Uncertainty, Calibration and Probability

The Statistics of Scientific and Industrial Measurement

Second Edition

The Adam Hilger Series on Measurement Science and Technology

Uncertainty, Calibration and Probability

The Statistics of Scientific and Industrial Measurement

Second Edition

C F Dietrich

BSc, PhD, FInstP, ChP

Adam Hilger
Bristol, Philadelphia and New York

British Library Cataloguing in Publication Data

Dietrich, C. F.
   Uncertainty, calibration and probability.—2nd ed.
   1. Quality control. Calibration & measurement
   I. Title
   658.562

   ISBN 0-7503-0060-4

Library of Congress Cataloging-in-Publication Data

Dietrich, C. F. (Cornelius Frank)
   Uncertainty, calibration, and probability: the statistics of scientific and and industrial measurement/C. F. Dietrich.—2nd ed.
      p.   cm.—(The Adam Hilger series on measurement science and technology)
   Includes bibliographical references and index.
   ISBN 0-7503-0060-4 (hbk.)
   1. Distribution (Probability theory)   2. Mensuration—Statistical methods.   I. Title.   II. Series.
QA273.6.D53   1991
519.2—dc20   90-38472
                                           CIP

First published, 1973
Second edition, 1991

Editorial panel: **M N Afsar** Tufts University
                 A E Bailey formerly National Physical Laboratory
                 R B D Knight National Physical Laboratory
                 R C Ritter University of Virginia

Published under the Adam Hilger imprint by IOP Publishing Ltd
Techno House, Redcliffe Way, Bristol BS1 6NX, England
335 East 45th Street, New York, NY 10017-3483, USA

US Editorial Office: 1411 Walnut Street, Philadelphia, PA 19102

Typeset by P&R Typesetters, Salisbury, Wiltshire
Printed in Great Britain by Galliard (Printers) Ltd, Great Yarmouth, Norfolk

To Drusilla

Contents

vii

9 Consistency and Significance Tests

10 Method of Least Squares

Appendix I Tables

Choice of table

Table I: Values of the density function of the normal distribution

$$f(c) = e^{-c^2/2} / \sqrt{(2\pi)}$$

Table II: Values of the integral

$$\frac{1}{\sqrt{(2\pi)}} \int_{-k_1}^{k_1} e^{-c^2/2} \, dc = \underset{-k_1 \text{ to } k_1}{P}$$

Table III: Values of k_1 corresponding to

$$\underset{-k_1 \text{ to } k_1}{P} = \frac{1}{\sqrt{(2\pi)}} \int_{-k_1}^{k_1} e^{-c^2/2} \, dc$$

Table IV: Student t distribution: μ and s known

Table V: Confidence limits for the mean μ

Table VI: Chi-square χ^2 distribution

Table VII: Confidence intervals for σ the standard deviation

Table VIII: F test. Upper limits for F. Probability of F exceeding these values $= 0.05$

Table IX: F test. Upper limits for F. Probability of F exceeding these values $= 0.01$

Table X: $\bar{x}$ and σ known. Tolerance limits for two tolerance probabilities

Table XI: $\bar{x}$ and s known. Tolerance limits for two tolerance probabilities

Table XII: $\bar{x}$ and σ known. Tolerance limits for three tolerance probabilities and two confidence probabilities

Table XIII: $\bar{x}$ and s known. Tolerance limits for four tolerance probabilities and three confidence probabilities

Table XIV: μ and s known. Tolerance limits for three tolerance probabilities and two confidence probabilities

Appendix II

Proof that $F_0(p; v_1; v_2) = \dfrac{1}{F_0(1 - p; v_2; v_1)}$

Appendix III
Proof that

$$\underbrace{\sum_{q=1}^{q=m} \sum_{r=1}^{r=n_q} \frac{w_{rq}(x_{rq} - \mu)^2}{\sigma^2}}_{\chi_0^2} = \underbrace{\sum_{q=1}^{q=m} \sum_{r=1}^{r=n_q} \frac{w_{rq}(x_{rq} - \bar{x}_q)^2}{\sigma^2}}_{\chi_1^2}$$

$$+ \underbrace{\sum_{q=1}^{q=m} \sum_{r=1}^{r=n_q} \frac{w_{rq}(x_{rq} - \bar{x})^2}{\sigma^2}}_{\chi_2^2}$$

$$+ \underbrace{\sum_{q=1}^{q=m} \sum_{r=1}^{r=n_q} \frac{w_{rq}(x_{rq} - \mu)^2}{\sigma^2}}_{\chi_3^2}$$

Preface

The purpose of this book is to provide a guide to the estimation of uncertainty in calibration and in scientific measurements, and also to the estimation of uncertainty in quality control of manufacturing processes. Some attention is made to the matching of graded parts after manufacture and to the estimation of the percentage of faulty parts accepted because of uncertainties in the sizing machines used to measure the parts.

Chapter 1 introduces the idea of uncertainty or error distributions, whilst Chapter 2 deals with the normal or Gaussian distribution and its properties. Chapter 3 is devoted mainly to the proof of a double integral theorem and to the derivation of some integral functions; the former being used to prove that certain of the properties proved in Chapter 2 for the Gaussian distribution are also true for any distribution or combinations thereof, whilst the latter are used to show how probability distributions may be convoluted or combined. In Chapter 4 the integral functions are applied to the combination of rectangular distributions, both with themselves and with a Gaussian distribution. The important point stressed in this chapter is the close approximation of the combination of several rectangular distributions to a Gaussian distribution, leading to an approximate value for the probability of an uncertainty of a given magnitude for the combination of rectangular distributions. Chapter 5 is now a new chapter dealing with applications to industry, such as the estimation of the percentage of faulty parts accepted and correct parts rejected because of uncertainty in the measuring equipment used. A table of how to estimate these percentages has been added (Table XXII in Appendix I). Chapter 6 deals with distributions ancillary to the Gaussian distribution because of their importance in consistency tests and in the case of the Student t distribution because of its use when a sample size is small. Chapter 7 is a new chapter and attempts to set out a basic methodology for the estimation of uncertainty founded on basic statistical theory. As before Chapter 8 deals with the estimation of the various components of uncertainty. Chapter 9 deals with consistency tests and with tests for the goodness of fit for a set of empirical points to a fitted or given curve. Chapter 10 deals with the method of least squares, including curve fitting. A new section has been added dealing with the use of orthogonal polynomials in curve fitting. This method avoids the solution of the linear normal equations containing as variables the constants of the various powers of the independent variable. If the curve to be fitted is of degree n, then this will involve calculating determinants of order $n + 1$, and if n is sizeable this can be very cumbersome and error prone. Further if curves of different degree are fitted, in order to find the best fit all the coefficients of the various powers of the independent variable have to be calculated afresh each time. The method of orthogonal polynomials avoids this, the coefficients of the various powers previously calculated remaining constant as each new coefficient of the next higher power is added. Chapter 11 deals with the theorems of Bernoulli, Stirling and the binomial theorem, and the distributions of Poisson and the hypergeometric function, together with examples. In the examples given in each chapter an attempt has been made to approximate the questions to actual problems encountered in calibration work.

Finally I would like to thank Dr J C E Jennings for all his advice and help and in particular for his help with the section on polynomial curve fitting, which is largely based on work done by him on this subject. I would also like to thank him for work he contributed to the statistical functions of Appendix VII.

C F Dietrich
Camberley
January 1990

1

Uncertainties and Frequency Distributions

Introduction: Calibration

1.01 It is only in this century and to a large extent in the last fifty years or so that the real importance of calibration measurements has come to be realized. In our industrialized society it is essential that instruments used in quality control and for the measurement of precision workpieces be properly calibrated. Interchangeability between mass produced parts, a need which is a cornerstone of present-day industry, can only be satisfactorily achieved if measuring equipment is properly calibrated.

1.02 The need for action at national level was realized some fifty years ago by Australia, with its National Australian Testing Authority (NATA), and a little later, after World War II, by the Japanese. In Great Britain the British Calibration Service (BCS) was set up in late 1966 with a remit to establish a national calibration service. This government-backed organization decided that it would be too expensive to set up a chain of government calibration laboratories throughout the country and chose instead to organize a scheme whereby existing laboratories in industry, the universities and in the government sector would be approved by BCS to issue authenticated calibration certificates, having gone through a strict assessment exercise. Measurements were divided between different fields e.g. electrical DC and LF mechanical, optical, flow, thermal, radiological etc. Each field was supervised by a committee of experts, who drew up a series of criteria documents covering each field. These criteria documents dealt with laboratory equipment, laboratory environment, laboratory personnel and also specific types of measurement within each field. The service was run by a headquarters staff of experts, who assessed each laboratory seeking approval, in conjunction with outside experts.

1.03 It was realized early on for a laboratory seeking approval that the most important factor governing its approval was proof that it could measure

1

to a given accuracy. This was achieved by submitting to each applicant laboratory a number of workpieces or instruments which had been previously calibrated at the National Physical Laboratory (NPL). The applicant laboratory's measurements were then compared with those of the NPL, and if the differences were sufficiently small, the laboratory was granted approval to issue authenticated certificates for those measurements. Further, it was also realized that an approved laboratory would need to be supervised, to see that its equipment was kept in calibration, its laboratory environment maintained etc., and most important of all that its measurement capability was maintained. This was achieved by setting up a so-called audit measurement scheme, which entailed the sending round to approved laboratories a series of workpieces and instruments, previously calibrated at the NPL, for measurement by the laboratories.

1.04   Another crucial factor was realized, namely that all measurements have a degree of uncertainty associated with them, and that if the measurement values given on certificates were to have any validity, the degree of this uncertainty would need to be carefully assessed and stated. From statistical theory, the error or particular uncertainty of measurement has a probability associated with it. Generally the larger the error, the smaller the probability of such an error occurring. It was decided that appropriate criteria for calibration should be that the correct value of the quantity measured should lie between equal plus and minus limits about a mean value with a probability of 0.954 or, put another way, that there was a 95.4% chance that the correct value should lie between the stated limits. Approximately, this means that there is a 1 in 20 chance of the correct value lying outside these limits. In special cases different probability limits may be chosen, for example 0.9973 which gives the probability of the correct value lying outside these limits as approximately 1 in 400.

1.05   The subject of uncertainty of measurement raises many problems and it is the purpose of this book to provide an understanding of these problems and how to deal with them.

1.06   The British Calibration Service (BCS) became part of the National Physical Laboratory in 1977 and has now been integrated with another NPL-based organization, the National Testing Laboratory Accreditation Scheme (NATLAS), to form a new organization known as the National Measurement Accreditation Service (NAMAS). The two sections of the service will cover calibration (BCS) and testing (NATLAS).

1.07   Many countries of the EC have now followed the example of the United Kingdom and have set up their own national calibration services, which in general are closely modelled on the British scheme. The arrival of national calibration schemes in other countries of Europe has now led to the signing of reciprocal agreements between some of these countries, which cover the recognition of each other's national calibration scheme certificates. This enables a purchaser of a piece of equipment or of an instrument to obtain an authenticated certificate, traceable to international standards, which gives a factual truthful assessment of the equipment's specification or measurement capability.

1.08 The origin of calibration goes back to the early eighteenth century, when the words 'calibre' and 'calliper' were different versions of the same word which described an instrument for measuring or comparing bores. The version 'calibre' became associated with guns and is now synonymous with the internal diameter of a gun barrel, whilst the transitive verb 'to calibrate' became associated with the actual measurement of gun bores and also with a so-called calibre scale. On this scale the distances of graduations from an initial line were proportional to the cube roots of the natural numbers. The first interval represented the diameter of a cannon ball of one pound weight, and successive intervals from zero thus gave the diameters of balls whose weights were successive multiples of one pound weight.

1.09 From its earliest known use 'to calibrate' was thus associated with measurement and with scales, and later in the mid-nineteenth century the term became associated with graduations on thermometer tubes, barometer tubes and pressure measuring devices, the graduations being spaced so as to correct for any irregularities in the bores of the tubes. The term had now become linked not only with measurement but with the idea of correcting errors, that is by making the graduation intervals of unequal size. In modern times, where the emphasis is on mass production and standardization, the term 'to calibrate' has become associated with instruments which already have scales, and a set of calibration measurements usually takes the form of a table of corrections to be applied to the instrument readings in order to allow for errors of one sort or another. In its modern use the term 'to calibrate' means to make measurements which are traceable to the national standards. Thus, if a piece of measuring equipment is used to measure a series of workpieces that have been previously measured on calibration equipment, that is equipment that has very small errors compared with the measuring equipment under test, then the differences between corresponding measured quantities will give the errors in the measuring equipment. The sizes of the calibrated workpieces, obtained from the calibration equipment, give a close approximation to the most probable values of these sizes. A set of calibration measurements thus gives what may be thought of as the most probable value of a measurement quantity to be associated either with a component or an instrument.

Uncertainty of Measurement

1.10 No measurement is perfect; a single measurement if repeated several times may give as many slightly differing results, whilst the results if repeated on another day may all show a slight shift. This idea of an uncertainty in measurements may seem disconcerting, but it is nevertheless something that has to be accepted and as far as possible allowed for. In fact it is the main aim of this book to show how the degree of uncertainty of calibration readings can be estimated. A general definition of uncertainty of measurement is that it is the residual error which may exist in an instrument or workpiece after calibration corrections have been made. In other words, all measurements have a certain indeterminancy associated with them, however small, and the phrase 'uncertainty of measurement' is used to describe the magnitude of indeterminancy or error which may be present in a set of calibration measurements. A much more precise definition will be given later after consideration has been given to the concept of distributed uncertainties.

Sources of Uncertainties

1.11 Before considering the nature of uncertainties more precisely, we shall consider the various types of uncertainty which may be present in any calibration measurement and which add their contribution to the total uncertainty. The following catalogue of uncertainties is not exhaustive but covers most of the principal sources of uncertainty:

1. Uncertainties in standards or in calibration equipment;
2. Uncertainties due to operator error;
3. Resolution or discrimination uncertainties;
4. Environmental uncertainties, including variation of temperature, of pressure, of flow rate, of power supplies, etc.;
5. Lack of repeatability–instability uncertainties;
6. Functional uncertainties, caused by the malfunctioning of equipment;
7. Uncertainties caused by lack of cleanliness;
8. Uncertainties due to poor quality surface texture and incorrect geometry;
9. Uncertainties associated with lapse of time, which produces changes in equipment or workpieces.

The last type of uncertainty can be assessed only from several sets of measurements taken over an extended period of time. When assessing the total uncertainty of measurement of a set of measurements every possible source of uncertainty should be accounted for, each source making its contribution to the total uncertainty.

Uncertainties as Distributions

1.12 We must examine more closely what we mean by an uncertainty. One very common concept is that an uncertainty is the amount by which a measurement differs from some true value. This uncertainty is often thought of as having a maximum value, so that we speak of the magnitude of any measurement plus or minus some figure which specifies the uncertainty of measurement or error. It might also be thought that if the uncertainties from different sources are known, then the total uncertainty is the sum of the component uncertainties. As we shall see, this simple idea of uncertainties, if adopted, would lead to completely erroneous results.

1.13   Let us consider a simple example of a source of error. Consider an angle measuring device such as a rotary table, fitted with a precise circular scale and microscope capable of reading the angular orientation of the table top to a second of arc. A piece of glass with a straight line engraved on it is mounted horizontally on the table top so that the prolongation of the line passes through the axis of rotation of the table top. A second piece of glass, with two parallel lines engraved on it, spaced about three times the width of the first line, is mounted just above the first piece of glass with its two lines radiating approximately from the centre of rotation of the table. The second piece of glass is mounted rigidly and independently of the table top. If both sets of lines are now viewed through a microscope mounted above the table top, the single line can be moved, so as to fit centrally between the two lines, by rotating the fine rotation screw of the table. The orientation of the table top in degrees, minutes and seconds is now noted. If the table top is now rotated a little and the single line again positioned centrally between the two lines it is very unlikely that the second angular orientation of the table top will be the same as the first. If a fairly large number of such readings are taken, say one hundred and ten, then many of these readings will be different.

1.14   Let us now consider these one hundred and ten readings and attempt to analyse them. Firstly we take the sum of all the readings and divide by the number of readings to obtain the mean. The readings are now arranged in ascending order of size and the difference between the largest and smallest readings is divided into eleven equal sub-ranges. In general the number of sub-ranges is usually taken between eight and sixteen, depending on the number of readings taken and their distribution. A table can now be made of the number of readings occurring in each sub-range, and a diagram constructed, consisting of a series of contiguous rectangles, the base of each rectangle being equal to a sub-range, whilst the height is made proportional to the number of readings falling in that particular range. The diagram obtained would look rather like that shown in Figure 1.14.

Histograms

1.15   The diagram shown in Figure 1.14 is called a *histogram*. As can be seen, the histogram looks fairly symmetrical and, had a much larger number of readings been taken, it would have become even more so. Let it now be

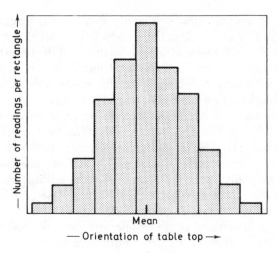

Figure 1.14   Frequency histogram

assumed that the diagram is in fact symmetrical. We now plot a series of points whose co-ordinates are the height of each rectangle (ordinates) and the position of the centre of the corresponding base (abscissae). If a curve is now drawn through these points we obtain a graph similar to that shown in Figure 1.15. Such a curve is known as a *frequency diagram.*

Frequency Distributions

1.16   The curve obtained is very similar to the one we should have obtained had we taken a very large number of readings and divided the abscissa axis into a very large number of equal parts. From a consideration of Figure 1.14 it is manifest that the area under the curve of Figure 1.15 is proportional to the total number of readings, and that the area contained between two lines parallel to the ordinate axis (see shaded area in Figure 1.15) is proportional to the number of readings in the range covered by the difference in the abscissae of these two lines. If the shaded area described above is divided by the total area under the curve the quotient obtained gives the fraction of readings which lie in the shaded area. This number is equal to the probability of a reading lying in the measurement range between the two points defined by the parallel lines. Often the ordinates of the rectangles of Figure 1.14 are plotted as the ratio of the number of readings in the rectangle to the total number of readings, and it is easily seen that if this is done the area under the curve, which is now called a *probability curve*, is equal to *unity*, which is only another way of saying that the probability of a reading occurring over the whole range of measurement is equal to unity. Frequency distributions

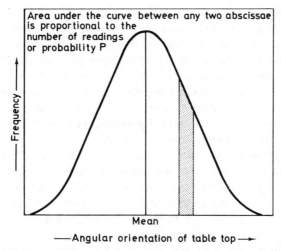

Figure 1.15 Frequency distribution

(see Figure 1.15) are usually plotted as probability curves, as it is generally more convenient to do so, but the ordinate at any point on a probability curve is often called the *density function* of the curve rather than the probability function of the curve. The integral of the *density function* from minus infinity to x_1, i.e. $\int_{-\infty}^{x_1} f(x)\, dx$, is defined as the *distribution function* of the probability distribution.

Probability

1.17 The idea of the probability of a measurement lying within a given range has thus been introduced, and from a consideration of Figure 1.15 it can be seen that the probability of a measurement lying within a given range is equal to the area contained under the curve between the limits of the range, provided of course that the curve is a probability curve, that is that the total area under the curve is equal to unity. This latter statement is a useful definition of a probability curve. Since the curve of Figure 1.15 is symmetrical it is seen that the maximum probability of occurrence of a reading coincides with the position of the mean value of the set of readings, and this is one of the reasons why the mean value of a set of readings is usually selected as the best estimate of a variable being measured.

Asymmetry

1.18 If the frequency distribution had been asymmetrical, the mean would not have coincided with the maximum of the curve. In calibration measurements, however, frequency or probability curves corresponding to an infinite number of readings are usually symmetrical—although of course the curves obtained in practice from a limited number of readings are not—so that the problem of deciding whether to take the arithmetic mean or the maximum of a frequency curve as the best value of a set of measurements rarely occurs. It should, however, be borne in mind that most frequency curves obtained from a limited set of readings are asymmetrical to a degree, but if a very large number of readings be taken the curve would be found to be symmetrical, so that unless the frequency curve is definitely known to be asymmetrical the mean should always be taken as the best value of a set of measurements, even though the frequency curve of a limited number of readings, from which the mean is derived, is asymmetrical. In the case of real asymmetry the most probable value of a set of readings will coincide with the maximum or *mode* of the curve as it is known. In these circumstances it is necessary to know the exact shape of the curve in order to determine the mode.

1.19   It should be said, however, that there is a very important frequency curve, known as the Bernoulli or binomial curve, which can be symmetrical or asymmetrical, and even in its asymmetrical form the mean is very close indeed to the maximum or mode of the curve, and so to a very close approximation the mean can be taken as the most probable value of the maximum of the curve.

Gaussian or Normal Distribution

1.20   The particular shape of the curve shown in Figure 1.15 is one that occurs very frequently in calibration measurements and, in fact, its shape can be predicted from quite simple assumptions as we shall see later. (See paragraphs 2.01 to 2.10 and 2.12 to 2.21.) The curve is often spoken of as a normal or Gaussian distribution. An uncertainty is in this case defined as the difference between any measurement and the corresponding mean measurement, represented by the position of the maximum of the curve. The normal distribution was originally discovered by De Moivre in 1733 but the discovery seems to have passed unnoticed: it was rediscovered independently by Gauss in 1809 and Laplace in 1812.

Uncertainty of Measurement: Definition

1.21   We are now in a position to define the uncertainty of measurement of a set of measurements. The uncertainty of measurement is defined as the range on either side of the mean value of a set of measurements between which a given fraction of readings would lie if a very large number of observations were made. Another way of saying this is to define the uncertainty of measurement as the probability of a reading lying in a given range on either side of the mean value. Usually the two ranges are equal and the probability is given for the complete range, that is from the limit point on one side of the mean to the limit point on the other side of the mean. The important point to grasp here is that there is no one uncertainty; all that can be said is that if limit points are selected on either side of the mean value the probability of a reading lying within the range so defined can be predicted. The greater the certainty required for a reading or observation to occur within a range, the greater that range has to be made. In the case of the Gaussian distribution, which is of course an idealized one, the range associated with certainty, that is a probability of unity, is from minus infinity to plus infinity.

2

The Gaussian Distribution

Simple Derivation of the Gaussian Distribution

2.01 The normal or Gaussian type distribution curve is one that is met with very frequently, and is the idealized curve which would occur if we had a truly random variable. We shall now derive the equation for the Gaussian distribution, using a simple method of derivation based on several simple assumptions.

2.02 Let us suppose that n measures $x_1, x_2, \ldots, x_n$ of a variable x have been made, and that the probability curve or density function of x is $y = f(x)$. We shall also assume that all the measures conform to the law of probability expressed by the above curve. Let us now move the origin of our co-ordinates to a point μ on the x axis. Our co-ordinates now become

$$\beta_1 = x_1 - \mu, \, \beta_2 = x_2 - \mu, \ldots, \beta_n = x_n - \mu \qquad 2.02(1)$$

2.03 The probability that each of the measurements $\beta_1, \beta_2, \ldots, \beta_n$ occurs separately is

$$f(\beta_1), \, f(\beta_2), \, f(\beta_3), \ldots, f(\beta_n)$$

whilst the compound probability that just these particular n measurements occur, out of all the possible measurements that might be obtained, is given by the product

$$P = f(\beta_1)f(\beta_2) \cdots f(\beta_n) \qquad 2.03(1)$$

2.04 We shall now make the following assumptions:

1. That P, regarded as a function of $\beta_1, \beta_2, \beta_3, \ldots, \beta_n$ and μ, is continuous and differentiable over the whole range.
2. There is a single greatest value of P in the range which is also a maximum of $f(x)$.
3. That the sum of the squares of the quantities $\beta_1, \beta_2, \beta_3, \ldots, \beta_n$, that is $\beta_1^2 + \beta_2^2 + \cdots + \beta_n^2$, is a minimum.

10

This last assumption is known as the principle of least squares (see Chapter 8) which plays a very important part in probability theory.

Linking of mean with most probable value

2.05 Let us now take logarithms of both sides of equation 2.03(1) and differentiate it with respect to μ. Equating $dP/d\mu$ to zero, we obtain the condition for maximum P as

$$\frac{f'(\beta_1)}{f(\beta_1)}\frac{d\beta_1}{d\mu} + \frac{f'(\beta_2)}{f(\beta_2)}\frac{d\beta_2}{d\mu} + \cdots + \frac{f'(\beta_n)}{f(\beta_n)}\frac{d\beta_n}{d\mu} = 0 \qquad 2.05(1)$$

Let the sum of the squares of the quantities $\beta_1, \beta_2, \beta_3, \ldots, \beta_n$ be equal to Q, thus

$$Q = \beta_1^2 + \beta_2^2 + \beta_3^2 + \cdots + \beta_n^2 \qquad 2.05(2)$$

If we differentiate this expression with respect to μ and equate to zero we then have the condition for Q to be a minimum or a maximum. Thus

$$\frac{1}{2}\frac{dQ}{d\mu} = \beta_1\frac{d\beta_1}{d\mu} + \beta_2\frac{d\beta_2}{d\mu} + \cdots + \beta_n\frac{d\beta_n}{d\mu} = 0 \qquad 2.05(3)$$

But from equations 2.02(1) we have

$$\frac{d\beta_1}{d\mu} = -1, \frac{d\beta_2}{d\mu} = -1, \ldots, \frac{d\beta_n}{d\mu} = -1 \qquad 2.05(4)$$

Therefore substituting in 2.05(3) we have

$$\beta_1 + \beta_2 + \beta_3 + \cdots + \beta_n = 0 \qquad 2.05(5)$$

or substituting for the βs in terms of x and μ from equation 2.02(1) we have

$$\sum_{r=1}^{r=n} x_r = n\mu \qquad 2.05(6)$$

Thus μ is the mean of xs. That the condition given by 2.05(5) corresponds to a minimum is easily shown, thus

$$\frac{1}{2}\frac{d^2Q}{d\mu^2} = \sum_{r=1}^{r=n}\left\{\beta_r\frac{d^2\beta_r}{d\mu^2} + \left(\frac{d\beta_r}{d\mu}\right)^2\right\} \qquad 2.05(7)$$

and since $d\beta_r/d\mu = -1$ a constant, $d^2\beta_r/d\mu^2 = 0$, therefore

$$\frac{1}{2}\frac{d^2Q}{d\mu^2} = \sum_{r=1}^{r=n}\left(\frac{d\beta_r}{d\mu}\right)^2 = n$$

which is positive. Therefore the condition expressed by equation 2.05(5) corresponds to a minimum.

2.06 Returning now to equations 2.05(1) and 2.05(3) and combining them, using the Lagrange method of undetermined multipliers, we have

$$\sum_{r=1}^{r=n} \left\{ \frac{f'(\beta_r)}{f(\beta_r)} - \lambda\beta_r \right\} \frac{d\beta_r}{d\mu} = 0 \qquad\qquad 2.06(1)$$

where λ is a constant. Since all the β_r are independent, this equation is satisfied for all values of β_r if

$$\frac{f'(\beta_r)}{f(\beta_r)} - \lambda\beta_r = 0 \qquad\qquad 2.06(2)$$

Thus, integrating we have

$$\log f(\beta_r) = \frac{\lambda\beta_r^2}{2} + C \qquad\qquad 2.06(3)$$

or

$$f(\beta_r) = e^{\lambda\beta_r^2/2 + C}$$

$$= A\, e^{\lambda\beta_r^2/2} \qquad\qquad 2.06(4)$$

2.07 Now condition 2 of paragraph 2.04 specified that P of equation 2.03(1) must have a maximum value, and thus λ must be determined to make this true. For a maximum

$$\frac{d^2P}{d\mu^2} < 0 \qquad\qquad \text{for specified } \mu$$

From 2.03(1) and 2.06(4)

$$P = A^n\, e^{\lambda/2 \sum_1^n \beta_r^2}$$

Thus

$$\frac{dP}{d\mu} = \lambda A^n\, e^{\lambda/2 \sum_1^n \beta_r^2} \cdot \left(\sum \beta_r \frac{d\beta_r}{d\mu} \right) \qquad\qquad 2.07(1)$$

and thus

$$\frac{d^2P}{d\mu^2} = \lambda A^n\, e^{\lambda/2 \sum_1^n \beta_r^2} \cdot \left[\sum_1^n \left(\beta_r \frac{d^2\beta_r}{d\mu^2} + \left(\frac{d\beta_r}{d\mu} \right)^2 \right) \right]$$

$$+ \lambda^2 A^n\, e^{\lambda/2 \sum_1^n \beta_r^2} \cdot \left(\sum_1^n \beta_r \frac{d\beta_r}{d\mu} \right)^2 \qquad\qquad 2.07(2)$$

Now from 2.05(1) $\sum_1^n \beta_r = 0$ and also from 2.05(1)

$$\frac{d^2\beta_r}{d\mu^2} = 0 \qquad \text{and} \qquad \frac{d\beta_r}{d\mu} = -1$$

Therefore

$$\frac{d^2P}{d\mu^2} = \lambda A^n \ e^{\lambda/2 \sum_1^n \beta_r^2}[(-1)^2] + \lambda^2 A^n \ e^{\lambda/2 \sum_1^n \beta_r^2}[0]$$

$$= \lambda A^n \ e^{\lambda/2 \sum_1^n \beta_r^2} \qquad \qquad 2.07(3)$$

Thus for $d^2P/d\mu^2$ to be < 0, λ must be negative. Thus

$$f(\beta) = A \ e^{-\lambda \beta^2/2} \qquad \qquad 2.07(4)$$

Note that the λ in 2.07(4) is now a positive quantity.

Determination of Constants A and λ

2.08   Since $f(\beta)$ is a probability density function, it follows that

$$\int_{-\infty}^{\infty} f(\beta) \ d\beta = 1$$

that is

$$\int_{-\infty}^{\infty} A \ e^{-\lambda \beta^2/2} \ d\beta = 1 \qquad \qquad 2.08(1)$$

Writing $\lambda \beta^2/2 = x^2$ we have $d\beta = \sqrt{(2/\lambda)} \, dx$ and substituting for β in terms of x we have

$$\int_{-\infty}^{\infty} A \ e^{-x^2} \sqrt{(2/\lambda)} \ dx = 1 \qquad \qquad 2.08(2)$$

2.09   Consider the integral

$$\int_{-\infty}^{\infty} e^{-x^2} \ dx \int_{-\infty}^{\infty} e^{-y^2} \ dy \qquad \qquad 2.09(1)$$

where x and y are independent. Now the integral is the limit of

$$\sum_{r=-n}^{r=n} e^{-x_r^2} \ \delta x \sum_{r=-n}^{r=n} e^{-y_r^2} \ \delta y_r \qquad \qquad 2.09(2)$$

when $n \to \infty$ and δx and $\delta y \to 0$ where $x_r = (r+1)\delta x$ and $y_r = (r+1)\delta y$. Now since both series are absolutely convergent, the terms may be rearranged in any order. This means that we can choose a relationship between x and y and so long as every value of x and y is covered by the integral, this rearranged integral and 2.09(1) will be equal.

2.10   Let us write $x^2 + y^2 = r^2$, and replace the element of area $dx \, dy$ by the element $r \, dr \, d\theta$. Thus, if r ranges from 0 to infinity and θ from 0 to 2π, then every single value of x and y will be covered. Thus we may put

$$\int_{-\infty}^{\infty} e^{-x^2}\, dx \int_{-\infty}^{\infty} e^{-y^2}\, dy = \int_0^{\infty}\int_0^{2\pi} e^{-r^2} r\, dr\, d\theta \qquad 2.10(1)$$

Let us write

$$I = \int_{-\infty}^{\infty} e^{-x^2}\, dx = \int_{-\infty}^{\infty} e^{-y^2}\, dy \qquad 2.10(2)$$

Thus we have

$$I^2 = \int_0^{\infty}\int_0^{2\pi} r\, e^{-r^2}\, dr\, d\theta$$

$$= 2\pi \int_0^{\infty} r\, e^{-r^2}\, dr = 2\pi[-e^{-r^2}/2]_0^{\infty} = \pi \qquad 2.10(3)$$

Therefore

$$I = \sqrt{\pi} \qquad 2.10(4)$$

Thus, substituting in 2.08(2) for $\int_{-\infty}^{\infty} e^{-x^2}\, dx$, we have

$$A\sqrt{(2/\lambda)}\sqrt{\pi} = 1$$

or

$$A = \sqrt{(\lambda/2\pi)} \qquad 2.10(5)$$

and

$$f(\beta) = \sqrt{(\lambda/2\pi)}\, e^{-\lambda\beta^2/2} \qquad 2.10(6)$$

Standard Deviation of Gaussian Distribution

2.11 We still have to determine the constant λ, and in order to do this we define another constant σ, known as the standard deviation of $f(\beta)$. σ is defined as the square root of the average second moment about the mean. Thus

$$\sigma^2 = \int_{-\infty}^{\infty} \beta^2 f(\beta)\, d\beta \bigg/ \int_{-\infty}^{\infty} f(\beta)\, d\beta \qquad 2.11(1)$$

$$= \sqrt{(\lambda/2\pi)} \int_{-\infty}^{\infty} \beta^2\, e^{-\lambda\beta^2/2}\, d\beta$$

from 2.10(6), and since

$$\int_{-\infty}^{\infty} f(\beta)\, d\beta = 1 \qquad \text{(see 2.08(1))}$$

As before, putting $\lambda\beta^2/2 = x^2$, we have $d\beta = \sqrt{(2/\lambda)}\,dx$, and then on substituting for β the integral becomes

$$\sigma^2 = \frac{2}{\lambda\sqrt{\pi}} \int_{-\infty}^{\infty} x^2\,e^{-x^2}\,dx \qquad\qquad 2.11(2)$$

Integrating by parts, we have

$$\sigma^2 = \frac{2}{\lambda\sqrt{\pi}} \int_{-\infty}^{\infty} -\frac{x}{2}\frac{d(e^{-x^2})}{dx} = \frac{2}{\lambda\sqrt{\pi}} \left\{ \underbrace{\left[-\frac{x}{2}e^{-x^2} \right]_{-\infty}^{\infty}}_{0} + \underbrace{\frac{1}{2}\int_{-\infty}^{\infty} e^{-x^2}\,dx}_{\sqrt{\pi}/2} \right\}$$

$$2.11(3)$$

$$= 1/\lambda$$

Therefore, finally, we have

$$f(\beta) = \frac{e^{-\beta^2/2\sigma^2}}{\sigma\sqrt{(2\pi)}} \qquad\qquad 2.11(4)$$

which is the required expression for the Gaussian distribution. Since $\beta = x - \mu$ we have on substituting for β

$$f(x) = \frac{e^{-(x-\mu)^2/2\sigma^2}}{\sigma\sqrt{(2\pi)}} \qquad\qquad 2.11(5)$$

where μ is the mean value of x. It is to be noted that the standard deviation of $f(x)$ is derived from the integral

$$\sigma^2 = \int_{-\infty}^{\infty} \frac{(x-\mu)^2\,e^{-(x-\mu)^2/2\sigma^2}\,dx}{\sigma\sqrt{(2\pi)}} \qquad\qquad 2.11(6)$$

and *not* from

$$\sigma^2 = \int_{-\infty}^{\infty} \frac{x^2\,e^{-(x-\mu)^2/2\sigma^2}\,dx}{\sigma\sqrt{(2\pi)}} \qquad\qquad 2.11(7)$$

This follows from the definition of σ given at the beginning of this paragraph.

Alternative Derivation of Gaussian Function

2.12 Before proceeding with a detailed discussion of the Gaussian distribution, it is worth considering an alternative derivation which may help the reader to understand what causes a normal or Gaussian type distribution.

2.13 Let us make the simple assumption that the deviation of any one random reading or measurement from the mean is produced by the chance instantaneous value of each of many contributing small deviations. Further, let us assume that the mean value of a measurement is x_0, and that there exist n sources of influence, each of which can contribute a deviation of $\pm\delta$. Note that for the sake of simplicity we have assumed that all the deviations δ from the various sources are equal.

2.14   Now, the probability that any one deviation will be $+\delta$ or $-\delta$ is $\frac{1}{2}$. Now, according to Bernoulli's theorem (see paragraph 11.05) if we have a number of articles that can be divided into two classes, and the fraction in Class 1 is, say, p, giving the fraction in Class 2 as $1 - p$, the probability of finding r articles of Class 1 in a random choice of n articles ($n \geqslant r$) is

$$^nC_r p^r (1 - p)^{n-r} \qquad\qquad 2.14(1)$$

where p is the probability of selecting an article from Class 1.

2.15   In our case, let Class 1 denote the positive δs and Class 2 the negative ones, and let us calculate the probability of $n + r$ positive δs and $n - r$ negative ones, where the total number of δs is now taken as $2n$. In our case $p = \frac{1}{2}$ since there are equal numbers of positive and negative δs. Thus, replacing n by $2n$ and r by $n + r$ in 2.14(1), we have

$$P = {}^{2n}C_{n+r} \cdot 1/2^{2n} \qquad\qquad 2.15(1)$$

where P is the probability of an excess of $2r$ positive δs over the negative ones. The deviation from x_0 is thus $2r\delta$.

2.16   Let us now examine 2.15(1). Now

$$^{2n}C_{n+r} \equiv \frac{2n!}{(n + r)!(n - r)!} \qquad\qquad 2.16(1)$$

Since n is very large, we use Stirling's approximation for $n!$ (see paragraphs 11.15 to 11.17 and expression 11.17(1)), that is

$$n! \simeq \sqrt{(2\pi)} n^{n+1/2} e^{-n+1/12n} \qquad\qquad 2.16(2)$$

but since n is large we may neglect the $1/12n$ term. Thus

$$2n! = \sqrt{(2\pi)} (2n)^{2n+1/2} e^{-2n} \qquad\qquad 2.16(3)$$

$$(n + r)! = \sqrt{(2\pi)} (n + r)^{n+r+1/2} e^{-n-r} \qquad\qquad 2.16(4)$$

$$(n - r)! = \sqrt{(2\pi)} (n - r)^{n-r+1/2} e^{-n+r} \qquad\qquad 2.16(5)$$

and so

$$(n + r)!(n - r)! = 2\pi\, e^{-2n} [(n + r)(n - r)]^{n+1/2} \left(\frac{n + r}{n - r}\right)^r$$

$$= 2\pi\, e^{-2n} (n^2 - r^2)^{n+1/2} \left(\frac{n + r}{n - r}\right)^r$$

$$= 2\pi\, e^{-2n} n^{2n+1} \left(1 - \frac{r^2}{n^2}\right)^{n+1/2} \left(1 + \frac{r}{n}\right)^r \left(1 - \frac{r}{n}\right)^{-r}$$

Thus

$$P = (2n)^{2n+1/2}\,e^{-2n}/\sqrt{(2\pi)}e^{-2n}n^{2n+1}\left(1 - \frac{r^2}{n^2}\right)^{n+1/2}\left(1 + \frac{r}{n}\right)^{r}\left(1 - \frac{r}{n}\right)^{-r}2^{2n}$$

$$= 2^{1/2}/\sqrt{(2\pi)}n^{1/2}\left(1 - \frac{r^2}{n^2}\right)^{n}\left(1 + \frac{r}{n}\right)^{r+1/2}\left(1 - \frac{r}{n}\right)^{-r+1/2}$$

Now

$$\left(1 - \frac{r^2}{n^2}\right)^{n} = \left\{1 - \left(\frac{r^2}{n}\right)\cdot\frac{1}{n}\right\}^{n} \to e^{-r^2/n} \qquad 2.16(6)$$

for large n. Also

$$\left(1 + \frac{r}{n}\right)^{r+1/2} = \left\{\left(1 + \frac{r}{n}\right)^{n}\right\}^{(r+1/2)/n} \to e^{r(r+1/2)/n} \qquad 2.16(7)$$

for large n. Similarly

$$\left(1 - \frac{r}{n}\right)^{-r+1/2} \to e^{r(r-1/2)/n} \qquad 2.16(8)$$

for large values of n. Thus, finally

$$P = \frac{e^{-r^2/n}}{\sqrt{(\pi n)}} \qquad 2.16(9)$$

2.17 Let us put the deviation

$$2r\delta = z \qquad 2.17(1)$$

giving

$$r = \frac{z}{2\delta}$$

Now r increases by integers, and so

$$2(r + 1)\delta = z + \delta z$$

giving

$$\delta z = 2\delta \qquad 2.17(2)$$

If we now substitute in 2.16(9) for r, we obtain

$$P_z\,dz = \frac{e^{-z^2/4\delta^2 n}}{\sqrt{(\pi n)}}\cdot\left(\frac{\delta z}{2\delta}\right) \qquad 2.17(3)$$

where $P_z\,dz$ is the probability of a deviation from the mean having a value between z and $z + dz$. Let us now put $2\delta^2 n = q^2$ which gives

$$P_z\,dz = \frac{e^{-z^2/2q^2}\,dz}{q\sqrt{(2\pi)}} \qquad 2.17(4)$$

Standard deviation

2.18 Let us now determine the standard deviation of P_z. This is the mean value of $z^2 P_z$ (see 2.11(1)). Thus

$$\sigma^2 = \int_{-\infty}^{\infty} \frac{z^2 e^{-z^2/2q^2} dz}{q\sqrt{(2\pi)}} \qquad 2.18(1)$$

If we put $z/q\sqrt{2} = y$, giving $dz = q\sqrt{2}\, dy$, on substituting for z in terms of y in 2.18(1) we get

$$\sigma^2 = \int_{-\infty}^{\infty} \frac{2q^2 y^2 e^{-y^2} q\sqrt{2}\, dy}{q\sqrt{(2\pi)}}$$

$$= \frac{1}{\sqrt{\pi}} \int_{-\infty}^{\infty} 2q^2 y^2 e^{-y^2}\, dy$$

$$= \frac{2q^2}{\sqrt{\pi}} \int_{-\infty}^{\infty} y^2 e^{-y^2}\, dy$$

$$= \frac{2q^2}{\sqrt{\pi}} \cdot \frac{\sqrt{\pi}}{2} = q^2 \qquad 2.18(2)$$

since $\int_{-\infty}^{\infty} y^2 e^{-y^2}\, dy = \sqrt{\pi}/2$. (See 2.11(3) and 2.10(4).) Therefore

$$q = \sigma \qquad 2.18(3)$$

and so

$$P_z\, dz = \frac{e^{-z^2/2\sigma^2}\, dz}{\sigma\sqrt{(2\pi)}} \qquad 2.18(4)$$

Thus

$$P_z = \frac{e^{-z^2/2\sigma^2}}{\sigma\sqrt{(2\pi)}} \qquad 2.18(5)$$

If we measure our abscissae from the zero defined by x_0, then $x = x_0 + z$. Therefore, substituting for z in terms of x, we have

$$P_x = \frac{e^{-(x-x_0)^2/2\sigma^2}}{\sigma\sqrt{(2\pi)}} \qquad 2.18(6)$$

exactly the same as 2.11(5).

Mean value of x

2.19 It is easily shown that x_0 is the mean value of x, viz.

$$\bar{x} = \frac{\int_{-\infty}^{\infty} xP_x\,dx}{\int_{-\infty}^{\infty} P_x\,dx} = \int_{-\infty}^{\infty} xP_x\,dx \qquad\qquad 2.19(1)$$

$$= \int_{-\infty}^{\infty} \frac{x\,e^{-(x-x_0)^2/2\sigma^2}\,dx}{\sigma\sqrt{(2\pi)}} \qquad\qquad 2.19(2)$$

Put

$$(x - x_0)/\sqrt{2}\sigma = y$$

Therefore,

$$x = x_0 + \sqrt{2}\sigma y$$

and

$$dx = \sqrt{2}\sigma\,dy$$

therefore,

$$\bar{x} = \int_{-\infty}^{\infty} \frac{e^{-y^2}(\sqrt{2}\sigma)\,dy(x_0 + \sqrt{2}\sigma y)}{\sigma\sqrt{(2\pi)}}$$

$$= \int_{-\infty}^{\infty} \frac{x_0\,e^{-y^2}\,dy}{\sqrt{\pi}} + \int_{-\infty}^{\infty} \frac{\sqrt{2}\sigma y\,e^{-y^2}\,dy}{\sqrt{\pi}}$$

Or

$$\bar{x} = \frac{x_0}{\sqrt{\pi}}(\sqrt{\pi})\dagger + \sqrt{2}\sigma\underbrace{\left[\frac{e^{-y^2}}{2}\right]_{-\infty}^{\infty}}_{0}$$

$$= x_0$$

thus

$$x_0 \equiv \mu \equiv \text{mean value of } x \qquad\qquad 2.19(3)$$

Generalization of Proof

2.20 It may be objected that the proof is not very general since the deviations were taken as multiples of the same small quantity, δ. Let us consider s distributions, each formed in the same way as 2.18(5) from many small

† From 2.10(4).

deviations, but let the basic deviation δ be different in each case. Thus the qth density function is given by

$$P_{z_q} = \frac{e^{-z_q^2/2\sigma_q^2}}{\sigma_q\sqrt{(2\pi)}}$$ 2.20(1)

where $s \geqslant q \geqslant 1$ and

$$\sigma_q^2 = 2\delta_q^2 n$$ 2.20(2)

and where as in 2.17(1)

$$z_q = 2\delta_q r$$ 2.20(3)

Let us now consider

$$z = \sum_{q=1}^{q=s} z_q$$ 2.20(4)

If we can find the distribution corresponding to z, this will be a distribution made up of all the possible ways of combining s groups of different δs, and this will thus be quite general. We thus require to find $P_z \, dz$ when

$$z + \delta z \geqslant \sum_{q=1}^{q=s} z_q > z$$

2.21 Now it will be shown later (see paragraph 2.29) that the required density function for z is given by (see 2.29(3))

$$P_z = \frac{e^{-(z-\mu)/2\sigma_z^2}}{\sigma_z\sqrt{(2\pi)}}$$ 2.21(1)

where

$$\sigma_z^2 = \sum_{q=1}^{q=s} \sigma_{z_q}^2$$ 2.21(2)

and where the mean of z is given by

$$\bar{z} \equiv \mu$$ 2.21(3)

and z is given by 2.20(4), that is a_r has been put equal to unity in 2.29(1), and 2.29(2). A particular value of z may thus be considered as the resultant of many small random deviations of different size. It is to be noted that the density function is independent of the particular small deviations involved in producing a particular deviation $z - \mu$ from μ; that is, it is a function only of z, σ_z and μ, the latter two parameters being constants of the distribution.

Approximation to the Mean μ of a Distribution

2.22 We have seen that the mean value of x for the distribution given by the density function

$$P_x = \frac{e^{-(x-x_0)^2}}{2\sigma^2} \qquad 2.18(6)$$

is given by

$$x = x_0 = \mu \qquad 2.19(3)$$

But if P_x is determined from a number of readings our best estimate of x_0 is given by

$$\sum_1^n x_r/n \qquad 2.22(1)$$

determined from n readings. This estimate of μ is usually denoted by $\bar{x}$. Later we shall see how to estimate the uncertainty in this mean, since the correct value of x_0 is obtained only when $n \to \infty$.

Approximation to the Standard Deviation σ

2.23 From the manner in which σ^2 is defined it is clear that

$$\sigma^2 = \frac{\sum_1^n (x_r - \bar{x})^2}{n} \qquad 2.23(1)$$

for large n. If, however, n is not large, that is less than 200, then 2.23(1) is not the best estimate of σ^2, as will be demonstrated below. Let us define the mean of a series of observations as μ when the number of readings tends to infinity. The mean of a finite number of readings n we will call $\bar{x}$, and this will in general not coincide with μ. Let the readings be denoted by x_r, and let the differences between μ and the readings be defined as the *deviations* and the differences between $\bar{x}$ and the readings as the *residuals*.

2.24 Let the *residuals* be

$$r_1 = \bar{x} - x_1, r_2 = \bar{x} - x_2, \ldots, r_n = \bar{x} - x_n \qquad 2.24(1)$$

whilst the *deviations* are

$$\varepsilon_1 = \mu - x_1, \varepsilon_2 = \mu - x_2, \ldots, \varepsilon_n = \mu - x_n \qquad 2.24(2)$$

Adding the last equations we have

$$\sum_1^n \varepsilon_r = n\mu - \sum_1^n x_r = n\mu - n\bar{x}$$

therefore

$$\bar{x} = \mu - \sum_1^n \frac{\varepsilon_r}{n} \qquad\qquad 2.24(3)$$

Substituting in 2.24(1) for $\bar{x}$ and using 2.24(2) we have

$$r_1 = \mu - \sum_1^n \frac{\varepsilon_r}{n} - x_1 = \varepsilon_1 - \sum_1^n \frac{\varepsilon_r}{n}$$

or

$$r_1 = \frac{(n-1)}{n}\varepsilon_1 - \frac{\varepsilon_2}{n} - \frac{\varepsilon_3}{n} - \cdots - \frac{\varepsilon_n}{n} \qquad\qquad 2.24(4)$$

Now let σ_e denote the standard deviation of the deviations and σ_r the standard deviation of the residuals, then

$$\sigma_r^2 = \left(\frac{n-1}{n}\right)^2 \sigma_e^2 + (n-1)\frac{\sigma_e^2}{n^2}$$

from comparing 2.24(4) with 2.29(1) and 2.29(2)

$$= \left(\frac{n-1}{n}\right)\sigma_e^2 \qquad\qquad 2.24(5)$$

So the probability that a residual lies between r and $r + dr$ is

$$\frac{e^{-r^2/2\sigma_r^2}\,dr}{\sigma_r\sqrt{(2\pi)}} = \frac{1}{\sigma_e}\left(\frac{n}{n-1}\right)^{1/2}\frac{\exp[-r^2n/2\sigma_e^2(n-1)]\,dr}{\sqrt{(2\pi)}} \qquad\qquad 2.24(6)$$

Since

$$\sigma_r^2 = \sum_1^n \frac{r_r^2}{n}$$

then from 2.24(5)

$$\sigma_e^2 = \frac{n}{(n-1)}\sigma_r^2 \qquad\qquad 2.24(7)$$

that is

$$\sigma_e^2 = \frac{\sum r_r^2}{n-1} \qquad\qquad 2.24(8)$$

or

$$\sigma_e = \left(\sum_1^n \frac{r_r^2}{n-1}\right)^{1/2}$$

$$= \left\{\sum_1^n \frac{(x_1 - \bar{x})^2}{n-1}\right\}^{1/2} = s_v \qquad\qquad 2.24(9)$$

It is usual to call the expression in 2.24(8) an estimate of σ and this is usually denoted by s_v. σ^2 is often called the *variance* of a distribution. The suffix v of s denotes the number of degrees of freedom of s which, in the case of one variable, is equal to $n - 1$, where n is the number of readings. s without suffix denotes the standard deviation of a sample, whilst s^2 is known as the sample variance, where

$$s^2 = \sum_1^n \frac{(x_r - \bar{x})^2}{n} \qquad 2.24(10)$$

Mean Absolute Deviation η

2.25   It is sometimes quicker to calculate this quantity rather than σ, and then to express σ in terms of η. η is defined as follows:

$$\eta = \int_{-\infty}^{\infty} \frac{|x| e^{-x^2/2\sigma^2}}{\sigma\sqrt{(2\pi)}} \qquad 2.25(1)$$

Putting $x/\sqrt{2}\sigma = y$, giving $dx = \sqrt{2}\sigma\, dy$, we have

$$\eta = \int_{-\infty}^{\infty} \frac{\sqrt{2}\sigma |y| e^{-y^2} dy \sqrt{2}\sigma}{\sigma\sqrt{(2\pi)}}$$

$$= \sqrt{\left(\frac{2}{\pi}\right)}\sigma \int_{-\infty}^{\infty} |y| e^{-y^2}\, dy$$

$$= 2\sqrt{\left(\frac{2}{\pi}\right)}\sigma \int_0^{\infty} y\, e^{-y^2}\, dy$$

$$= 2\sqrt{\left(\frac{2}{\pi}\right)}\sigma \left[\frac{-e^{-y^2}}{2}\right]_0^{\infty}$$

$$= \sqrt{\left(\frac{2}{\pi}\right)}\sigma \qquad 2.25(2)$$

Now it can be shown that[†]

$$\eta = \frac{\sum |r_r|}{\sqrt{(n(n-1))}} \qquad 2.25(3)$$

where $r_r = r$th residual $\bar{x} - x_r$. Thus

$$\sigma = \sqrt{\left(\frac{\pi}{2}\right)}\eta = \sqrt{\left(\frac{\pi}{2n(n-1)}\right)}\sum_1^n |r_r| \qquad 2.25(4)$$

† *Ast Nach* **44** 29, 1856.

Combination of Two Gaussian Distributions

2.26 So far we have considered the distribution of a single variable but now we shall consider the case of two independent variables x and y, each of which has a Gaussian distribution. We shall derive an expression which represents the combined probability distribution of both variables. A simple example of this type of distribution is met with if we have two meters, one measuring voltage and the other measuring current, and we wish to know the uncertainty in the wattage.

2.27 Since the range of uncertainty is always small, it is always possible to represent the variation of one variable in terms of another by an expression such as $z = ax$, where a is constant for the particular points about which z and x are considered to vary. Now the probability of z lying in the range z to $z + dz$ is given by the expression

$$P_z \, dz = \frac{e^{-z^2/2\sigma_z^2} \, dz}{\sigma_z \sqrt{(2\pi)}} \qquad 2.27(1)$$

If we put

$$z = ax \qquad 2.27(2)$$

then

$$\delta z = a \, \delta x \qquad 2.27(3)$$

and

$$\delta z^2 = a^2 \, \delta x^2$$

Therefore

$$\sum \delta z^2 = a^2 \sum \delta x^2$$

and thus

$$\sigma_z^2 = \sum \frac{\delta z^2}{n-1} = a^2 \frac{\sum \delta x^2}{n-1} = a^2 \sigma_x^2 \qquad 2.27(4)$$

Thus, replacing σ_z by $a\sigma_x$ we have

$$P_z \, dz = \frac{e^{-z^2/2a^2\sigma_x^2} \, dz}{a\sigma_x \sqrt{(2\pi)}} \qquad 2.27(5)$$

Similarly, if

$$z = by \qquad 2.27(6)$$

$$P_z \, dz = \frac{e^{-z^2/2b^2\sigma_y^2} \, dz}{b\sigma_y \sqrt{(2\pi)}} \qquad 2.27(7)$$

If, however, $z = ax + by$, what now is the probability function for z? The probability of an uncertainty of δx in x is given by

$$\frac{e^{-x^2/2\sigma_x^2}\,dx}{\sigma_x\sqrt{(2\pi)}} \qquad\qquad 2.27(8)$$

Similarly the probability of an uncertainty of δy in y is given by

$$\frac{e^{-y^2/2\sigma_y^2}\,dy}{\sigma_y\sqrt{(2\pi)}} \qquad\qquad 2.27(9)$$

If

$$z = ax + by \qquad\qquad 2.27(10)$$

then we wish to find the probability of an uncertainty

$$z + \delta z \geqslant ax + by \geqslant z \qquad\qquad 2.27(11)$$

Now the probability of an uncertainty δx occurring at the same time as an error δy is given by

$$\frac{e^{-x^2/2\sigma_x^2}\,dx\,e^{-y^2/2\sigma_y^2}\,dy}{\sigma_y\sigma_x\cdot 2\pi} \qquad\qquad 2.27(12)$$

2.28 Now, in order to satisfy 2.27(11), we must integrate 2.27(12) with respect to y between the limits y and $y + \delta y$, and with respect to x between the limits $-\infty$ to $+\infty$. This is so because whatever value is chosen for y there is always a range of values for x which makes 2.27(11) true, that is, from

$$x = \frac{z - by}{a} \qquad \text{to} \qquad \frac{z + \delta z - by}{a}$$

Thus

$$P_z\,dz = \frac{1}{\sigma_x\sigma_y 2\pi}\int_{-\infty}^{\infty} e^{-x^2/2\sigma_x^2}\,dx \int_{y}^{y+\delta y} e^{-y^2/2\sigma_y^2}\,dy$$

$$= \frac{1}{\sigma_x\sigma_y 2\pi}\int_{-\infty}^{\infty} e^{-x^2/2\sigma_x^2}\,dx \int_{\frac{z-ax}{b}}^{\frac{z+\delta z-ax}{b}} e^{-y^2/2\sigma_y^2}\,dy \qquad\qquad 2.28(1)$$

where the limits for y have been obtained from 2.27(11) in terms of z and x. Now since

$$\int_{\frac{z-ax}{b}}^{\frac{z+\delta z-ax}{b}} e^{-y^2/2\sigma_y^2}\,dy = \exp\left\{-\frac{(z-ax)^2}{2b^2\sigma_y^2}\right\}\frac{\delta z}{b} \qquad\qquad 2.28(2)$$

we have

$$P_z \, dz = \frac{dz}{b\sigma_x\sigma_y 2\pi} \int_{-\infty}^{\infty} \exp\left\{-x^2/2\sigma_x^2 - \frac{(z-ax)^2}{2b^2\sigma_y^2}\right\} dx \qquad 2.28(3)$$

Now consider the exponent of e, that is

$$-\frac{x^2}{2\sigma_x^2} - \frac{(z-ax)^2}{2b^2\sigma_y^2}$$

If we are to evaluate the expression for $P \, dz$ we must be able to complete the square for x in the exponent of e. Now

$$\frac{x^2}{2\sigma_x^2} + \frac{(z-ax)^2}{2b^2\sigma_y^2} = \frac{x^2(a^2\sigma_x^2 + b^2\sigma_y^2) - 2a\sigma_x^2 xz + z^2\sigma_x^2}{2\sigma_x^2\sigma_y^2 b^2} \qquad 2.28(4)$$

and we may write

$$fx^2 + gx + c = f(x+h)^2 + k$$
$$= fx^2 + 2fxh + fh^2 + k \qquad 2.28(5)$$

Equating coefficients each side of 2.28(5) we have

$$g = 2fh$$

and

$$c = fh^2 + k$$

therefore

$$h = g/2f \qquad 2.28(6)$$

and

$$k = c - fh^2$$
$$= c - \frac{fg^2}{4f^2}$$
$$= c - \frac{g^2}{4f} \qquad 2.28(7)$$

Comparing the left-hand side of 2.28(5) with the right-hand side of 2.28(4) we have on equating corresponding coefficients of x

$$f = \frac{a^2\sigma_x^2 + b^2\sigma_y^2}{2\sigma_x^2\sigma_y^2 b^2}$$

$$g = -\frac{2a\sigma_x^2 z}{\sigma_x^2\sigma_y^2 b^2} = -\frac{az}{b^2\sigma_y^2}$$

$$c = \frac{z^2\sigma_x^2}{2\sigma_x^2\sigma_y^2 b^2} = \frac{z^2}{2b^2\sigma_y^2}$$

Thus from 2.28(6) and 2.28(7)

$$h = -\frac{a\sigma_x^2 z}{a^2\sigma_x^2 + b^2\sigma_y^2}$$

2.28(8)

and

$$k = \frac{z^2}{2b^2\sigma_y^2} - \frac{a^2\sigma_x^2 z^2}{2(a^2\sigma_x^2 + b^2\sigma_y^2)\sigma_y^2 b^2}$$

$$= \frac{z^2}{2(a^2\sigma_x^2 + b^2\sigma_y^2)}$$

2.28(9)

and so the exponent of e, using 2.28(5), is equal to

$$-\frac{a^2\sigma_x^2 + b^2\sigma_y^2}{2\sigma_x^2\sigma_y^2 b^2}\left(x - \frac{a\sigma_x^2 z}{(a^2\sigma_x^2 + b^2\sigma_y^2)}\right)^2 - \frac{z^2}{2(a^2\sigma_x^2 + b^2\sigma_y^2)}$$

Thus

$$P_z \, dz = \frac{\exp\left\{-\dfrac{z^2}{2(a^2\sigma_x^2 + b^2\sigma_y^2)}\right\}}{b\sigma_x\sigma_y 2\pi}$$

$$\times \int_{-\infty}^{\infty} \exp\left\{-\frac{a^2\sigma_x^2 + b^2\sigma_y^2}{2\sigma_x^2\sigma_y^2 b^2}\left(x - \frac{a z\sigma_x^2}{(a^2\sigma_x^2 + b^2\sigma_y^2)}\right)^2\right\} dx$$

Now the integral is

$$\int_{-\infty}^{\infty} e^{-f(x+h)^2} \, dx = I_z$$

2.28(10)

and so writing

$$\sqrt{f}(x + h) = \eta$$

we have

$$dx = d\eta/\sqrt{f}$$

giving, on substituting for x in terms of η

$$I_z = \int_{-\infty}^{\infty} \frac{e^{-\eta^2} \, d\eta}{\sqrt{f}} = \sqrt{(\pi/f)} \qquad \text{(see paragraph 2.10)}$$

$$= \frac{\sqrt{(2\pi)}\sigma_x\sigma_y b}{(a^2\sigma_x^2 + b^2\sigma_y^2)^{1/2}}$$

2.28(11)

Thus

$$P_z \, dz = \frac{e^{-z^2/2(a^2\sigma_x^2 + b^2\sigma_y^2)} \, dz}{\sqrt{\{2\pi(a^2\sigma_x^2 + b^2\sigma_y^2)\}}} \qquad 2.28(12)$$

Comparing with the standard form of the Gaussian distribution, that is

$$P_x \, dx = \frac{e^{-x^2/2\sigma^2} \, dx}{\sqrt{(2\pi)}\sigma}$$

we see that the standard deviation of the frequency function 2.28(12) is equal to

$$(a^2\sigma_x^2 + b^2\sigma_y^2)^{1/2} \qquad 2.28(13)$$

2.29 Thus we have proved that the probability function of $ax + by$ is that given by 2.28(12) and that its standard deviation is equal to the square root of the sum of the squares of the separate standard deviations, each multiplied by its appropriate constant, that is, a or b. By writing $ax + by = \chi$ and $cv + fv = \kappa$ it is easily seen that the standard deviation of the expression

$$\chi + \kappa = ax + by + cv + fv$$

is equal to

$$(a^2\sigma_x^2 + b^2\sigma_y^2 + c^2\sigma_v^2 + f^2\sigma_v^2)^{1/2}$$

Thus, if we have an expression of the form

$$z = \sum_{r=1}^{r=n} a_r x_r \qquad 2.29(1)$$

where x_r are independent variables, then the standard deviation is equal to

$$\left(\sum_{r=1}^{r=n} a_r^2 \sigma_r^2\right)^{1/2} = \sigma_z \qquad 2.29(2)$$

where σ_r is the standard deviation of x_r and

$$P_z \, dz = \frac{e^{-(z-\mu)^2/2\sigma_z^2}}{\sigma_z\sqrt{(2\pi)}} = \frac{\exp\{-(z-\mu)^2/2\sum_1^n a^2\sigma_r^2\}}{(\sum a^2\sigma_r^2)^{1/2}\sqrt{(2\pi)}} \qquad 2.29(3)$$

where $\mu \equiv$ mean of z. This is a very important theorem since it enables us to calculate the frequency distribution of a number of independent variables; it would for instance enable the calculation of the probability of an uncertainty in the wattage consumed by a circuit for which the current and voltage frequency distributions are known (see paragraph 2.26). Stated generally, the frequency distribution of a combination of Gaussian distributions is itself a Gaussian distribution, with a standard deviation equal to the square root of the sum of the squares of the component standard deviations.

Standard Deviation of the Mean or Standard Error

2.30 Consider equation 2.29(1). If we put $a_r = 1/n$ and the x_r are independent measures of the same quantity, then

$$z = \frac{1}{n} \sum_{r=1}^{r=n} x_r = \bar{x}$$

that is z is the mean of the xs. Thus using 2.29(2), the standard deviation of $\bar{x}$ is equal to

$$\left(\sum_1^n \frac{\sigma_r^2}{n^2} \right)^{1/2} = \left(\frac{n\sigma^2}{n^2} \right)^{1/2} = \frac{\sigma}{\sqrt{n}} \qquad 2.30(1)$$

where $\sigma_r = \sigma$, $r = 1$ to n, since all σ_r are equal, because all the x_r are independent measures of the same quantity. Thus we have shown that the standard deviation of the mean of a variable x is equal to the standard deviation of x divided by the square root of the number of measures of x. The frequency function of $\bar{x}$ is of course

$$P_{\bar{x}} \, d\bar{x} = \frac{1}{\sigma} \sqrt{\left(\frac{n}{2\pi} \right)} e^{-(\bar{x} - \mu)^2 n / 2\sigma^2} \, d\bar{x} \qquad 2.30(2)$$

The quantity $\sigma/\sqrt{n}$ is often called the standard error of a variable. An important point to note is that the standard deviation of a variable is a constant, but that the standard error of a variable diminishes by $1/\sqrt{n}$ as n increases.

Standard Deviation of the Standard Deviation

2.31 When σ is calculated from the formula

$$\sigma^2 = \sum_1^n \frac{(x_r - \bar{x})^2}{n} \qquad 2.23(1)$$

the value obtained is only correct if n is very large (200 or more). It was also shown that if n is small the best estimate of σ^2 was given by

$$\sigma_e^2 = s_v^2 = \sum \frac{(x_r - \bar{x})^2}{n - 1} \qquad \text{(see 2.24(9))}$$

But this is only an estimate and represents the most probable value of σ; the correct value may be either smaller or larger than this.

2.32 We shall now show that the estimate of σ, σ_e or s has its own probability distribution, which under the conditions we shall consider is approximately Gaussian, and which can be used to calculate the probability of a deviation in σ from the estimated value with fair accuracy provided that n is larger than about 25. When n is smaller than this recourse must be made to a more accurate derivation of the probability distribution for σ_e or s (see paragraph 6.16).

2.33   Let us consider n deviations $\varepsilon_1, \varepsilon_2, \ldots, \varepsilon_n$ from the mean value μ, and let the standard deviation have a value σ_1. The probability of just this set of deviations is thus

$$P = \frac{\exp(-\sum_1^n \varepsilon_r^2 / 2\sigma_1^2)}{\sigma_1^n (2\pi)^{n/2}} \qquad 2.33(1)$$

If on the other hand the standard deviation were $\sigma_1 + \gamma$, the probability of the same set of deviations would be

$$\frac{\exp\{-\sum_1^n \varepsilon_r^2 / 2(\sigma_1 + \gamma)^2\}}{(\sigma_1 + \gamma)^n (2\pi)^{n/2}} \qquad 2.33(2)$$

Thus the ratio of the probability that $\sigma_1 + \gamma$ is the true value of σ to the probability that σ_1 is the true value of σ is equal to the ratio of expression 2.33(2) to 2.33(1); that is

$$
\begin{aligned}
Q &= \left(1 + \frac{\gamma}{\sigma_1}\right)^{-n} \exp\left[\frac{1}{2} \sum_1^n \varepsilon_r^2 \left\{\frac{1}{\sigma_1^2} - \frac{1}{(\sigma_1 + \gamma)^2}\right\}\right] \\
&= \left(1 + \frac{\gamma}{\sigma_1}\right)^{-n} \exp\left\{\frac{1}{2} \sum_1^n \varepsilon_r^2 \frac{(2\gamma\sigma_1 + \gamma^2)}{\sigma_1^2 (\sigma_1 + \gamma)^2}\right\} \\
&= \exp\left\{\frac{1}{2} \sum_1^n \varepsilon_r^2 \frac{(2\gamma\sigma_1 + \gamma^2)}{\sigma_1^2 (\sigma_1 + \gamma)^2} - n \log_e\left(1 + \frac{\gamma}{\sigma_1}\right)\right\} \qquad 2.33(3)
\end{aligned}
$$

since

$$a^{-n} = e^{-n \log_e a}$$

2.34   Now suppose that σ_1 is the value which makes 2.33(1) a maximum, that is that σ_1 is the most probable value of σ. Thus taking the log of 2.33(1), we have

$$\log P = -n \log \sigma_1 - \sum_1^n \varepsilon_r^2 / 2\sigma_1^2 - \frac{n}{2} \log 2\pi$$

and, differentiating with respect to σ, we have

$$\frac{1}{P} \frac{dP}{d\sigma_1} = \frac{-n}{\sigma_1} - \frac{\sum_1^n \varepsilon_r^2 (-2)}{2\sigma_1^3} = 0$$

for a maximum value for σ_1, and so

$$\sigma_1^2 = \sum_1^n \varepsilon_r^2 / n \qquad 2.34(1)$$

or

$$\sigma_1 = \left(\sum_1^n \varepsilon_r^2/n \right)^{1/2} \qquad 2.34(2)$$

Thus, substituting for $\sum_1^n \varepsilon_r^2$ in 2.33(3) we have

$$Q = \exp\left\{ \frac{1}{2} n \frac{(2\gamma\sigma_1 + \gamma^2)}{(\sigma_1 + \gamma)^2} - n \log\left(1 + \frac{\gamma}{\sigma_1} \right) \right\}$$

The exponent of e can be expanded to give

$$Q = \exp\left\{ +\frac{1}{2} \frac{n(2\gamma\sigma_1 + \gamma^2)}{\sigma_1^2(1 + \gamma/\sigma_1)^2} - n\left(\frac{\gamma}{\sigma_1} - \frac{\gamma^2}{2\sigma_1^2} + \frac{\gamma^3}{3\sigma_1^3} - \cdots \right) \right\}$$

$$= \exp\left\{ +\left(\frac{n\gamma}{\sigma_1} + \frac{1}{2} n \frac{\gamma^2}{\sigma_1^2} \right)\left(1 - \frac{2\gamma}{\sigma_1} + \frac{3\gamma^2}{\sigma_1^2} - \frac{4\gamma^3}{\sigma_1^3} + \cdots \right) \right.$$

$$\left. - n\left(\frac{\gamma}{\sigma_1} - \frac{\gamma^2}{2\sigma_1^2} + \frac{\gamma^3}{3\sigma_1^3} - \cdots \right) \right\}$$

which on simplification and neglecting terms of a higher power than γ^2, on the assumption that γ is small compared with σ_1, gives

$$Q = e^{-n\gamma^2/\sigma_1^2} \qquad 2.34(3)$$

since the first powers of γ cancel. Thus the probability that the value of σ_1 lies between $\sigma_1 + \gamma$ and $\sigma_1 + \gamma + d\gamma$ is approximately equal to

$$Ke^{-n\gamma^2/\sigma_1^2} \, d\gamma \qquad 2.34(4)$$

where K is a constant. Now

$$\int_{-\infty}^{\infty} Ke^{-n\gamma^2/\sigma_1^2} \, d\gamma = 1 \qquad 2.34(5)$$

since 2.34(4) is a probability density function. Thus, putting $\sqrt{n}\gamma/\sigma_1 = x$, giving $d\gamma = \sigma_1/\sqrt{n} \, dx$ we have, on substituting in 2.34(5),

$$K \frac{\sigma_1}{\sqrt{n}} \int_{-\infty}^{\infty} e^{-x^2} \, dx = K \frac{\sigma_1}{\sqrt{n}} \sqrt{\pi} = 1$$

and therefore

$$K = \frac{\sqrt{n}}{\sigma_1\sqrt{\pi}} \qquad 2.34(6)$$

Thus the density function for σ_1 is

$$\frac{\sqrt{n}}{\sigma_1\sqrt{\pi}} e^{-n\gamma^2/\sigma_1^2} \, d\gamma \qquad 2.34(7)$$

which, on comparison with the standard form for a Gaussian distribution, that is

$$\frac{e^{-x^2/2\sigma^2}}{\sigma\sqrt{(2\pi)}} dx$$

shows that the standard deviation of the density function for σ_1 is equal to

$$\sigma_1/\sqrt{(2n)} \qquad\qquad 2.34(8)$$

It should be noted that we have assumed γ small compared with σ_1 and also n large in order to make the expression for $\sum \varepsilon_r^2$, that is 2.34(1), a reasonable approximation to σ^2. In the next section we shall consider the use of some of the expressions so far derived.

Tolerance Intervals and Tolerance Probabilities

2.35 Let us consider the expression for the Gaussian distribution, that is

$$P = \frac{e^{-z^2/2\sigma^2}}{\sigma\sqrt{(2\pi)}} \qquad\qquad 2.11(4)$$

If we write $z/\sigma = c$, then $dz = \sigma\, dc$ and the expression $P\, dz$ becomes

$$P\, dz = \frac{e^{-c^2/2}}{\sqrt{(2\pi)}} dc \qquad\qquad 2.35(1)$$

$$P_n = \frac{e^{-c^2/2}}{\sqrt{(2\pi)}} \qquad\qquad 2.35(2)$$

is the normalized frequency distribution since it corresponds to a frequency distribution whose standard deviation is unity. Table I in Appendix I gives values of P_n for a number of values of c at appropriate intervals up to $c = 4$, a range which covers most practical cases.

2.36 It will be seen from 2.18(5) that as σ increases the curve for P becomes flatter. See Figure 2.36.

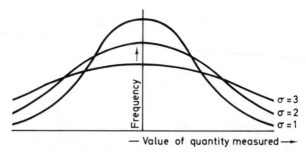

Figure 2.36 Graph showing the effect of increasing σ, the standard deviation, and the flattening of the Gaussian curves which is produced

2.37 Now the area under the curve formed by the function P, taken between the limits z_1 and z_2 (see Figure 2.37a) represents the probability of occurrence of an uncertainty between these limits (see paragraph 1.16 and 1.17). When dealing with uncertainties it is usual to measure the limits from the mean value μ of the distribution, thus if the distribution is given in the form

$$P = \frac{e^{-(x-\mu)^2/2\sigma^2}}{\sigma\sqrt{(2\pi)}} \qquad 2.11(5)$$

the limits will be given as $x_1 - \mu$ and $x_2 - \mu$ (see Figure 2.37b). When we are dealing with uncertainties we are usually interested in the probability of an uncertainty in a given range on either side of the mean value, and in most cases these two ranges are taken as equal.

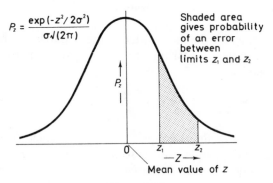

Figure 2.37a

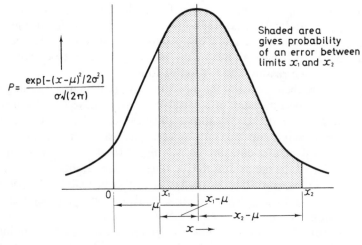

Figure 2.37b

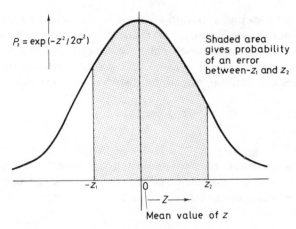

Figure 2.38

2.38   If we arrange for the mean value of the variable z to be at the zero of our co-ordinate system by writing $z = x - \mu$, then our probability density function is given by

$$P\,dz = \frac{e^{-z^2/2\sigma^2}}{\sigma\sqrt{(2\pi)}}$$ 2.18(4)

Thus in most cases the limits z_1 and z_2 will be on either side of the ordinate axis, and in the great majority of these cases $z_1 = -z_2$, z_2 being positive. If we regard the limit points z as positive, the two limits become $-z_1$ and z_2. See Figure 2.38.

2.39   The interval $-z_1$ to z_2 is spoken of as a *tolerance interval* and represents the uncertainty in the value of z. Associated with the tolerance interval is the corresponding probability, that is the probability that the value of z will lie between z_1 and z_2. This probability is known as the *tolerance probability*. Thus if the standard deviation of a Gaussian distribution has been found using 2.24(8) the uncertainties in the measurements can be expressed in a form that can be readily understood as follows.

2.40   First the mean value of the measurements (calculated by using 2.22(1)) is taken as the best measure of a variable. Now we have seen from what has been said so far that there is no *one* uncertainty to be associated with the mean or best value, but that any uncertainty has a known probability of occurrence. In these circumstances it is usual to choose two positions, one on either side of the mean, and to calculate the probability of an uncertainty between these two positions. Note that when this probability has been found it represents the probability of any one measurement lying between the two limit positions chosen.

2.41   Now it would not be of much use to choose the two limit points (these are usually taken as being equally disposed on either side of the mean) so that the probability of an uncertainty lying between them is small, that is 0.1, 0.2 or even 0.5, since even in the latter case there is the probability that every other reading will be outside the chosen range. It is for this reason that the uncertainty of measurement is usually expressed for certain specified values of probability; these are usually 0.95, 0.9545, 0.99 and 0.9973. These values may appear rather odd at first but the reason for their choice will be explained later. What should be noted is that the probability of an error lying *outside* the range represented by the first two values is about 1/20, whilst the corresponding figures for the last two values are 1/100 and 1/370. The first two values are used for measurements where a reasonably small probability of the chosen range being exceeded is sufficient, in this case about 1/20. The third and fourth values are usually associated with measurements where it is necessary to feel very confident that the quantity measured will not lie outside the chosen range, and in particular the last value is often selected for dimensional metrology measurements.

2.42 We shall now find the values of the limit points of the ranges expressed by the probabilities quoted in paragraph 2.41. If we consider the integral

$$\frac{1}{\sigma\sqrt{(2\pi)}} \int_{-z_1}^{z_1} e^{-z^2/2\sigma^2} \, dz \qquad 2.42(1)$$

the solution of this will give us the probability of an uncertainty lying in the range $-z_1$ to z_1. Writing $z/\sigma = c$, then $dz = \sigma \, dz$ and the integral becomes

$$\underset{-k_1 \text{ to } k_1}{P} = \frac{1}{\sqrt{(2\pi)}} \int_{-z_1/\sigma}^{z_1/\sigma} e^{-c^2/2} \, dc \qquad 2.42(2)$$

Table II in Appendix I gives values of this integral at appropriate intervals of $z_1/\sigma = k_1$ up to $k_1 = 3.891$.

2.43 Thus, to discover the probability associated with any selected uncertainty range $-z_1$ to z_1, z_1 is divided by the value of the standard deviation σ obtained from a set of measurements and the value of k_1 is found. The value of the integral $\underset{-k \text{ to } k}{P}$ is then read off against the appropriate entry for k_1. If k_1 falls between two entries of k_1 it will be necessary to interpolate, but linear interpolation will usually be quite sufficient for the accuracy required.

2.44 Table III in Appendix I is the inverse of Table II and gives the values of k_1 for given values of $\underset{-k_1 \text{ to } k_1}{P}$ at intervals of 0.01. This table is useful when it is desired to find the value of k_1 corresponding to a given probability, that is if we select a probability it enables us to find the corresponding range. Referring now to the four probability values given in paragraph 2.40, namely, 0.95, 0.9545, 0.99 and 0.9973, we find that the corresponding values of k_1 are 1.96, 2.00, 2.576 and 3.00. The second and fourth figures are more easily seen from Table II. Since $k_1 = z_1/\sigma$, the corresponding values of z_1 are 1.96σ,

2.00σ, 2.576σ and 3.00σ. Thus the four values of the probability we have considered are associated with uncertainty ranges equal to twice the values quoted above, or plus the above values on one side of the mean and minus the above values on the other side. It is of interest to note that the probability associated with an uncertainty range of one standard deviation σ on either side of the mean is equal to 0.683.

2.45 If for some reason an asymmetrical uncertainty range is required, say $-z_1$ to z_2, the probability of the range $-z_1$ to 0 for one side of the mean is found by dividing z_1 by σ and looking up the value of $P_{-k_1 \text{to} k_1}$ corresponding to this value of $k_1 = z_1/\sigma$, and then dividing by 2. Similarly the probability of the range 0 to z_2 is found from $k_1 = z_2/\sigma$ and the value of $P_{-k_1 \text{to} k_1}$ obtained is divided by 2. Thus the value of the *tolerance probability* associated with the tolerance range $-z_1$ to z_2 is given by

$$\left[\left(P_{-k_1 \text{to} k_1}\right)_{z_1} + \left(P_{-k_1 \text{to} k_1}\right)_{z_2}\right]/2 \qquad 2.45(1)$$

Thus, provided that we know σ, the standard deviation, and μ the mean value, we can find the *tolerance probability* for a given *tolerance interval* using Table II, or find the *tolerance interval* given the *tolerance probability* using Table III.

Alternative expression of tolerance probabilities

2.46 Tolerance probabilities are usually given for an uncertainty to lie *inside* the quoted range, but they are sometimes quoted as the probability of an uncertainty being found *outside* the range. The latter probability is always unity minus the first. For example, the probability of an uncertainty in the range -2σ to $+2\sigma$ is 0.9545, or the probability of an uncertainty occurring outside this range is $1 - 0.9545 = 0.0455 \simeq 1/20$.

Confidence Intervals and Confidence Probabilities with Reference to the Mean Value

2.47 Let us now turn our attention to the mean value of a set of readings. In many cases of measurement it is not the probability of a single reading that is required but the probability of an error in the mean of a number of readings. For instance, if a large number of readings of the position of a measuring machine carriage or of the angular disposition of the face of a given workpiece have been made we may then wish to know how reliable the mean value of these readings is.

2.48 Having found the standard deviation σ and the mean μ of the readings, using respectively 2.24(8) and 2.22(1), we know that the standard deviation of the mean is $\sigma/\sqrt{n}$, and that the probability distribution of the mean is thus

$$P_m \, dx = \frac{\sqrt{n}}{\sigma\sqrt{(2\pi)}} \, e^{-(x-\mu)^2 n/2\sigma^2} \, dx \qquad 2.48(1)$$

on substituting for the standard deviation and mean in 2.30(2).

2.49 Suppose we choose limit points x_1 and x_2 on either side of the mean value, then we require to know the probability that the *mean* value shall lie between these limit points. The range between x_1 and x_2 is called the *confidence interval* for the mean and the associated probability of the mean lying in this range is called the *confidence probability*. We find the latter quantity, having chosen the *confidence interval*, as follows.

2.50 The standard deviation of the mean is equal to $\sigma/\sqrt{n} = \sigma_m$. Thus if we require the probability of an uncertainty in the mean, between two values of x equally disposed on either side of the mean, then if the larger limit point is x_1, $x_1 > \mu$, then the other limit point will be $\mu - (x_1 - \mu) = 2\mu - x_1$. That this is the second or lower limit point is easily seen from Figure 2.50. Thus the *confidence probability* of an uncertainty between the two *confidence limit* points $2\mu - x_1$ and x_1 is given by

$$\int_{2\mu-x_1}^{x_1} P_m \, dx = \frac{1}{\sigma} \sqrt{\left(\frac{n}{2\pi}\right)} \int_{2\mu-x_1}^{x_1} e^{-(x-\mu)^2 n/2\sigma^2} \, dx \qquad 2.50(1)$$

If we put $(x-\mu)\sqrt{n}/\sigma = c$, then $dz = \sigma/\sqrt{n} \, dc$ and the limit points become

$$c_1 = (2\mu - x_1 - \mu)\frac{\sqrt{n}}{\sigma} = -(x_1 - \mu)\frac{\sqrt{n}}{\sigma}$$

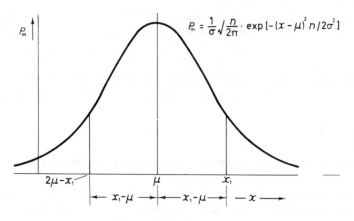

Figure 2.50

and

$$c_2 = (x_1 - \mu) \frac{\sqrt{n}}{\sigma}$$

Thus the probability integral becomes

$$\underset{-k_1 \text{ to } k_1}{P} = \frac{1}{\sqrt{(2\pi)}} \int_{-(x_1-\mu)(\sqrt{n})/\sigma}^{(x_1-\mu)(\sqrt{n})/\sigma} e^{-c^2/2} \, dc \qquad 2.50(2)$$

where

$$k_1 = (x_1 - \mu) \frac{\sqrt{n}}{\sigma} = \frac{(x_1 - \mu)}{\sigma_m} \qquad 2.50(3)$$

2.51 Thus the procedure is exactly the same as in the case of the probability of a single reading lying between specified limits, that is the limit semi-range $x_1 - \mu$ is divided by the standard deviation of the mean, giving k_1, and the corresponding value $\underset{-k_1 \text{ to } k_1}{P}$ from Table II gives the required *confidence probability* for the uncertainty in the mean, that is the probability of the mean lying within the range $x_1 - \mu$ on either side of the mean value. The confidence range is given by $k_1 \sigma/\sqrt{n}$, whilst $k_1/\sqrt{n}$ is known as the *confidence factor*.

2.52 It should be noted that the tolerance intervals and confidence intervals and their related probabilities, which we have discussed above, are reliable only if the value of the standard deviation σ and of the mean value μ are known accurately. Generally σ and μ are derived from several readings which may not often exceed ten or twenty, and in these circumstances the values we obtain for σ and μ, in terms of their estimates s and $\bar{x}$, may be in error. This could mean that the tolerance interval or confidence interval should be larger or smaller for the same tolerance probability or confidence probability. Both possibilities exist but it is usual to consider the possibility of the intervals being larger and a confidence probability is then associated with the tolerance probability of a tolerance interval. In the case of a confidence interval, the interval is smallest for a chosen confidence probability when σ is known accurately, and increases as n decreases, where σ is estimated from n readings as s_y. We shall deal with this in more detail in Chapter 6, paragraph 6.21 onwards.

Difference between a tolerance interval and a confidence interval

2.53 The reader may be a little puzzled over the use of the terms tolerance interval and confidence interval since they may seem to be synonymous. This is not the case, however, since a *tolerance interval* applies to the range over which it is expected that a single reading will lie, whilst a *confidence interval* is the range over which the *mean* of a set of readings is expected to lie. There is a further important difference: as the number of readings n increases, a tolerance interval converges to a finite value, whilst under the same circumstances a confidence interval leads to zero. See also paragraphs 6.21 and 6.22.

Weighted Mean

2.54 Sometimes a particular quantity, x, may be measured by several different methods, leading to differing values of x and different standard deviations. Suppose under these circumstances we wish to ascertain the best value to take for the quantity in question. We proceed as follows. Suppose that n measurements are made of a quantity x, leading to $x_1, x_2, x_3, \ldots, x_n$ different values for x. Let the standard deviation of each of these values of x be $\sigma_1, \sigma_2, \ldots, \sigma_n$. Now let the most probable value of x be q, then the deviations from q are given by $q - x_1, q - x_2, \ldots, q - x_n$. The probability of precisely this set of deviations is therefore the product

$$\frac{e^{-(x_1-q)^2/2\sigma_1^2}}{\sigma_1\sqrt{(2\pi)}} \cdot \frac{e^{-(x_2-q)^2/2\sigma_2^2}}{\sigma_2\sqrt{(2\pi)}} \cdots \frac{e^{-(x_n-q)^2/2\sigma_n^2}}{\sigma_n\sqrt{(2\pi)}} = \frac{\exp\{-\sum_{r=1}^{r=n}(x_r-q)^2/2\sigma_r^2\}}{(2\pi)^{n/2}\prod_1^n \sigma_r}$$

$$2.54(1)$$

where

$$\prod_1^n \sigma_r = \sigma_1 \cdot \sigma_2 \cdot \sigma_3 \cdots \sigma_n$$

Now the most probable value of this quantity is given for that value of q which makes this expression a maximum, that is it is the value of q which makes the exponent of e a minimum. Thus we require the value of q which makes

$$\sum_{r=1}^{r=n}(x_r - q)^2/2\sigma_r^2 \quad \text{a minimum.}$$

Differentiating this expression with respect to q and equating to zero we obtain

$$-\sum_{r=1}^{r=n} 2\frac{(x_r - q)}{2\sigma_r^2} = 0$$

or

$$q\sum_{r=1}^{r=n}\frac{1}{\sigma_r^2} = \sum_{r=1}^{r=n}\frac{x_r}{\sigma_r^2}$$

giving

$$q = \frac{\sum_1^n x_r/\sigma_r^2}{\sum_1^n 1/\sigma_r^2} \qquad 2.54(2)$$

That this value of q represents a minimum is easily demonstrated since the second derivative of the exponent of e in 2.54(1) is

$$\sum_{r=1}^{r=n} \frac{1}{\sigma_r^2} \quad \text{which is } positive.$$

2.55 Thus the best value for the variable x is given by 2.54(2). It is seen that each value of x_r is weighted by the reciprocal of the square of the standard deviation of each of the x_r.

2.56 If the x_r in 2.54(2) are replaced by means the standard deviations must be replaced by the standard errors, and thus 2.54(2) becomes

$$\bar{q} = \frac{\sum_1^n \bar{x}_r/\bar{\sigma}_r^2}{\sum_1^n 1/\bar{\sigma}_r^2} \qquad 2.56(1)$$

Now each $\bar{x}_r$ is the mean of n_r readings and so

$$\bar{\sigma}_r = \frac{\sigma_r}{\sqrt{n_r}} \qquad 2.30(1)$$

where n_r is the number of observations made for $\bar{x}_r$, and thus substituting for $\bar{\sigma}_r$ we have

$$\bar{q} = \sum_1^n \frac{\bar{x}_r n_r}{\sigma_r^2} \Big/ \sum_1^n \frac{n_r}{\sigma_r^2} \qquad 2.56(2)$$

2.57 If all the σ_r of 2.54(2) are equal, that is they all belong to the same set of measurements, then 2.54(2) reduces to

$$q = \sum x_r/n \qquad 2.57(1)$$

which is identical with 2.22(1). If all the σ_r of 2.56(2) are equal this reduces to

$$\bar{q} = \sum_1^n n_r \bar{x}_r \Big/ \sum n_r \qquad 2.57(2)$$

wherein it is seen that each $\bar{x}_r$ is weighted in proportion to the number of readings producing it. It is easily seen that 2.57(1) and 2.57(2) are identical because if we put $\bar{x}_r = \sum_{p=1}^{p=n_r} x_{rp}$ we have that

$$\bar{q} = \sum_{r=1}^{r=n} \left(n_r \sum_{p=1}^{p=n_r} \frac{x_{rp}}{n_r} \right) \Big/ \sum_1^n n_r$$

$$= \sum \frac{x}{n} \qquad \text{since } \sum_1^n n_r = n \qquad \text{and } \sum_{r=1}^{r=n} \sum_{p=1}^{p=n_r} x_{rp} = \text{sum of all the } xs$$

2.58 It is seen that in forming the weighted mean (see 2.54(2) and 2.56(1)), each value of the variable x_r or $\bar{x}_r$ is weighted by the reciprocal of the corresponding variance. 2.54(2) and 2.56(2) are often written respectively as

$$q = \frac{\sum_1^n w_r x_r}{\sum_1^n w_r} = \frac{\sum_1^n w_r x_r}{W} \qquad 2.58(1)$$

and

$$\bar{q} = \frac{\sum_1^n \bar{w}_r \bar{x}_r}{\sum_1^n \bar{w}_r} = \frac{\sum \bar{w}_r \bar{x}_r}{\bar{W}} \qquad 2.58(2)$$

where

$$w_r = k^2/\sigma_r^2 \qquad \text{and} \qquad W = \sum_1^n w_r \qquad 2.58(3)$$

and

$$\bar{w}_r = k^2/\bar{\sigma}_r^2 \qquad \text{and} \qquad \bar{W} = \sum_1^n \bar{w}_r \qquad 2.58(4)$$

w_r and $\bar{w}_r$ are known as the weights of the variables x_r and $\bar{x}_r$ respectively whilst k is constant. The analogy with the calculation of the centre of gravity of a system of weights is obvious, hence the name weight for the w_r.

Standard Deviation of the Weighted Mean

2.59 Referring to $z = \sum_1^n a_r x_r$ (2.29(1)) we see that the standard deviation of z is given by

$$\sigma_z = \left(\sum_1^n a_r^2 \sigma_r^2 \right)^{1/2} \qquad 2.29(2)$$

where σ_r is the standard deviation of x_r. Comparing 2.29(1) with 2.54(2), we see that

$$a_r \equiv 1/\sigma_r^2 \sum_1^n 1/\sigma_r^2 \qquad 2.59(1)$$

Thus the standard deviation of 2.54(2) or 2.58(1) is given by

$$\sigma_q = \left\{ \sum_1^n (1/\sigma_r^2)^2 \cdot \sigma_r^2 \middle/ \left(\sum_1^n 1/\sigma_r^2 \right)^2 \right\}^{1/2}$$

using 2.29(2)

$$= \left(1 \middle/ \sum_1^n 1/\sigma_r^2 \right)^{1/2} \qquad 2.59(2)$$

or

$$\frac{1}{\sigma_q^2} = \sum_1^n 1/\sigma_r^2 \qquad\qquad 2.59(3)$$

2.60 Similarly, the standard deviation of 2.56(1), that is of

$$\bar{q} = \sum_1^n \bar{x}_r/\bar{\sigma}_r^2 \sum_1^n 1/\bar{\sigma}_r^2 \qquad\qquad 2.56(1)$$

is easily seen to be given by

$$1/\bar{\sigma}_q^2 = \sum_1^n 1/\bar{\sigma}_r^2 \qquad\qquad 2.60(1)$$

or

$$= \sum_1^n n_r/\sigma_r^2 \qquad\qquad 2.60(2)$$

From 2.58(3) and 2.59(3) or 2.58(4) and 2.60(1) it is easily seen that

$$W = k^2/\sigma_q^2 = k^2 \sum 1/\sigma_r^2 \qquad\qquad 2.60(3)$$

and

$$\bar{W} = k^2/\bar{\sigma}_q^2 = k^2 \sum 1/\bar{\sigma}_r^2 \qquad\qquad 2.60(4)$$

Thus W is not only the sum of the weights, but is itself the weight to be associated with the mean, that is the weight of q (2.58(1) is thus W and the weight of $\bar{q}$ (2.58(2) is $\bar{W}$). Thus the fact noted in paragraph 2.55, that the weight of each variable x_r is inversely proportional to the square of the associated standard deviation, is also true of the mean formed from these variables, since $\bar{W}$ is proportional to the reciprocal of the square of the standard deviation of the mean.

Estimation of the Standard Deviation of the Weighted Mean from Observations Made

2.61 Let us consider n groups of observations where each group consists of n_q observations, q taking the values 1 to m, and where the n_q are not necessarily equal. Let each observation be denoted by x_{rq}, the standard deviation of each variable x_{rq} being σ_{rq}, with r taking the values from 1 to n_q. Let the most probable value of x be $\bar{x}$, where the meaning of $\bar{x}$ is as yet undefined. Thus the *a priori* probability of occurrence of the observation x_{rq} is equal to

$$\frac{e^{-(x_{rq} - \bar{x})^2/2\sigma_{rq}^2}}{\sigma_{rq}(2\pi)^{1/2}} \qquad\qquad 2.61(1)$$

The probability of occurrence of the set of observations x_{rq} ($r = 1, 2, \ldots, n_q$; and $q = 1, 2, \ldots, m$) is thus

$$Q = \frac{\exp\{-\sum_{q=1}^{q=m}\sum_{r=1}^{r=n_q}(x_{rq} - \bar{x})^2/2\sigma_{rq}^2\}}{(\prod_{q=1}^{q=m}\prod_{r=1}^{r=n_q}\sigma_{rq})(2\pi)^{N/2}} \qquad 2.61(2)$$

where $\sum_{q=1}^{q=m} n_q = N$, the total number of observations. Let us now put

$$w_{rq} = k^2/\sigma_{rq}^2 \qquad 2.61(3)$$

where w_{rq} is the weight of the observation x_{rq} (see 2.58(2)). Thus, substituting for the σ_{rq} in 2.61(2) we have

$$Q = \frac{\prod_{q=1}^{q=m}\prod_{r=1}^{r=n_q}(w_{rq})^{1/2}\exp\{-\sum_{q=1}^{q=m}\sum_{r=1}^{r=n_q}w_{rq}(x_{rq} - \bar{x})^2/2k^2\}}{(k)^N(2\pi)^{N/2}} \qquad 2.61(4)$$

Taking logarithms of both sides of 2.61(4) and putting $\log Q = T$, we have

$$T = \frac{1}{2}\sum_{q=1}^{q=m}\sum_{r=1}^{r=n_q}\log w_{rq} - N\log k - \sum_{q=1}^{q=m}\sum_{r=1}^{r=n_q}\frac{w_{rq}(x_{rq} - \bar{x})^2}{2k^2} - \frac{N}{2}\log 2\pi \quad 2.58(6)$$

2.62 Now, the most likely values for $\bar{x}$ and k are those which make Q a maximum. This is known as the principle of maximum likelihood. The required values of σ and $\bar{x}$ are thus given by solving

$$\frac{\partial Q}{\partial \bar{x}} = 0 \qquad \text{and} \qquad \frac{\partial Q}{\partial k} = 0$$

or what is the same thing

$$\frac{\partial T}{\partial \bar{x}} = 0 \qquad\qquad 2.62(1)$$

and

$$\frac{\partial T}{\partial k} = 0 \qquad\qquad 2.62(2)$$

Thus

$$\frac{\partial T}{\partial \bar{x}} = 0 = \sum_{q=1}^{q=m}\sum_{r=1}^{r=n_q}\frac{w_{rq}(x_{rq} - \bar{x})}{k^2}$$

Therefore

$$\bar{x} = \frac{\sum_{q=1}^{q=m}\sum_{r=1}^{r=n_q}w_{rq}x_{rq}}{\sum_{q=1}^{q=m}\sum_{r=1}^{r=n_q}w_{rq}} \qquad 2.62(3)$$

$\bar{x}$ is thus the weighted mean of all the x_{rq} values of x.

2.63 Now

$$\frac{\partial T}{\partial k} = 0 = \frac{-N}{k} + \sum_{q=1}^{q=m} \sum_{r=1}^{r=n_q} \frac{w_{rq}(x_{rq} - \bar{x})^2}{k^3}$$

therefore

$$k^2 = \sum_{q=1}^{q=m} \sum_{r=1}^{r=n_q} \frac{w_{rq}(x_{rq} - \bar{x})^2}{N} \qquad 2.63(1)$$

Now by 2.59(3), 2.58(1) and 2.54(2) we know that the standard deviation of the mean $\bar{x}$ (see 2.62(3)) is given by

$$\frac{1}{\sigma_{\bar{x}}^2} = \sum_{q=1}^{q=m} \sum_{r=1}^{r=n_q} \frac{1}{\sigma_{rq}^2} = \sum_{q=1}^{q=m} \sum_{r=1}^{r=n_q} \frac{w_{rq}}{k^2} \qquad 2.63(2)$$

Thus

$$\sigma_{\bar{x}}^2 = \frac{k^2}{\sum_{q=1}^{q=m} \sum_{r=1}^{r=n_q} w_{rq}} = \frac{\sum_{q=1}^{q=m} \sum_{r=1}^{r=n_q} w_{rq}(x_{rq} - \bar{x})^2}{N \sum_{q=1}^{q=m} \sum_{r=1}^{r=n_q} w_{rq}} \qquad 2.63(3)$$

using 2.63(1).

2.64 The best estimate $s_{\bar{x}}^2$ of $\sigma_{\bar{x}}^2$ from a finite number of readings is obtained by multiplying $\bar{\sigma}_{\bar{x}}^2$ by $N/(N-1)$ (see 2.24(7)). Thus

$$s_{\bar{x}}^2 = \frac{\sum_{q=1}^{q=m} \sum_{r=1}^{r=n_q} w_{rq}(x_{rq} - \bar{x})^2}{(N-1) \sum_{q=1}^{q=m} \sum_{r=1}^{r=n_q} w_{rq}} \qquad 2.64(1)$$

The best estimate of the weighted standard deviation of the x_{rq} is obtained by multiplying $s_{\bar{x}}$ by $\sqrt{N}$, where N is the total number of readings, giving the weighted deviation as

$$s_x^2 = \frac{\sum_{q=1}^{q=m} \sum_{r=1}^{r=n_q} N w_{rq}(x_{rq} - \bar{x})^2}{(N-1) \sum_{q=1}^{q=m} \sum_{r=1}^{r=n_q} w_{rq}} \qquad 2.64(2)$$

where $\bar{x}$ is given by 2.62(3) and is the weighted mean.

Standard Error of the Mean by External Consistency

2.65 Let us again consider the distribution of paragraph 2.61 and find the mean of each of the m groups of n_q observations. Thus, simplifying 2.62(3), we have

$$\bar{x}_q = \frac{\sum_{r=1}^{r=n_q} w_{rq} x_{rq}}{\sum_{r=1}^{r=n_q} w_{rq}} \qquad 2.65(1)$$

Similarly, the standard deviation of each of these means is given by

$$\sigma_{\bar{x}_q}^2 = \frac{\sum_{r=1}^{r=n_q} w_{rq}(x_{rq} - \bar{x}_q)^2}{n_q \sum_{r=1}^{r=n_q} w_{rq}} \qquad 2.65(2)$$

The best estimate of the standard deviation of each mean for a finite number of readings is thus

$$s_{\bar{x}_q}^2 = \frac{\sum_{r=1}^{r=n_q} w_{rq}(x_{rq} - \bar{x}_q)^2}{(n_q - 1) \sum_{r=1}^{r=n_q} w_{rq}} \qquad 2.65(3)$$

2.66 Now by 2.58(1), the weight of $\bar{x}_q$ is

$$\sum_{r=1}^{r=n_q} w_{rq} = w_q \qquad 2.66(1)$$

Thus the weighted mean of the m values of $\bar{x}_q$ is given by

$$\bar{x} = \frac{\sum_{q=1}^{q=m} w_q \bar{x}_q}{\sum_{q=1}^{q=m} w_q} \qquad 2.66(2)$$

Substituting for w_q from 2.66(1) and $\bar{x}_q$ from 2.65(1), we have

$$\bar{x} = \frac{\sum_{q=1}^{q=m} \left\{ \sum_{r=1}^{r=n_q} w_{rq} x_{rq} \cdot \sum_{r=1}^{r=n_q} w_{rq} / \sum_{r=1}^{r=n_q} w_{rq} \right\}}{\sum_{q=1}^{q=m} \sum_{r=1}^{r=n_q} w_{rq}}$$

$$= \frac{\sum_{q=1}^{q=m} \sum_{r=1}^{r=n_q} w_{rq} x_{rq}}{\sum_{q=1}^{q=m} \sum_{r=1}^{r=n_q} w_{rq}} \qquad 2.66(3)$$

Thus the weighted mean of the intermediate m means is equal to the weighted mean of all the observations taken together, since 2.66(3) is identical with 2.62(3).

2.67 The standard deviation of the mean of the m means is given by

$$_2\sigma_{\bar{x}}^2 = \frac{\sum_{q=1}^{q=m} w_q (\bar{x}_q - \bar{x})^2}{m \sum_{q=1}^{q=m} w_q} \qquad 2.67(1)$$

using 2.63(3) and putting $x_{nq} = \bar{x}_q$. The number of groups is m and thus m replaces N. Substituting for w_q from 2.66(1), we have

$$_2\sigma_{\bar{x}}^2 = \frac{\sum_{q=1}^{q=m} \sum_{r=1}^{r=n_q} w_{rq} (\bar{x}_q - \bar{x})^2}{m \sum_{q=1}^{q=m} \sum_{r=1}^{r=n_q} w_{rq}} \qquad 2.67(2)$$

where the suffix 2 of $_2\sigma_{\bar{x}}^2$ denotes the standard deviation of the mean by *external consistency*.

2.68 The best estimate $s_2(\bar{x})$ of the standard deviation of the mean by external consistency is thus given by

$$s_2^2(\bar{x}) = \frac{\sum_{q=1}^{q=m} \sum_{r=1}^{r=n_q} w_{rq}(\bar{x}_q - \bar{x})^2}{(m-1)\sum_{q=1}^{q=m} \sum_{r=1}^{r=n_q} w_{rq}} \qquad 2.68(1)$$

See 2.24(7), there being m groups of $\bar{x}_q$.

Standard Error of the Mean by Internal Consistency

2.69 Let us again consider the groups of observations dealt with in paragraph 2.61. Let $\bar{x}_q$ be the mean of each group of n_q observations. The probability of occurrence of the observation x_{rq} is equal to

$$\frac{e^{-(x_{rq} - \bar{x}_q)^2/2\sigma_{rq}^2}}{\sigma_{rq}(2\pi)^{1/2}} \qquad 2.69(1)$$

The probability of occurrence of the set of observations x_{rq} ($r = 1, 2, \ldots, n_q$; and $q = 1, 2, \ldots, m$) is thus

$$Q = \frac{\exp\{-\sum_{q=1}^{q=m} \sum_{r=1}^{r=n_q}(x_{rq} - \bar{x}_q)^2/2\sigma_{rq}^2\}}{(\prod_{q=1}^{q=m} \prod_{r=1}^{r=n_q} \sigma_{rq})(2\pi)^{N/2}} \qquad 2.69(2)$$

where $\sum_{q=1}^{q=m} n_q = N$ the total number of observations. It is to be noted that 2.69(2) does not use the common mean of all the observations, but that of the n_q observations only.

2.70 As before let us put the weight of each observation x_{qr} equal to

$$w_{rq} = k^2/\sigma_{rq}^2 \qquad 2.61(3)$$

Thus, substituting for σ_{rq} in 2.69(2) and taking logarithms of both sides, we have

$$\log Q = T = \frac{1}{2} \sum_{q=1}^{q=m} \sum_{r=1}^{r=n_q} \log w_{rq} - N \log k - \sum_{q=1}^{q=m} \sum_{r=1}^{r=n_q} \frac{w_{rq}(x_{rq} - \bar{x}_q)^2}{2k^2} - \frac{N}{2} \log 2\pi \qquad 2.70(1)$$

For just this set of observations to occur the probability Q must be a maximum, and thus Q or T must assume its maximum value for this to happen. The only variable at our disposal is k, and so, differentiating T with respect to k, we have

$$-\frac{N}{k} + \sum_{q=1}^{q=m} \sum_{r=1}^{r=n_q} \frac{w_{rq}(x_{rq} - \bar{x}_q)^2}{k^3} = 0$$

Thus

$$k^2 = \sum_{q=1}^{q=m} \sum_{r=1}^{r=n_q} \frac{w_{rq}(x_{rq} - \bar{x}_q)^2}{N}$$ 2.70(2)

If σ is the standard deviation of the mean given by 2.62(3), we know that

$$\frac{1}{\sigma^2} = \sum_{q=1}^{q=m} \sum_{r=1}^{r=n_q} \frac{1}{\sigma_{rq}^2} = \sum_{q=1}^{q=m} \sum_{r=1}^{r=n_q} \frac{w_{rq}}{k^2}$$ 2.63(2)

and therefore

$$\sigma^2 = \frac{k^2}{\sum_{q=1}^{q=m} \sum_{r=1}^{r=n_q} w_{rq}}$$

$$= \sum_{q=1}^{q=m} \sum_{r=1}^{r=n_q} \frac{w_{rq}(x_{rq} - \bar{x}_q)^2}{N \sum_{q=1}^{q=m} \sum_{r=1}^{r=n_q} w_{rq}}$$ 2.70(3)

Notice that this assumes that $\sum_{q=1}^{q=m} \sum_{r=1}^{r=n_q} w_{rq}(x_{rq} - \bar{x}_q)^2$ is a valid approximation for $\sum_{q=1}^{q=m} \sum_{r=1}^{r=n_q} w_{rq}(x_{rq} - \bar{x})^2$ (see paragraph 9.29 and expression 9.29(5)). If the observations are consistent then the approximation is valid.

2.71 The best estimate of σ^2 from a finite number of readings is obtained by multiplying σ^2 by $N/(N - m)$ (see 2.24(7)), since $\sum_{q=1}^{q=m} \sum_{r=1}^{r=n_q} w_{rq}(x_{rq} - \bar{x}_q)^2$ has $N - m$ degrees of freedom, that is, m values of $\bar{x}_q$ are calculated. Thus

$$s_1^2 = \sum_{q=1}^{q=m} \sum_{r=1}^{r=n_q} \frac{w_{rq}(x_{rq} - \bar{x}_q)^2}{(N - m) \sum_{q=1}^{q=m} \sum_{r=1}^{r=n_q} w_{rq}}$$ 2.71(1)

This estimate of the standard deviation of the mean is known as the standard deviation of the mean by internal consistency. For a discussion of how to use the two standard deviations s_1 and s_2 (2.71(1) and 2.68(1)), see paragraph 9.12 et seq., and in particular equations 9.29(6) and 9.30(2).

Examples on Chapter 2

Example 2.1 If twelve readings have been taken at each end of a workpiece using a toolmaker's microscope or a similar device, determine the mean length of the workpiece, the confidence interval for a confidence probability of 0.954, and the tolerance interval for the length of the workpiece, if found from one reading at either end, for a tolerance probability of 0.99.

Readings in inches   End 1   1.0246, 1.0248, 1.0247, 1.0244, 1.0242, 1.0245, 1.0249, 1.0248, 1.0249, 1.0246, 1.0248, 1.0245.

End 2   4.4623, 4.4622, 4.4625, 4.4626, 4.4623, 4.4627, 4.4625, 4.4628, 4.4624, 4.4623, 4.4628, 4.4626.

(Assume distribution to be Gaussian and use Tables II and III of Appendix I.)

Answer: The mean reading at End 1 is given by

$$\sum \frac{x_r}{n} = \bar{x}_1 = 1.024\,64$$

where $n = 12$. The estimate of the standard deviation of these readings is given by using

$$\sigma_1^2 = \sum_1^n (x_r - \bar{x})^2/(n - 1) \qquad\qquad 2.24(8)$$

giving

$$\sigma_1 = 0.000\,24$$

Similarly the mean reading at the other end is found from the second block of readings, yielding

$$\bar{x}_2 = 4.462\,50 \qquad \text{and} \qquad \sigma_2 = 0.000\,20$$

The mean length is thus $\bar{x}_2 = \bar{x}_1 = 3.437\,86$ which we will round to 3.4379. The standard error or standard deviation of the mean at End 1 is given by

$$\sigma_1/\sqrt{n} = 0.000\,065 = \bar{\sigma}_1$$

and the standard error or the standard deviation of the mean at End 2 is likewise

$$\sigma_2/\sqrt{n} = 0.000\,058 = \bar{\sigma}_2$$

by 2.30(1). The standard deviation of the mean length $(\bar{x}_2 - \bar{x}_1)$ or standard error of the length is given by

$$\bar{\sigma} = (\bar{\sigma}_1^2 + \bar{\sigma}_2^2)^{1/2}$$

by 2.29(2) with $a_r = 1$, and $n = 2$. Thus

$$\bar{\sigma} = 0.000\,087$$

Referring to Table III we see that for $\underset{-k_1\,\text{to}\,k_1}{P}$ equal to 0.954, $k_1 = 2.0$. Thus the required confidence interval is

$$\pm 2.0\bar{\sigma} = \pm 2.0 \times 0.000\,087 = 0.000\,176$$

k_1, the coefficient of $\bar{\sigma}$ is in this instance called the *confidence factor* of the *confidence interval*, that is the number by which the standard error or standard deviation of the mean is multiplied by to give the required confidence interval. The standard deviation of the length is likewise given by

$$\sigma = (\sigma_1^2 + \sigma_2^2)^{1/2} = 0.000\,31.$$

Referring to Table III again, we find the *tolerance factor* for a tolerance probability of 0.99 equal to 2.576. Thus the tolerance interval or uncertainty in the length of the workpiece, for a single reading taken at each end, is equal to

$$k_1\sigma = 2.576 \times 0.000\,31$$

$$= \underline{0.000\,80}$$

with a tolerance probability of 0.99. Had the tolerance probability been 0.954, the same as for the error in the mean, the tolerance interval would have been $2.0 \times 0.000\,31 = \underline{0.000\,61}$. The advantage of taking several readings is clearly shown in this example where the uncertainty in the length has been reduced by a factor of 3.5 by so doing.

Example 2.2 If the standard deviation of a measuring microscope has been established as 0.0001 in by taking many readings, and the length of a workpiece is required to an uncertainty of ± 0.0001 in, how many readings at each end of the workpiece need to be taken, if the confidence probability required is 0.99?

Answer: Since the standard deviation of the microscope is given as $\sigma = 0.0001$ in, the standard deviation of the length of a workpiece is equal to $(\sigma^2 + \sigma^2)^{1/2} = \sqrt{2}\sigma$, and the standard error of this is thus $\sqrt{2}\sigma/\sqrt{n}$ where n is the number of readings at each end. As the required *confidence probability* is 0.99, the *confidence factor* associated with this value, as given by Table III, is equal to $k_1 = 2.576$. Thus the *confidence interval* associated with this *confidence probability* for n readings is equal to $\sqrt{2}\sigma/\sqrt{n} \times 2.576$. Now this is required to be equal to 0.0001 in, thus with σ given as 0.0001, we have $\sqrt{2} \times 0.0001 \times 2.576/\sqrt{n} = 0.0001$, giving

$$n = \left[\frac{\sqrt{2} \times 0.001 \times 2.576}{0.0001} \right]^2$$

$$= 13.2$$

Thus about thirteen readings at each end would be required. The 0.99 confidence probability means that if a hundred groups of thirteen readings each were taken, then on average, in ninety-nine of these, the mean length would lie within the confidence limits calculated in the example.

Example 2.3 If a number of 1 lb weights have been calibrated and the standard deviation of each weight is estimated as 10^{-4} lb, what is the fractional uncertainty in the weight of a test piece likely to be, when weighed against sixteen of these weights and other sources of error are neglected, the tolerance probability required being 0.95?

Answer: If σ_w is the standard deviation of each weight then the total standard deviation is given by $\sigma = \sqrt{(n\sigma_w^2)}$ from 2.29(2) with $a_r = 1$.

If k_1 is the tolerance factor corresponding to the tolerance probability, then the uncertainty in the weight of the test piece is $k_1\sigma = k_1(n\sigma_w^2)^{1/2}$. The fractional uncertainty is thus

$$\frac{k_1\sigma}{nw} = \frac{k_1\sigma_w}{w\sqrt{n}}$$

where w is the weight of each calibrated weight. Since $\sigma_w = 10^{-4}$ and from Table III, $k_1 = 1.96$, whilst $w = 1$ and $n = 16$, the required fractional uncertainty is

$$\frac{k_1\sigma_w}{w\sqrt{n}} = \frac{1.96 \times 10^{-4}}{1 \times 4} = 0.49 \times 10^{-4}$$

It is to be noticed that the fractional uncertainty of each weight to the same tolerance probability is $k_1\sigma_w/w$. Thus the fractional uncertainty of the combination of n weights is less than that of a single weight in the ratio $1/\sqrt{n}$.

Example 2.4 The standard deviation of the variation in temperature of a standards laboratory is $1°C$ and the difference in the coefficient of expansion between slip gauges used to measure a workpiece and the workpiece itself is 10^{-5} in $°C^{-1}$ in^{-1}. The standard deviation of the difference in size between the workpiece and slip gauges is 4.0×10^{-5} in, as found from ten readings. What is the fractional uncertainty in the workpiece to a tolerance probability of 0.954, if the nominal size is 2.5 in? If the standard deviation of the difference in size between the workpiece and the gauges is assumed known at 4×10^{-5} in, and only two readings are taken of the difference, what now is the fractional uncertainty in the size of the workpiece to the same tolerance probability of 0.954?

Answer: 2.24×10^{-5}, 3.02×10^{-5}.

Example 2.5 The results of the measurements of six different laboratories on a standard resistance are, in ohms, 1.00007, 1.00001, 1.00009, 1.00005, 1.00006, 1.00004. The corresponding standard deviations of the uncertainties in the measurements of each laboratory are respectively proportional to 5, 6, 2, 7, 4, 3. Find the weighted mean value of the resistance, and also the unweighted mean value.

Answer: 1.000068, 1.000053.

Example 2.6 Shafts are manufactured such that the standard deviation from the mean $\bar{x}_1$ is σ_1. Likewise the corresponding bores are manufactured such that the standard deviation from the mean $\bar{x}_2$ is σ_2, $\bar{x}_2 > \bar{x}_1$. If bores and shafts are mated at random, derive an expression for the proportion of matings in which the bore will not enter the shaft†.

Answer:

$$\int_{-\infty}^{-(\bar{x}_2 - \bar{x}_1)} \frac{\exp\{-y^2/2(\sigma_1^2 + \sigma_2^2)\}\, dy†}{[2\pi(\sigma_1^2 + \sigma_2^2)]^{1/2}} \qquad \text{or} \qquad \left[1 - \frac{P}{-k_1 \text{ to } k_1}\right]\bigg/2$$

where

$$k_1 = (x_2 - x_1)/(\sigma_1^2 + \sigma_2^2)^{1/2}$$

If $\bar{x}_2 - \bar{x}_1 = 0.001$ in and $\sigma_1 = \sigma_2 = 0.0003$ in find the proportion of bores which will fail to mate with randomly chosen shafts.

Answer: 0.0093.

Show also that the answer to the first part of the question is also the proportion of matings in which the clearance will exceed $2(\bar{x}_2 - \bar{x}_1)$.

Example 2.7 Two voltmeters are known to have standard deviations of ± 0.1 and ± 0.2 V. Four readings using the first voltmeter yield a mean of 3.50 V across a resistor, whilst the mean of six readings using the second voltmeter yields a mean of 3.45 V. Find the mean voltage across the resistor, and the confidence limits for a confidence probability of 0.95.

Answer: Mean 3.486 V, confidence limits ± 0.084 V about mean value.

† Hint: consider the combination of the two distributions

$$e^{-(x - \bar{x}_1)^2/2\sigma_1^2}/\sigma_1\sqrt{(2\pi)} \qquad \text{and} \qquad e^{-(y - \bar{x}_2)^2/2\sigma_2^2}/\sigma_2\sqrt{(2\pi)}$$

and see paragraphs 2.27 to 2.28.

3

General Distributions

3.01 In this chapter we shall deal with general distributions, of which Gaussian distributions are a particular case. The results we shall obtain will apply quite generally to Gaussian and non-Gaussian distributions unless the contrary is stated.

A Double Integral Theorem

3.02 Before proceeding further we shall need to prove a theorem concerning a double integral, which will be needed in dealing with general distributions, and later on, in deriving the frequency distribution of two combined or convoluted distributions.

3.03 Let us consider two functions $f(x)$ and $\psi(z)$. The only restriction we place on these two functions is that they should be bounded in the ranges x_c to x_d and z_e to z_f, and continuous except for a finite number of discontinuities. Let us divide x_c to x_d into p intervals

$$x_{c+1} - x_c, x_{c+2} - x_{c+1}, \ldots, x_{c+p-1} - x_{c+p-2}, x_d - x_{c+p-1}$$

where p is an integer. Let $f(x)_r$ be the mean value of $f(x)$ in the interval $x_{c+r} - x_{c+r-1}$, and let us consider the sum

$$\sum_{r=0}^{r=p} (x_{c+r} - x_{c+r-1}) f(x)_r \qquad 3.03(1)$$

where $x_{c+p} = x_d$. As $f(x)$ is bounded in the interval x_c to x_d and continuous except for a finite number of discontinuities, the limit of the above, when the intervals $x_{c+1} - x_{c+r-1}$ are made vanishingly small and p tends to infinity, is the integral of $f(x)$ between x_c and x_d.

3.04 Similarly, if we consider the sum

$$\sum_{r=0}^{r=q} (z_{e+r} - z_{e+r-1}) \psi(z)_r \qquad 3.04(1)$$

(Note: $z_{e+q} = z_f$) where $\psi(z)$ is bounded in the range z_e to z_f and continuous in this range except for a finite number of discontinuities, and the range has been divided up into q equal parts, with $\psi(z)_r$ the mean values of $\psi(z)$ for each range $z_{e+r} - z_{e+r-1}$, then as q tends to infinity and $z_{e+r} - z_{e+r-1}$ becomes vanishingly small, the sum tends to the integral of $\psi(z)$ over the range z_e to z_f. It is to be noted that since both series are absolutely convergent, the two series can be multiplied together term by term and the resultant series of terms arranged in any desired order without changing the value of the sum.

3.05 Let us now suppose that x and z are connected by the relation

$$\mu = ax + bz \qquad 3.05(1)$$

where μ is also a variable and a and b are positive. Then

$$z = \frac{\mu - ax}{b} \qquad 3.05(2)$$

and thus

$$\psi(z) \equiv \psi\left(\frac{\mu - ax}{b}\right) \qquad 3.05(3)$$

We will now consider $\psi(z)$ between z_e and z_f and $f(x)$ between x_c and x_d, where $z_f > z_e$ and $x_d > x_c$. Let

$$f(x) = f(r) = f(x_c + r\varepsilon) \qquad 3.05(4)$$

where

$$x = x_c + r\varepsilon \qquad 3.05(5)$$

and where

$$0 \leqslant r \leqslant p = \frac{x_d - x_c}{\varepsilon} \qquad 3.05(6)$$

r and p being integers. The range $x_d - x_c$ has thus been divided into p equal steps, each equal to ε. Similarly let

$$\psi(z) \equiv \psi\left(\frac{\mu - ax}{b}\right) \equiv \psi(r + g)$$

$$\equiv \psi\left\{\frac{(\mu_1 + g\varepsilon a) - a(x_c + r\varepsilon)}{b}\right\} \qquad 3.05(7)$$

where

$$\mu = \mu_1 + g\varepsilon a \qquad 3.05(8)$$

and where g is an integer and εa is the increase in μ for an increment ε in x. μ_1 will be defined later (see 3.07(1)).

3.06 Let us now consider the array

$$x_d$$

$$f(x_c), f(x_c + \varepsilon), f(x_c + 2\varepsilon), \ldots, f(x_c + (p-1)\varepsilon), f(x_c + p\varepsilon)$$

$$\mu_2 \qquad\qquad\qquad\qquad\qquad\qquad\qquad\qquad\qquad \mu_1$$

$$\psi(z_f), \psi\left(z_f - \frac{a\varepsilon}{b}\right), \psi\left(z_f - \frac{2a\varepsilon}{b}\right), \ldots, \psi\left(z_f - \frac{a}{b}(x_d - x_c)\right), \ldots, \psi\left(z_e + \frac{a}{b}(x_d - x_c)\right), \ldots, \psi\left(z_e + \frac{a\varepsilon}{b}\right), \psi(z_e)$$

The number of terms in the top row is equal to $p + 1$ whilst the number in the bottom row is equal to $q + 1$ where

$$q = \frac{(z_f - z_e)}{\varepsilon} \frac{b}{a} \qquad\qquad 3.06(1)$$

It is to be noted that the first line of the array gives $f(x)$ at intervals of $\delta x = \varepsilon$ for increasing x, whilst the second line gives $\psi(z)$ at intervals of $\delta z = -a\varepsilon/b$, that is for decreasing z. The term $a\varepsilon/b$ for decreasing z is obtained by differentiating 3.05(2), giving

$$\delta z = \frac{-a\delta x}{b} \qquad\qquad 3.06(2)$$

and putting $\delta x = \varepsilon$.

3.07 Let us now assume that there are more $\psi(z)$ terms than $f(x)$ terms in the array. If we put $x = x_d$ and $z = z_e$ and substitute in 3.05(1) we have

$$\mu_1 = ax_d + bz_e \qquad\qquad 3.07(1)$$

Similarly, if we put $x = x_c$ and $z = z_f$, and again substitute in 3.05(1) we obtain

$$\mu_2 = ax_c + bz_f \qquad\qquad 3.07(2)$$

Now the number of terms in the first line of the array is given by $(x_d - x_c)/\varepsilon + 1$, and the number of terms in the second line by

$$\frac{(z_f - z_e)}{\varepsilon a/b} + 1$$

Thus if the number of terms in line two of the array is to exceed the number of terms in the first line, then

$$(z_f - z_e)b/\varepsilon a > (x_d - x_c)/\varepsilon$$

or

$$b(z_f - z_e) > a(x_d - x_c) \qquad\qquad 3.07(3)$$

Thus

$$\mu_2 - \mu_1 = b(z_f - z_e) - a(x_d - x_c) > 0 \qquad 3.07(4)$$

by 3.07(3) and thus

$$\mu_2 > \mu_1 \qquad 3.07(5)$$

3.08 Let us now consider the sum

$$_1\eta_g = \sum f(x_c + r\varepsilon)\psi\left[\frac{(\mu_1 + g\varepsilon a) - a(x_c + r\varepsilon)}{b}\right]\frac{\varepsilon a}{b} \qquad 3.08(1)$$

$$\equiv \sum f(r)\psi(r + g)\cdot\frac{\varepsilon a}{b}$$

where the summation is made with respect to r, where r goes from 0 to $(x_d - x_c)/\varepsilon$, and g, which is constant for any one summation in r, can vary between 0 and $(\mu_2 - \mu_1)/\varepsilon a = \{b(z_f - z_e) - a(x_d - x_c)\}/\varepsilon q$ in unit steps. If $g = 0$, that is $\mu = \mu_1$ (from 3.05(8)), then

$$_1\eta_{g=0} = \sum_{r=0}^{r=\frac{x_d - x_c}{\varepsilon}} f(x_c + r\varepsilon)\psi\left\{\frac{\mu_1 - a(x_c + r\varepsilon)}{b}\right\}\frac{\varepsilon a}{b} \qquad 3.08(2)$$

$$\equiv \sum_{r=0}^{r=\frac{x_d - x_c}{\varepsilon}} f(x_c + r\varepsilon)\psi\left\{z_e + \frac{a}{b}(x_d - x_c) - \frac{r\varepsilon a}{b}\right\}\frac{\varepsilon a}{b} \qquad 3.08(3)$$

using 3.07(1) to express μ_1 in terms of z and x. Thus $r = 0$ gives the term $f(x_c)$ multiplied by the term $\psi[z_e + a(x_d - x_c)/b]$ or p terms to the left of $\psi(z_e)$ in the second line of the array. When $r = p = (x_d - x_c)/\varepsilon$ we obtain the last term of $_1\eta_{g=0}$, namely $f(x_d)$ multiplied by the term $\psi(z_e)$. $_1\eta_{g=0}$ thus represents the sum of the series obtained by multiplying the terms joined by the broken lines in the array.

3.09 Similarly if

$$g = \frac{\mu_2 - \mu_1}{\varepsilon a} = \frac{b(z_f - z_e) - a(x_d - x_c)}{\varepsilon a} = \tau \qquad 3.09(1)$$

$$_1\eta_{g=\tau} = \sum_{r=0}^{r=\frac{x_d - x_c}{\varepsilon}} f(x_c + r\varepsilon)\psi\left(z_f - \frac{r\varepsilon a}{b}\right)\frac{\varepsilon a}{b} \qquad 3.09(2)$$

$$\equiv \sum_{r=0}^{r=\frac{x_d - x_c}{\varepsilon}} f(x_c + r\varepsilon)\psi\left\{\frac{\mu_2 - a(x_c + r\varepsilon)}{b}\right\}\frac{\varepsilon a}{b} \qquad 3.09(3)$$

If $r = 0$ the first term is $f(x_c)\psi(z_f)$, whilst if $r = (x_d - x_c)/\varepsilon$ the last term is $f(x_d)\psi\{z_f - a(x_d - x_c)/b\}$. The series $_1\eta_{g=\tau}$ thus represents the sum of the series obtained by multiplying the terms joined by the full lines in the array.

Thus as g varies from 0 to $(\mu_2 - \mu_1)/\varepsilon a$ the series ${}_1\eta_g$ represents the terms of the top row of the array successively multiplied by successive terms in the second row, where the first term used in the second row begins with $\psi[z_e + a(x_d - x_c)/b]$ for $g = 0$ and moves one term to the left for each unit increase in g until $g = \tau$ when the first term of the second row is $\psi(z_f)$.

3.10   Thus when $r \to \infty, g \to \infty, \delta x \to \delta r\varepsilon = \varepsilon \to 0$. Since $\delta r = 1$, and the sum ${}_1\eta_g$ thus tends to the integral

$$ {}_1\bar\eta_\mu = \frac{a}{b} \int_{x_c}^{x_d} f(x)\psi\left(\frac{\mu - ax}{b}\right) dx \qquad \text{3.10(1)} $$

valid for $\mu_2 \geqslant \mu \geqslant \mu_1$ where the limits are obtained as follows

$$ x = x_c + r\varepsilon \qquad \text{3.05(5)} $$

thus when $r = 0, x = x_c$ and when $r = (x_d - x_c)/\varepsilon, x = x_d$. Since the series ${}_1\eta_g$ is valid from $g = 0$ to

$$ g = \tau = \frac{\mu_2 - \mu_1}{\varepsilon a} \qquad \text{3.09(1)} $$

the integral is valid for μ from μ_1 to μ_2; by substitution of g in 3.05(8).

3.11   Let us now consider the sum

$$ \sum_{g=0}^{g=\tau} {}_1\eta_g \varepsilon = \sum_{g=0}^{g=\tau} \sum_{r=0}^{r=\frac{x_d - x_c}{\varepsilon}} f(x_c + r\varepsilon)\psi\left\{\frac{(\mu_1 + g\varepsilon a) - \mu(x_c + r\varepsilon)}{b}\right\}\frac{\varepsilon a}{b} \qquad \text{3.11(1)} $$

Now from equation 3.05(8), that is $\mu = \mu_1 + g\varepsilon q$, we have

$$ \delta\mu = \delta g\varepsilon q = \varepsilon q \qquad \text{3.11(2)} $$

since $\delta g = 1$. Thus equation 3.11(1) can be written

$$ \frac{1}{a}\sum_{g=0}^{g=\tau} {}_1\eta_g \overbrace{\varepsilon a}^{\delta\mu} = \frac{1}{b}\sum_{g=0}^{g=\tau} \overbrace{\varepsilon a}^{\delta\mu} \sum_{r=0}^{r=\frac{x_d - x_c}{\varepsilon}} f(x_c + r\varepsilon)\psi\left\{\frac{(\mu_1 + g\varepsilon a) - a(x_c + r\varepsilon)}{b}\right\}\delta x \qquad \text{3.11(3)} $$

In the limit when g and r tend to infinity and $\varepsilon \to 0$ the sums in equation 3.11(3) tend to the integrals

$$ \frac{1}{a}\int_{\mu_1}^{\mu_2} {}_1\bar\eta_\mu \, d\mu = \frac{1}{b}\int_{\mu_1}^{\mu_2} d\mu \int_{x_c}^{x_d} f(x)\psi\left(\frac{\mu - ax}{b}\right) dx \qquad \text{3.11(4)} $$

The limits for μ are obtained as follows. From 3.05(8), $\mu = \mu_1 + g\varepsilon a$, so when $g = 0, \mu = \mu_1$. When $g = \tau = (\mu_2 - \mu_1)/\varepsilon$ (3.09(1)), then substituting for g in 3.05(8) gives $\mu = \mu_2$.

3.12 Now let us consider the sums involved when the terms of the top row of the array are multiplied by terms in the second row to the right of those shown by the broken lines. It is seen that as the sets of terms in the second row, which are used, start farther and farther to the right, so more and more terms at the *end* of the first row are left out of the sum.

3.13 Let us consider the sum

$$_2\eta_g = \sum f(x_c + r\varepsilon)\psi \left\{ \frac{(\mu_1 - g\varepsilon a) - a(x_c + r\varepsilon)}{b} \right\} \frac{\varepsilon a}{b} \qquad 3.13(1)$$

where

$$\mu = \mu_1 - g\varepsilon a \qquad 3.13(2)$$

and

$$\mu_1 = ax_d + bz_e \qquad 3.07(1)$$

$g = 0$ gives the series $_1\eta_{g=0}$. As g increases, the maximum value of r in $_2\eta_g$, for given g, decreases as more terms of the top line of the array are left out of the sum. The maximum value of r for given g is obtained by putting

$$\psi \left\{ \frac{(\mu_1 - g\varepsilon a) - a(x_c + r\varepsilon)}{b} \right\} = \psi(z_e)$$

Therefore

$$\frac{\mu_1 - g\varepsilon a - ax_c - r\varepsilon a}{b} = z_e$$

Thus substituting for μ_1 from 3.07(1) and rearranging we obtain

$$r_{max} = \frac{x_d - x_c}{\varepsilon} - g \qquad 3.13(3)$$

Substituting from 3.13(2) for g and using 3.07(1) we get

$$r_{max} = \frac{\mu - ax_c - bz_e}{\varepsilon a} \qquad 3.13(4)$$

The maximum value of g is attained when $r_{max} = 0$ giving

$$\mu = ax_c + bz_e = \mu_3 \qquad 3.13(5)$$

The maximum value of g is thus obtained by putting $\mu = \mu_3$ in 3.13(2), giving

$$g = \frac{\mu_1 - \mu_3}{\varepsilon a} = \frac{(x_d - x_c)}{\varepsilon} \qquad 3.13(6)$$

3.14 Thus the required limits for the sum $_2\eta_g$ are

$$\frac{\mu_1 - \mu_3}{\varepsilon a} = \frac{x_d - x_c}{\varepsilon} \geqslant g \geqslant 1$$

and

$$\frac{\mu - ax_c - bz_e}{\varepsilon a} \geqslant r \geqslant 0$$

Thus

$$_2\eta_g = \sum_{r=0}^{r=\frac{\mu - ax_c - bz_e}{\varepsilon a}} f(x_c + r\varepsilon)\psi\{(\mu_1 - g\varepsilon a) - a(x_c + r\varepsilon)\}\frac{\varepsilon a}{b} \qquad 3.14(1)$$

valid for

$$\frac{x_d - x_c}{\varepsilon} \geqslant g \geqslant 1$$

As before when $r \to \infty, g \to \infty$, and $\varepsilon \to 0, \delta x \to \delta r\varepsilon = \varepsilon \to 0$; then the sum represented by $_2\eta_g$ tends to the integral

$$_2\bar{\eta}_\mu = \frac{a}{b}\int_{x_c}^{\frac{\mu - bz_e}{a}} f(x)\psi\left(\frac{\mu - ax}{b}\right)dx \qquad 3.14(2)$$

valid for $\mu_1 \geqslant \mu \geqslant \mu_3$, where the limits for x are obtained from

$$x = x_c + r\varepsilon \qquad 3.05(5)$$

When $r = 0, x = x_c$ and when $r = (\mu - ax_c - bz_e)/\varepsilon a$, then $x = (\mu - bz_e)/a$; μ_1 and μ_3 correspond to $g = 1$ and $g = (\mu_1 - \mu_3)/\varepsilon a$, obtained by substituting in $\mu = \mu_1 - g\varepsilon a$. Therefore when $g = 1, \mu \to \mu_1 - \varepsilon a \to \mu_1$ as $\varepsilon \to 0$. When $g = (\mu_1 - \mu_3)/\varepsilon a, \mu = \mu_1 - [(\mu_1 - \mu_3)/\varepsilon a]\varepsilon a = \mu_3$.

3.15 Let us now consider the sum

$$\sum_{g=1}^{g=\frac{x_d - x_c}{\varepsilon}} {}_2\eta_g\varepsilon = \sum_{g=1}^{g=\frac{x_d - x_c}{\varepsilon}} \varepsilon \sum_{r=0}^{r=\frac{\mu - ax_c - bz_e}{\varepsilon a}} f(x_c + r\varepsilon)\psi\{(\mu_1 - g\varepsilon a) - a(x_c + r\varepsilon)\}\frac{\varepsilon a}{b}$$

$$3.15(1)$$

Now from 3.13(2) $\delta\mu = -\delta g\varepsilon a = -\varepsilon a$ since $\delta g = 1$ and $\varepsilon = \delta x$. 3.15(1) can then be rewritten as

$$\frac{1}{a}\sum_{g=1}^{g=\frac{x_d - x_c}{\varepsilon}} {}_2\eta_g\widetilde{\varepsilon a} = $$

$$\frac{1}{b}\sum_{g=1}^{g=\frac{x_d - x_c}{\varepsilon}} \widetilde{\varepsilon a} \sum_{r=0}^{r=\frac{\mu - ax_c - bz_e}{\varepsilon a}} f(x_c + r\varepsilon)\psi\{(\mu_1 - g\varepsilon a) - a(x_c + r\varepsilon)\}\widetilde{\varepsilon}$$

$$3.15(2)$$

In the limit when $g \to \infty$, $r \to \infty$ and $\varepsilon \to 0$ the sums in equation 3.15(2) tend to the integrals

$$\frac{1}{a} \int_{\mu_3}^{\mu_1} {}_2\bar{\eta}_\mu \, d\mu = \frac{1}{b} \int_{\mu_3}^{\mu_1} d\mu \int_{x_c}^{\frac{\mu - bz_e}{a}} f(x)\psi\left(\frac{\mu - ax}{b}\right) dx \qquad 3.15(3)$$

It should be noted that $\mu_1 > \mu_3$ (see 3.13(6) and note also that $x_d > x_c$) and that the limits for $d\mu$ have been inverted for this reason: the sum implied this anyway since on substituting for εa a minus sign was introduced.

3.16 Finally let us consider the case when the terms of the top row of the array are multiplied by terms in the second row to the left of those shown by the full lines of the array. It is seen that as the set of terms used in the second row starts farther and farther to the left then more terms at the *beginning* of the first row are left out of the sum.

3.17 Let us consider the sum

$$_3\eta_g = \sum f(x_c + r\varepsilon)\psi\left\{\frac{(\mu_2 + g\varepsilon a) - a(x_c + r\varepsilon)}{b}\right\}\frac{\varepsilon a}{b} \qquad 3.17(1)$$

where

$$\mu = \mu_2 + g\varepsilon a \qquad 3.17(2)$$

and

$$\mu_2 = ax_c + bz_f \qquad 3.17(3)$$

$g = 0$ gives the series $_1\eta_{g=\tau}$ since $\tau = (\mu_2 - \mu_1)/\varepsilon a$ (see 3.09(1)). As g increases, the minimum value of r in $_3\eta_g$ increases as more terms to the left of the top row are left out of the sum. The minimum value of r for given g is obtained by putting

$$\psi\left\{\frac{(\mu_2 + g\varepsilon a) - a(x_c + r\varepsilon)}{b}\right\} = \psi(z_f)$$

Therefore

$$\frac{\mu_2 + g\varepsilon a - ax_c - r\varepsilon a}{b} = z_f \qquad 3.17(4)$$

Thus substituting for μ_2 from 3.07(2) we obtain

$$r_{\min} = g = \frac{\mu - \mu_2}{\varepsilon a} \qquad \text{from 3.17(2)} \qquad 3.17(5)$$

$$= \frac{\mu - ax_c - bz_f}{\varepsilon a} \qquad 3.17(6)$$

on substituting for μ_2 from 3.17(3). The maximum value of r which is always attained in the summation is given by $x_d = x_c + r\varepsilon$ or $r_{max} = (x_d - x_c)/\varepsilon$. The maximum value of g is given when r_{min} is equal to $(x_d - x_c)/\varepsilon$. The maximum value of g is thus given when r_{min} attains its maximum value and thus

$$g_{max} = \frac{x_d - x_c}{\varepsilon} \qquad \qquad 3.17(7)$$

Thus the maximum value of μ is given by substituting g_{max} in 3.17(2) giving

$$\mu_{max} = \mu_2 + a(x_d - x_c) = ax_d + bz_f = \mu_4 \qquad 3.17(8)$$

and $_3\eta_g$ can be written as

$$_3\eta_g = \sum_{r=\frac{\mu - ax_c - bz_f}{\varepsilon a}}^{r = \frac{x_d - x_c}{\varepsilon}} f(x_c + r\varepsilon)\psi\left\{ \frac{(\mu_2 + g\varepsilon a) - a(x_c + r\varepsilon)}{b} \right\} \frac{\varepsilon a}{b} \qquad 3.17(9)$$

valid for $(x_d - x_c)/\varepsilon \geqslant g \geqslant 1$. As before, when $r \to \infty$, $g \to \infty$ and $\delta x \to \delta r\varepsilon = \varepsilon \to 0$ then the sum represented by $_3\eta_g$ tends to the integral

$$_3\bar{\eta}_\mu = \frac{a}{b} \int_{\frac{\mu - bz_f}{a}}^{x_d} f(x)\psi\left(\frac{\mu - ax}{b} \right) dx \qquad 3.17(10)$$

valid for $\mu_4 \geqslant \mu \geqslant \mu_2$. The limits for x are obtained from

$$x = x_c + r\varepsilon \qquad \qquad 3.05(5)$$

Upper limit $x_{max} = x_c + ((x_d - x_c)/\varepsilon)\varepsilon = x_d$ from $r = (x_d - x_c)/\varepsilon$, the upper limit of r in 3.17(9).

Lower limit $x_{min} = x_c + ((\mu - ax_c - bz_f)/\varepsilon a)\varepsilon = (\mu - bz_f)/a$ from $r = (\mu - ax_c - bz_f)/\varepsilon a$, the lower limit of r in 3.17(9).
The limits for μ are obtained from

$$\mu = \mu_2 + g\varepsilon a \qquad \qquad 3.17(2)$$

When

$$g = g_{max} = \frac{x_d - x_c}{\varepsilon}$$

$$\mu = \mu_2 + \frac{(x_d - x_c)}{\varepsilon}\varepsilon a = ax_c + bz_f + ax_d - ax_c$$

$$= ax_d + bz_f = \mu_4 \qquad \text{upper limit} \qquad 3.17(11)$$

When

$$g = 1, \quad \mu = \mu_2 + \varepsilon a \to \mu_2 \quad \text{as} \quad \varepsilon \to 0 \qquad \text{lower limit} \qquad 3.17(12)$$

3.18   Let us now consider the sum

$$\frac{1}{a} \sum_{g=1}^{g=\frac{x_d - x_c}{\varepsilon}} {}_3\eta_g \overbrace{\varepsilon a}^{\delta\mu} =$$

$$\frac{1}{b} \sum_{g=1}^{g=\frac{x_d - x_c}{\varepsilon}} \overbrace{\varepsilon a}^{\delta\mu} \sum_{r=\frac{\mu - ax_c - bz_f}{\varepsilon a}}^{r=\frac{x_d - x_c}{\varepsilon}} f(x_c + r\varepsilon)\psi\left\{\frac{(\mu_2 + g\varepsilon a) - (ax_c + r\varepsilon)}{b}\right\}\overbrace{\frac{\delta x}{\varepsilon}}$$

$$3.18(1)$$

As before, $\delta\mu = \delta g\varepsilon a = \varepsilon a$ since $\delta g = 1$. Thus in the limit when $r \to \infty, g \to \infty$ and $\varepsilon \to 0$, the sums in 3.18(1) tend to the integrals

$$\frac{1}{a}\int_{\mu_2}^{\mu_4} {}_3\bar\eta_\mu \, d\mu = \frac{1}{b}\int_{\mu_2}^{\mu_4} d\mu \int_{\frac{\mu - bz_f}{a}}^{x_d} f(x)\psi\left[\frac{\mu - ax}{b}\right] dx \qquad 3.18(2)$$

where the limits are obtained as at the end of paragraph 3.17, that is, the limits for x are obtained from 3.05(5) whilst those of μ are obtained from 3.17(2).

3.19   It is to be noted that when

$$\mu = ax + bz \qquad 3.05(1)$$

and

$$b(z_f - z_c) \geqslant a(x_d - x_c) \qquad 3.07(3)$$

with $x_d > x_c, z_f > z_e$, and a and b are positive, that

$$\mu_1 = ax_d + bz_e \qquad 3.07(1)$$

$$\mu_2 = ax_c + bz_f \qquad 3.07(2)$$

$$\mu_3 = ax_c + bz_e \qquad 3.13(5)$$

$$\mu_4 = ax_d + bz_f \qquad 3.17(8)$$

giving

$$\mu_4 > \mu_2 > \mu_1 > \mu_3 \qquad 3.19(1)$$

and that

$$_1\bar\eta_\mu = \frac{a}{b}\int_{x_c}^{x_d} f(x)\psi\left(\frac{\mu - ax}{b}\right) dx \qquad 3.10(1)$$

valid for $\mu_2 \geqslant \mu \geqslant \mu_1$

$$_2\bar\eta_\mu = \frac{a}{b}\int_{x_c}^{\frac{\mu - bz_e}{a}} f(x)\psi\left(\frac{\mu - ax}{b}\right) dx \qquad 3.14(2)$$

valid for $\mu_1 \geqslant \mu \geqslant \mu_3$

$$_3\bar{\eta}_\mu = \frac{a}{b} \int_{\frac{\mu - bz_f}{a}}^{x_d} f(x)\psi\left(\frac{\mu - ax}{b}\right) dx \qquad 3.17(10)$$

valid for $\mu_4 \geqslant \mu \geqslant \mu_2$.

3.20　If we now consider x and z related by

$$\mu = ax - bz \qquad 3.20(1)$$

where once again $b(z_f - z_e) > a(x_d - x_c), x_d > x_c, z_f > z_e$, a and b are positive, and we repeat the arguments of paragraphs 3.05 to 3.18, we obtain an analogous set of integrals, namely

$$_1\bar{\eta}'_\mu = \frac{a}{b} \int_{x_c}^{x_d} f(x)\psi\left(\frac{ax - \mu}{b}\right) dx \qquad 3.20(2)$$

valid for $\mu'_3 \geqslant \mu \geqslant \mu'_4$

$$_2\bar{\eta}'_\mu = \frac{a}{b} \int_{x_c}^{\frac{\mu + bz_f}{a}} f(x)\psi\left(\frac{ax - \mu}{b}\right) dx \qquad 3.20(3)$$

valid for $\mu'_4 \geqslant \mu \geqslant \mu'_2$

$$_3\bar{\eta}'_\mu = \frac{a}{b} \int_{\frac{\mu + bz_e}{a}}^{x_d} f(x)\psi\left(\frac{ax - \mu}{b}\right) dx \qquad 3.20(4)$$

valid for $\mu'_1 \geqslant \mu \geqslant \mu'_3$ where

$$\mu'_1 \geqslant \mu'_3 \geqslant \mu'_4 \geqslant \mu'_2 \qquad 3.20(5)$$

and

$$\mu'_1 = ax_d - bz_e \qquad 3.20(6)$$

$$\mu'_2 = ax_c - bz_f \qquad 3.20(7)$$

$$\mu'_3 = ax_c - bz_e \qquad 3.20(8)$$

$$\mu'_4 = ax_d - bz_f \qquad 3.20(9)$$

As before

$$\frac{1}{a} \int_{\mu'_4}^{\mu'_3} {}_1\bar{\eta}'_\mu \, d\mu = \frac{1}{b} \int_{\mu'_4}^{\mu'_3} d\mu \int_{x_c}^{x_d} f(x)\psi\left(\frac{ax - \mu}{b}\right) dx \qquad 3.20(10)$$

$$\frac{1}{a} \int_{\mu'_2}^{\mu'_4} {}_2\bar{\eta}'_\mu \, d\mu = \frac{1}{b} \int_{\mu'_2}^{\mu'_4} d\mu \int_{x_c}^{\frac{\mu + bz_f}{a}} f(x)\psi\left(\frac{ax - \mu}{b}\right) dx \qquad 3.20(11)$$

$$\frac{1}{a} \int_{\mu'_3}^{\mu'_1} {}_3\bar{\eta}'_\mu \, d\mu = \frac{1}{b} \int_{\mu'_3}^{\mu'_1} d\mu \int_{\frac{\mu + bz_e}{a}}^{x_d} f(x)\psi\left(\frac{ax - \mu}{b}\right) dx \qquad 3.20(12)$$

It is to be especially noted that when $ax - bz$ is used instead of $ax + bz$, that the corresponding integrals for $ax - bz$ are *not* obtained by changing the sign of b in the set for $ax + bz$. Note that the integral expressions for $_1\bar{\eta}_\mu$ and $_1\bar{\eta}'_\mu$ are both integrated between x_c and x_d and are valid between the second largest and third largest value of μ, where the μ' expressions are obtained from the μ expressions by writing $-bz$ for bz in 3.07(1), 3.07(2), 3.13(5) and 3.17(8). In the expressions for $_2\bar{\eta}_\mu$ and $_2\bar{\eta}'_\mu$ the lower limit of each integral is x_c, but the upper limit of $_2\bar{\eta}_\mu$ is $\mu - bz_e$ whilst that of $_2\bar{\eta}'_\mu$ is $\mu + bz_f$. Each expression is valid between the third and fourth largest values of μ for each expression, and the upper limit of each integral contains a z term, common to the two values of μ between which each integral is valid, but with its sign reversed. Finally $_3\bar{\eta}_\mu$ and $_3\bar{\eta}'_\mu$ each have x_d as their upper limit, but the lower limit of $_3\bar{\eta}_\mu$ is $(\mu - bz_f)/a$ whilst that of $_3\bar{\eta}'_\mu$ is $(\mu + bz_e)/a$. Each expression is valid between the largest and second largest values of μ for each expression, and the lower limit of each integral contains a z term common to the two values of μ between which each integral is valid, once again with its sign reversed, i.e. bz_f for $_3\bar{\eta}_\mu$ and $-bz_e$ for $_3\bar{\eta}'_\mu$.

3.21 If we again consider the relationship

$$\mu = ax + bz \qquad 3.05(1)$$

but this time for

$$b(z_f - z_e) \leqslant a(x_d - x_c) \qquad 3.21(1)$$

with $z_f > z_e, x_d > x_c$, with a and b positive as before, then $\psi(z)$ should be placed at the top of the array and there there will be more terms in $f(x)$ than in $\psi(z)$. In the expressions for the η_μ

$$f(x)\psi\left(\frac{\mu - ax}{b}\right) dx$$

should be replaced by

$$\psi(z)f\left(\frac{\mu - bz}{a}\right) dz$$

and the other constants replaced as follows: b by a, a by b, z_e by x_c, x_c by z_e, z_f by x_d and x_d by z_f, yielding

$$_1\bar{\eta}_\mu = \frac{b}{a}\int_{z_e}^{z_f} \psi(z)f\left(\frac{\mu - bz}{a}\right) dz \qquad 3.21(2)$$

valid for $\mu_1 \geqslant \mu \geqslant \mu_2$

$$_2\bar{\eta}_\mu = \frac{b}{a}\int_{z_e}^{\frac{\mu - ax_c}{b}} \psi(z)f\left(\frac{\mu - bz}{a}\right) dz \qquad 3.21(3)$$

valid for $\mu_2 \geqslant \mu \geqslant \mu_3$

$$_3\bar{\eta}_\mu = \frac{b}{a} \int_{\frac{\mu - ax_d}{b}}^{z_f} \psi(z) f\left(\frac{\mu - bz}{a}\right) dz \qquad 3.21(4)$$

valid for $\mu_4 \geqslant \mu \geqslant \mu_1$. Note that

$$\mu_4 \geqslant \mu_1 \geqslant \mu_2 \geqslant \mu_3 \qquad 3.21(5)$$

The values of μ_1, μ_2, μ_3 and μ_4 are given by 3.07(1), 3.07(2), 3.13(5) and 3.17(8). The new μ are obtained by interchanging the constants as prescribed above in the previous μ and using 3.21(1). If we consider the relationship

$$\mu = -ax + bz \qquad 3.21(6)$$

with

$$b(z_f - z_e) \leqslant a(x_d - x_c) \qquad 3.21(1)$$

then using the rules given at the end of paragraph 3.20 and operating on the $\bar{\eta}_\mu$s of 3.21(1), 3.21(2) and 3.21(3) we have the new $\bar{\eta}_\mu$s and limits as follows

$$\mu_1 = -ax_d + bz_e \qquad 3.21(7)$$

$$\mu_2 = -ax_c + bz_f \qquad 3.21(8)$$

$$\mu_3 = -ax_c + bz_e \qquad 3.21(9)$$

$$\mu_4 = -ax_d + bz_f \qquad 3.21(10)$$

These four expressions yield

$$\mu_2 > \mu_3 > \mu_4 > \mu_1 \qquad 3.21(11)$$

using 3.21(1), and thus

$$_1\bar{\eta}_\mu = \frac{b}{a} \int_{z_e}^{z_f} \psi(z) f\left(\frac{bz - \mu}{a}\right) dz \qquad 3.21(12)$$

valid for $\mu_3 \geqslant \mu \geqslant \mu_4$

$$_2\bar{\eta}_\mu = \frac{b}{a} \int_{z_e}^{\frac{\mu + ax_d}{b}} \psi(z) f\left(\frac{bz - \mu}{a}\right) dz \qquad 3.21(13)$$

valid for $\mu_4 \geqslant \mu \geqslant \mu_1$

$$_3\bar{\eta}_\mu = \frac{b}{a} \int_{\frac{\mu + ax_c}{b}}^{z_f} \psi(z) f\left(\frac{bz - \mu}{a}\right) dz \qquad 3.21(14)$$

valid for $\mu_2 \geqslant \mu \geqslant \mu_3$. Let us now consider the relationship

$$\mu = ax - bz \qquad 3.20(1)$$

with

$$b(z_f - z_e) \leqslant a(x_d - x_c) \qquad 3.21(1)$$

then we obtain the $\bar{\eta}_\mu$ and their limits by considering the expressions 3.20(2) to 3.20(4) and 3.20(6) to 3.20(9) and by using the rules given at the beginning of paragraph 3.21. Thus we have

$$\mu'_1 = ax_d - bz_e \qquad 3.20(5)$$

$$\mu'_2 = ax_c - bz_f \qquad 3.20(6)$$

$$\mu'_3 = ax_c - bz_e \qquad 3.20(7)$$

$$\mu'_4 = ax_d - bz_f \qquad 3.20(8)$$

and $\mu'_1 \geqslant \mu'_4 \geqslant \mu'_3 \geqslant \mu'_2$

$$_1\bar{\eta}'_\mu = \frac{b}{a} \int_{z_e}^{z_f} \psi(z) f\left(\frac{bz - \mu}{a}\right) dz \qquad 3.21(15)$$

valid for $\mu'_4 \geqslant \mu \geqslant \mu'_3$

$$_2\bar{\eta}'_\mu = \frac{b}{a} \int_{z_e}^{\frac{\mu + ax_d}{b}} \psi(z) f\left(\frac{bz - \mu}{a}\right) dz \qquad 3.21(16)$$

valid for $\mu'_3 \geqslant \mu \geqslant \mu'_2$

$$_3\bar{\eta}'_\mu = \frac{b}{a} \int_{\frac{\mu + ax_c}{b}}^{z_f} \psi(z) f\left(\frac{bz - \mu}{a}\right) dz \qquad 3.21(17)$$

valid for $\mu'_1 \geqslant \mu \geqslant \mu'_4$. Finally if we consider the relationship

$$\mu = -ax + bz \qquad 3.21(6)$$

with

$$b(z_f - z_e) \geqslant a(x_d - x_c) \qquad 3.07(3)$$

we obtain the η_μs and their limits by considering the expressions 3.21(7) to 3.21(10) and 3.21(12) to 3.21(14) and by using the rules given at the beginning of paragraph 3.21, whence

$$\mu_1 = -ax_d + bz_e \qquad 3.21(7)$$

$$\mu_2 = -ax_c + bz_f \qquad 3.21(8)$$

$$\mu_3 = -ax_c + bz_e \qquad 3.21(9)$$

$$\mu_4 = -ax_d + bz_f \qquad 3.21(10)$$

These four expressions yield $\mu_2 \geqslant \mu_4 \geqslant \mu_3 \geqslant \mu_1$ using 3.07(3). Thus we have

$$_1\bar{\eta}_\mu = \frac{a}{b} \int_{x_c}^{x_d} f(x)\psi\left(\frac{ax + \mu}{b}\right) dx \qquad 3.21(18)$$

valid for $\mu_4 \geqslant \mu \geqslant \mu_3$

$$_2\bar{\eta}_\mu = \frac{a}{b} \int_{x_c}^{\frac{\mu + bz_f}{a}} f(x)\psi\left(\frac{ax + \mu}{b}\right) dx \qquad 3.21(19)$$

valid for $\mu_3 \geqslant \mu \geqslant \mu_1$

$$_3\bar{\eta}_\mu = \frac{a}{b} \int_{\frac{\mu + bz_e}{a}}^{x_d} f(x)\psi\left(\frac{ax + \mu}{b}\right) dx \qquad 3.21(20)$$

valid for $\mu_2 \geqslant \mu \geqslant \mu_4$. One final word, it is *permissible* to change the sign of μ; thus if $\mu = ax - bz$, and $b(z_f - z_e) \geqslant a(x_d - x_c)$, then if we change the sign of μ, we set $\mu = -ax + bz$, again for $z_f - z_c \geqslant a/b(x_d - x_c)$. The size sequence of the various μ will also be reversed. To see this, compare 3.20(2) to 3.20(4) with 3.21(18) to 3.21(20), where the only difference is the sign of μ in the variable ψ and the interchange of the μ in the limits, since for $ax - bz$, $\mu_1 \geqslant \mu_3 \geqslant \mu_4 \geqslant \mu_2$ whilst for $-ax + bz$, $\mu_2 > \mu_4 > \mu_3 > \mu_1$, and the limits for $_1\eta_\mu$, $_2\eta_\mu$ and $_3\eta_\mu$ are thus reversed, that is respectively to $\mu_3 \geqslant \mu_4$, $\mu_3 \geqslant \mu_1$ and $\mu_2 \geqslant \mu_4$.

Expressions for the Double Integral Theorem

3.22   Let us now consider the sum

$$S_1 = \sum {_1\eta_g\varepsilon} + \sum {_2\eta_g\varepsilon} + \sum {_3\eta_g\varepsilon} \qquad 3.22(1)$$

where for the sake of clarity the limits have been omitted. A little consideration of the array and the relevant sums involved, that is equations 3.11(1), 3.15(1) and 3.18(1), shows that the sum S_1 is made up of the sum of the products of each term of the top row with each term of the bottom row. This can be also written

$$\sum_{r=0}^{r = \frac{x_d - x_c}{\varepsilon}} f(r)\varepsilon \cdot \sum_{k=0}^{k = (z_f - z_e)\frac{b}{\varepsilon a}} \psi(r)\frac{\varepsilon a}{b} = S_2 \qquad 3.22(2)$$

where

$$z = z_e + k\varepsilon a/b \qquad 3.22(3)$$

and

$$x = x_c + r\varepsilon \qquad 3.05(5)$$

Now $\delta z = \delta k \varepsilon a / b = \varepsilon a / b$ since $\delta k = 1$ and $z_f > z_e$. Thus when k and $r \to \infty$ and $\varepsilon \to 0$, S_2 tends to

$$\int_{x_c}^{x_d} f(x)\, dx \cdot \int_{z_e}^{z_f} \psi(z)\, dz \qquad 3.22(4)$$

Now we know that when g and r tend to infinity and ε tends to zero the sum S_1 tends to

$$\frac{1}{a}\left[\int_{\mu_3}^{\mu_1} {}_2\bar{\eta}_\mu\, d\mu + \int_{\mu_1}^{\mu_2} {}_1\bar{\eta}_\mu\, d\mu + \int_{\mu_2}^{\mu_4} {}_3\bar{\eta}_\mu\, d\mu \right] \qquad 3.22(5)$$

Also, since all the series involved are absolutely convergent, $S_1 = S_2$, since S_1 is the same as S_2 with the terms rearranged. Thus since $f(x)$ and $\psi(z)$ are bounded and continuous except for a finite number of discontinuities, the limit $S_1 = $ limit S_2 and, finally,

$$\int_{x_c}^{x_d} f(x)\, dx \int_{z_e}^{z_f} \psi(z)\, dz = \frac{1}{b} \int_{\overbrace{bz_e + ax_c}^{\mu_3}}^{\overbrace{bz_e + ax_d}^{\mu_1}} d\mu \int_{x_c}^{\overbrace{\frac{\mu - bz_e}{a}}} f(x)\psi\left(\frac{\mu - ax}{b}\right) dx$$

$$+ \frac{1}{b} \int_{\overbrace{bz_e + ax_d}^{\mu_1}}^{\overbrace{bz_f + ax_c}^{\mu_2}} d\mu \int_{x_c}^{x_d} f(x)\psi\left(\frac{\mu - ax}{b}\right) dx$$

$$+ \frac{1}{b} \int_{\overbrace{bz_f + ax_c}^{\mu_2}}^{\overbrace{bz_f + ax_d}^{\mu_4}} d\mu \int_{\underbrace{\frac{\mu - bz_f}{a}}}^{x_d} f(x)\psi\left(\frac{\mu - ax}{b}\right) dx$$

$$3.22(6)$$

This identity is valid for

$$b(z_f - z_e) \geq a(x_d - x_c) \qquad 3.07(3)$$

and

$$\mu = ax + bz \qquad 3.05(1)$$

where μ_1, μ_2, μ_3 and μ_4 are given by 3.07(1), 3.07(2), 3.13(5) and 3.17(8), and $\mu_4 \geq \mu_2 \geq \mu_1 \geq \mu_3$.

3.23 If $b(z_f - z_e) \leq a(x_d - x_c)$, then $\psi(z)$ should be placed on the top line of the array, and

$$f(x)\psi\left(\frac{\mu - ax}{b}\right) dx$$

should be replaced by

$$\psi(z)f\left(\frac{\mu - bz}{a}\right)dz$$

the other constants being interchanged as follows: b by a, a by b, z_e by x_c, x_c by z_e, z_f by x_d, and x_d by z_f, as described at the beginning of paragraph 3.21 yielding

$$\int_{x_c}^{x_d} f(x)\,dx \int_{z_e}^{z_f} \psi(z)\,dz = \frac{1}{a}\int_{bz_e + ax_c}^{bz_f + ax_c} d\mu \int_{z_e}^{\frac{\mu - ax_c}{b}} \psi(z)f\left(\frac{\mu - bz}{a}\right)dz$$

$$+ \frac{1}{a}\int_{bz_e + ax_c}^{bz_e + ax_d} d\mu \int_{z_e}^{z_f} \psi(z)f\left(\frac{\mu - bz}{a}\right)dz$$

$$+ \frac{1}{a}\int_{bz_e + ax_d}^{bz_f + ax_d} d\mu \int_{\frac{\mu - ax_d}{b}}^{z_f} \psi(z)f\left(\frac{\mu - bz}{a}\right)dz$$

3.23(1)

valid for

$$b(z_f - z_e) \leqslant a(x_d - x_c) \qquad\qquad 3.21(1)$$

with

$$\mu = ax + bz \qquad\qquad 3.05(1)$$

and where μ_1, μ_2, μ_3 and μ_4 are given by 3.07(1), 3.07(2), 3.13(5) and 3.17(8), and $\mu_4 \geqslant \mu_1 \geqslant \mu_2 \geqslant \mu_3$.

Similar identities exist for the other functions of μ, that is for $\mu = ax - bz$, etc., which have been discussed in the previous paragraphs.

Alternative Forms of Theorem

3.24 The identity 3.22(6) can be written in alternative form by using equation 3.05(1), solving for $x = (\mu - bz)/a$ and substituting for x in terms of z. This gives

$$\int_{x_c}^{x_d} f(x)\,dx \int_{z_e}^{z_f} \psi(z)\,dz = \frac{1}{a}\int_{bz_e + ax_c}^{bz_e + ax_d} d\mu \int_{z_e}^{\frac{\mu - ax_c}{b}} \psi(z)f\left(\frac{\mu - bz}{a}\right)dz$$

$$+ \frac{1}{a}\int_{bz_e + ax_d}^{bz_f + ax_c} d\mu \int_{\frac{\mu - ax_d}{b}}^{\frac{\mu - ax_e}{b}} \psi(z)f\left(\frac{\mu - bz}{a}\right)dz$$

$$+ \frac{1}{a}\int_{bz_f + ax_c}^{bz_f + ax_d} d\mu \int_{\frac{\mu - ax_d}{b}}^{z_f} \psi(z)f\left(\frac{\mu - bz}{a}\right)dz$$

3.24(1)

valid for

$$b(z_f - z_e) \geqslant a(x_d - x_c) \qquad \text{3.07(3)}$$

3.25 Similarly, the identity 3.23(1) can be written in alternative form by using equation 3.05(2), that is $z = (\mu - ax)/b$ and substituting for z in terms of x. This gives

$$\int_{x_c}^{x_d} f(x)\, dx \int_{z_e}^{z_f} \psi(z)\, dz = \frac{1}{b} \int_{bz_e + ax_c}^{bz_f + ax_c} d\mu \int_{x_c}^{\frac{\mu - bz_e}{a}} f(x)\psi\left(\frac{\mu - ax}{b}\right) dx$$

$$+ \frac{1}{b} \int_{bz_e + ax_c}^{bz_e + ax_d} d\mu \int_{\frac{\mu - bz_f}{a}}^{\frac{\mu - bz_e}{a}} f(x)\psi\left(\frac{\mu - ax}{b}\right) dx$$

$$+ \frac{1}{b} \int_{bz_e + ax_d}^{bz_e + ax_d} d\mu \int_{\frac{\mu - bz_f}{a}}^{x_d} f(x)\psi\left(\frac{\mu - ax}{b}\right) dx$$

$$\text{3.25(1)}$$

valid for $b(z_f - z_e) \leqslant a(x_d - x_c)$.

3.26 It should be noted that the two forms in terms of x, that is 3.22(6) and 3.25(1) have *different* limits, the former applies to the case $b(z_f - z_e) \geqslant a(x_d - x_c)$ and the latter to the case $b(z_f - z_e) \leqslant a(x_d - x_c)$. The other two forms which are expressed in terms of z, that is 3.23(1) and 3.24(1), also have different limits and it is to be noted that 3.23(1) covers the case when $b(z_f - z_e) \leqslant a(x_d - x_c)$ whilst 3.24(1) covers the case $b(z_f - z_e) \geqslant a(x_d - x_c)$.

Integral Functions

3.27 Let

$$\mu = ax + bz \qquad \text{3.05(1)}$$

and

$$b(z_f - z_e) \geqslant a(x_d - x_c) \qquad \text{3.07(3)}$$

The integral function of μ made up of

$$P(\mu) = \frac{1}{a}({}_2\bar{\eta}_\mu + {}_1\bar{\eta}_\mu + {}_3\bar{\eta}_\mu) = \frac{1}{b}\left\{ \int_{x_c}^{\frac{\mu - bz_e}{a}} f(x)\psi\left(\frac{\mu - ax}{b}\right) dx \right.$$

$$\text{valid for } \mu_1 \geqslant \mu \geqslant \mu_3$$

$$\left. + \int_{x_c}^{x_d} f(x)\psi\left(\frac{\mu - ax}{b}\right) dx + \int_{\frac{\mu - bz_f}{a}}^{x_d} f(x)\psi\left(\frac{\mu - ax}{b}\right) dx \right\}$$

$$\text{valid for } \mu_2 \geqslant \mu \geqslant \mu_1 \qquad \text{valid for } \mu_4 \geqslant \mu \geqslant \mu_2 \qquad \text{3.27(1)}$$

where

$$\mu_1 = ax_d + bz_e \qquad\qquad 3.27(2a)$$

$$\mu_2 = ax_c + bz_f \qquad\qquad 3.27(2b)$$

$$\mu_3 = ax_c + bz_e \qquad\qquad 3.27(2c)$$

$$\mu_4 = ax_d + bz_f \qquad\qquad 3.27(2d)$$

and

$$\mu_4 \geqslant \mu_2 \geqslant \mu_1 \geqslant \mu_3$$

is of interest since, as we shall see later, it represents the combined or convoluted probability density function of $f(x)$ and $\psi(z)$, where these are themselves probability density functions, and x and z are related by the equation $\mu = ax + bz$ (3.05(1)).

It is to be noted that if $\bar{x}$ and $\bar{z}$ are the mean values of x and z, then the mean value of μ is given by

$$\bar{\mu} = a\bar{x} + b\bar{z} \qquad\qquad 3.27(2e)$$

If

$$\mu = ax + bz \qquad\qquad 3.05(1)$$

but $b(z_f - z_e) \leqslant a(x_d - x_c)$ (3.21(1)), then the appropriate integral is

$$P(\mu) = \frac{1}{b}(_2\bar{\eta}_\mu + {}_1\bar{\eta}_\mu + {}_3\bar{\eta}_\mu) = \frac{1}{a}\left\{ \int_{z_e}^{\frac{\mu - ax_c}{a}} \psi(z)f\left(\frac{\mu - bz}{a}\right) dz \right.$$

$$\text{valid for } \mu_2 \geqslant \mu \geqslant \mu_3$$

$$+ \int_{z_e}^{z_f} \psi(z)f\left(\frac{\mu - bz}{a}\right) dz + \left. \int_{\frac{\mu - ax_d}{b}}^{z_c} \psi(z)f\left(\frac{\mu - bz}{a}\right) dz \right\}$$

$$\text{valid for } \mu_1 \geqslant \mu \geqslant \mu_2 \qquad \text{valid for } \mu_4 \geqslant \mu \geqslant \mu_1 \qquad 3.27(3)$$

where $\mu_4 \geqslant \mu_1 \geqslant \mu_2 \geqslant \mu_3$ and the μ are given by 3.27(2a), (2b), (2c) and (2d). The integral is obtained by operating on 3.27(1), using the rules given at the beginning of paragraph 3.21.

3.28 Integral 3.27(1) can be expressed in terms of z, and integral 3.27(3) in terms of x, by using $\mu = ax + bz$ (3.05(1)) and recalculating the limits for x or z for the integral signs. Note that the limits for the μ remain unaltered when x and z are interchanged. $P(\mu)$ (3.27(1)) as a function of z is

$$P(\mu) = \frac{1}{a}\left\{ \int_{z_e}^{\frac{\mu - ax_e}{b}} \psi(z)f\left(\frac{\mu - bz}{a}\right) dz + \int_{\frac{\mu - ax_d}{b}}^{\frac{\mu - ax_e}{b}} \psi(z)f\left(\frac{\mu - bz}{a}\right) dz \right.$$

$\qquad\qquad$ valid for $\mu_1 \geqslant \mu \geqslant \mu_3$ $\qquad$ valid for $\mu_2 \geqslant \mu \geqslant \mu_1$

$$\left. + \int_{\frac{\mu - ax_d}{b}}^{z_f} \psi(z)f\left(\frac{\mu - bz}{a}\right) dz \right\} \qquad\qquad 3.28(1)$$

$\qquad\qquad$ valid for $\mu_4 \geqslant \mu \geqslant \mu_2$

where $\mu_4 \geqslant \mu_2 \geqslant \mu_1 \geqslant \mu_3$.

$\quad P(\mu)$ (3.27(3)) as a function of x is

$$P(\mu) = \frac{1}{b}\left\{ \int_{x_c}^{\frac{\mu - bz_e}{a}} f(x)\psi\left(\frac{\mu - ax}{b}\right) dx + \int_{\frac{\mu - bz_f}{a}}^{\frac{\mu - bz_e}{a}} f(x)\psi\left(\frac{\mu - ax}{b}\right) dx \right.$$

$\qquad\qquad$ valid for $\mu_2 \geqslant \mu \geqslant \mu_3$ $\qquad$ valid for $\mu_1 \geqslant \mu \geqslant \mu_2$

$$\left. + \int_{\frac{\mu - bz_f}{a}}^{x_d} f(x)\psi\left(\frac{\mu - ax}{b}\right) dx \right\} \qquad\qquad 3.28(2)$$

$\qquad\qquad$ valid for $\mu_4 \geqslant \mu \geqslant \mu_1$

where $\mu_4 \geqslant \mu_1 \geqslant \mu_2 \geqslant \mu_3$.

$\quad$ Similar functions exist for the other functions of μ. For

$$\mu = ax - bz \qquad\qquad 3.20(1)$$

and

$$b(z_f - z_e) \geqslant a(x_d - x_c) \qquad\qquad 3.07(3)$$

then

$$P(\mu) = \frac{1}{a}({}_2\bar{\eta}_\mu + {}_1\bar{\eta}_\mu + {}_3\bar{\eta}_\mu) = \frac{1}{b}\left\{ \int_{x_c}^{\frac{\mu + bz_f}{a}} f(x)\psi\left(\frac{ax - \mu}{b}\right) dx \right.$$

$\qquad\qquad$ valid for $\mu_4 \geqslant \mu \geqslant \mu_2$

$$\left. + \int_{x_c}^{x_d} f(x)\psi\left(\frac{ax - \mu}{b}\right) dx + \int_{\frac{\mu + bz_e}{a}}^{x_d} f(x)\psi\left(\frac{ax - \mu}{b}\right) dx \right\} \qquad 3.28(3)$$

$\qquad\qquad$ valid for $\mu_3 \geqslant \mu \geqslant \mu_4$ $\qquad$ valid for $\mu_1 \geqslant \mu \geqslant \mu_3$

where $\mu_1 \geqslant \mu_3 \geqslant \mu_4 \geqslant \mu_2$.

For

$$\mu = ax - bz \tag{3.26(1)}$$

but with

$$b(z_f - z_e) \leqslant a(x_d - x_c) \tag{3.21(1)}$$

then

$$P(\mu) = \frac{1}{b}({}_2\bar\eta_\mu + {}_1\bar\eta_\mu + {}_3\bar\eta_\mu) = \frac{1}{a}\left\{ \int_{z_e}^{\frac{\mu + ax_d}{b}} \psi(z) f\left(\frac{bz - \mu}{a}\right) dz \right.$$

valid for $\mu_3 \geqslant \mu \geqslant \mu_2$

$$+ \int_{z_e}^{z_f} \psi(z) f\left(\frac{bz - \mu}{a}\right) dz + \left. \int_{\frac{\mu + ax_c}{b}}^{z_f} \psi(z) f\left(\frac{bz - \mu}{a}\right) dz \right\} \tag{3.28(4)}$$

valid for $\mu_4 \geqslant \mu \geqslant \mu_3$ valid for $\mu_1 \geqslant \mu \geqslant \mu_4$

where $\mu_1 \geqslant \mu_4 \geqslant \mu_3 \geqslant \mu_2$ and where

$$\mu_1 = ax_d - bz_e \tag{3.20(5)}$$
$$\mu_2 = ax_c - bz_f \tag{3.20(6)}$$
$$\mu_3 = ax_c - bz_e \tag{3.20(7)}$$
$$\mu_4 = ax_d - bz_f \tag{3.20(8)}$$

For

$$\mu = -ax + bz \tag{3.21(6)}$$

and

$$b(z_f - z_e) \geqslant a(x_d - x_c) \tag{3.07(3)}$$

then

$$P(\mu) = \frac{1}{a}({}_2\bar\eta_\mu + {}_1\bar\eta_\mu + {}_3\bar\eta_\mu) = \frac{1}{b}\left\{ \int_{x_c}^{\frac{\mu + bz_f}{a}} f(x) \psi\left(\frac{ax + \mu}{b}\right) dx \right.$$

valid for $\mu_3 \geqslant \mu \geqslant \mu_1$

$$+ \int_{x_c}^{x_d} f(x) \psi\left(\frac{ax + \mu}{b}\right) dx + \left. \int_{\frac{\mu + bz_e}{a}}^{x_d} f(x) \psi\left(\frac{ax + \mu}{b}\right) dx \right\} \tag{3.28(5)}$$

valid for $\mu_4 \geqslant \mu \geqslant \mu_3$ valid for $\mu_2 \geqslant \mu \geqslant \mu_4$

where $\mu_2 \geqslant \mu_4 \geqslant \mu_3 \geqslant \mu_1$ and μ_1, μ_2, μ_3 and μ_4 are given by 3.21(7), 3.21(8), 3.21(9) and 3.21(10).

For $\mu = -ax + bz$ (3.21(6)), but with

$$b(z_f - z_e) \leqslant (x_d - x_c) \qquad\qquad 3.21(1)$$

then

$$P(\mu) = \frac{1}{b}(_2\bar{\eta}_\mu + _1\bar{\eta}_\mu + _3\bar{\eta}_\mu) = \frac{1}{a}\left\{ \int_{z_e}^{\frac{\mu + ax_d}{b}} \psi(z)f\left(\frac{bz - \mu}{a}\right) dz \right.$$

$$\text{valid for } \mu_4 \geqslant \mu \geqslant \mu_1$$

$$+ \int_{z_e}^{z_f} \psi(z)f\left(\frac{bz - \mu}{a}\right) dz + \left. \int_{\frac{\mu + ax_c}{b}}^{z_f} \psi(z)f\left(\frac{bz - \mu}{a}\right) dz \right\} \qquad 3.28(6)$$

$$\text{valid for } \mu_3 \geqslant \mu \geqslant \mu_4 \qquad \text{valid for } \mu_2 \geqslant \mu \geqslant \mu_3$$

where $\mu_2 \geqslant \mu_3 \geqslant \mu_4 \geqslant \mu_1$ and where

$$\mu_1 = -ax_d + bz_e \qquad\qquad 3.21(7)$$

$$\mu_2 = -ax_c + bz_f \qquad\qquad 3.21(8)$$

$$\mu_3 = -ax_c + bz_e \qquad\qquad 3.21(9)$$

$$\mu_4 = -ax_d + bz_f \qquad\qquad 3.21(10)$$

Range of P(μ)

3.29 It is interesting to note that in all six cases of $P(\mu)$ considered, namely $\mu = ax + bz$, $\mu = ax - bz$, and $\mu = -ax + bz$, with $b(z_f - z_e) \geqslant a(x_d - x_c)$ or $b(z_f - z_e) \leqslant a(x_d - x_c)$, for each value of μ the total range of the convolution function $P(\mu)$ is equal to $b(z_f - z_e) + a(x_d - x_c)$. Likewise the range of $_2\bar{\eta}_\mu$ or of $_3\bar{\eta}_\mu$ for $b(z_f - z_e) \geqslant a(x_d - x_c)$ is always $a(x_d - x_c)$ whilst for $b(z_f - z_e) \leqslant a(x_d - x_c)$ it is always $b(z_f - z_e)$. The range of $_1\bar{\eta}_\mu$ is always $b(z_f - z_e) - a(x_d - x_c)$ when $b(z_f - z_e) \geqslant a(x_d - x_c)$ and is always $a(x_d - x_c) - b(z_f - z_e)$ when $b(z_f - z_e) \leqslant a(x_d - x_c)$.

General Case: Combination of Uncertainty Populations or Frequency Distributions

3.30 Let $\alpha(x)$ be a bounded function between $-c$ and d, and since it is to be considered as a probability density function it must also be positive between these two limits. If a function is to be a density function between the limits $-c$ to d its integral between these limits must be equal to unity. Manifestly

$$P_x = \frac{\alpha(x)}{\int_{-c}^{d} \alpha(x)\, dx} \qquad\qquad 3.30(1)$$

is such a function (see Figure 3.30).

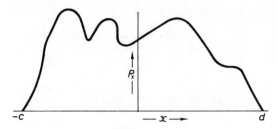

Figure 3.30

3.31　We first assume, as a matter of convenience†, that $x = 0$ gives the mean value of $\alpha(x)$ between the limits $-c$ to d. Now

$$\bar{x} \int_{-c}^{d} \alpha(x)\,dx = \int_{-c}^{d} x\alpha(x)\,dx \qquad\qquad 3.31(1)$$

and $\int_{-c}^{d} \alpha(x)\,dx > 0$, thus for $\bar{x}$ to be zero

$$\int_{-c}^{d} x\alpha(x)\,dx = 0 \qquad\qquad 3.31(2)$$

The standard deviation σ_x of P_x, which is defined as the square root of the mean second moment about the mean value of x, is thus given by

$$\sigma_x^2 = \int_{-c}^{d} x^2 P_x\,dx$$

$$= \frac{\int_{-c}^{d} x^2 \alpha(x)\,dx}{\int_{-c}^{d} \alpha(x)\,dx} = \int_{-c}^{d} \frac{x^2 \alpha(x)\,dx}{A} \qquad\qquad 3.31(3)$$

where

$$A = \int_{-c}^{d} \alpha(x)\,dx \qquad\qquad 3.31(4)$$

3.32　Similarly, if $\beta(z)$ is another function, bounded and positive between $-e$ to f, with its mean at zero, then, as before,

$$\int_{-e}^{f} z\beta(z)\,dz = 0 \qquad\qquad 3.32(1)$$

and the associated density function is given by

$$P_z = \frac{\beta(z)}{\int_{-e}^{f} \beta(z)\,dz} \qquad\qquad 3.32(2)$$

† This is merely a question of choice of co-ordinates.

whilst its standard deviations σ_z is given by the expression

$$\sigma_z^2 = \int_{-e}^{f} z^2 P_z \, dz = \frac{\int_{-e}^{f} z^2 \beta(z) \, dz}{\int_{-e}^{f} \beta(z) \, dz} = \int_{-e}^{f} \frac{z^2 \beta(z) \, dz}{B} \qquad 3.32(3)$$

where

$$B = \int_{-e}^{f} \beta(z) \, dz \qquad 3.32(4)$$

3.33 Let us now consider the compound probability of an uncertainty $ax + bz$, where the density functions for x and z are given by equations 3.30(1) and 3.32(2). Let

$$\mu + \delta\mu \geqslant ax + bz \geqslant \mu \qquad 3.33(1)$$

Then the probability of an uncertainty lying between the limits specified by the above inequality is $P_x P_z \, dx \, dz$, integrated over the range specified by the inequality and by the limits of the functions P_x and P_z, that is $-c$ to d, and $-e$ to f respectively.

3.34 Let us first specify that

$$f + e \geqslant \frac{a}{b}(d + c) \qquad 3.34(1)$$

and integrate P_z over the range

$$\frac{\mu + \delta\mu - ax}{b} \geqslant z \geqslant \frac{\mu - ax}{b} \qquad 3.34(2)$$

Thus

$$\int_{\frac{\mu - ax}{b}}^{\frac{\mu + \delta\mu - ax}{b}} P_z \, dz = \int \frac{\beta(z)}{B} \, dz = \frac{\beta((\mu - ax)/b)}{B} \frac{\delta\mu}{b} \qquad 3.34(3)$$

from the mean value theorem and where $\delta z = \delta\mu/b$ from 3.34(2). Thus the required probability $P_\mu \, d\mu$ is given by

$$\alpha(x)\beta\left(\frac{\mu - ax}{b}\right) dx \, d\mu/bAB$$

integrated over the range specified by 3.33(1) and the limits of the functions $\alpha(x)$ and $b(z)$. Now the limits of μ are given by using the expression

$$\mu = ax + bz \qquad 3.34(4)$$

Thus the minimum value of μ is given by $x = -c$ and $z = -e$, and so $\mu_{min} = -(ac + be)$. The maximum value of μ is given by $x = +d$, and $z = f$, giving $\mu_{max} = ad + bf$. But we cannot just integrate this expression between these limits, because $\alpha(x)$ is defined only between $-c$ to d and $\beta(z)$ is defined only between $-e$ to f; outside these ranges they are to be considered zero but, since neither will in fact be zero outside these limits, means must be found of circumventing this problem. The functions $P(x)$ and $P(z)$ may be specified by algebraic expressions such that at $-c$ and d, $P(x)$ is zero and at $-e$ and f, $P(z)$ is zero and, in between these values, the algebraic expressions give a close enough fit to $P(x)$ and $P(z)$. However, outside these limits the expressions will have finite values and thus integrating the expression for $P(\mu)$ between the limits $\mu_{min} = -(ac + be)$ and $\mu_{max} = ad + bf$ will lead to an incorrect value for the integral $P(\mu)$.

3.35   If we now consider the array in paragraph 3.06 we see that the integrals $_1\bar{\eta}_\mu$, $_2\bar{\eta}_\mu$ and $_3\bar{\eta}_\mu$ which we obtained from the array satisfy the conditions we have stated in the previous paragraph. The two functions $f(x)$ and $\psi(z)$ used in conjunction with the array were defined between set finite limits, were zero outside these limits and, further, were related by precisely the same relation as we are now considering, that is $\mu = ax + bz$. In the case of the array we obtained three values for the function $P(\mu)$ of μ, each valid over a certain range of μ. Thus the function $P(\mu)$ we require is given by considering the function $P(\mu)$ given by 3.27(1).

3.36   If we compare the two functions of x and z we see that $f(x)$ of 3.27(1) is equivalent to $P_x \equiv \alpha(x)/A$ and that $\psi(z)$ of 3.27(1) is equivalent to $P_z \equiv \beta(z)/B$. The relationship between z and x is the same, that is $\mu = ax + bz$. Thus all we have to do is to compare limits. This leads to

$$x_c = -c, \quad x_d = d, \quad z_e = -e, \quad \text{and} \quad z_f = f$$

Thus, inserting the expressions for P_x and P_z in 3.27(1) we have

$$P(\mu)\,d\mu = \left\{ \overbrace{\left(\frac{1}{bAB} \int_{-c}^{\frac{\mu+be}{a}} \alpha(x)\beta\left(\frac{\mu-ax}{b}\right) dx \right)}^{_1P(\mu)} \right.$$

$$\text{valid for } \mu_1 \geqslant \mu \geqslant \mu_3$$

$$+ \overbrace{\left(\frac{1}{bAB} \int_{-c}^{d} \alpha(x)\beta\left(\frac{\mu-ax}{b}\right) dx \right)}^{_2P(\mu)}$$

$$\text{valid for } \mu_2 \geqslant \mu \geqslant \mu_1$$

$$+ \left. \overbrace{\left(\frac{1}{bAB} \int_{\frac{\mu-bf}{a}}^{d} \alpha(x)\beta\left(\frac{\mu-ax}{b}\right) dx \right)}^{_3P(\mu)} \right\}\,d\mu$$

$$\text{valid for } \mu_4 \geqslant \mu \geqslant \mu_2$$

3.36(1)

where

$$\mu_1 = ad - be, \qquad \mu_2 = -ac + bf \Big\}$$
$$\mu_3 = -ac - be, \qquad \mu_4 = ad + bf \Big\} \qquad 3.36(2)$$

Thus the integral $_1P(\mu)$ gives the value of $P(\mu)$, the required density function over the range $\mu_1 \geqslant \mu \geqslant \mu_3$. Similarly $_2P(\mu)$ and $_3P(\mu)$ give $P(\mu)$ over the ranges given in 3.36(1). The probability of an uncertainty in a given range of μ is obtained by integrating the appropriate functions, $_1P(\mu)$, $_2P(\mu)$ and $_3P(\mu)$, over the required range of μ. If desired, of course, $P(\mu)$ can be written in the form

$$P(\mu)\, d\mu = \frac{1}{AB} \Bigg\{ \int_{-c}^{\frac{\mu+be}{a}} \alpha(x)\, dx \int_{\frac{\mu-ax}{b}}^{\frac{\mu+\delta\mu-ax}{b}} \beta(z)\, dz$$

$$+ \int_{-c}^{d} \alpha(x)\, dx \int_{\frac{\mu-ax}{b}}^{\frac{\mu+\delta\mu-ax}{b}} \beta(z)\, dz$$

$$+ \int_{\frac{\mu-bf}{a}}^{d} \alpha(x)\, dx \int_{\frac{\mu-ax}{b}}^{\frac{\mu+\delta\mu-ax}{b}} \beta(z)\, dz \Bigg\} d\mu \qquad 3.36(3)$$

Standard Deviation of Combined or Convoluted Distribution

3.37 Suppose that we now calculate the standard deviation of $P(\mu)$. In order to do this we multiply $P(\mu)$ by μ^2 and integrate $\mu^2\,_1P(\mu)$, $\mu^2\,_2P(\mu)$ and $\mu^2\,_3P(\mu)$ over its appropriate range in μ. Thus using 3.36(1), multiplying by μ^2 and integrating we have

$$\sigma_\mu^2 = \frac{1}{bAB} \Bigg\{ \int_{\mu_2}^{\mu_1} \mu^2\, d\mu \int_{-c}^{\frac{\mu+be}{a}} \alpha(x)\beta\left(\frac{\mu-ax}{b}\right) dx$$

$$+ \int_{\mu_1}^{\mu_2} \mu^2\, d\mu \int_{-c}^{d} \alpha(x)\beta\left(\frac{\mu-ax}{b}\right) dx$$

$$+ \int_{\mu_2}^{\mu_4} \mu^2\, d\mu \int_{\frac{\mu-bf}{a}}^{d} \alpha(x)\beta\left(\frac{\mu-ax}{b}\right) dx \Bigg\} \qquad 3.37(1)$$

where μ_1, μ_2, μ_3 and μ_4 are given by 3.36(2). Notice that the integrands of each term are similar. Now we can write

$$\mu^2 = a^2 x^2 + b^2\left(\frac{\mu - ax}{b}\right)^2 + 2abx\left(\frac{\mu - ax}{b}\right) \qquad 3.37(2)$$

and so the integral of each of the terms 3.37(1) can now be written

$$\frac{1}{bAB}\int \delta\mu \int \left\{ a^2 x^2 \alpha(x)\beta\left(\frac{\mu - ax}{b}\right) + b^2 \alpha(x)\left(\frac{\mu - ax}{b}\right)^2 \beta\left(\frac{\mu - ax}{b}\right) \right.$$
$$\left. + 2abx\alpha(x)\left(\frac{\mu - ax}{b}\right)\beta\left(\frac{\mu - ax}{b}\right) \right\} dx$$

$$3.37(3)$$

Let us now write

$$x^2 \alpha(x) = K(x) \qquad 3.37(4)$$

$$\left(\frac{\mu - ax}{b}\right)^2 \beta\left(\frac{\mu - ax}{b}\right) = L\left(\frac{\mu - ax}{b}\right) \qquad 3.37(5)$$

$$x\alpha(x) = M(x) \qquad 3.37(6)$$

and

$$\left(\frac{\mu - ax}{b}\right)\beta\left(\frac{\mu - ax}{b}\right) = N\left(\frac{\mu - ax}{b}\right) \qquad 3.37(7)$$

The integral of each of the three terms of 3.37(1) now becomes

$$\frac{1}{bAB}\int \delta\mu \int \left\{ a^2 K(x)\beta\left(\frac{\mu - ax}{b}\right) + b^2 \alpha(x)L\left(\frac{\mu - ax}{b}\right) \right.$$
$$\left. + 2abM(x)N\left(\frac{\mu - ax}{b}\right) \right\} dx$$

$$3.37(8)$$

3.38 Let us now compare each of the terms of 3.37(8), each of which is integrated between three sets of limits given by 3.37(1), with the right-hand side of equation 3.22(6). If we write $-c = x_c$, $d = x_d$, $-e = z_e$ and $f = z_f$, then we see that 3.37(8) represents three integrals of the type shown on the right-hand side of 3.22(6), where each of the terms of 3.37(8) represents

$$f(x)\psi\left(\frac{\mu - ax}{b}\right)$$

of 3.22(6). Thus each of the three integrals represented by 3.37(8) may be put in the form represented by the left-hand side of 3.22(6). We can therefore write

$$\sigma_\mu^2 = \frac{a^2}{AB} \int_{-c}^{d} K(x)\,dx \int_{-e}^{f} \beta(z)\,dz + \frac{b^2}{AB} \int_{-c}^{d} \alpha(x)\,dx \int_{-e}^{f} L(z)\,dz$$

$$+ \frac{2ab}{AB} \int_{-c}^{d} M(x)\,dx \int_{-e}^{f} N(z)\,dz \qquad 3.38(1)$$

3.39 Now the third term of 3.38(1) is zero since

$$\int_{-c}^{d} M(x)\,dx = \int_{-c}^{d} x\alpha(x)\,dx = 0$$

(see equation 3.31(2)) and

$$\int_{-e}^{f} N(z)\,dz = \int_{-e}^{f} z\beta(z)\,dz = 0$$

(see equation 3.32(1)). Now

$$\int_{-e}^{f} \beta(z)\,dz = B \qquad \text{and} \qquad \int_{-c}^{d} \alpha(x)\,dx = A$$

(see equations 3.32(4) and 3.31(4) respectively). Thus, finally

$$\sigma_\mu^2 = \frac{a^2}{A} \int_{-c}^{d} x^2\alpha(x)\,dx + \frac{b^2}{B} \int_{-e}^{f} z^2\beta(z)\,dz \qquad 3.39(1)$$

But this equation can be written

$$\sigma_\mu^2 = a^2\sigma_x^2 + b^2\sigma_z^2 \qquad 3.39(2)$$

by virtue of equations 3.31(3) and 3.32(3) respectively. Thus the standard deviation of the variable $\mu = ax + bz$ is given by the expression 3.39(2) in terms of the variances of x and z. If we compare 3.39(2) with 2.28(13) we see that the standard deviation of the expression $\mu = ax + bz$, given by 3.39(2), is true not only for the special case when the density functions of x and z have a Gaussian form but also when each has any form.

Rule for Compounding Variances of Uncertainty Distributions

3.40 Let u and w be two other independent variables with frequency distributions $\eta(u)$ and $\xi(w)$. Let

$$\tau = cu + dw \qquad 3.40(1)$$

Thus the standard deviation of τ is related to that of u and w by the expression

$$\sigma_\tau^2 = c^2\sigma_u^2 + d^2\sigma_w^2 \qquad 3.40(2)$$

which follows by comparison with 3.39(2). Now

$$\mu = ax + bz \qquad 3.05(1)$$

and the standard deviation of μ is given in terms of σ_x and σ_y by

$$\sigma_\mu^2 = a^2 \sigma_x^2 + b^2 \sigma_z^2 \qquad 3.39(2)$$

The standard deviation of $\mu + \tau$ is given by

$$\sigma_{\mu\tau}^2 = \sigma_\mu^2 + \sigma_\tau^2 \qquad 3.40(3)$$

by comparing with 3.39(2) and putting $a = b = 1$. Thus, substituting for σ_τ and σ_μ from 3.40(2) and 3.39(2), we have

$$\sigma_{\mu\tau}^2 = \sigma_{xzuw}^2 = a^2 \sigma_x^2 + b^2 \sigma_z^2 + c^2 \sigma_u^2 + d^2 \sigma_w^2 \qquad 3.40(3)$$

which gives the standard deviation of $ax + bz + cu + dw$ in terms of σ_x, σ_z, σ_u and σ_w. It is a simple matter to extend the argument by repetition of the above and to show that the standard deviation of

$$K = \sum_1^n a_r x_r \qquad 3.40(4)$$

where the a_r are constants and the x_r variables, is given by

$$\sigma_K^2 = \sum_1^n a_r^2 \sigma_r^2 \qquad 3.40(5)$$

where σ_r is the standard deviation of x_r. Comparing 3.40(5) with 2.29(2), we see that the latter which was proved for Gaussian distributions is true for all distributions, including the case when the separate distributions compounded are different from each other.

Standard Deviation of $\Sigma_1^n x_r$

3.41 If we put the a_r in 3.40(4) and 3.40(5) equal to unity, the standard deviation of the sum of n independent variables is given by

$$\sigma_K^2 = \sum_1^n \sigma_r^2 \qquad 3.41(1)$$

Standard Deviation of Arithmetic Mean or Standard Error

3.42 If we take equation 3.40(4) and put $a_r = 1/n$, then

$$K = \frac{1}{n} \sum_1^n x_r \qquad 3.42(1)$$

and K is thus the mean value of the x_r. Further, if the x_r belong to the same uncertainty population, then all the σ_r are equal. Let us therefore put $\sigma_r = \sigma$. Thus, making these substitutions in 3.40(5) we obtain

$$\sigma_K^2 = \sum_1^n \frac{\sigma^2}{n^2} = \frac{n\sigma^2}{n^2} = \frac{\sigma^2}{n} \qquad 3.42(2)$$

and so

$$\sigma_K = \frac{\sigma}{\sqrt{n}} \qquad 3.42(3)$$

that is, the standard error is equal to the standard deviation of a probability distribution of a variable, divided by the square root of the number of readings taken. This result is of course true for probability distributions of any form. Thus once again we see that an equation that was initially proved for a Gaussian distribution (see 2.30(1)) is true for any distribution.

Combination of a Stochastic Distribution with a Gaussian Distribution

3.43 By a stochastic distribution we mean a distribution of uncertainties which is given in the form of a number of discrete values of the variable concerned. Such a distribution is called a stochastic one because it is not continuous, and is zero between one value of the variable and the next.

3.44 Such distributions are of interest in calibration work because it is sometimes useful when dealing with estimated uncertainties to assume that each estimated uncertainty can assume either a plus or a minus maximum value.

3.45 Let us first consider the case of two estimated uncertainties, one of which can assume the values A and $-A$, and the other which can assume the values B and $-B$, both sets being measured from a mean value. It is easily seen that the uncertainties which can occur are $A + B$, $-A - B$, $A - B$ and $-A + B$. If we now consider three distributions with components A and $-A$, B and $-B$, and C and $-C$, then the resultant uncertainties which can occur are

$$\left.\begin{array}{llll} A + B + C, & A - B - C, & A + B - C, & A - B + C \\ -A - B - C, & -A + B - C, & -A - B + C, & -A + B - C \end{array}\right\}$$

$$3.45(1)$$

3.46 Likewise, if four distributions with components A and $-A$, B and $-B$, C and $-C$, and D and $-D$ are considered, the resultant uncertainties which can occur are

$$
\left.
\begin{array}{llll}
A + B + C + D, & A + B - C - D, & A - B + C + D, & A - B - C + D \\
-A - B - C - D, & -A - B + C + D, & -A + B - C - D, & -A + B + C - D \\
A + B + C - D, & A + B - C + D, & A - B + C - D, & A - B - C - D \\
-A - B - C + D, & -A - B + C - D, & -A + B - C + D, & -A + B + C + D
\end{array}
\right\}
$$

$$3.46(1)$$

It is seen from above that:
  combining two distributions gives four possible uncertainties;
  combining three distributions gives eight possible uncertainties;
  combining four distributions gives sixteen possible uncertainties.
Also it is to be noted that each positive uncertainty has a negative counterpart of equal magnitude.

3.47 By induction it is easily seen that if n distributions were combined this would lead to 2^n possible resultant uncertainties. Since to each possible uncertainty there is a corresponding negative uncertainty of equal magnitude, the number of uncertainties of differing magnitude is 2^{n-1}. The probability of an uncertainty of any one magnitude occurring is thus $1/2^{n-1}$. Let each uncertainty in the combination be denoted by ε_r, where r assumes all integral values from -2^{n-1} to 2^{n-1}, excluding zero, thus making 2^n values of ε_r. Thus the probability of an uncertainty lying between $\pm \varepsilon_r$ is $2r/2^n = r/2^{n-1}$, where r can take the values 1 to 2^{n-1} in integral steps. Also $\varepsilon_r = -\varepsilon_{-r}$ and $\varepsilon_r \geqslant \varepsilon_{r-1}$, whilst $|\varepsilon_{-r}| \geqslant |\varepsilon_{-r+1}|$.

3.48 Suppose we wish to combine the above stochastic uncertainty population with a Gaussian one whose standard deviation is σ. This is likely to be the case if we have a number of calibration uncertainty sources, some of which are Gaussian and the others plus or minus estimated ones as described above.

3.49 The density function for the Gaussian distribution is

$$e^{-z^2/2\sigma^2}/\sigma\sqrt{(2\pi)}$$

$$3.49(1)$$

Let us put the density function for the stochastic distribution equal to $Q(x)$.

  Then the density function $P(\mu)$ for the convolution of the two distributions is given by substitution in 3.27(1), where $Q(x) \equiv f(x)$ and $e^{-z^2/2\sigma^2}/\sigma\sqrt{(2\pi)} \equiv \psi(z)$. In this case the limits of $\psi(z)$, i.e. z_e and z_f are respectively equal to $-\infty$ and $+\infty$, whilst those of $Q(x)$ we will put equal to x_c and x_d. We require the density function of $\mu = x + y$. Thus we put $a = b = 1$ in 3.27(1) and substitute in 3.27(2a) to 3.27(2d) for z_e and z_f, giving

$$\mu_1 = x_d - \infty = -\infty \qquad\qquad 3.49(2a)$$

$$\mu_2 = x_c + \infty = +\infty \qquad\qquad 3.49(2b)$$

$$\mu_3 = x_c - \infty = -\infty \qquad\qquad 3.49(2c)$$

$$\mu_4 = x_d + \infty = +\infty \qquad\qquad 3.49(2d)$$

Now from 3.27(1), $_2\bar{\eta}_\mu$ is valid over $\mu_1 \geqslant \mu \geqslant \mu_3$, that is between $-\infty$ and $-\infty$. The integral for $_2\bar{\eta}_\mu$ is thus zero. $_1\bar{\eta}_\mu$ is valid over $\mu_2 \geqslant \mu \geqslant \mu_1$, that is between $-\infty$ and $+\infty$. Thus $_1\bar{\eta}_\mu$ can be integrated between x_c to x_d. $_3\bar{\eta}_\mu$ is valid over $\mu_4 \geqslant \mu \geqslant \mu_2$, that is between $+\infty$ and $+\infty$. The integral for $_3\bar{\eta}_\mu$ is thus zero.

Thus the only integral to survive is $_1\bar{\eta}_\mu$. The required integral for $P(\mu)$ is thus equal to

$$P(\mu) = {}_1\bar{\eta}_\mu = \frac{1}{\sigma\sqrt{(2\pi)}} \int_{x_c}^{x_d} Q(x) e^{-(\mu-x)^2/2\sigma^2} \, dx \qquad 3.49(3)$$

It is easily shown by a similar argument that if either of the two functions $f(x)$ or $\psi(x)$ have infinite range limits, then only the integral for $_1\bar{\eta}_\mu$ survives. If both functions have infinite range limits then x_c and x_d just become $-\infty$ and $+\infty$ respectively, and the integral for $P(\mu)$ again reduces to $_1\bar{\eta}_\mu$.

3.50 Now in the case under consideration, $Q(y) \, dy$ is discontinuous and has the value $1/2^n$ at values of y equal to ε_r where ε_r is an uncertainty of the discontinuous population as described in paragraph 3.47, and is zero elsewhere. Therefore we replace the integral sign in front of the integrand of 3.49(3) by a sum, and the density function for $P(\mu)$ becomes

$$P(\mu) = \frac{1}{\sigma\sqrt{(2\pi)}} \sum_{r=-2^{n-1}}^{r=2^{n-1}} \frac{e^{-(\mu-\varepsilon_r^2)/2\sigma^2}}{2^n} \qquad 3.50(1)$$

The probability of an uncertainty P between $x \geqslant \mu \geqslant -x$ is thus given by

$$P = \frac{1}{\sigma\sqrt{(2\pi)2^n}} \sum_{r=-2^{n-1}}^{r=2^{n-1}} \int_{-x}^{x} e^{-(\mu-\varepsilon_r)^2/2\sigma^2} \, d\mu \qquad 3.50(2)$$

Put $(\mu - \varepsilon_r)/\sigma = y$, then $d\mu = \sigma dy$. The limits of integration $-x$ and $+x$ thus become

$$y = -(x + \varepsilon_r)/\sigma \qquad \text{and} \qquad (x - \varepsilon_r)/\sigma$$

and the integral becomes

$$\frac{1}{\sqrt{(2\pi)}.2^n} \sum_{r=-2^{n-1}}^{r=2^{n-1}} \int_{-(x+\varepsilon_r)/\sigma}^{(x-\varepsilon_r)/\sigma} e^{-y^2/2} \, dy \qquad 3.50(3)$$

As the integral is symmetrical it can be written as

$$\frac{1}{\sqrt{(2\pi)}.2^n} \sum_{r=-2^{n-1}}^{r=2^{n-1}} \left\{ \int_0^{(x-\varepsilon_r)/\sigma} e^{-y^2/2} \, dy + \int_0^{(x+\varepsilon_r)/\sigma} e^{-y^2/2} \, dy \right\} \qquad 3.50(4)$$

Referring to Table II of Appendix I we see that the complete integral can now be written as

$$\frac{1}{2^{n+1}} \sum_{r=-2^{n-1}}^{r=2^{n-1}} \left(P_{\frac{-(x-\varepsilon_r)}{\sigma} \text{ to } \frac{(x-\varepsilon_r)}{\sigma}} + P_{\frac{-(x+\varepsilon_r)}{\sigma} \text{ to } \frac{(x+\varepsilon_r)}{\sigma}} \right) \qquad 3.50(5)$$

Since $\varepsilon_r = -\varepsilon_{-r}$ the above summation can be modified to become

$$P = \frac{1}{2^n} \sum_{r=1}^{r=2^{n-1}} \left(\underset{\frac{-(x-\varepsilon_r)}{\sigma} \text{ to } \frac{(x-\varepsilon_r)}{\sigma}}{P} + \underset{\frac{-(x+\varepsilon_r)}{\sigma} \text{ to } \frac{(x+\varepsilon_r)}{\sigma}}{P} \right) \qquad 3.50(6)$$

where the ε_r are the *positive* values of the uncertainties of the discontinuous distribution. It is to be noted that if $(x - \varepsilon_r)$ is negative it should be treated as *positive*, but the value of $\underset{-k \text{ to } k}{P}$ obtained should be treated as *negative*.

Standard Deviation of the Combination of a Stochastic Distribution with a Gaussian Distribution (For Alternative see Paragraphs 3.56 and 3.57)

3.51   Let us now find the standard deviation of the combination given in paragraph 3.50. For the stochastic distribution consisting of the ε_r, the standard deviation squared is given by

$$\frac{1}{2^n} \sum_{r=-2^{n-1}}^{r=2^{n-1}} \varepsilon_r^2 \qquad 3.51(1)$$

where the number of ε_r is equal to 2^n. Since the standard deviation of the Gaussian system is σ, if we combine the stochastic distribution with the Gaussian one we obtain, using 3.41(1) with $n = 2$, the square of the standard deviation σ_c of the combination as

$$\sigma_c^2 = \sigma^2 + \frac{1}{2^n} \sum_{r=-2^{n-1}}^{r=2^{n-1}} \varepsilon_r^2 \qquad 3.51(2)$$

or, alternatively,

$$\sigma_c^2 = \sigma^2 + \frac{1}{2^{n-1}} \sum_{r=1}^{r=2^{n-1}} \varepsilon_r^2 \qquad 3.51(3)$$

where only the positive ε_r are used.

3.52   This result is easily verified as follows. Consider the distribution function 3.50(2) for P, then the square of the standard deviation σ_c is given by

$$\sigma_c^2 = \int P\mu^2 \, d\mu = \frac{1}{\sigma\sqrt{(2\pi)2^n}} \sum_{r=-2^{n-1}}^{r=2^{n-1}} \int_{-\infty}^{\infty} \mu^2 e^{-(\mu-\varepsilon_r)^2/2\sigma^2} \, d\mu \qquad 3.52(1)$$

Put $(\mu - \varepsilon_r)/\sigma = y$, giving $d\mu = \sigma \, dy$ and $\mu = \sigma y + \varepsilon_r$, and therefore

$$\sigma_c^2 = \frac{1}{\sqrt{(2\pi)2^n}} \sum_{r=-2^{n-1}}^{r=2^{n-1}} \int_{-\infty}^{\infty} (\sigma y + \varepsilon_r)^2 e^{-y^2/2} \, dy \qquad 3.52(2)$$

The integral becomes, on multiplying out the squared term,

$$\int_{-\infty}^{\infty} (\sigma^2 y^2 + \varepsilon_r^2 + 2\sigma\varepsilon_r y)e^{-y^2/2} \, dy$$

Now the last term is zero since y is an odd function and $e^{-y^2/2}$ is an even one. Also, putting the integral produced by the first term equal to I, we have

$$I = \int_{-\infty}^{\infty} \sigma^2 y^2 e^{-y^2/2} \, dy = \int_{-\infty}^{\infty} \sigma^2(-y) d(e^{-y^2/2})$$

Therefore, integrating by parts we have

$$I = \sigma^2 [-y e^{-y^2/2}]_{-\infty}^{\infty} + \int_{-\infty}^{\infty} \sigma^2 e^{-y^2/2} \, dy$$
$$\downarrow$$
$$0$$

$$= \sigma^2 \sqrt{(2\pi)} \qquad\qquad 3.52(3)$$

since $\int_{-\infty}^{\infty} e^{-y^2/2} \, dy = \sqrt{(2\pi)}$ (see paragraph 2.10).

Thus the integral becomes $(\varepsilon_r^2 + \sigma^2)\sqrt{(2\pi)}$ and the standard deviation σ_c is given by

$$\sigma_c^2 = \sigma^2 + \frac{1}{2^n} \sum_{r=-2^{n-1}}^{r=2^{n-1}} \varepsilon_r^2 \qquad\qquad 3.52(4)$$

in agreement with 3.51(2).

Relation between ε_r and a_r of a Stochastic Distribution

3.53 Let there be n systematic uncertainties $a_1, a_2, a_3, \ldots, a_n$. There will thus be 2^n values of ε_r, where the ε_r are made up of sums, each of which contain all the a_r, wherein each a_r has either a plus or a minus sign.

Let $n = 3$. There are thus $2^3 = 8$ values of ε_r, given as follows:

$$\varepsilon_4 = a_1 + a_2 + a_3$$
$$\varepsilon_3 = a_1 - a_2 + a_3$$
$$\varepsilon_2 = a_1 + a_2 - a_3$$
$$\varepsilon_1 = a_1 - a_2 - a_3$$
$$\varepsilon_{-1} = -\varepsilon_1$$
$$\varepsilon_{-2} = -\varepsilon_2$$
$$\varepsilon_{-3} = -\varepsilon_3$$
$$\varepsilon_{-4} = -\varepsilon_4$$

Thus

$$\sum_{r=-4=-(2^2)}^{r=4=2^2} \varepsilon_r^2 = 2 \begin{pmatrix} a_1^2 + a_2^2 + a_3^2 + 2a_1a_2 + 2a_1a_3 + 2a_2a_3 \\ a_1^2 + a_2^2 + a_3^2 - 2a_1a_2 + 2a_1a_3 - 2a_2a_3 \\ a_1^2 + a_2^2 + a_3^2 - 2a_1a_2 - 2a_1a_3 - 2a_2a_3 \\ a_1^2 + a_2^2 + a_3^2 - 2a_1a_2 - 2a_1a_3 + 2a_2a_3 \end{pmatrix}$$

$$= 8(a_1^2 + a_2^2 + a_3^2) = 8 \sum_{r=1}^{r=3} a_r^2$$

$$= 2^3 \sum_{r=1}^{r=3} a_r^2 \qquad\qquad 3.53(1)$$

It is easily shown that if $n = 4$ then

$$\sum_{r=-8=-(2)^3}^{r=8=2^3} \varepsilon_r^2 = 2^4 \sum_{r=1}^{r=4} a_r^2 \qquad\qquad 3.53(2)$$

3.54   This immediately suggests that if we have n values of a_r, then

$$\sum_{r=-2^{n-1}}^{r=2^{n-1}} \varepsilon_r^2 = 2^n \sum_{r=1}^{r=n} a_r^2 \qquad\qquad 3.54(1)$$

We will now prove 3.54(1) by induction. Let us assume that 3.54(1) is true for n values of a_r, and let us suppose that there are now $n + 1$ values of a_r, then the new ε_q' are given by $\varepsilon_r \pm a_{n+1}$, that is each ε_r gives rise to new ε_q', q varying from -2^n to 2^n, and therefore

$$\sum_{q=-2^n}^{q=2^n} (\varepsilon_q')^2 = \sum_{r=-2^{n-1}}^{r=2^{n-1}} \{(\varepsilon_r + a_{n+1})^2 + (\varepsilon_r - a_{n+1})^2\}$$

$$= 2 \sum_{r=-2^{n-1}}^{r=2^{n-1}} (\varepsilon_r^2 + a_{n+1}^2)$$

$$= 2 \times 2^n \sum_{r=1}^{r=n} a_r^2 + 2 \times 2^n (a_{n+1})^2$$

from 3.54(1)

$$= 2^{n+1} \sum_{r=1}^{r=n+1} a_r^2 \qquad\qquad 3.54(2)$$

But we know that this formula is true if $n = 3$ and 4; thus it is true for any value of n, and so 3.54(1) is true.

Alternative Formula for Standard Deviation of the Combination of a Stochastic and a Gaussian Distribution

3.55 We shall have occasion to use the relationship 3.54(1) in the next chapter, but it is worth noticing here that the standard deviation of the combination 3.52(4) can be written as

$$\sigma_c^2 = \sigma^2 + \sum_{r=1}^{r=n} a_r^2 \qquad\qquad 3.55(1)$$

by substituting for $\sum \varepsilon_r^2$ from 3.54(1).

3.56 This relationship also follows by considering each stochastic pair as a separate distribution. The standard deviation of each pair is thus $\{a_r^2 + (-a_r)^2\}/2 = a_r^2$, using the definition of standard deviation given in 2.23(1), where $n = 2$. Thus the standard deviation σ_{st} of the stochastic distribution is given by

$$\sigma_{st}^2 = \sum_{r=1}^{r=n} a_r^2 = \frac{1}{2^n} \sum_{r=-2^{n-1}}^{r=2^{n-1}} \varepsilon_r^2 = \frac{1}{2^{n-1}} \sum_{r=1}^{r=2^{n-1}} \varepsilon_r^2 \qquad 3.56(1)$$

and 3.55(1) follows when the Gaussian variance is added to give the variance of the combination.

Example

3.57 We will now consider an example, using the formula we have just developed. Suppose that we have a set of Gaussian uncertainties, with a standard deviation of ten units, and four non-Gaussian uncertainties, which we shall assume can take only plus or minus specified values. Let these be ± 3, ± 2, ± 4, and ± 1 units. Referring to 3.46(1) we see that the ε_r, of which there are $2^4 = 16$, are given by

$$
\left.
\begin{aligned}
4 + 3 + 2 + 1 &= 10 \\
4 + 3 + 2 - 1 &= 8 \\
4 + 3 - 2 - 1 &= 4 \\
4 + 3 - 2 + 1 &= 6 \\
4 - 3 + 2 + 1 &= 4 \\
4 - 3 + 2 - 1 &= 2 \\
4 - 3 - 2 + 1 &= 0 \\
-4 + 3 + 2 + 1 &= 2
\end{aligned}
\right\}
$$

positive ε_r; the rest of the ε_r are of similar magnitude but of negative sign.

Thus

$$\varepsilon_1 = 0, \quad \varepsilon_2 = 2, \quad \varepsilon_3 = 2, \quad \varepsilon_4 = 4, \quad \varepsilon_5 = 4, \quad \varepsilon_6 = 6, \quad \varepsilon_7 = 8, \quad \varepsilon_8 = 10$$

Using 3.56(1) the standard deviation of the ε_r is thus given by

$$\left(\sum_{r=1}^{r=4} a_r^2 \right)^{1/2} = (4^2 + 3^2 + 2^2 + 1^2)^{1/2} = \sqrt{30} = 5.48$$

3.58 If the ε_r had been a Gaussian distribution, the probability of an uncertainty between $\pm 2\sigma = 11.96$ units would have been 95.4 per cent. Since the largest ε_r is 10, the probability of an uncertainty between these limits for the stochastic ε_r distribution is 100 per cent.

3.59 Let us now combine the stochastic ε_r distribution with the Gaussian distribution. Using 3.55(1) the standard deviation of the combined system is given by

$$\sigma_c^2 = 10^2 + 30$$

giving

$$\sigma_c = \sqrt{(130)} = 11.40$$

Let us now find the probability of an uncertainty between plus and minus two standard deviations of the combination, that is between ± 22.80 units. Thus, using 3.50(6) we proceed as follows: $\sigma = 10$, the standard deviation of the Gaussian distribution

$$\varepsilon_1 = 0, \quad \varepsilon_2 = 2, \quad \varepsilon_3 = 2, \quad \varepsilon_4 = 4, \quad \varepsilon_5 = 4, \quad \varepsilon_6 = 6, \quad \varepsilon_7 = 8, \quad \varepsilon_8 = 10$$

Putting $x = 22.8$, we now calculate $(x - \varepsilon_r)/\sigma$ and $(x + \varepsilon_r)/\sigma$ in units of σ.

$$\frac{x - \varepsilon_1}{\sigma} = \frac{22.8 - 0}{10} = 2.28 \qquad \frac{x + \varepsilon_1}{\sigma} = \frac{22.8 + 0}{10} = 2.28$$

$$\frac{x - \varepsilon_2}{\sigma} = \frac{22.8 - 2.0}{10} = 2.08 \qquad \frac{x + \varepsilon_2}{\sigma} = \frac{22.8 + 2.0}{10} = 2.48$$

$$\frac{x - \varepsilon_3}{\sigma} = \frac{22.8 - 2.0}{10} = 2.08 \qquad \frac{x + \varepsilon_3}{\sigma} = \frac{22.8 + 2.0}{10} = 2.48$$

$$\frac{x - \varepsilon_4}{\sigma} = \frac{22.8 - 4.0}{10} = 1.88 \qquad \frac{x + \varepsilon_4}{\sigma} = \frac{22.8 + 4.0}{10} = 2.68$$

$$\frac{x - \varepsilon_5}{\sigma} = \frac{22.8 - 4.0}{10} = 1.88 \qquad \frac{x + \varepsilon_5}{\sigma} = \frac{22.8 + 4.0}{10} = 2.68$$

$$\frac{x - \varepsilon_6}{\sigma} = \frac{22.8 - 6.0}{10} = 1.68 \qquad \frac{x + \varepsilon_6}{\sigma} = \frac{22.8 + 6.0}{10} = 2.88$$

$$\frac{x - \varepsilon_7}{\sigma} = \frac{22.8 - 8.0}{10} = 1.48 \qquad \frac{x + \varepsilon_7}{\sigma} = \frac{22.8 + 8.0}{10} = 3.08$$

$$\frac{x - \varepsilon_8}{\sigma} = \frac{22.8 - 10}{10} = 1.28 \qquad \frac{x + \varepsilon_8}{\sigma} = \frac{22.8 + 10}{10} = 3.28$$

Using Table II of Appendix I, the respective $\underset{-k_1 \text{ to } k_1}{P}$ are

0.977 28,   0.977 28,   0.962 32,   0.986 78,   0.962 32,   0.986 78,

0.939 68,   0.992 59,   0.939 68,   0.992 59,   0.906 78,   0.995 99,

0.860 81,   0.997 908,   0.799 09,   0.998 95

Thus

$$P = \frac{1}{2^4} \sum_{r=1}^{r=2^3=8} \left(\underset{-\left(\frac{x-\varepsilon_r}{\sigma}\right) \text{ to } \left(\frac{x-\varepsilon_r}{\sigma}\right)}{P} + \underset{-\left(\frac{x+\varepsilon_r}{\sigma}\right) \text{ to } \left(\frac{x+\varepsilon_r}{\sigma}\right)}{P} \right)$$

$$= \underline{0.9548}$$

3.60 It is interesting to note that the probability of an uncertainty occurring within the range of plus and minus two standard deviations of our combination is slightly more than the probability of an uncertainty occurring between plus and minus two standard deviations of a Gaussian distribution.

3.61 Let us now consider the uncertainty range formed by plus and minus the addition of the magnitudes of the four non-Gaussian uncertainties plus twice the standard deviation of the Gaussian distribution, that is

$$\pm(1 + 2 + 3 + 4 + 2 \times 10) = \pm 30 \text{ units}$$

If we put $x = 30$ and repeat our previous calculation, we find the probability of an uncertainty lying between ± 30 units is equal to 0.992. Expressing the 30 uncertainty units in terms of the standard deviation of the combination, we find that 30 units is equal to $30/11.4$ standard deviations $= 2.6316$. Assuming a Gaussian distribution, and using Table II of Appendix I, we find the probability of an uncertainty lying between ± 2.6316 standard deviations is 0.9914. Thus once again we see that the probability for the combination is slightly more than for a Gaussian distribution with the same standard deviation when both have the same limits.

Skewness

3.62 We have seen how the first moment of a distribution leads to the mean value if it is equated to zero, and how the second moment taken about the mean leads to the standard deviation. We will now consider the significance of the third moment about the mean. This is equal to

$$\int_b^a (x - \mu)^3 f(x)\, dx \qquad\qquad 3.62(1)$$

for a continuous probability distribution $f(x)$ which exists between the limits a and b where μ is the mean value. If only discrete readings are available the third moment is given by

$$\sum_{i=1}^{i=n} \frac{(x_1 - \bar{x})^3}{n} \qquad 3.62(2)$$

or for a very large number of readings, the working mean $\bar{x}$ is replaced by μ the true mean, that is

$$\sum_{i=1}^{i=n} \frac{(x_i - \mu)^3}{n} \qquad 3.62(3)$$

The so-called coefficient of skewness is defined as the skewness which is given by 3.62(1), (2) or (3) divided by the cube of the standard deviation or its estimate, as appropriate. The resulting coefficient is dimensionless and independent of the scale used. For a continuous distribution it is

$$\int_a^b \frac{(x - \mu)^3 f(x) \, dx}{\sigma^3} \qquad 3.62(4)$$

For a discrete experimental distribution derived from n readings it is equal to

$$\sum_{i=1}^{i=n} \frac{(x_i - \bar{x})^3}{ns^3} \qquad 3.62(5)$$

where s is the estimate of σ, given by

$$s^2 = \sum_{i=1}^{i=n} \frac{(x_i - \bar{x})^2}{n - 1} \qquad 2.24(10)$$

If n is very large the $\bar{x}$ is replaced by μ the true mean and s by σ the standard deviation of the distribution rather than its estimate, leading to

$$\sum_{i=1}^{i=n} \frac{(x_i - \mu)^3}{n\sigma^3} \qquad 3.62(6)$$

for the skewness.

3.63 When the skewness is worked out for any discrete distribution, and a positive value obtained, it means that more than half the deviations are on the left of the mean, but that the large deviations are on the right. (See Figure 3.63.) The large deviations although small in number have a large effect because of the cubic factor.

If the skewness coefficient is negative, the large deviations lie on the left of the mean but the majority lie on the right. For continuous distributions, the outline of the curve in Figure 3.63 gives the appearance for positive skewness, that is the area to the left of the mean is greatest, and vice versa for negative skewness.

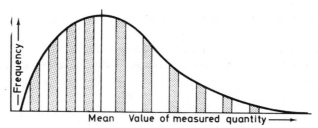

Figure 3.63   Showing positive skewness

3.64   The coefficient of skewness is sensitive to variations in the number of large deviations, and so must be used with caution, that is a large number of readings must be obtained before a stable value of the skewness can be realized.

Coefficient of Kurtosis or Peakedness of a Distribution

3.65   Analogous to the coefficient of skewness, the coefficient of peakedness or Kurtosis is defined as the normalized fourth moment about the mean value. For continuous distributions the coefficient is equal to

$$\frac{\int_a^b (x - \mu)^4 f(x)\, dx}{\sigma^4} \qquad\qquad 3.65(1)$$

whilst for discrete distributions it is equal to

$$\frac{\sum_{i=1}^{i=n} (x_i - \bar{x})^4}{ns^4} \qquad\qquad 3.65(2)$$

for a limited number of readings, whilst it is equal to

$$\sum_{i=1}^{i=n} \frac{(x_i - \mu)^4}{n\sigma^4} \qquad\qquad 3.65(3)$$

for a large number of readings, when μ and σ are known, rather than their estimates. The fourth moment is even more sensitive to large deviations, and is thus of restricted use, and should not be relied upon unless a very large number of observations have been made.

Distribution of Difference between Two Distributions

3.66   Two sets of parts which have to fit together are made, and the mean value of the smaller mating part is $\bar{x}_1$, whilst that of the larger is $\bar{x}_2$. The corresponding standard deviations of the two sets are σ_1 and σ_2. If the sets of parts each have a Gaussian distribution, find the density function of the difference of any two parts picked at random, one from one set and one from the other.

The required density function is that corresponding to the variable $x_2 - x_1$. Now we know that the density functions of the variables x_2 and x_1 are Gaussian, and so

$$P(x_1) = \frac{e^{-(x_1 - \bar{x}_1)^2/2\sigma_1^2}}{\sigma_1\sqrt{(2\pi)}} \qquad\qquad 3.66(1)$$

and

$$P(x_2) = \frac{e^{-(x_2 - \bar{x}_2)^2/2\sigma_2^2}}{\sigma_2\sqrt{(2\pi)}} \qquad\qquad 3.66(2)$$

The limits of both functions are from $-\infty$ to $+\infty$. Let us now put $\mu = x_2 - x_1$.

Now consider 3.28(5) where the value of the density function $P(\mu)$ is obtained from the combination of two density distributions $f(x)$ and $\psi(z)$, and where $\mu = -ax + bz$, $b(z_f - z_e) \geqslant a(x_d - x_c)$ and z_f, z_e and x_d, x_c are the upper and lower limits of the density functions $\psi(z)$ and $f(x)$ respectively. Thus we put

$$f(x) = \frac{e^{-(x_1 - \bar{x}_1)^2/2\sigma_1^2}}{\sigma_1\sqrt{(2\pi)}} \equiv P(x_1) \qquad\qquad 3.66(3)$$

and

$$\psi(z) = \frac{e^{-(x_2 - \bar{x}_2)^2/2\sigma_2^2}}{\sigma_2\sqrt{(2\pi)}} \equiv P(x_2) \qquad\qquad 3.66(4)$$

Comparing

$$\mu = -ax + bz \qquad\qquad 3.21(6)$$

with

$$\mu = x_2 - x_1 = -x_1 + x_2$$

we see that

$$a = 1 \qquad \text{and} \qquad b = 1, \qquad \text{and} \qquad x \equiv x_1 \qquad \text{and} \qquad z \equiv x_2$$

Let us put the limits of x_1 equal to $-c$ and d, and the limits of x_2 equal to $-e$ and f respectively, where c, d, e and f are positive quantities. Comparing with 3.21(7) to 3.21(10) we see that $x_c = -c$, $x_d = d$, $z_e = -e$ and $z_f = f$.

Therefore

$$\mu_1 = -d - e \qquad\qquad 3.66(5)$$

$$\mu_2 = c + f \qquad\qquad 3.66(6)$$

$$\mu_3 = c - e \qquad\qquad 3.66(7)$$

$$\mu_4 = -d + f \qquad\qquad 3.66(8)$$

where $\mu_2 \geqslant \mu_4 \geqslant \mu_3 \geqslant \mu_1$ from 3.07(3).

Let us now let the limits of x_2 and x_1 tend to infinity, i.e. c, d, e and f $\to \infty$. Then

$$\left.\begin{array}{c} \mu_1 \to -\infty \\ \mu_2 \to \infty \\ \mu_3 \to -\infty \\ \mu_4 \to \infty \end{array}\right\} \qquad 3.66(9)$$

Thus $_1\bar{\eta}_\mu$, which is valid between μ_3 to μ_4 (see 3.21(18)), is thus valid for μ between $-\infty$ to $+\infty$.

$_2\bar{\eta}_\mu$, which is valid between μ_1 to μ_3 (see 3.21(19)), is thus valid for μ between $-\infty$ and $-\infty$. Thus the integral for $_2\bar{\eta}_\mu$ is zero. Similarly $_3\bar{\eta}_\mu$, which is valid between μ_4 to μ_2 (see 3.21(20)), is thus valid between $+\infty$ to $+\infty$. Thus the integral for $_3\bar{\eta}_\mu$ is zero. Thus only $_1\bar{\eta}_\mu$ survives. If we now let c $\to$ d $\to \infty$, the limits for the integral of $_1\bar{\eta}_\mu$ tend to $-\infty$ and $+\infty$. Finally

$$P(\mu) = {}_1\bar{\eta}_\mu = \int_{-\infty}^{\infty} f(x)\psi\left(\frac{ax + \mu}{b}\right) dx$$

$$= \int_{-\infty}^{\infty} \frac{e^{-(x_1 - \bar{x}_1)^2/2\sigma_1^2} \cdot e^{-(x_2 - \bar{x}_2)^2/2\sigma_2^2}\, dx_1}{\sigma_1 \sigma_2 2\pi} \qquad 3.66(10)$$

Compare this result with the general statement given in paragraph 3.49 that if either or both functions convoluted are valid over an infinite range, i.e. $-\infty$ to $+\infty$ then only the integral for $_1\bar{\eta}_\mu$ survives.

The exponent of e in this is thus

$$\frac{-(x_1 - \bar{x}_1)^2}{2\sigma_1^2} - \frac{(x_1 + \mu - \bar{x}_2)^2}{2\sigma_2^2}$$

$$= -\left\{ \frac{x_1^2}{2\sigma_1^2} - \frac{2x_1\bar{x}_1}{2\sigma_1^2} + \frac{\bar{x}_1^2}{2\sigma_1^2} + \frac{x_1^2}{2\sigma_2^2} - \frac{2x_1(\mu - \bar{x}_2)}{2\sigma_2^2} + \frac{(\mu - \bar{x}_2)^2}{2\sigma_2^2} \right\}$$

$$= -\left\{ x_1^2\left(\frac{1}{2\sigma_1^2} + \frac{1}{2\sigma_2^2}\right) - 2x_1\left(\frac{\bar{x}_1}{2\sigma_1^2} + \frac{\mu - \bar{x}_2}{2\sigma_2^2}\right) + \frac{\bar{x}_1^2}{2\sigma_1^2} + \frac{(\mu - \bar{x}_2)^2}{2\sigma_2^2} \right\}$$

$$= -\left[\left(\frac{1}{2\sigma_1^2} + \frac{1}{2\sigma_2^2}\right)\left\{ x_1^2 - 2x_1\left(\frac{\bar{x}_1}{2\sigma_1^2} + \frac{\mu - \bar{x}_2}{2\sigma_2^2}\right) \Big/ \left(\frac{1}{2\sigma_1^2} + \frac{1}{2\sigma_2^2}\right) \right\} + \frac{\bar{x}_1^2}{2\sigma_1^2} + \frac{(\mu - \bar{x})^2}{2\sigma_2^2} \right]$$

$$= -\left[\underbrace{\left(\frac{1}{2\sigma_1^2} + \frac{1}{2\sigma_2^2}\right)\left\{ x_1 - \left(\frac{\bar{x}_1}{2\sigma_1^2} + \frac{\mu - \bar{x}_2}{2\sigma_2^2}\right) \Big/ \left(\frac{1}{2\sigma_1^2} + \frac{1}{2\sigma_2^2}\right) \right\}^2}_{\alpha} \right.$$

$$\left. \underbrace{- \left(\frac{\bar{x}_1}{2\sigma_1^2} + \frac{\mu - \bar{x}_2}{2\sigma_2^2}\right)^2 \Big/ \left(\frac{1}{2\sigma_1^2} + \frac{1}{2\sigma_2^2}\right) + \frac{\bar{x}_1^2}{2\sigma_1^2} + \frac{(\mu - \bar{x}_2)}{2\sigma_2^2}}_{\beta} \right]$$

$$3.66(11)$$

Dealing first with β we have

$$\beta = \frac{2\sigma_1^2\sigma_2^2}{\sigma_1^2 + \sigma_2^2}\left\{\frac{-\bar{x}_1^2}{4\sigma_1^4} - \frac{(\mu - \bar{x}_2)^2}{4\sigma_2^4} + \frac{2\bar{x}_1(\mu - \bar{x}_2)}{4\sigma_1^2\sigma_2^2} + \frac{\bar{x}_1^2(\sigma_1^2 + \sigma_2^2)}{2\sigma_1^2 . 2\sigma_1^2\sigma_2^2} + \frac{(\mu - \bar{x}_2)^2(\sigma_1^2 + \sigma_2^2)}{4\sigma_1^2\sigma_2^4}\right\}$$

$$= \frac{2\sigma_1^2\sigma_2^2}{\sigma_1^2 + \sigma_2^2}\left\{\frac{\bar{x}_1^2}{4\sigma_1^2\sigma_2^2} + \frac{(\mu - \bar{x}_2)^2}{4\sigma_1^2\sigma_2^2} + \frac{2\bar{x}_1(\mu - \bar{x}_2)}{4\sigma_1^2\sigma_2^2}\right\}$$

$$= \frac{2\sigma_1^2\sigma_2^2}{(\sigma_1^2 + \sigma_2^2)} \times \frac{1}{4\sigma_1^2\sigma_2^2}\{\bar{x}_1 + \mu - \bar{x}_2\}^2$$

$$= \frac{1}{2(\sigma_1^2 + \sigma_2^2)}\{\mu - (\bar{x}_2 - \bar{x}_1)\}^2 \qquad\qquad 3.66(12)$$

Taking α, we put $\sqrt{\alpha} = y$. Thus

$$\left(\frac{1}{2\sigma_1^2} + \frac{1}{2\sigma_2^2}\right)^{1/2}\left\{x_1 - \left(\frac{\bar{x}_1}{2\sigma_1^2} + \frac{\mu - \bar{x}_2}{2\sigma_2^2}\right)\Big/\left(\frac{1}{2\sigma_1^2} + \frac{1}{2\sigma_2^2}\right)\right\} = y \quad 3.66(13)$$

and

$$dx_1 = dy\Big/\left(\frac{1}{2\sigma_1^2} + \frac{1}{2\sigma_2^2}\right)^{1/2} = \frac{\sqrt{2}\sigma_1\sigma_2\,dy}{(\sigma_1^2 + \sigma_2^2)^{1/2}} \qquad 3.66(14)$$

When $x_1 = -\infty$, $y = -\infty$, and when $x_1 = \infty$, $y = \infty$.
    Thus $P(\mu)$ reduces to

$$P(\mu) = e^{-\{\mu - (\bar{x}_2 - \bar{x}_1)\}^2/2(\sigma_1^2 + \sigma_2^2)}\int_{-\infty}^{\infty}\frac{e^{-y^2}\,dy}{(\sigma_1^2 + \sigma_2^2)^{1/2}\sqrt{(2\pi)}} \qquad 3.66(15)$$

Now from 2.10(4)

$$\int_{-\infty}^{\infty} e^{-y^2}\,dy = \sqrt{\pi}$$

and so, finally,

$$P(\mu) = \frac{e^{-\{\mu - (\bar{x}_2 - \bar{x}_1)\}^2/2(\sigma_1^2 + \sigma_2^2)}}{\sqrt{(2\pi)}(\sigma_1^2 + \sigma_2^2)^{1/2}} \qquad\qquad 3.66(16)$$

which is the density function of a Gaussian distribution with a standard deviation of $(\sigma_1^2 + \sigma_2^2)^{1/2}$ and a mean value at $\bar{x}_2 - \bar{x}_1$. The most probable value of the difference between two parts picked at random, one from each of the two sets, is thus equal to the difference between the means of the two sets.

The Fitting Together of Two Sets of Parts with Standard Deviations of σ_1 and σ_2 and Means of $\bar{x}_1$ and $\bar{x}_2$

3.67 The problem given below is of common occurrence in engineering. Two sets of parts are required to fit, one into the other, the standard deviation of the smaller part being σ_1 with mean value $\bar{x}_1$, whilst the standard deviation of the larger part is σ_2 with mean value $\bar{x}_2$. Suppose that mating parts can have clearances varying between a maximum value 'a' and a minimum value 'b'. We will now find the proportion of parts that will fit if selective assembly is used, assuming that the distributions are in each case Gaussian.

Consider Figure 3.67a; the two density distributions are given by

$$f_1 = \frac{e^{-(x_1 - \bar{x}_1)^2/2\sigma_1^2}}{\sigma_1 \sqrt{(2\pi)}} \qquad\qquad 3.67(1)$$

and

$$f_2 = \frac{e^{-(x_2 - \bar{x}_2)^2/2\sigma_2^2}}{\sigma_2 \sqrt{(2\pi)}} \qquad\qquad 3.67(2)$$

The mean clearance, that is difference between the means, is given by $\bar{x}_2 - \bar{x}_1$. Now the maximum clearance allowed is 'a'. Thus we put

$$\bar{x}_2 - \bar{x}_1 + \alpha = a \qquad\qquad 3.67(3)$$

and similarly for the minimum clearance 'b' we put

$$\bar{x}_2 - \bar{x}_1 + \beta = b \qquad\qquad 3.67(4)$$

Note that α and β can be negative or positive, depending on a, b, x_2 and x_1. Let us now superimpose f_2 on f_1, that is move the mean of f_2 to coincide with the mean of f_1. Our figure will now appear as shown in Figure 3.67b.

Let us consider the point $_1x_\beta$ for the distribution f_1 and $_2x_\beta$ for the distribution f_2 where $_2x_\beta - {}_1x_\beta = \beta$, then since $_2x_\beta$ is to the left of $_1x_\beta$, a part taken from distribution f_2 at $_2x_\beta$ and matched with a part taken from distribution f_1 at $_1x_\beta$ will give a clearance of $\bar{x}_2 - \bar{x}_1 + \beta$, that is the smallest clearance allowable. As we match corresponding parts by moving along each curve from left to right, the distance between $_1x_\beta$ and $_2x_\beta$ decreases, provided that $_1x_\beta$ and $_2x_\beta$ have been chosen correctly, that is $f_2(_2x_\beta) \leqslant f_1(_1x_\beta)$, otherwise the distance between the new positions of $_1x_\beta$ and $_2x_\beta$ would

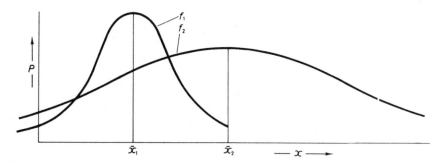

Figure 3.67a

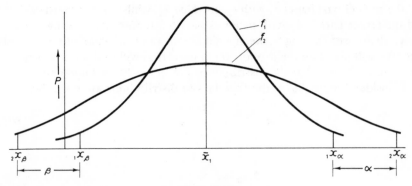

Figure 3.67b

increase momentarily and make the clearance less than the minimum value specified. Similarly, if $_2x_\alpha - _1x_\alpha = \alpha$, where α is the final distance between $_2x$ and $_1x$ then $_1x_\alpha$ and $_2x_\alpha$ give the positions of the final pieces to be matched. We now have to find $_1x_\beta$ and $_1x_\alpha$. We require the area under f_1 between these points to be a maximum and equal to the area under f_2 between $_2x_\beta$ and $_2x_\alpha$; thus we require

$$\frac{1}{\sigma_1\sqrt{(2\pi)}} \int_{_1x_\beta}^{_1x_\alpha} e^{-(x-\bar{x}_1)^2/2\sigma_1^2} \, dx = \frac{1}{\sigma_2\sqrt{(2\pi)}} \int_{_2x_\beta}^{_2x_\alpha} e^{-(x-\bar{x}_1)^2/2\sigma_2^2} \, dx$$

3.67(5)

This equality may be written

$$\int_{\frac{_1x_\beta-\bar{x}_1}{\sigma_1}}^{\frac{_1x_\alpha-\bar{x}_1}{\sigma_1}} e^{-c^2/2} \, dc = \int_{\frac{_2x_\beta-\bar{x}_1}{\sigma_2}}^{\frac{_2x_\alpha-\bar{x}_1}{\sigma_2}} e^{-c^2/2} \, dc$$

3.67(6)

by writing $c = (x - \bar{x}_1)/\sigma_1$ in the first integral and $c = (x - \bar{x}_1)/\sigma_2$ in the second integral. Since the function is the same for both integrals one solution is obtained by equating the limits, that is

$$\frac{_1x_\alpha - \bar{x}_1}{\sigma_1} = \frac{_2x_\alpha - \bar{x}_1}{\sigma_2} = \frac{_1x_\alpha - \bar{x}_1 + \alpha}{\sigma_2}$$

and

$$\frac{_1x_\beta - \bar{x}_1}{\sigma_1} = \frac{_2x_\beta - \bar{x}_1}{\sigma_2} = \frac{_1x_\beta - \bar{x}_1 + \beta}{\sigma_2}$$

whence

$$_1x_\alpha - \bar{x}_1 = \frac{\alpha\sigma_1}{\sigma_2 - \sigma_1}$$

3.67(7)

and

$$_1x_\beta - \bar{x}_1 = \frac{\beta\sigma_1}{\sigma_2 - \sigma_1}$$
3.67(8)

and also

$$_2x_\alpha - \bar{x}_1 = \frac{\alpha\sigma_2}{\sigma_2 - \sigma_1}$$
3.67(9)

and

$$_2x_\beta - \bar{x}_1 = \frac{\beta\sigma_2}{\sigma_2 - \sigma_1}$$
3.67(10)

This solution is shown in Figure 3.67c, where β is negative.

3.68 We will now consider another solution, since the one given is generally not the best, that is the maximum number of possible parts are not used.

Consider Figure 3.68. Let us now move each of the points $_1x_\beta$ and $_2x_\beta$ a distance p to the right, and each of the points $_1x_\alpha$ and $_2x_\alpha$ a distance q to

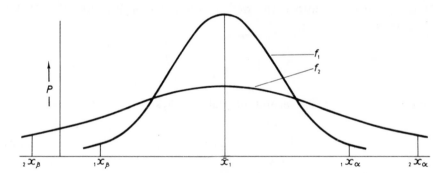

Figure 3.67c

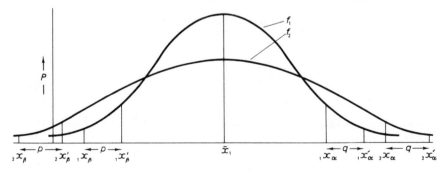

Figure 3.68

the right. Thus the area under f_1 between the new limit points is

$$\int_{1x_\beta + p}^{1x_\alpha + q} f_1(x)\, dx = \frac{1}{\sqrt{(2\pi)}} \int_{\frac{1x_\beta - \bar{x}_1 + p}{\sigma_1}}^{\frac{1x_\alpha - \bar{x}_1 + q}{\sigma_1}} e^{-c^2/2}\, dc \qquad 3.68(1)$$

whilst the new area under f_2 is now

$$\int_{2x_\beta + p}^{2x_\alpha + q} f_2\, dx = \frac{1}{\sqrt{(2\pi)}} \int_{\frac{2x_\beta - \bar{x}_1 + p}{\sigma_2}}^{\frac{2x_\alpha - \bar{x}_1 + q}{\sigma_2}} e^{-c^2/2}\, dc \qquad 3.68(2)$$

where $_1x'_\beta - {}_1x_\beta = {}_2x'_\beta - {}_2x_\beta = p$ and $_1x'_\alpha - {}_1x_\alpha = {}_2x'_\alpha - {}_2x_\alpha = q$. Now for parts to be matched we require these two areas equal, that is

$$\int_{\frac{1x_\beta - \bar{x}_1 + p}{\sigma_1}}^{\frac{1x_\alpha - \bar{x}_1 + q}{\sigma_1}} e^{-c^2/2}\, dc = \int_{\frac{2x_\beta - \bar{x}_1 + p}{\sigma_2}}^{\frac{2x_\alpha - \bar{x}_1 + q}{\sigma_2}} e^{-c^2/2}\, dc \qquad 3.68(3)$$

If we equate the top limits and bottom limits as before and substitute for the xs, we find that $p = q = 0$, that is, this is the same solution as before. However, there is another way we can equate limits. We write the second integral as

$$-\int_{\frac{2x_\alpha - \bar{x}_1 + q}{\sigma_2}}^{\frac{2x_\beta - \bar{x}_1 + p}{\sigma_2}} e^{-c^2/2}\, dc$$

and writing $c = -r$, the second integral becomes

$$\int_{\frac{-(2x_\alpha - \bar{x}_1 + q)}{\sigma_2}}^{\frac{-(2x_\beta - \bar{x}_1 + p)}{\sigma_2}} e^{-r^2/2}\, dr \qquad 3.68(4)$$

since the integrand is a symmetric function. Once again, equating the limits of the first integral of 3.68(3) with those of 3.68(4), we have

$$\frac{1x_\alpha - \bar{x}_1 + q}{\sigma_1} = \frac{-(2x_\beta - \bar{x}_1 + p)}{\sigma_2} \qquad 3.68(5)$$

and

$$\frac{1x_\beta - \bar{x}_1 + p}{\sigma_1} = \frac{-(2x_\alpha - \bar{x}_1 + q)}{\sigma_2} \qquad 3.68(6)$$

Thus, substituting for the xs from equations 3.67(7), 3.67(8), 3.67(9) and 3.67(10), we have

$$\frac{\alpha}{\sigma_2 - \sigma_1} + \frac{q}{\sigma_1} = \frac{-\beta}{\sigma_2 - \sigma_1} - \frac{p}{\sigma_2} \qquad 3.68(7)$$

and

$$\frac{\beta}{\sigma_2 - \sigma_1} + \frac{p}{\sigma_1} = \frac{-\alpha}{\sigma_2 - \sigma_1} - \frac{q}{\sigma_2} \qquad 3.68(8)$$

Solving these equations for p and q we obtain

$$p = \frac{-(\alpha + \beta)\sigma_1\sigma_2}{\sigma_2^2 - \sigma_1^2} \qquad 3.68(9)$$

and

$$q = \frac{-(\alpha + \beta)\sigma_1\sigma_2}{\sigma_2^2 - \sigma_1^2} \qquad 3.68(10)$$

The new limits are now

$$_1x'_\beta = {}_1x_\beta + p \qquad 3.68(11)$$

to

$$_1x'_\alpha = {}_1x_\alpha + q \qquad 3.68(12)$$

and

$$_2x'_\beta = {}_2x_\beta + p \qquad 3.68(13)$$

to

$$_2x'_\alpha = {}_2x_\alpha + q \qquad 3.68(14)$$

and therefore

$$_1x'_\alpha - \bar{x}_1 = {}_1x_\alpha - \bar{x}_1 + q = \frac{\alpha\sigma_1}{\sigma_2 - \sigma_1} - \frac{(\alpha + \beta)\sigma_1\sigma_2}{\sigma_2^2 - \sigma_1^2}$$

$$= \frac{\sigma_1(\alpha\sigma_1 - \beta\sigma_2)}{\sigma_2^2 - \sigma_1^2} \qquad 3.68(15)$$

using 3.68(12) and 3.68(10). Similarly, using 3.68(11) and 3.68(9)

$$_1x'_\beta - \bar{x}_1 = \frac{-\sigma_1(\alpha\sigma_2 - \beta\sigma_1)}{\sigma_2^2 - \sigma_1^2} \qquad 3.68(16)$$

and using 3.68(14) and 3.68(10)

$$_2x'_\alpha - \bar{x}_1 = \frac{\sigma_2(\alpha\sigma_2 - \beta\sigma_1)}{\sigma_2^2 - \sigma_1^2} \qquad 3.68(17)$$

and using 3.68(13) and 3.68(9)

$$_2x'_\beta - \bar{x}_1 = \frac{-\sigma_2(\alpha\sigma_1 - \beta\sigma_2)}{\sigma_2^2 - \sigma_1^2} \qquad 3.68(18)$$

or

$$\frac{_1x'_\alpha - \bar{x}_1}{\sigma_1} = \frac{-(_1x'_\beta - \bar{x}_1)}{\sigma_2} = \frac{\alpha\sigma_1 - \beta\sigma_2}{\sigma_2^2 - \sigma_1^2} \qquad 3.68(19)$$

and

$$\frac{_2x'_\alpha - \bar{x}_1}{\sigma_2} = \frac{-(_1x'_\beta - \bar{x}_1)}{\sigma_1} = \frac{\alpha\sigma_2 - \beta\sigma_1}{\sigma_2^2 - \sigma_1^2} \qquad 3.68(20)$$

using 3.68(5) and 3.68(6) and substituting x' for the x, using 3.68(1) to 3.68(14) and 3.68(15) to 3.68(18).

This solution is better than the original one because although the new spreads equal the old ones, that is

$$\underbrace{_1x_\alpha - {}_1x_\beta}_{\text{para. 3.67}} = \underbrace{_1x'_\alpha - {}_1x'_\beta}_{\text{para. 3.68}} = \frac{\sigma_1(\alpha - \beta)}{\sigma_2 - \sigma_1} \qquad 3.68(21)$$

and

$$\underbrace{_2x_\alpha - {}_2x_\beta}_{\text{para. 3.67}} = \underbrace{_2x'_\alpha - {}_2x'_\beta}_{\text{para. 3.68}} = \frac{\sigma_2(\alpha - \beta)}{\sigma_2 - \sigma_1} \qquad 3.68(22)$$

the mean of $_1x_\alpha$ and $_1x_\beta$ is

$$\frac{_1x_\alpha + {}_1x_\beta}{2} = \frac{\sigma_1(\alpha + \beta)}{2(\sigma_2 - \sigma_1)} + \bar{x}_1 \qquad 3.68(23)$$

whilst the mean of $_1x'_\alpha$ and $_1x'_\beta$ is

$$\frac{_1x'_\alpha + {}_1x'_\beta}{2} = \frac{-\sigma_1(\alpha + \beta)}{2(\sigma_1 + \sigma_2)} + \bar{x}_1 \qquad 3.68(24)$$

Similarly, the mean of $_2x_\alpha$ and $_2x_\beta$ is

$$\frac{\sigma_2(\alpha + \beta)}{2(\sigma_2 - \sigma_1)} + \bar{x}_1 \qquad 3.68(25)$$

whilst the mean of $_2x'_\alpha$ and $_2x'_\beta$ is

$$\frac{\sigma_2(\alpha + \beta)}{2(\sigma_1 + \sigma_2)} + \bar{x}_1 \qquad 3.68(26)$$

Thus the area under the x' solutions is larger, because these means are always smaller, since σ_1 and σ_2 are always positive. Whilst the solution given does not give the maximum, it is a near approximation to the maximum value for the parts which can be matched.

3.69 It can be shown that the position of $_1x_\beta$ and $_2x_\beta$, where $_2x_\beta - {_1x_\beta} = \beta$ and where the initial rates of increase of f_1 and f_2 are equal, is given by

$$_1x_\beta'' = \frac{\beta\sigma_1^2}{\sigma_2^2 - \sigma_1^2} + \bar{x}_1 \pm \frac{\sigma_1\sigma_2}{(\sigma_2^2 - \sigma_1^2)}\left[\beta^2 + 2(\sigma_2^2 - \sigma_1^2)\log_e\frac{\sigma_2}{\sigma_1}\right]^{1/2}$$

3.69(1)

Note: It is as well to check that $|_1x_\beta' - \bar{x}_1|$ is less than the above value of $|_1x_\beta'' - \bar{x}_1|$, otherwise the minimum clearance $\bar{x}_2 - \bar{x}_1 + \beta$ will initially decrease. $_2x_\beta'' = {_1x_\beta''} + \beta$, and it is easily shown that the $_1x_\alpha''$ and $_2x_\alpha''$ limits are given by

$$\frac{1}{\sqrt{(2\pi)}}\int_{\frac{_1x_\beta'' - \bar{x}_1}{\sigma_1}}^{\frac{_1x_\beta'' - \bar{x}_1 + \beta}{\sigma_2}} e^{-c^2/2}\,dc = R = \frac{1}{\sqrt{(2\pi)}}\int_{\frac{_1x_\alpha'' - \bar{x}_1}{\sigma_1}}^{\frac{_1x_\alpha'' - \bar{x}_1}{\sigma_2} + \frac{\alpha}{\sigma_2}} e^{-c^2/2}\,dc$$

$$= \frac{1}{\sqrt{(2\pi)}}\int_{A}^{\frac{\sigma_1}{\sigma_2}A + \frac{\alpha}{\sigma_2}} e^{-c^2/2}\,dc \qquad 3.69(2)$$

The second integral has to be found by means of a few trials and interpolation until the correct value of $A = ({_1x_\alpha''} - \bar{x}_1)/\sigma_1$ is found, whence $_1x_\alpha'' \cdot {_2x_\alpha''}$ is found from $_2x_\alpha'' = {_1x_\alpha''} + \alpha$. An integral of the form

$$\frac{1}{\sqrt{(2\pi)}}\int_{T}^{S} e^{-c^2/2}\,dc$$

is found from Table II, Appendix I, as

$$\frac{1}{2}\frac{1}{\sqrt{(2\pi)}}\left[\int_{-S}^{S} e^{-c^2/2}\,dc - \int_{-T}^{T} e^{-c^2/2}\,dc\right] = \frac{1}{2}\left[\underset{-S\text{ to }S}{P} - \underset{-T\text{ to }T}{P}\right]$$

Note that if S or T is negative the *sign* of P is reversed. The proportion of parts which it is possible to mate is of course given by

$$\frac{1}{\sqrt{(2\pi)}}\int_{\frac{_1x_\beta - \bar{x}_1}{\sigma_1}}^{\frac{_1x_\alpha - \bar{x}_1}{\sigma_1}} e^{-c^2/2}\,dc \qquad \text{or} \qquad \frac{1}{\sqrt{(2\pi)}}\int_{\frac{_2x_\beta - \bar{x}_1}{\sigma_2}}^{\frac{_2x_\alpha - \bar{x}_1}{\sigma_2}} e^{-c^2/2}\,dc$$

3.69(3)

Note: If $({_1x_\alpha} - \bar{x}_1)/\sigma_1 < ({_1x_\beta} - \bar{x}_1)/\sigma_1$ then the limits would be reversed.

Recapitulation

$\sigma_1 =$ standard deviation of smaller part;
$\sigma_2 =$ standard deviation of larger part;
$\bar{x}_1 =$ mean of smaller part;
$\bar{x}_2 =$ mean of larger part.

If the maximum clearance is a and the minimum clearance is b, then

$$\bar{x}_2 - \bar{x}_1 + \alpha = a$$

and

$$\bar{x}_2 - \bar{x}_1 + \beta = b$$

whence $_1x''_\alpha \cdot _2x''_\alpha$ is found from $_2x''_\alpha = {}_1x''_\alpha + \alpha$. The x of 3.69(3) should of course be replaced by the x''.

Method 1   Non-preferred solution

3.70 This gives

$$_1x_\alpha - \bar{x}_1 = \frac{\alpha\sigma_1}{\sigma_2 - \sigma_1} \qquad \text{or} \qquad \frac{_1x_\alpha - \bar{x}_1}{\sigma_1} = \frac{\alpha}{\sigma_2 - \sigma_1} = \frac{_2x_\alpha - \bar{x}_1}{\sigma_2}$$

$$\text{3.70(1)}$$

$$_1x_\beta - \bar{x}_1 = \frac{\beta\sigma_1}{\sigma_2 - \sigma_1} \qquad \text{or} \qquad \frac{_1x_\beta - \bar{x}_1}{\sigma_1} = \frac{\beta}{\sigma_2 - \sigma_1} = \frac{_2x_\beta - \bar{x}_1}{\sigma_2}$$

$$\text{3.70(2)}$$

and

$$_2x_\alpha = {}_1x_\alpha + \alpha \qquad\qquad\qquad \text{3.70(3)}$$

and

$$_2x_\beta = {}_1x_\beta + \beta \qquad\qquad\qquad \text{3.70(4)}$$

where $_1x_\beta$ and $_2x_\beta$ are the starting points on the two distribution curves, and $_1x_\alpha$ and $_2x_\alpha$ are the end points. The proportion of parts that can be matched is given by either of the integrals of 3.69(3), which are equal. This is the non-preferred solution, except for the case of $\alpha = 0$. Note that for this solution, matching can commence from the maximum of each distribution and proceed in either direction.

Method 2   Preferred solution

This is denoted by the primed xs, and the limits are given by

$$_1x'_\alpha - \bar{x}_1 = \frac{\sigma_1(\alpha\sigma_1 - \beta\sigma_2)}{\sigma_2^2 - \sigma_1^2} \qquad\qquad \text{3.68(15)}$$

$$_1x'_\beta - \bar{x}_1 = \frac{-\sigma(\alpha\sigma_2 + \beta\sigma_1)}{\sigma_2^2 - \sigma_1^2} \qquad\qquad \text{3.68(16)}$$

and

$$_2x'_\alpha - \bar{x}_1 = \frac{\sigma_2(\alpha\sigma_2 - \beta\sigma_1)}{\sigma_2^2 - \sigma_1^2} \qquad 3.68(17)$$

$$_2x'_\beta - \bar{x}_1 = \frac{-\sigma_2(\alpha\sigma_1 - \beta\sigma_2)}{\sigma_2^2 - \sigma_1^2} \qquad 3.68(18)$$

The proportion of matching parts is again given by either of the two integrals of 3.69(3), and where the x are replaced by x'. This solution should *not* be applied when $\alpha = \infty$ since the limits may be exceeded within the calculated ranges. When using this solution check that $|_1x''_\beta - \bar{x}_1|$ is less than $|_1x''_\beta - \bar{x}_1|$ where both quantities are on the same side of $\bar{x}_1$.

Method 3

This is denoted by double primed x and the limits are given by

$$_1x''_\beta - \bar{x}_1 = \frac{\beta\sigma_1^2}{\sigma_2^2 - \sigma_1^2} \pm \frac{\sigma_1\sigma_2}{\sigma_2^2 - \sigma_1^2}\left(\beta^2 + 2(\sigma_2^2 - \sigma_1^2)\log_e\frac{\sigma_2}{\sigma_1}\right)^{1/2}$$

$$3.69(1)$$

and $_2x''_\beta = {}_1x''_\beta + \beta$, whence $_2x''_\beta - \bar{x}_1$.

The other two limits, $_1x''_\alpha - \bar{x}_1$ and $_2x''_\alpha - \bar{x}_1$, are given by finding A, where $A = ({}_1x''_\alpha - \bar{x}_1)/\sigma_1$ and A is found as follows. First R is found from

$$R = \frac{1}{\sqrt{(2\pi)}} \int_{\frac{1x''_\beta - \bar{x}_1}{\sigma_1}}^{\frac{1x''_\beta - \bar{x}_1 + \beta}{\sigma_2}} e^{-c^2/2}\, dc$$

$$= \frac{1}{\sqrt{(2\pi)}} \int_B^{\frac{\sigma_1}{\sigma_2}B + \frac{\beta}{\sigma_2}} e^{-c^2/2}\, dc \qquad 3.70(5)$$

$$= \tfrac{1}{2}\left(\underset{-C\text{ to }C}{P} - \underset{-B\text{ to }B}{P} \right) \qquad 3.70(6)$$

where

$$C = \frac{\sigma_1}{\sigma_2}B + \frac{\beta}{\sigma_2} \qquad 3.70(7)$$

and

$$B = ({}_1x''_\beta - \bar{x}_1)/\sigma_1 \qquad 3.70(8)$$

A is then found from

$$R = \frac{1}{\sqrt{(2\pi)}} \int_{A}^{\frac{\sigma_1}{\sigma_2}A + \frac{\alpha}{\sigma_2}} e^{-c^2/2}\, dc$$

$$= \tfrac{1}{2}\left(\underset{-E\text{ to }E}{P} - \underset{-A\text{ to }A}{P} \right) \qquad\qquad 3.70(9)$$

where

$$E = \frac{\sigma_1}{\sigma_2}A + \frac{\alpha}{\sigma_2} \qquad\qquad 3.70(10)$$

by assuming a few trial values for *A* and then interpolating.

The proportion of matching parts is again given by the integral

$$\frac{1}{\sqrt{(2\pi)}} \int_{\frac{1 x''_\beta - \bar{x}_1}{\sigma_1}}^{\frac{1 x''_\alpha - \bar{x}_1}{\sigma_1}} e^{-c^2/2}\, dc \qquad \text{or} \qquad \frac{1}{\sqrt{(2\pi)}} \int_{\frac{2 x''_\beta - \bar{x}_1}{\sigma_2}}^{\frac{2 x''_\alpha - \bar{x}_1}{\sigma_2}} e^{-c^2/2}\, dc$$

Method 2 is much the easier solution to use, and will generally give a satisfactory result. $_1 x''_\alpha - \bar{x}_1$ may of course be used by replacing β by α.

3.71 The following gives a worked example based on the forgoing theory. Two sets of matching parts have been made and the standard deviation of the smaller part is 0.0011 in, whilst that of the larger part is 0.001 52 in. The mean of the larger parts $\bar{x}_2$ is 2.0014 in whilst the mean of the smaller parts $\bar{x}_1$ is 1.9997 in and the minimum clearance *b* allowable is 0.0003 in whilst the maximum allowable clearance is 0.0010 in. Find the fraction of parts which can be matched using methods 1, 2 and 3 of the previous example.

Method 1

Calculating the necessary limits (in inches), we have

$$\beta = b - \bar{x}_2 + \bar{x}_1 = 0.003 - 2.0014 + 1.9997 = -0.001\,40$$

$$\alpha = a - \bar{x}_2 + \bar{x}_1 = 0.0010 - 2.0014 + 1.9997 = -0.000\,70$$

therefore

$$\frac{_1 x_\alpha - \bar{x}_1}{\sigma_1} = \frac{\alpha}{\sigma_2 - \sigma_1} = \frac{-0.0007}{0.000\,42} = -1.666$$

$$\frac{_1 x_\beta - \bar{x}_1}{\sigma_1} = \frac{\beta}{\sigma_2 - \sigma_1} = \frac{0.0014}{0.000\,42} = -3.333$$

whence the proportion of matching parts is given by

$$\frac{1}{\sqrt{(2\pi)}} \int_{-3.333}^{-1.666} e^{-c^2/2}\, dc = \tfrac{1}{2}(-0.904\,32 + 0.999\,14)$$

$$= 0.0474$$

The various constants associated with this method are given, in inches, below

$$_1x_\alpha = \bar{x}_1 + \frac{\alpha\sigma_1}{\sigma_2 - \sigma_1} = 1.9997 - 1.111\,61 \times 0.0011 = 1.9979$$

$$_2x_\alpha = \bar{x}_1 + \frac{\alpha\sigma_2}{\sigma_2 - \sigma_1} = 1.9997 - 1.111\,61 \times 0.00152 = 1.9972$$

$$_1x_\beta = \bar{x}_1 + \frac{\beta\sigma_1}{\sigma_2 - \sigma_1} = 1.9997 - 3.333\,33 \times 0.0011 = 1.9960$$

$$_2x_\beta = \bar{x}_1 + \frac{\beta\sigma_2}{\sigma_2 - \sigma_1} = 1.9997 - 3.333\,33 \times 0.001\,52 = 1.994\,633$$

$$_1x_\alpha - \bar{x}_1 = -0.001\,83, \qquad _2x_\alpha - \bar{x}_1 = 0.002\,53,$$

$$_1x_\beta - \bar{x}_1 = -0.003\,700, \qquad _2x_\beta - \bar{x}_1 = -0.0051$$

Only a small proportion of parts are fitted by this method. The appearance of the answer is given in Figure 3.71.

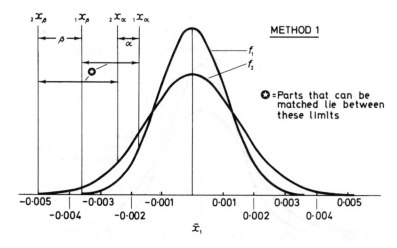

Figure 3.71

3.72

Method 2

$$_1x'_\alpha - \bar{x}_1 = \frac{\sigma_1(\alpha\sigma_1 - \beta\sigma_2)}{\sigma_2^2 - \sigma_1^2} = 0.0011 \times 1.234\,096 = 0.001\,357$$

$$_1x'_\beta - \bar{x}_1 = \frac{-\sigma_1(\alpha\sigma_2 - \beta\sigma_1)}{\sigma_2^2 - \sigma_1^2} = -0.0011 \times 0.432\,569 = -0.000\,4758$$

$$_2x'_\alpha - \bar{x}_1 = 0.000\,6570, \qquad _2x'_\beta - \bar{x}_1 = -0.001\,8758$$

The limit points of the required area integral are thus

$$\frac{_1x'_\alpha - \bar{x}_1}{\sigma_1} = 1.2341 \qquad \text{and} \qquad \frac{_1x'_\beta - \bar{x}_1}{\sigma_1} = -0.432\,57$$

whence the required fraction of parts which can be matched is

$$\frac{1}{\sqrt{(2\pi)}} \int_{-0.432\,57}^{1.2341} e^{-c^2/2}\, dc = \tfrac{1}{2}\{0.782\,82 - (-0.334\,65)\}$$

$$= 0.5587$$

It is seen that the number of parts matchable has risen to over fifty per cent. The corresponding graph is shown in Figure 3.72.

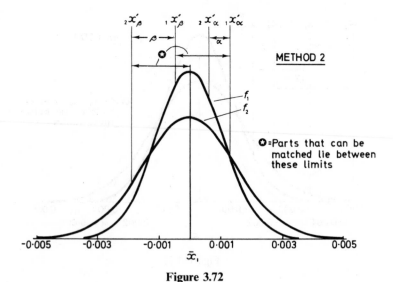

Figure 3.72

3.73

Method 3

$$_1x''_\beta - \bar{x}_1 = \frac{\beta\sigma_1^2}{\sigma_2^2 - \sigma_1^2} \pm \frac{\sigma_2\sigma_1}{(\sigma_2^2 - \sigma_1^2)}\left[\beta^2 + 2(\sigma_2^2 - \sigma_1^2)\log_e\frac{\sigma_2}{\sigma_1}\right]^{1/2}$$
$$= -0.001\,539 \pm 0.002\,48$$

that is

$$= -0.004\,02 \qquad \text{or} \qquad +0.000\,94$$

(See Figure 3.73.)

Solution

$$_1x''_\beta - \bar{x}_1 = -0.004\,02, \qquad B = (_1x''_\beta - \bar{x}_1)/\sigma_1 = -3.6545$$

$$C = \frac{\sigma_1}{\sigma_2}B + \frac{\beta}{\sigma_2} = -3.5657, \qquad \text{giving} \qquad R = \tfrac{1}{2}[-0.999\,62 + 0.999\,73]$$

$$= 0.000\,05$$

Thus we require

$$0.000\,05 = \frac{1}{\sqrt{(2\pi)}}\int_A^{\frac{\sigma_1}{\sigma_2}A + \frac{\alpha}{\sigma_2}} e^{-c^2/2}\,dc = \frac{1}{\sqrt{(2\pi)}}\int_A^{0.7237A - 0.000\,46} e^{-c^2/2}\,dc$$

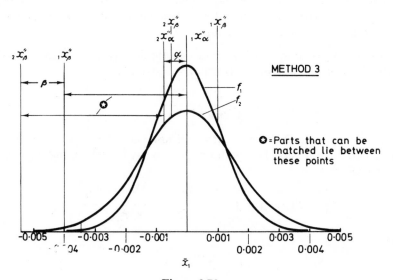

Figure 3.73

where

$$A = (_1x_\alpha'' - \bar{x}_1)/\sigma_1$$

The solution of the integral is

$$A = -0.001\,86 \qquad \text{giving} \qquad _1x_\alpha'' - \bar{x}_1 = -0.000\,002\,05,$$

$$\text{whilst} \qquad _2x_\beta'' - \bar{x}_1 = -0.005\,42$$

from $_2x_\beta'' = _1x_\beta'' + \beta$ and the value of $_1x_\beta'' - \bar{x}_1$.

Lastly, $_2x_\alpha'' - \bar{x}_1 = -0.000\,702$ from $_2x_\alpha'' = _1x_\alpha'' + \alpha$ and the value of $_1x_\alpha'' - \bar{x}_1$. The proportion of parts that can be matched is

$$= \frac{1}{\sqrt{(2\pi)}} \int_{\frac{_1x_\beta'' - \bar{x}_1}{\sigma_1}}^{\frac{_1x_\alpha'' - \bar{x}_1}{\sigma_1}} e^{-c^2/2}\, dc = \frac{1}{\sqrt{(2\pi)}} \int_{-3.6545}^{-0.001\,86} e^{-c^2/2}\, dc$$

$$= \tfrac{1}{2}(-0.001\,48 + 0.999\,73)$$

$$= 0.4991$$

It is to be noted that the second solution, that is $_1x_\beta'' - \bar{x}_1 = +0.000\,94$, cannot be used since it is easily seen that $|_1x_\beta'' - _2x_\beta''|$ increases in either direction from this solution.

Examples on Chapter 3

Example 3.1 Using the reasoning of paragraph 3.66, find an expression for the fraction of parts which will not mate. If $\bar{x}_1 = 2.001$ in and $\bar{x}_2 = 2.002$ in, and $\sigma_1 = \sigma_2 = 0.0005$ in, find the fraction of parts which will not mate, when any two parts are picked at random.

Answer: From paragraph 3.66 we know that the density function corresponding to the difference in size of the two parts is given by

$$P(\mu) = \frac{e^{-\{\mu - (\bar{x}_2 - \bar{x}_1)\}^2/2(\sigma_1^2 + \sigma_2^2)}}{\sqrt{(2\pi)}(\sigma_1^2 + \sigma_2^2)^{1/2}} \qquad\qquad 3.66(16)$$

where $\mu = x_2 - x_1$. Now two parts will not mate if μ is less than 0. Thus the required expression for the fraction of parts for which mating will not occur is given by

$$\int_{-\infty}^{0} P(\mu)\, d\mu = \int_{-\infty}^{0} \frac{e^{-\{\mu - (\bar{x}_2 - \bar{x}_1)\}^2/2(\sigma_1^2 + \sigma_2^2)}\, d\mu}{\sqrt{(2\pi)}(\sigma_1^2 + \sigma_2^2)^{1/2}} = I_\mu$$

Putting

$$\frac{\mu - (\bar{x}_2 - \bar{x}_1)}{(\sigma_1^2 + \sigma_2^2)^{1/2}} = c$$

we have

$$d\mu = (\sigma_1^2 + \sigma_2^2)^{1/2} \, dc$$

therefore

$$I_\mu = \int_{-\infty}^{-k} \frac{e^{c^2/2} \, dc}{\sqrt{(2\pi)}} = \frac{1}{2} \int_{-\infty}^{\infty} \frac{e^{-c^2/2} \, dc}{\sqrt{(2\pi)}} + \int_{0}^{-k} \frac{e^{-c^2/2} \, dc}{\sqrt{(2\pi)}}$$

$$= \frac{1}{2} \int_{-\infty}^{\infty} \frac{e^{-c^2/2} \, dc}{\sqrt{(2\pi)}} - \frac{1}{2} \int_{-k}^{k} \frac{e^{-c^2/2} \, dc}{\sqrt{(2\pi)}}$$

where

$$k = \frac{\bar{x}_2 - \bar{x}_1}{(\sigma_1^2 + \sigma_2^2)^{1/2}}$$

Thus

$$I_\mu = \tfrac{1}{2}\left(1 - \underset{-k \text{ to } k}{P}\right)$$

by Table II of Appendix I. Substituting in the expression for k the values given, we have

$$k = \frac{0.001}{10^{-4}(25 + 25)^{1/2}} = 1.4142$$

therefore

$$I_\mu = \tfrac{1}{2}(1 - 0.843\,05)$$

using Table II

$$= 0.0785$$

Example 3.2 If in the previous example, $\bar{x}_1$ and $\bar{x}_2$ remain the same, but $\sigma_1 = 0.0005$ in and $\sigma_2 = 0.0008$ in, find the fraction of parts which will not mate when picked at random.

Answer: The required fraction by Example 3.1 is given by

$$I_\mu = \tfrac{1}{2}\left(1 - \underset{-k \text{ to } k}{P}\right) \qquad \text{where} \qquad k = \frac{\bar{x}_2 - \bar{x}_1}{(\sigma_1^2 + \sigma_2^2)^{1/2}}$$

Thus

$$k = \frac{0.001}{10^{-4}(25 + 64)^{1/2}} = 0.9434$$

and

$$I_\mu = \tfrac{1}{2}(1 - 0.6560) = 0.1720$$

With the increase in σ_2, the fraction of non-mating parts, when mating pairs are picked at random, is seen to have increased considerably.

Example 3.3   Two independent variables x_1 and x_2 are related by the expression $y = a_1 x_1 + a_2 x_2$; the mean value of x_1 is $\bar{x}_1$ and its standard deviation is σ_1, whilst the corresponding quantities for x_2 are $\bar{x}_2$ and σ_2 respectively. Find an expression for the density function of y and find between what limits ninety per cent of the values of y will lie if $a_1 = 3, a_2 = 2, \sigma_1 = 0.20$, $\sigma_2 = 0.30, x_1 = 4$ and $x_2 = 6$. Assume the limits are equally distributed about the mean value of y, given by $y = a_1 x_1 + a_2 x_2$. (See paragraph 3.66.)

Answer: Density distribution $P(y)$ of y is

$$\frac{e^{-(y - a_1\bar{x}_1 - a_2\bar{x}_2)^2/2(a_1^2\sigma_1^2 + a_2^2\sigma_2^2)}}{\sqrt{(2\pi)}(a_1^2\sigma_1^2 + a_2^2\sigma_2^2)^{1/2}}$$

The 90% limits of $y \equiv 36 \pm 1.396$, where 36 is the mean value of y.

Example 3.4   Apply Method 1 to Example 3.2.

Hint: Put $a = \infty$ and $b = 0$. This may appear odd, but the number of parts for which the clearance is large will be very small. $\sigma_1 = 0.0005, \sigma_2 = 0.0008$.

Answer: Proportion of non-matching parts $= 0.0478$. Note that as $\sigma_2 - \sigma_1$ decreases the proportion of parts which cannot be matched falls. It is seen that the proportion of parts which can be matched is greater than in the random matching used in Example 3.2 for the same data.

Example 3.5   Two sets of parts which are to be matched have been made. The average value of the smaller part is 3.001 in with a standard deviation of 0.0009. The average value of the larger part is 3.0018 in with a standard deviation of 0.0005. If the clearances can vary between 0.0005 and 0.0012, find the proportion of parts that can be fitted, using Method 2 of paragraph 3.70.

Answer: 0.6181.

Example 3.6   A set of matching parts has been made, and the standard deviation of the smaller part is σ_1, whilst that of the larger is σ_2. If the maximum clearance is a and the minimum clearance is b, show that the proportion of parts which will fit if chosen at random is

$$\int_b^a \frac{e^{-\{\mu - (\bar{x}_2 - \bar{x}_1)\}^2/2(\sigma_1^2 + \sigma_2^2)} \, d\mu}{\sqrt{(2\pi)}(\sigma_1^2 + \sigma_2^2)^{1/2}}$$

Example 3.7   Using the above expression calculate the fraction of mating parts that will be obtained when random matching is employed for the data of Example 3.5.

Answer: 0.265 80. As one would expect, this is considerably less than the selective assembly method of Example 3.5.

Example 3.8 The Gaussian uncertainty contributions to the uncertainty of measurement of a quantity have standard deviations of 0.1, 0.3, 0.15 and 0.1 units. There are also three stochastic uncertainty contributions which can assume the values ±0.1, ±0.18 and ±0.2 units. Find an approximate value for the uncertainty for a tolerance probability of 0.954.

Answer: 0.927 units.

Example 3.9 If four stochastic uncertainties are possible with values of ±0.1, ±0.15, ±0.17 and ±0.20, write down the uncertainties that are possible. Calculate the probability of an uncertainty lying outside ±0.52, and ±0.37. Find the standard deviation of the uncertainties, and assuming that the resultant distribution is Gaussian, calculate the probability of an uncertainty lying outside the same limits as above.

Answer:

0.62	-0.12	0.32	-0.02
-0.62	0.12	-0.32	0.02
0.22	0.28	-0.08	-0.42
-0.22	-0.28	0.08	0.42

The probability of an uncertainty outside ±0.52 is 0.125;
The probability of an uncertainty outside ±0.37 is 0.25;
Standard deviation $= 0.3184$.
Probability of an uncertainty outside ±0.52, assuming a Gaussian distribution $= 0.102$.
Probability of an uncertainty outside ±0.37, assuming a Gaussian distribution $= 0.255$.
Note the approximate agreement between the answers for corresponding uncertainties.

4

Rectangular Distributions

4.01 In this chapter we shall discuss in some detail the properties of rectangular probability distributions which are of importance because of their application to the assessment of systematic uncertainties.

4.02 Where estimated uncertainties have to be considered, and this is almost always the case when instruments, components, workpieces, etc., are calibrated, the question of how to deal with these uncertainties needs to be investigated.

4.03 Estimated uncertainties are essentially uncertainties about which little can be known. Unlike a Gaussian type uncertainty, whose frequency distribution can be determined by repeated measurement, and whose standard deviation can be found with increasing accuracy as the number of readings is increased, the frequency distribution of an estimated uncertainty remains unknown.

4.04 Let us now consider what can reasonably be assumed about such uncertainty distributions. It is usually possible to assume an upper limit to the magnitude of an estimated uncertainty distribution, but nothing can usually be said about its distribution; under these circumstances it is reasonable to assume that any uncertainty between the plus and minus limits, set by the assumed maximum magnitude of the uncertainties, is equi-probable. It is easily seen that the frequency distribution of the uncertainty population described by our assumptions is a rectangular one, and so in the next pages we will consider the properties of rectangular distributions, how they combine with one another and how they combine with Gaussian distributions.

Rectangular Distribution

4.05 Let us consider a rectangular distribution which has a semi-range (magnitude) of plus or minus h about a mean value. Now since the area under a probability distribution must be unity, the height of the rectangle must be equal to $1/2h$ (see Figure 4.05). Now the height or ordinate of the

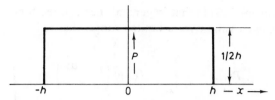

Figure 4.05 Rectangular distribution, range $\pm h$

rectangle is equal to the density function P of the distribution, which in this case is a constant.

4.06 The standard deviation of the above rectangular distribution is thus given by

$$\sigma_{\text{rec}}^2 = \int_{-h}^{h} x^2 P \, dx$$

$$= \int_{-h}^{h} \frac{x^2 \, dx}{2h} \qquad \text{since} \qquad P = \frac{1}{2h}$$

$$= \left[\frac{x^3}{6h} \right]_{-h}^{h}$$

$$= \frac{h^2}{3}$$

whence $\sigma_{\text{rec}} = h/\sqrt{3} = 0.5770h$. The magnitude h may be expressed as $\sqrt{3}\sigma_{\text{rec}} = 1.732\sigma_{\text{rec}}$. The probability of an uncertainty lying between the limits $\pm \sigma_{\text{rec}}$ for a rectangular distribution is equal to

$$\int_{-\sigma_{\text{rec}}}^{\sigma_{\text{rec}}} P \, dx = \int_{-\sigma_{\text{rec}}}^{\sigma_{\text{rec}}} \frac{dx}{2h}$$

$$= \int_{-\sigma_{\text{rec}}}^{\sigma_{\text{rec}}} \frac{dx}{2\sqrt{3}\sigma_{\text{rec}}}$$

$$= \frac{1}{\sqrt{3}} = 0.5770$$

This result follows from the constancy of P, the area under the rectangle thus being proportional to the abscissa x.

Compounding or Convoluting Two Rectangular Distributions

4.07 Let one rectangular distribution have a semi-range of $\pm k$, whilst the other has a semi-range of $\pm h$, each having a mean of $x = 0$. Also let $k \geqslant h$.

Now we know from 3.27(1) that in general the required function is given by three integrals, each valid over a limited range, that is

$$P_1(\mu) = \frac{1}{b} \int_{x_c}^{\frac{\mu - bz_e}{a}} f(x)\psi\left(\frac{\mu - ax}{b}\right) dx$$

valid for $\mu_1 \geqslant \mu \geqslant \mu_3$

$$P_2(\mu) = \int_{x_c}^{x_d} f(x)\psi\left(\frac{\mu - ax}{b}\right) dx \qquad \qquad 3.27(1)$$

valid for $\mu_2 \geqslant \mu \geqslant \mu_1$

$$P_3(\mu) = \int_{\frac{\mu - bz_f}{a}}^{x_d} f(x)\psi\left(\frac{\mu - ax}{b}\right) dx$$

valid for $\mu_4 \geqslant \mu \geqslant \mu_2$

where

$$\left.\begin{aligned}\mu_1 = bz_e + ax_d, \quad \mu_2 = bz_f + ax_c\\ \mu_3 = bz_e + ax_c, \quad \mu_4 = bz_f + ax_d\end{aligned}\right\} \qquad 4.07(1)$$

and where $f(x)$ and $\psi(z)$ are probability density functions and x and z are related by

$$\mu = ax + bz \qquad \qquad 4.07(2)$$

Note also that the range of ψ, that is $z_f - z_e$, and the range of f, that is $x_d - x_c$, must satisfy the relation

$$z_f - z_e \geqslant \frac{a}{b}(x_d - x_c) \qquad \qquad 4.07(3)$$

for the set of equations given by 3.27(1).

4.08   In the case we are considering we require the probability of occurrence of $\mu = x + z$, and thus we put $a = b = 1$ in 4.07(2). Now the range of the h rectangle is $2h$, whilst that of the k rectangle is $2k$ and since $k \geqslant h$ it follows that $2k \geqslant 2h$. Thus we put $f(x) = 1/2h$ and $\psi(z) = 1/2k$ in 3.27(1). The limits are seen by inspection to be as follows:

$$\left.\begin{aligned}x_c = -h, \quad x_d = h\\ z_e = -k, \quad z_f = k\end{aligned}\right\} \qquad 4.08(1)$$

Thus the three integrals forming the three parts of the required probability density function become

$$P_1(\mu) = \int_{-h}^{\mu + k} \frac{dx}{4hk} = \frac{\mu + h + k}{4hk} \qquad \qquad 4.08(2)$$

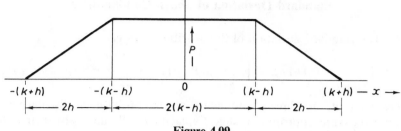

Figure 4.09

valid for $-(k - h) \geqslant \mu \geqslant -(k + h)$

$$P_2(\mu) = \int_{-h}^{h} \frac{dx}{4hk} = \frac{1}{2k}$$

4.08(3)

valid for $k - h \geqslant \mu \geqslant -(k - h)$

$$P_3(\mu) = \int_{\mu - k}^{h} \frac{dx}{4hk} = \frac{h + k - \mu}{4hk}$$

4.08(4)

valid for $k + h \geqslant \mu \geqslant k - h$.

4.09 It is easily seen that $P_1(\mu)$ is a part of the straight line between the co-ordinates $-(h + k)$, 0, and $-(k - h)$, $1/2k$, whilst $P_2(\mu)$ is equal to $1/2k$ over the range $-(k - h)$ to $k - h$, and finally $P_3(\mu)$ is a part of the straight line between the co-ordinates $k - h$, $1/2k$ and $k + h$, 0. The distribution is shown in Figure 4.09.

Probability Distribution Formed by Combining Two Rectangular Distributions Whose Semi-ranges are $\pm h$ and $\pm k$, $k \geqslant h$

4.10 It is easily seen that the distribution given by 4.08(2) to 4.08(4) is a probability distribution since the area under the distribution is

$$\frac{2h}{2} \times \frac{1}{2k} + \frac{2(k - h)}{2k} + \frac{2h}{2} \times \frac{1}{2k} = \frac{h}{2k} + 1 - \frac{k}{h} + \frac{h}{2k}$$

$$= 1 \text{ as required}$$

If $k = h$, the distribution becomes an isosceles triangle, with $P_{\max} = 1/2k$. A point to be noted is that even in this simple case, the probability of large uncertainties is reduced because of the smaller value of the density function for the large values of $\mu = x + z$.

Standard Deviation of above Combination

4.11 The standard deviation of the distribution is given by

$$\sigma^2 = \int_{-(k+h)}^{-(k-h)} \mu^2 P_1(\mu)\, d\mu + \int_{-(k-h)}^{k-h} \mu^2 P_2(\mu)\, d\mu + \int_{k-h}^{k+h} \mu^2 P_3(\mu)\, d\mu \quad 4.11(1)$$

where each of the three components of the probability distribution is integrated over its appropriate range. Replacing P_1, P_2 and P_3 by their values from 4.08(2), 4.08(3) and 4.08(4) we have

$$\sigma^2 = \int_{-(k+h)}^{-(k-h)} \frac{\mu^2(\mu + h + k)}{4hk}\, d\mu + \int_{-(k-h)}^{k-h} \frac{\mu^2\, d\mu}{2k} + \int_{k-h}^{k+h} \frac{\mu^2(h + k - \mu)}{4hk}\, d\mu$$

$$= \left[\left\{\frac{\mu^4}{4} + \frac{\mu^3}{3}(h + k)\right\}\Big/ 4hk\right]_{-(k+h)}^{-(k-h)} + \left(\frac{\mu^3}{6k}\right)_{-(k-h)}^{k-h}$$

$$+ \left[\left\{\frac{\mu^3}{3}(h + k) - \frac{\mu^4}{4}\right\}\Big/ 4hk\right]_{k-h}^{k+h}$$

$$= \left\{\frac{(k - h)^4}{4} - \frac{(k + h)^4}{4} - \frac{(k - h)^3}{3}(h + k) + \frac{(k + h)^3}{3}(h + k)\right\}\Big/ 4hk$$

$$+ \{(k - h)^3 + (k - h)^3\}/6k$$

$$+ \left\{\frac{(k + h)^3}{3}(h + k) - \frac{(k - h)^3(h + k)}{3} - \frac{(k + h)^4}{4} + \frac{(k - h)^4}{4}\right\}\Big/ 4hk$$

$$= \left\{\frac{(k - h)^4}{2} - \frac{(k + h)^4}{2} - \frac{2(k - h)^3(h + k)}{3} + \frac{2(k + h)^3(h + k)}{3}\right\}\Big/ 4hk$$

$$+ (k - h)^3/3k$$

$$= \{(k - h)^2 + (k + h)^2\}\{(k - h)^2 - (k + h)^2\}/8hk$$

$$+ (h + k)2h\{(k + h)^2 + (k - h)^2 + (k - h)(k + h)\}/6hk + (k - h)^3/3k$$

$$= 2\{(k^2 + h^2)(-4hk)/8hk + (h + k)(3k^2 + h^2)\}/3k + (k - h)^3/3k$$

$$= -(k^2 + h^2) + k^2 + h^2/3 + hk + h^3/3k + k^2/3 - hk + h^2 - h^3/3k$$

$$= (h^2 + k^2)/3$$

$$4.11(2)$$

4.12 The above integration could have been somewhat shortened by noticing that the expression was symmetrical about the ordinate axis, thus σ^2 could have been written as

$$\sigma^2 = 2\int_0^{k-h} \frac{\mu^2\, d\mu}{2k} + \int_{k-h}^{k+h} \frac{\mu^2(h + k - \mu)\, d\mu}{2hk} \qquad 4.12(1)$$

It is left as an exercise to the reader to show that it reduces to the same answer as that given by 4.11(2).

4.13 It is to be noted that the standard deviation of the combination of the two rectangular probability distributions is equal to

$$\sigma = \{(h^2 + k^2)/3\}^{1/2} \qquad\qquad 4.13(1)$$

a result which is in agreement with 3.41(1) since the standard deviation of the rectangular distribution of semi-range k is equal to $k/\sqrt{3}$ and that of the one with semi-range h is equal to $h/\sqrt{3}$; thus using 3.41(1) and taking the square root we arrive at 4.13(1).

4.14 It is interesting to consider one or two practical examples of the combination of two rectangular distributions, and to compare the probability of an uncertainty between stated limits with the corresponding probability for a Gaussian distribution taken between the same limits.

Combination of Two Rectangular Distributions with Semi-ranges of ±*a* and ±2*a*

4.15 If we put $k = 2a$ and $h = a$, we obtain a distribution similar to that shown in Figure 4.09, with a standard deviation of

$$[(2a)^2 + a^2]^{1/2}/\sqrt{3} = \sqrt{(5/3)}a = 1.291a.$$

Alternatively $a = 0.7746\sigma$. The semi-range of the combination which is $3a$ is thus equal to 2.3238σ. Figure 4.15 shows a normal distribution of the same standard deviation superimposed on the combination of the two rectangles we are considering. The figures on the drawing give the probability of an uncertainty greater than two standard deviations for the combination of the

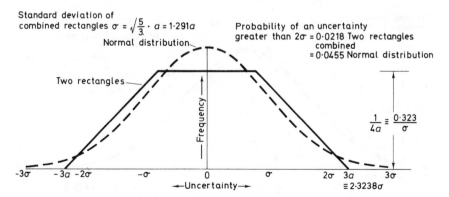

Figure 4.15 Combination of two rectangular distributions with semi-ranges of a and $2a$, with a normal distribution of equal standard deviation superposed

two rectangles and also for the Gaussian or normal distribution. It is to be noted that the probability for the combination of the two rectangles is the smaller of the two. Even in this simple example the fit between the Gaussian and the combined rectangular distributions is distinctly noticeable.

Combination of Two Rectangular Distributions each with a Semi-range of $\pm a$

4.16   Figure 4.16 shows a comparison between two distributions of equal standard deviation, a Gaussian distribution and a combination of two rectangles of equal semi-range, that is $h = k = a$. The standard deviation of the combination is equal to $\sqrt{(a^2 + a^2)} = \sqrt{(2/3)}a = 0.816a$. Alternatively $a = 1.2247\sigma$. The semi-range of the combination which is $2a$ is thus equal to 2.4494σ. Once again it is seen that the probability of an uncertainty greater than $\pm 2\sigma$ is smaller for the combination of the two rectangular distributions than for the superimposed normal distribution. In this example the fit between the combination of the two rectangular distributions is closer than in the previous example.

Compounding of Three Rectangular Distributions with Semi-ranges of $\pm j$, $\pm k$, $\pm h$. General Case $j \leqslant h + k, j \geqslant k \geqslant h$

4.17   We shall now consider the combination of three rectangular distributions. Let the semi-ranges of the three distributions that are to be combined be h, k and j, where $j \geqslant k \geqslant h$. There are, however, two possibilities: one, that $h + k \geqslant j$ or the other, that $j \geqslant h + k$. We shall begin by considering $h + k \geqslant j$.

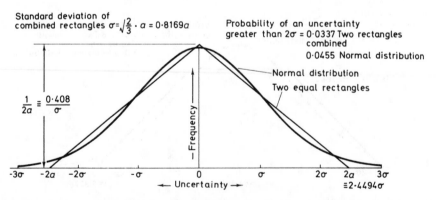

Figure 4.16   Combination of two rectangular distributions, each with a semi-range of a, with a normal distribution of equal standard deviation superposed

4.18 Now we have already combined the h and k distributions, the resultant combination being given by 4.08(2), 4.08(3) and 4.08(4), and we now combine the j rectangular distribution with each of these components to form the new hkj distribution, which will thus be made up of nine components.

4.19 Dealing first with 4.08(2), which is valid between $-(k+h)$ to $-(k-h)$, we have that the range of 4.08(2) is equal to $2h$, that is $-(k-h)$ to $-(k+h)$. Now the range of the j distribution is $2j$, that is $-j$ to j, and as $j \geqslant k \geqslant h$, thus $2j \geqslant 2h$. Since $\mu = x + z$ and thus $a = b = 1$, we require $z_f - z_e \geqslant x_d - x_c$ (see 3.27(2)). Thus in 3.27(1) we put the j distribution in place of ψ and the hk combined distribution 4.08(2) in place of f. Therefore we write

$$f(x) = \frac{x + h + k}{4hk} \qquad\qquad 4.19(1)$$

$$\psi(z) = \frac{1}{2j} \qquad\qquad 4.19(2)$$

Also

$$x_c = -(k+h), \quad x_d = -(k-h) \quad \text{from range of } 4.08(2)$$

$$z_e = -j, \qquad\quad z_f = j$$

$$\left.\begin{array}{ll} \mu_1 = -j - k + h, & \mu_2 = j - k - h \\ \mu_3 = -j - k - h, & \mu_4 = j - k + h \end{array}\right\} \quad \text{from } 4.07(1)$$

4.20 Thus substituting in 3.27(1) we have

$$P_1(\mu) = \int_{-(k+h)}^{\mu+j} \frac{(x+h+k)}{4hk} \cdot \left(\frac{1}{2j}\right) \cdot dx = \left\{ \frac{(x+h+k)^2}{16hkj} \right\}_{-(k+h)}^{\mu+j}$$

$$= \frac{1}{16hkj}(\mu + h + k + j)^2 \qquad\qquad 4.20(1)$$

valid for $-j - k + h \geqslant \mu \geqslant -j - k - h$ (see 3.27(1))

$$P_2(\mu) = \int_{-(k+h)}^{-(k-h)} \frac{(x+h+k)\,dx}{8hkj} = \left\{ \frac{(x+h+k)^2}{16hkj} \right\}_{-(k+h)}^{-(k-h)}$$

$$= \frac{1}{16hkj}(2h)^2 \qquad\qquad 4.20(2)$$

valid for $j - k - h \geqslant \mu \geqslant -j - k + h$ (see 3.27(1))

$$P_3(\mu) = \int_{\mu-j}^{-(k-h)} \frac{(x+h+k)\,dx}{8hkj} = \left\{ \frac{(x+h+k)^2}{16hkj} \right\}_{\mu-j}^{-(k-h)}$$

$$= \frac{1}{16hkj}\{(2h)^2 - (\mu - j + h + k)^2\} \qquad\qquad 4.20(3)$$

valid for $j - k + h \geqslant \mu \geqslant j - k - h$ (see 3.27(1)).

4.21 Now we deal with 4.08(3) and the j rectangle. The range of 4.08(3) is $2(k - h)$, that is $k - h - (-(k - h))$. Thus since $j \geqslant k \geqslant h$, $2j \geqslant 2(k - h)$, and so referring to 3.27(1) again, we put $\psi(z)$ equal to the j distribution density function and $f(x)$ equal to the function 4.08(3). Thus we write

$$f(x) = 1/2k \qquad\qquad 4.21(1)$$

$$\psi(z) = 1/2j \qquad\qquad 4.21(2)$$

$$\mu = x + z \qquad \text{and} \qquad a = b = 1$$

Also

$$x_c = -(k - h), \quad x_d = k - h \quad \text{from range of 4.08(3)}$$
$$z_e = -j, \qquad\qquad z_f = j$$
$$\left.\begin{array}{ll} \mu_1 = -j + k - h, & \mu_2 = j - k + h \\ \mu_3 = -j - k + h, & \mu_4 = j + k - h \end{array}\right\} \text{ from 4.07(1)}$$

4.22 Substituting in 3.27(1) we have

$$P_4(\mu) = \int_{-(k-h)}^{\mu+j} \frac{dx}{4kj} = \frac{\mu + j + k - h}{4kj} \qquad\qquad 4.22(1)$$

valid for $-j + k - h \geqslant \mu \geqslant -j - k + h$

$$P_5(\mu) = \int_{-(k-h)}^{k-h} \frac{dx}{4kj} = \frac{(k - h)}{2kj} \qquad\qquad 4.22(2)$$

valid for $j - k + h \geqslant \mu \geqslant -j + k - h$

$$P_6(\mu) = \int_{\mu-j}^{k-h} \frac{dx}{4kj} = \frac{k - h + j - \mu}{4kj} \qquad\qquad 4.22(3)$$

valid for $j + k - h \geqslant \mu \geqslant j - k + h$.

4.23 Finally we combine the 4.08(4) function and the j rectangular distribution. The range of 4.08(4) is $2h$, that is $k + h - (k - h)$. Now $2j \geqslant 2h$, and so on referring to 3.27(1) again we put $\psi(z)$ equal to the j distribution and $f(x)$ equal to function 4.08(4). Therefore we write

$$f(x) = \frac{h + k - x}{4hk} \qquad\qquad 4.23(1)$$

$$\psi(z) = 1/2j \qquad\qquad 4.23(2)$$

$$\mu = x + z \qquad \text{and} \qquad a = b = 1$$

Also

$$x_c = k - h, \quad x_d = k + h \quad \text{from range of 4.08(4)}$$
$$z_e = -j, \quad z_f = j$$
$$\left.\begin{array}{ll} \mu_1 = -j + k + h, & \mu_2 = j + k - h \\ \mu_3 = -j + k - h, & \mu_4 = j + k + h \end{array}\right\} \quad \text{from 4.07(1)}$$

4.24 Thus, substituting in 3.27(1) we have

$$P_7(\mu) = \int_{k-h}^{\mu+j} \frac{(h + k - x)\, dx}{8hkj} = \left\{\frac{-(h + k - x)^2}{16hkj}\right\}_{k-h}^{\mu+j}$$

$$= \frac{(2h)^2 - (h + k - j - \mu)^2}{16hkj} \qquad\qquad 4.24(1)$$

valid for $-j + k + h \geqslant \mu \geqslant -j + k - h$

$$P_8(\mu) = \int_{k-h}^{k+h} \frac{(h + k - x)\, dx}{8hkj} = \left\{\frac{-(h + k - x)^2}{16hkj}\right\}_{k-h}^{k+h}$$

$$= \frac{(2h)^2}{16hkj} \qquad\qquad 4.24(2)$$

valid for $j + k - h \geqslant \mu \geqslant -j + k + h$

$$P_9(\mu) = \int_{\mu-j}^{k+h} \frac{(h + k - x)\, dx}{8hk} = \left\{\frac{-(h + k - x)^2}{16hkj}\right\}_{\mu-j}^{k+h}$$

$$= \frac{(h + k + j - \mu)^2}{16hkj} \qquad\qquad 4.24(3)$$

valid for $j + k + h \geqslant \mu \geqslant j + k - h$.

4.25 We now have nine separate functions making up the complete density function of the three combined rectangular distributions, but some of the intervals over which the functions are valid overlap. In the overlapping regions the functions should be added to form a single function for each overlapping region.

4.26 The sorting out of the new intervals can be achieved by writing down the valid interval for each function in a table, and proceeding to write down the intervals beginning with the one with the smallest limit points and proceeding in ascending order of limit points. When this is done the appropriate functions in the overlapping regions can be added and so the final form of the composite function arrived at. Table 4.26A below gives the list of functions we have found, together with the interval limits and ranges. Table 4.26B gives the limits and corresponding intervals in ascending order, with the numbered functions at the side.

Table 4.26A

| | Interval | | |
Function	Lower limit	Upper limit	Range
P_1	$-j-k-h \rightarrow$	$-j-k+h$	$2h$
P_2	$-j-k+h \rightarrow$	$j-k-h$	$2(j-h)$
P_3	$j-k-h \rightarrow$	$j-k+h$	$2h$
P_4	$-j-k+h \rightarrow$	$-j+k-h$	$2(k-h)$
P_5	$-j+k-h \rightarrow$	$j-k+h$	$2(j-k+h)$
P_6	$j-k+h \rightarrow$	$j+k-h$	$2(k-h)$
P_7	$-j+k-h \rightarrow$	$-j+k+h$	$2h$
P_8	$-j+k+h \rightarrow$	$j+k-h$	$2(j-h)$
P_9	$j+k-h \rightarrow$	$j+k+h$	$2h$

Table 4.26B

| | Interval | | |
Function	Lower limit	Upper limit	Range
$P'_1 = P_1$	$-j-k-h \rightarrow$	$-j-k+h$	$2h$
$P'_2 = P_2 + P_4$	$-j-k+h \rightarrow$	$-j+k-h$	$2(k-h)$
$P'_3 = P_2 + P_5 + P_7$	$-j+k-h \rightarrow$	$j-k-h$	$2(j-k)$
$P'_4 = P_3 + P_5 + P_7$	$j-k-h \rightarrow$	$-j+k+h$	$2(k+h-j)$
$P'_5 = P_3 + P_5 + P_8$	$-j+k+h \rightarrow$	$j-k+h$	$2(j-k)$
$P'_6 = P_6 + P_8$	$j-k+h \rightarrow$	$j+k-h$	$2(k-h)$
$P'_7 = P_9$	$j+k-h \rightarrow$	$j+k+h$	$2h$
			$\sum \text{range} = 2(j+h+k)$

Reminder:

$$j \geqslant k \geqslant h \qquad \text{and} \qquad h+k \geqslant j \qquad\qquad 4.26(1)$$

4.27 We have thus ended up with nine functions and if we proceed to add the functions as shown in the first column of Table 4.26B, we arrive finally at the seven functions and seven ranges which represent the combination of three rectangular distributions, subject to the restrictions imposed by 4.26(1). The seven functions are

$$P'_1 = \frac{1}{16hkj} \cdot (h+k+j+\mu)^2 \qquad\qquad 4.27(1)$$

valid for $-j-k-h \leqslant \mu \leqslant -j-k+h$, range $= 2h$

$$P'_2 = \frac{k+j+\mu}{4kj} \qquad\qquad 4.27(2)$$

valid for $-j - k + h \leqslant \mu \leqslant -j + k - h$, range $2(k - h)$

$$P'_3 = \frac{8hk - (h + k - j - \mu)^2}{16hkj}$$ 4.27(3)

valid for $-j + k - h \leqslant \mu \leqslant j - k - h$, range $= 2(j - k)$

$$P'_4 = \frac{4hk - \mu^2 - (j - h - k)^2}{8hkj}$$ 4.27(4)

valid for $j - k - h \leqslant \mu \leqslant -j + k + h$, range $= 2(k + h - j)$

$$P'_5 = \frac{8hk - (h + k - j + \mu)^2}{16hkj}$$ 4.27(5)

valid for $-j + k + h \leqslant \mu \leqslant j - k + h$, range $= 2(j - k)$

$$P'_6 = \frac{k + j - \mu}{4kj}$$ 4.27(6)

valid for $j - k + h \leqslant \mu \leqslant j + k - h$, range $= 2(k - h)$

$$P'_7 = \frac{(h + k + j - \mu)^2}{16hkj}$$ 4.27(7)

valid for $j + k - h \leqslant \mu \leqslant j + k + h$, range $= 2h$.

It should be noted that the complete function, represented by the seven components, is symmetrical about the ordinate axis. This is seen thus: P'_4 the middle component is symmetrical, and has a range equally disposed about the ordinate axis, about which it is symmetrical. Pairing the functions on either side of P'_4, that is P'_3 with P'_5, P'_2 with P'_6 and P'_1 with P'_7, we see that the members of each pair are anti-symmetrical with one another, that is changing the sign of μ transforms one member into the other. Also the ranges of the paired functions are equal and equally disposed about the ordinate axis. Hence the complete function is symmetrical about the ordinate axis.

Combination of Three Rectangular Distributions of Equal Semi-range $\pm a$

4.28 Before proceeding to the case when $j \geqslant h + k$, we will insert particular values for h, k and j and plot the shape of the probability distributions obtained. First we put $h = k = j = a$. Inserting these values in the ranges of P'_1 to P'_7, we see that P'_2, P'_3, P'_5 and P'_6 have ranges of zero and thus disappear. The remaining functions P'_1, P'_4 and P'_7 with their ranges are given as follows:

$$P'_1 = (3a + \mu)^2/16a^3 \qquad\qquad 4.28(1)$$

valid for μ from $-3a$ to $-a$, range $= 2a$

$$P'_4 = (3a^2 - \mu^2)/8a^3 \qquad\qquad 4.28(2)$$

valid for μ from $-a$ to a, range $= 2a$

$$P'_7 = (3a - \mu)^2/16a^3 \qquad\qquad 4.28(3)$$

valid for μ from a to $3a$, range $= 2a$.

4.29   The standard deviation of the combination is given by

$$\sigma^2 = \int_{-3a}^{-a} \mu^2 P'_1 \, d\mu + \int_{-a}^{a} \mu^2 P'_4 \, d\mu + \int_{a}^{3a} \mu^2 P'_7 \, d\mu \qquad 4.29(1)$$

This is readily shown to be $a^2 \times 2/5 + a^2/5 + a^2 \times 2/5 = a^2$ giving $\sigma = a$, which is seen to be in agreement with 3.41(1), which gives

$$\sigma = \surd(a^2/3 + a^2/3 + a^2/3) = a.$$

4.30   Figure 4.30 gives a graphical representation of the combination given by 4.28(1), 4.28(2) and 4.28(3). The broken curve gives the normal or Gaussian distribution of the same standard deviation. The similarity between the two is seen to be very close. If the probability of an error lying outside $\pm 2\sigma$ is calculated for the combination the answer is found to be 0.0417, which is slightly smaller than in the Gaussian case which yields a value of 0.0455. This answer shows the same trend as that shown in Figure 4.16, that is that the probability of an error outside $\pm 2\sigma$ is smaller for a combination of rectangles than for the corresponding Gaussian distribution.

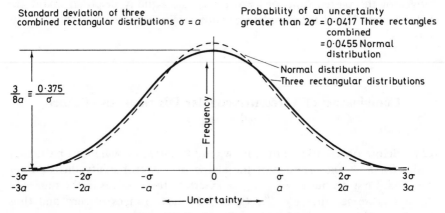

Figure 4.30   Combination of three rectangular distributions, each with a semi-range of a, with a normal distribution of equal standard deviation superposed

Combination of Three Rectangular Distributions with Semi-ranges of $\pm a$, $\pm 2a$ and $\pm 3a$

4.31 As a further example we will put $h = a$, $k = 2a$ and $j = 3a$. Putting in these values for the validity limits in 4.27(1) to 4.27(7) we find that the range of P'_4 is zero and this function thus drops out. The remaining functions and their ranges are given as follows:

$$P'_1 = (6a + \mu)^2/96a^3 \qquad\qquad 4.31(1)$$

valid for μ from $-6a$ to $-4a$, range $= 2a$

$$P'_2 = (5a + \mu)/24a^2 \qquad\qquad 4.31(2)$$

valid for μ from $-4a$ to $-2a$, range $= 2a$

$$P'_3 = (16a^2 - \mu^2)/96a^3 \qquad\qquad 4.31(3)$$

valid for μ from $-2a$ to 0, range $= 2a$

$$P'_5 = (16a^2 - \mu^2)/96a^3 \qquad\qquad 4.31(4)$$

valid for μ from 0 to $2a$, range $= 2a$

$$P'_6 = (5a - \mu)/24a^2 \qquad\qquad 4.31(5)$$

valid for μ from $2a$ to $4a$, range $= 2a$

$$P'_7 = (6a - \mu)^2/96a^3 \qquad\qquad 4.31(6)$$

valid for μ from $4a$ to $6a$, range $= 2a$.

4.32 The standard deviation of the combination is

$$\sigma = \{(3a)^2/3 + (2a)^2/3 + a^2/3\}^{1/2} = \sqrt{(14/3)} \cdot a = 2.16a$$

Figure 4.32 shows a graph of the combination with a Gaussian curve of equal standard deviation superposed. The probability of an error larger than

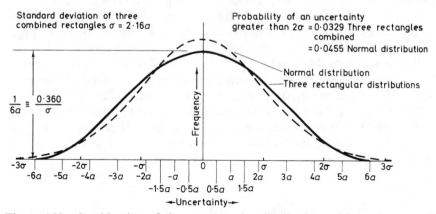

Figure 4.32 Combination of three rectangular distributions with semi-ranges of $3a$, $2a$ and a, with a normal distribution of equal standard deviation superposed

$\pm 2\sigma$ for the combination is 0.0329 compared with 0.0455 for a Gaussian distribution.

Compounding of Three Rectangular Distributions. General Case
$j \geqslant h + k, j \geqslant k \geqslant h$

4.33 We will now consider the alternative general case of the combination of three rectangular distributions, when as before the three semi-ranges h, k and j are related by the expression $j \geqslant k \geqslant h$, but with the second condition reversed, that is $j \geqslant h + k$. As before (see paragraph 4.18) we combine the three components of the combination of two rectangular distributions separately with the third rectangular distribution, which will once again lead to nine resultant components.

4.34 Starting as before with the function 4.08(2), which is valid for $-(k + h) \leqslant \mu \leqslant -(k - h)$, that is a range of $2h$, we combine this with the rectangular distribution which has a range of $2j$. Since $\mu = z + x$, $a = b = 1$, we require $z_f - z_e \geqslant x_d - x_c$ (see 3.27(2)). Thus as before we put the hk distribution in place of f and the j distribution in place of ψ, that is

$$f(x) = \frac{x + h + k}{4hk} \qquad\qquad 4.19(1)$$

$$\psi(z) = \frac{1}{2j} \qquad\qquad 4.19(2)$$

On substituting in 3.27(1) we arrive at the same three functions as before, that is 4.20(1), 4.20(2) and 4.20(3) valid between the same limits.

4.35 Similarly on combining 4.08(3) and 4.08(4) respectively with the j rectangular distribution, we obtain exactly the same functions as before, that is 4.22(1), 4.22(2), 4.22(3) and 4.24(1), 4.24(2) and 4.24(3) which are valid between the same limits as previously. As in paragraph 4.25 we now have the same nine functions, some of which overlap, and we shall now proceed as before to sort out the functions by addition into a series of non-overlapping contiguous functions. It is in this sorting out that the final composite function will be found to be different from 4.27(1) to 4.27(7) because of the condition $j \geqslant h + k$ rather than $j \leqslant h + k$ which appertained in the latter distribution.

4.36 The first table for our nine functions will be exactly the same as before (see Table 4.26A) but the second table, giving the list of limits in ascending order with the corresponding functions at the side, will be different because of the new condition, that is $j \geqslant h + k$ rather than $j \leqslant h + k$. The results are given in Table 4.36 below.

Reminder:

$$j \geqslant k \geqslant h \qquad \text{and} \qquad j \geqslant h + k \qquad\qquad 4.36(1)$$

Table 4.36

Function	Interval		Range
	Lower limit	*Upper limit*	
$P'_1 = P_1$	$-j - k - h \rightarrow$	$-j - k + h$	$2h$
$P'_2 = P_2 + P_4$	$-j - k + h \rightarrow$	$-j + k - h$	$2(k - h)$
$P'_3 = P_2 + P_5 + P_7$	$-j + k - h \rightarrow$	$-j + k + h$	$2h$
$P'_4 = P_2 + P_5 + P_8$	$-j + k + h \rightarrow$	$j - k - h$	$2(j - k - h)$
$P'_5 = P_3 + P_5 + P_8$	$j - k - h \rightarrow$	$j - k + h$	$2h$
$P'_6 = P_6 + P_8$	$j - k + h \rightarrow$	$j + k - h$	$2(k - h)$
$P'_7 = P_9$	$j + k - h \rightarrow$	$j + k + h$	$2h$
			$\sum \text{range} = 2(j + h + k)$

4.37 Once again we have arrived finally with seven functions, and if we proceed to add the functions as indicated in Table 4.36 we obtain the composite function which represents the combination of three rectangular distributions, subject to the conditions imposed by 4.36(1). The seven functions are

$$P'_1 = (h + k + j + \mu)^2 / 16hkj \qquad 4.37(1)$$

valid for $-j - k - h \leqslant \mu \leqslant -j - k + h$, range $= 2h$

$$P'_2 = (h + j + \mu) / 4kj \qquad 4.37(2)$$

valid for $-j - k + h \leqslant \mu \leqslant -j + k - h$, range $= 2(k - h)$

$$P'_3 = \{8hk - (h + k - j - \mu)^2\} / 16hkj \qquad 4.37(3)$$

valid for $-j + k - h \leqslant \mu \leqslant -j + k + h$, range $= 2h$

$$P'_4 = 1/2j \qquad 4.37(4)$$

valid for $-j + k + h \leqslant \mu \leqslant j - k - h$, range $= 2(j - k - h)$

$$P'_5 = \{8hk - (h + k - j + \mu)^2\} / 16hkj \qquad 4.37(5)$$

valid for $j - k - h \leqslant \mu \leqslant j - k + h$, range $= 2h$

$$P'_6 = (k + j - \mu) / 4kj \qquad 4.37(6)$$

valid for $j - k + h \leqslant \mu \leqslant j + k - h$, range $= 2(k - h)$

$$P'_7 = (h + k + j - \mu)^2 / 16hkj \qquad 4.37(7)$$

valid for $j + k - h \leqslant \mu \leqslant j + k + h$, range $= 2h$.

4.38 Comparing with the corresponding functions of paragraph 4.27 it is seen that the only different function is P_4, which in this group is a constant. Whilst the other functions are identical, it should be noted that ranges 3 and 5 are different whilst 4 is reversed in sign.

Combination of Three Rectangular Distributions with Semi-ranges of $\pm a$, $\pm 1.5a$ and $\pm 3a$

4.39 We will consider one case of the use of the general formula set out in paragraph 4.37, that is when $h = a$, $k = 1.5a$ and $j = 3a$. The composite function is obtained by substitution of these values, and leads to the following results

$$P'_1 = (5.5a + \mu)^2/72a^3 \qquad\qquad 4.39(1)$$

valid for $-5.5a \leqslant \mu \leqslant -3.5a$, range $= 2a$

$$P'_2 = (4.5a + \mu)/18a^2 \qquad\qquad 4.39(2)$$

valid for $-3.5a \leqslant \mu \leqslant -2.5a$, range $= a$

$$P'_3 = \{12a^2 - (\mu + 0.5)^2\}/72a^3 \qquad\qquad 4.39(3)$$

valid for $-2.5a \leqslant \mu \leqslant -0.5a$, range $= 2a$

$$P'_4 = 1/6a \qquad\qquad 4.39(4)$$

valid for $-0.5a \leqslant \mu \leqslant 0.5a$, range $= a$

$$P'_5 = \{12a^2 - (\mu - 0.5a)^2\}/72a^3 \qquad\qquad 4.39(5)$$

valid for $0.5a \leqslant \mu \leqslant 2.5a$, range $= 2a$

$$P'_6 = (4.5a - \mu)/18a^2 \qquad\qquad 4.39(6)$$

valid for $2.5a \leqslant \mu \leqslant 3.5a$, range $= a$

$$P_7 = (5.5a - \mu)^2/72a^3 \qquad\qquad 4.39(7)$$

valid for $3.5a \leqslant \mu \leqslant 5.5a$, range $= 2a$.

4.40 The standard deviation of the combination is

$$\sigma = [\{(3a)^2 + (1.5a)^2 + a^2\}/3]^{1/2} = \sqrt{(12.25/3)}\, a = 2.021a$$

Figure 4.40 shows a graph of the combination, with a Gaussian curve of equal standard deviation superposed. The probability of an error larger than $\pm 2\sigma$ for the combination is 0.0287 compared with 0.0455 for a Gaussian distribution.

Compounding of Four Rectangular Distributions each with a Semi-range of $\pm a$

4.41 Before leaving the subject of the combination of rectangular distributions, we will consider one further case, that of the combination of four equal rectangular distributions, each having a semi-range of $\pm a$. We take as our

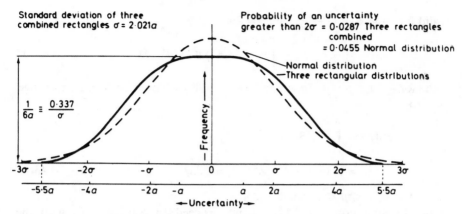

Figure 4.40 Combination of three rectangular distributions with semi-ranges of $3a$, $1.5a$ and a, with a normal distribution of equal standard deviation superposed

starting point the functions P_1, P_4 and P_7 of paragraph 4.28 which represent the composite function for the combination of three rectangular distributions each having a semi-range of $\pm a$.

4.42 Starting first with 4.28(1) we combine this with a rectangular distribution of semi-range $\pm a$. As we are again considering the probability of occurrence of $\mu = x + z$, we put $a = b = 1$ in 4.07(2). Since the range of function 4.28(1) is $2a$, equal to that of the rectangular distribution we are combining with it, it does not matter which function we choose to substitute for ψ or f in 3.27(1), since 4.07(3) is satisfied either way round in this special case of equality. Let us substitute 4.28(1) for $f(x)$ and the rectangular distribution of semi-range a for ψ.

4.43 Thus we put

$$f(x) = (3a + x)^2/16a^3 \qquad\qquad 4.28(1)$$

and

$$\psi(z) = 1/2a \qquad\qquad 4.43(1)$$

By inspection, the limits are

$$x_c = -3a, \quad x_d = -a \quad \text{(see 4.28(1))}$$

$$z_e = -a, \quad z_f = +a$$

Thus substituting in 3.27(1) we obtain the following components

$$P_1(\mu) = \int_{-3a}^{\mu+a} \{(3a + x)^2/32a^4\}\, dx$$

$$= \{(3a + x)^3/96a^4\}_{-3a}^{\mu+a} = (4a + \mu)^3/96a^4 \qquad\qquad 4.43(2)$$

valid for $-4a \leqslant \mu \leqslant -2a$, range $= 2a$.

$$P_2(\mu) = \int_{-3a}^{-a} \{(3a + x)^2/32a^4\} \, dx \qquad 4.43(3)$$

valid for $-2a \leqslant \mu \leqslant -2a$, range $= 0$. $P_2(\mu)$ is thus omitted because its range is zero.

$$P_3(\mu) = \int_{\mu-a}^{-a} \{(3a + x)^2/32a^4\} \, dx$$

$$= \{(3a + x)^3/96a^4\}_{\mu-a}^{-a} = \{8a^3 - (2a + \mu)^3\}/96a^4 \qquad 4.43(4)$$

valid for $-2a \leqslant \mu \leqslant 0$, range $= 2a$.

4.44 Dealing now with 4.28(2), we proceed to combine this with the rectangular distribution of semi-range $\pm a$. Proceeding as in paragraph 4.42, since the range of 4.28(2) is also equal to $2a$, we can again choose 4.28(2) or the a rectangular distribution to substitute in place of ψ or f in 3.27(1). Let us in this case substitute 4.28(2) for $\psi(z)$ and the a rectangular distribution for $f(x)$.

4.45 Thus we put

$$f(x) = 1/2a \qquad 4.45(1)$$

and

$$\psi(z) = (3a^2 - z^2)/8a^3 \qquad 4.28(2)$$

By inspection, the limits are

$$x_c = -a, \quad x_d = +a$$

$$z_e = -a, \quad z_f = +a \quad \text{(see 4.28(2))}$$

Substituting in 3.27(1) we obtain the next three components of the composite probability distribution as

$$P_4(\mu) = \int_{-a}^{\mu+a} [\{3a^2 - (\mu - x)^2\}/16a^4] \, dx$$

using $\mu = x + z$, that is, 4.07(2) with $a = b = 1$

$$= \{3a^2x + (\mu - x)^3/3\}_{-a}^{\mu-a}/16a^4$$

$$= \{-(\mu + a)^3/3 + 3a^2\mu + 17a^3/3\}/16a^4 \qquad 4.45(2)$$

valid for $-2a \leqslant \mu \leqslant 0$.

$$P_5(\mu) = \int_{-a}^{a} \{3a^2 - (\mu - x)^2\}/16a^4 \qquad 4.45(3)$$

valid for $0 \leqslant \mu \leqslant 0$. $P_5(\mu)$ is thus omitted because its range is zero.

$$P_6(\mu) = \int_{\mu-a}^{a} \{3a^2 - (\mu - x)^2\}/16a^4$$

$$= [\{3a^2 x - (\mu - x)^3/3\}/16a^4]_{\mu-a}^{a}$$

$$= \{(\mu - a)^3/3 - 3a^2\mu + 17a^3/3\}/16a^4 \qquad 4.45(4)$$

valid for $0 \leqslant \mu \leqslant 2a$.

4.46 Finally we combine the rectangular distribution of semi-range $\pm a$ with distribution 4.28(3). As previously, since the range of 4.28(3) is also $2a$ it can be substituted for either $f(x)$ or $\psi(z)$ in 3.27(1). Let us substitute 4.28(3) for $f(x)$ and the a rectangular distribution for $\psi(z)$.

Thus we put

$$f(x) = (3a - x)^2/16a^3 \qquad 4.28(3)$$

and

$$\psi(z) = 1/2a \qquad 4.46(1)$$

By inspection, the limits are

$$x_c = a, \qquad x_d = 3a \quad \text{(see 4.28(3))}$$

$$z_e = -a, \qquad z_f = a$$

Substituting in 3.27(1) we obtain the final three terms of the composite distribution we require, giving

$$P_7(\mu) = \int_{a}^{\mu+a} \{(3a - x)^2/16a^3\} \frac{dx}{2a}$$

$$= \left\{ -\frac{(3a - x)^3}{3} \cdot \frac{1}{32a^4} \right\}_{a}^{\mu+a}$$

$$= -\frac{1}{96a^4} \{(2a - \mu)^3 - 8a^3\} \qquad 4.46(2)$$

valid for $0 \leqslant \mu \leqslant 2a$.

$$P_8(\mu) = \int_{a}^{3a} \frac{1}{32a^4} (3a - x)^2 \, dx \qquad 4.46(3)$$

valid for $2a \leqslant \mu \leqslant 2a$. P_8 is thus omitted as its range is zero.

$$P_9(\mu) = \int_{\mu-a}^{3a} \frac{1}{32a^4} (3a - x)^2 \, dx$$

$$= \left\{ -\frac{1}{96a^4} (3a - x)^3 \right\}_{\mu-a}^{3a}$$

$$= (4a - \mu)^3/96a^4 \qquad 4.46(4)$$

valid for $2a \leqslant \mu \leqslant 4a$.

4.47   Where overlapping of functions occurs it is complete, and so makes the task of sorting into ascending intervals easier. Doing this, and adding we get the final components of the combination of these rectangles as follows:

$$P'_1 = (4a + \mu)^3/96a^4 \qquad\qquad 4.43(2)$$

valid for $-4a \leqslant \mu \leqslant -2a$, range $= 2a$

$$P'_2 = \frac{1}{16a^2} \left\{ \frac{-(2a + \mu)^3}{6a^2} - \frac{(a + \mu)^3}{3a^2} + 7a + 3\mu \right\} \qquad 4.47(1)$$

valid for $-2a \leqslant \mu \leqslant 0$, range $= 2a$

$$P'_3 = \frac{1}{16a^2} \left\{ \frac{-(2a - \mu)^3}{6a^2} - \frac{(a - \mu)^3}{3a^2} + 7a - 3\mu \right\} \qquad 4.47(2)$$

valid for $0 \leqslant \mu \leqslant 2a$, range $= 2a$

$$P'_4 = (4a - \mu)^3/96a^4 \qquad\qquad 4.46(4)$$

valid for $2a \leqslant \mu \leqslant 4a$, range $= 2a$.

4.48   The standard deviation of the combination is given by 3.41(1)

$$\sigma = \sqrt{\{4(a^2/3)\}} = 1.154a$$

Figure 4.48 shows a graph of this combination, with a Gaussian distribution of equal standard deviation superposed. Since the range of the combination is $\pm 4a$, its range exceeds $\pm 3\sigma = 3.462a$, and so besides the probability of an uncertainty exceeding $\pm 2\sigma$, the probability of an uncertainty exceeding $\pm 3\sigma$ is given, and compared with the corresponding probabilities for a Gaussian distribution. Note the very close correspondence between the combination distribution and the Gaussian distribution.

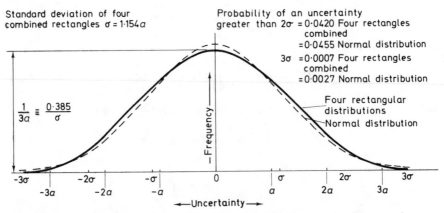

Figure 4.48   Combination of four rectangular distributions each with a semi-range of a, with a normal distribution of equal standard deviation superposed

Combination of a Gaussian and a Rectangular Distribution†

4.49 Before leaving the discussion of the combination of rectangular distributions, we investigate the convolution of a rectangular distribution with a Gaussian one. Let the Gaussian distribution be

$$P_k = \frac{e^{-z^2/2\sigma_k^2}}{\sigma_k\sqrt{(2\pi)}} \qquad 4.49(1)$$

where σ_k is the standard deviation. If the standard deviation of the rectangular distribution is σ_h, then its range is $\pm\sqrt{3}\cdot\sigma_h$ and it will be represented by

$$P_h = \frac{1}{2\sqrt{3}\cdot\sigma_h} \qquad 4.49(2)$$

that is the reciprocal of twice the range.

4.50 Referring to paragraph 4.07, $a = b = 1$ since we require the density function of $\mu = x + z$ (see 4.07(2)). Thus, in order to satisfy 4.07(3), we substitute P_k in place of $\psi(z)$ and P_h in place of $f(x)$. By inspection $x_c = -\sqrt{3}\sigma_h$, $x_d = \sqrt{3}\sigma_h$, $z_e = -\infty$, $z_f = \infty$. Thus substituting in 4.07(1) we have

$$\mu_1 = -\infty, \quad \mu_2 = \infty$$

$$\mu_3 = -\infty, \quad \mu_4 = \infty$$

4.51 Referring now to the validity limits of the three integrals of 3.27(1), we see that only $P_2(\mu)$ survives since $\mu_1 = \mu_3$ and $\mu_2 = \mu_4$. Thus substituting in $P_2(\mu)$ of 3.27(1) for $\psi(z)$ and $f(x)$ we have as the density function of our combination

$$P_{hk}(\mu) = \int_{-\sqrt{3}\sigma_h}^{\sqrt{3}\sigma_h} \frac{e^{-(\mu-x)^2/2\sigma_k^2}\,dx}{2\sqrt{3}\sigma_h\sigma_k\sqrt{(2\pi)}} \qquad 4.51(1)$$

4.52 The standard deviation of this function is, according to 3.41(1), given by

$$\sigma = \sqrt{(\sigma_h^2 + \sigma_k^2)} \qquad 4.52(1)$$

This is easily proved as follows. The standard deviation of 4.51(1) is given by

$$\sigma^2 = \int_{-\infty}^{\infty} \mu^2\,d\mu \int_{-\sqrt{3}\sigma_h}^{\sqrt{3}\sigma_h} \frac{e^{-(\mu-x)^2/2\sigma_k^2}\,dx}{2\sqrt{3}\sigma_h\sigma_k\sqrt{(2\pi)}} \qquad 4.52(2)$$

† Table XXIII in Appendix I gives tabulated values of k for indexed values of P between limits $(-k\sigma$ to $k\sigma)$, for a range of values of $\eta = \sigma_G/\sigma_R$ where $\sigma_G \equiv$ standard deviation of Gaussian distribution and $\sigma_R \equiv$ standard deviation of rectangular distribution. Values of P for indexed values of k are also given for a range of values of η, where P is the probability of an uncertainty between the limits $(-k\sigma$ to $k\sigma)$.

Since $\mu^2 e^{-(\mu-x)^2/2\sigma_k^2}$ is a continuous function we can reverse the order of integration. Thus

$$\sigma^2 = \int_{-\sqrt{3\sigma_h}}^{\sqrt{3\sigma_h}} dx \int_{-\infty}^{\infty} \frac{\mu^2 e^{-(\mu-x)^2/2\sigma_k^2} d\mu}{2\sqrt{3\sigma_h}\sigma_k\sqrt{(2\pi)}} \qquad 4.52(3)$$

Put $(\mu - x)/\sqrt{2}\sigma_k = y$. Thus $d\mu = dy\sqrt{2}\sigma_k$ and

$$\mu^2 = (\sqrt{2}\sigma_k y + x)^2 = x^2 + 2\sigma_k^2 y^2 + 2\sqrt{2}yx.$$

Substituting for μ in terms of y in 4.52(3) we have

$$\sigma^2 = \int_{-\sqrt{3\sigma_h}}^{\sqrt{3\sigma_h}} \frac{dx}{\sqrt{6\sigma_h}\sqrt{(2\pi)}} \int_{-\infty}^{\infty} (x^2 + 2\sigma_k^2 y^2 + 2\sqrt{2}\sigma_k xy) \, e^{-y^2} \, dy \qquad 4.52(4)$$

Now

$$\int_{-\infty}^{\infty} e^{-y^2} \, dy = \sqrt{\pi} \qquad 2.10(4)$$

$$\int_{-\infty}^{\infty} y^2 e^{-y^2} \, dy = -\int_{-\infty}^{\infty} \frac{y d(e^{-y^2})}{2}$$

$$= \left(\frac{-y e^{-y^2}}{2}\right)_{-\infty}^{\infty} + \int_{-\infty}^{\infty} \frac{e^{-y^2} dy}{2}$$

$$= \frac{\sqrt{\pi}}{2} \qquad 4.52(5)$$

and

$$\int_{-\infty}^{\infty} y e^{-y^2} \, dy = -\int_{-\infty}^{\infty} \frac{d(e^{-y^2})}{2}$$

$$= (-e^{-y^2})_{-\infty}^{\infty} = 0 \qquad 4.52(6)$$

Thus 4.52(4) reduces to

$$\sigma^2 = \int_{-\sqrt{3\sigma_h}}^{\sqrt{3\sigma_h}} \frac{dx}{\sqrt{6\sigma_h}} \frac{\sqrt{\pi}}{\sqrt{(2\pi)}} (x^2 + \sigma_k^2)$$

$$= \frac{1}{2\sqrt{3\sigma_h}} \left(\frac{x^3}{3} + \sigma_k^2 x\right)_{-\sqrt{3\sigma_h}}^{\sqrt{3\sigma_h}}$$

$$= \frac{1}{2\sqrt{3\sigma_h}} \left(\frac{2.3\sqrt{3\sigma_h^3}}{3} + 2\sqrt{3}\sigma_k^2\sigma_h\right)$$

$$= \sigma_h^2 + \sigma_k^2$$

in agreement with 4.52(1).

4.53 $P_{hk}(\mu)$ of 4.51(1) may be transformed by writing $(\mu - x)/\sigma_k = y$, giving $dx = -\sigma_k\, dy$, with limits

$$y_1 = (\mu - \sqrt{3}\sigma_h)/\sigma_k$$

$$y_2 = (\mu + \sqrt{3}\sigma_h)/\sigma_k$$

Thus on substituting in 4.51(1) we have

$$P_{hk}(\mu) = \frac{1}{2\sqrt{3}\sigma_h} \int_{\frac{\mu-\sqrt{3}\sigma_h}{\sigma_k}}^{\frac{\mu+\sqrt{3}\sigma_h}{\sigma_k}} \frac{e^{-y^2/2}\, dy}{\sqrt{(2\pi)}} \qquad 4.53(1)$$

If we write $\sigma_k = \alpha\sigma$ where $1 \geqslant \alpha \geqslant 0$, then $\sigma_h = \sigma\sqrt{(1 - \alpha^2)}$ from 4.52(1). Substituting in 4.53(1) for σ_k and σ_h and writing $\mu = q\sigma$ where $q \geqslant 0$ we have

$$P_{hk} = \frac{1}{2\sqrt{\{3(1 - \alpha^2)\}}\sigma} \int_{\frac{q-\sqrt{\{3(1-\alpha^2)\}}}{\alpha}}^{\frac{q+\sqrt{\{3(1-\alpha^2)\}}}{\alpha}} \frac{e^{-y^2/2}\, dy\dagger}{\sqrt{(2\pi)}} \qquad 4.53(2)$$

Values of the integral of P may be obtained from Table II, Appendix I, and thus P_{hk} can be plotted for any selected value of α. Figure 4.53 shows P_{hk} plotted for the combination of a rectangular distribution and a Gaussian distribution of equal standard deviation. The broken curve shows a Gaussian distribution of equal standard deviation to the combination.

4.54 The expression given by 4.53(2), whilst convenient for plotting P_{hk}, is not convenient for the calculation of probabilities, that is the integral of P_{hk} between limits, since no simple analytical solution of this integral, valid for all values of μ, exists. This difficulty can be overcome by using a function which is an approximation to a rectangular distribution, and combining this with a Gaussian distribution.

† It can be shown that

$$\int_{-\infty}^{z} \frac{e^{-y^2/2}\, dy}{\sqrt{(2\pi)}} = \frac{1}{2} + \frac{e^{-z^2/2}}{\sqrt{(2\pi)}} \sum_{r=0}^{r=\infty} \frac{z^{2r+1} r!\, 2^r}{(2r + 1)!},$$

whence

$$\int_{z_1}^{z_2} \frac{e^{-y^2/2}}{\sqrt{(2\pi)}} e^{-z^2/2}\, dy = \frac{e^{-z_2^2/2}}{\sqrt{(2\pi)}} \sum_{r=0}^{r=\infty} \frac{z_2^{2r+1} r!\, 2^r}{(2r + 1)!} - \frac{e^{-z_1^2/2}}{\sqrt{(2\pi)}} \sum_{r=0}^{r=\infty} \frac{z_1^{2r+1} r!\, 2^r}{(2r + 1)!}$$

Alternatively

$$\int_{z_1}^{z_2} P\, dy = \int_{z_1}^{0} P\, dy + \int_{0}^{z_2} P\, dy = \int_{0}^{z_2} P\, dy - \int_{0}^{z_1} P\, dy = \frac{1}{2}\left[\int_{-z_2}^{z_2} P\, dy - \int_{-z_1}^{z_1} P\, dy\right]$$

where $P = e^{-y^2/2}/\sqrt{(2\pi)}$

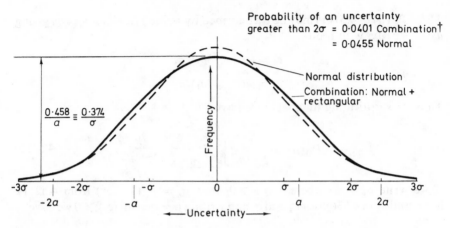

Figure 4.53 Combination of a rectangular distribution with a standard deviation of $\sigma_R = a/\sqrt{3}$ with a normal distribution of equal standard deviation, with a normal distribution of standard deviation, equal to the combination, superposed

Approximate Rectangular Density Function

4.55 The function which approximates to a rectangular distribution is

$$P_{\sigma_s} = \sum_{r=-n}^{r=n} \frac{e^{-(x-rk\sigma_s)^2/2\sigma_s^2}}{(2n+1)\sigma_s\sqrt{(2\pi)}} \qquad 4.55(1)$$

where σ_s is the standard deviation of the component distributions, k is a constant approximately equal to 2.10 and n is an integer. The larger the value of n chosen the more vertical are the sides of the approximate rectangle. The top of the rectangle has $2n+1$ maxima, which give it the appearance of a sine wave. The depth of modulation of the top is approximately $\pm 2.25\%$ of the average height, and this is independent of the value of n.

4.56 The general appearance of the function is shown in Figure 4.56. The standard deviation of 4.55(1) is given by

$$\int_{-\infty}^{\infty} x^2 P_{\sigma_s} \, dx\ddagger$$

Now

$$\int_{-\infty}^{\infty} x^2 \, e^{-(x+\beta)^2/2\sigma^2} \, dx$$

where $\beta = rk\sigma_s$ and $\sigma = \sigma_s$ can be written as

$$\int_{-\infty}^{\infty} \sqrt{2} \, e^{-y^2}(2\sigma^2 y^2 + \beta^2 - 2\sqrt{2}\sigma\beta y) \, dy$$

† Calculated from equations 4.61(5), 4.65(6), 4.65(7) and 4.65(8). See paragraph 4.65.
‡ Since the mean of P_{σ_s} is zero.

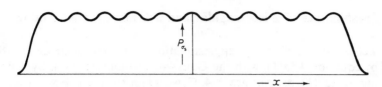

Figure 4.56 Approximation to a rectangular distribution given by 4.55(1)

by putting $(x + \beta)/\sqrt{2}\sigma = y$, and it is seen by referring to 2.10(4), 4.52(5) and 4.52(6) that this integral is equal to

$$\sigma^3 \sqrt{(2\pi)} + \beta^2 \sigma \sqrt{(2\pi)}$$

It is thus readily seen that the standard deviation of P_{σ_s} is given by

$$\sigma'^2 = \frac{1}{(2n + 1)\sigma_s \sqrt{(2\pi)}} \sum_{r=-n}^{r=n} (\sigma_s^3 \sqrt{(2\pi)} + k^2 r^2 \sigma_s^3 \sqrt{(2\pi)})$$

$$= \frac{\sigma_s^2}{(2n + 1)} \left(2n + 1 + k^2 \sum_{r=-n}^{r=n} r^2 \right) \qquad 4.56(1)$$

Now

$$r^2 = \frac{(r + 2)(r + 1)(r) - (r + 1)(r)(r - 1)}{3} - \frac{(r + 1)r - r(r - 1)}{2}$$

and therefore

$$\sum_{r=-n}^{r=n} r^2 = 2 \sum_{r=1}^{r=n} r^2 = 2\left\{ \frac{(n + 2)(n + 1)n}{3} - \frac{(n + 1)n}{2} \right\}$$

$$= 2\left\{ \frac{(n + 1)n}{6} (2n + 4 - 3) \right\}$$

$$= \frac{n(n + 1)(2n + 1)}{3} \qquad 4.56(2)$$

Thus

$$\sigma'^2 = \sigma_s^2 \left\{ 1 + \frac{k^2 n(n + 1)}{3} \right\}\dagger \qquad 4.56(3)$$

† The semi-range is approximately $(kn + 1)\sigma_s$, and thus the ratio of the standard deviation to the semi-range is equal to

$$[1 + \{k^2 n(n + 1)\}/3]^{1/2}/(kn + 1) = (1/\sqrt{3})[1 + \{2 + nk(k - 2)\}/(kn + 1)^2]^{1/2}.$$

Since $k = 2.1$, this ratio reduces approximately to $[1 + 2/(kn + 1)^2]^{1/2}/\sqrt{3} \to 1/\sqrt{3}$ with increasing n, as it should.

Combination of a Rectangular and a Gaussian Distribution

4.57 Let us now combine the approximation to a rectangular distribution given by equation 4.55(1), with the Gaussian distribution given by 4.49(1). Referring to 3.27(1), see paragraph 4.07, we again require $\mu = x + y$ and so $a = b = 1$. Now the range of 4.55(1) is from $-\infty$ to $+\infty$, the same as that of the Gaussian distribution 4.49(1), and so we have a choice as to which of them we substitute for $f(x)$ or $\psi(z)$. Let us substitute the rectangular approximation for $f(x)$ and the Gaussian distribution for $\psi(z)$. The limits are

$$x_c = -\infty, \quad x_d = +\infty$$

$$z_e = -\infty, \quad z_f = +\infty$$

Since both sets of limits are infinite, then the only integral of 3.27(1) to survive is $P_2(\mu)$ with limits x_c to x_d. (See paragraph 3.50.) The combination is represented from $-\infty$ to $+\infty$ by

$$P_3 = \frac{1}{2\pi\sigma_s\sigma_k(2n+1)} \int_{-\infty}^{\infty} e^{-(\mu-x)^2/2\sigma_k^2} \left(\sum_{r=-n}^{r=n} e^{-(x-rk\sigma_s)^2/2\sigma_s^2} \right) dx \quad 4.57(1)$$

4.58 The integration of this expression which is a little involved is accomplished as follows. The index of each exponential term of the integral is of the form

$(x - rk\sigma_s)^2/2\sigma_s^2 + (\mu - x)^2/2\sigma_k^2$

$= (x^2 + r^2k^2\sigma_s^2 - 2xrk\sigma_s)/2\sigma_s^2 + (\mu^2 + x^2 - 2\mu x)/2\sigma_k^2$

$= \{x^2(\sigma_k^2 + \sigma_s^2) - 2x(rk\sigma_s\sigma_k^2 + \mu\sigma_s^2) + \mu^2\sigma_s^2 + r^2k^2\sigma_s^2\sigma_k^2\}/2\sigma_s^2\sigma_k^2$

$= \{x^2 - 2x(rk\sigma_s\sigma_k^2 + \mu\sigma_s^2)/(\sigma_k^2 + \sigma_s^2)$

$\quad + (\mu^2\sigma_s^2 + r^2k^2\sigma_s^2\sigma_k^2)/(\sigma_k^2 + \sigma_s^2)\} \times (\sigma_k^2 + \sigma_s^2)/2\sigma_s^2\sigma_k^2$

$= \underbrace{\{x - (rk\sigma_s\sigma_k^2 + \mu\sigma_s^2)/(\sigma_k^2 + \sigma_s^2)\}^2(\sigma_k^2 + \sigma_s^2)/2\sigma_s^2\sigma_k^2}_{A}$

$\quad + (\mu^2\sigma_s^2 + r^2k^2\sigma_s^2\sigma_k^2)/2\sigma_s^2\sigma_k^2 - (rk\sigma_s\sigma_k^2 + \mu\sigma_s^2)^2/2\sigma_s^2\sigma_k^2(\sigma_k^2 + \sigma_s^2)$

$= A + \{\mu^2(\sigma_s^4 + \sigma_s^2\sigma_k^2 - \sigma_s^4) + r^2k^2\sigma_s^2\sigma_k^2(\sigma_k^2 + \sigma_s^2 - \sigma_k^2)\}/2\sigma_s^2\sigma_k^2(\sigma_k^2 + \sigma_s^2)$

$\quad - 2rk\sigma_s^3\sigma_k^2/2\sigma_s^2\sigma_k^2(\sigma_k^2 + \sigma_s^2)$

$= A + (\mu^2\sigma_s^2\sigma_k^2 + r^2k^2\sigma_s^4\sigma_k^2 - 2rk\sigma_s^3\sigma_k^2)/2\sigma_s^2\sigma_k^2(\sigma_k^2 + \sigma_s^2)$

$= A + (\mu - rk\sigma_s)^2/2(\sigma_k^2 + \sigma_s^2)$

The complete index can thus be written as

$\{x - (rk\sigma_s\sigma_k^2 + \mu\sigma_s^2)/(\sigma_k^2 + \sigma_s^2)\}^2(\sigma_k^2 + \sigma_s^2)/2\sigma_s^2\sigma_k^2 + (\mu - rk\sigma_s)^2/2(\sigma_k^2 + \sigma_s^2)$

$$4.58(1)$$

4.59 Equation 4.57(1) can now be written

$$P_3 = \frac{1}{2\pi\sigma_s\sigma_k(2n+1)} \sum_{r=-n}^{r=n} \exp - \left\{ \frac{(\mu - rk\sigma_s)^2}{2(\sigma_k^2 + \sigma_s^2)} \right\}$$

$$\times \int_{-\infty}^{\infty} \exp - \left\{ x - \frac{(rk\sigma_s\sigma_k^2 + \mu\sigma_s^2)^2}{\sigma_k^2 + \sigma_s^2} \right\} \left\{ \frac{1}{2} \left(\frac{1}{\sigma_k^2} + \frac{1}{\sigma_s^2} \right) \right\} dx \qquad 4.59(1)$$

Let us put the index of the exponential term under the integral equal to y^2, that is

$$\frac{1}{2} \left(\frac{1}{\sigma_k^2} + \frac{1}{\sigma_s^2} \right) \left(x - \frac{rk\sigma_s\sigma_k^2 + \mu\sigma_s^2}{(\sigma_k^2 + \sigma_s^2)} \right)^2 = y^2$$

or $a(x - b)^2 = y^2$, where

$$a = \frac{1}{2} \left(\frac{1}{\sigma_k^2} + \frac{1}{\sigma_s^2} \right)$$

and

$$b = \frac{rk\sigma_s\sigma_k^2 + \mu\sigma_s^2}{(\sigma_k^2 + \sigma_s^2)}$$

Thus

$$\sqrt{a}(x - b) = y \qquad \text{and} \qquad dx = \frac{dy}{\sqrt{a}} = \frac{dy\sigma_k\sigma_s\sqrt{2}}{\sqrt{(\sigma_k^2 + \sigma_s^2)}}$$

Therefore the integral becomes on substituting for x in terms of y

$$\int_{-\infty}^{\infty} \frac{e^{-y^2} dy\sigma_k\sigma_s\sqrt{2}}{\sqrt{(\sigma_k^2 + \sigma_s^2)}}$$

Now

$$\int_{-\infty}^{\infty} e^{-y^2} dy = \sqrt{\pi} \quad \text{(see 2.10(4))}$$

Thus finally

$$P_3 = \sum_{r=-n}^{r=+n} \frac{e^{-(\mu - rk\sigma_s)^2/2(\sigma_k^2 + \sigma_s^2)}}{(2n+1)\sqrt{(\sigma_k^2 + \sigma_s^2)}\sqrt{(2\pi)}} \qquad 4.59(2)$$

4.60 The standard deviation of P_3 is given by

$$\sigma^2 = \int_{-\infty}^{\infty} x^2 P_3 \, dx\dagger = \int_{-\infty}^{\infty} \sum_{r=-n}^{r=n} \frac{x^2 e^{-(x - rk\sigma_s)^2/2(\sigma_k^2 + \sigma_s^2)} dx}{(2n+1)\sqrt{\{2\pi(\sigma_k^2 + \sigma_s^2)\}}} \qquad 4.60(1)$$

† Since the mean of P_3 is zero.

Referring to paragraph 4.56, it is seen that this integral when evaluated gives

$$\sigma^2 = \sigma_k^2 + \sigma_s^2 \left\{ 1 + \frac{k^2 n(n+1)}{3} \right\}$$ 4.60(2)

in agreement with 3.41(1).

4.61   The probability P of an uncertainty between the limits $-p\sigma$ to $p\sigma$ for the function P_3 (4.59(2)), where σ is given by 4.60(2) and p is an integer, is given by

$$P = \int_{-p\sigma}^{p\sigma} \sum_{r=-n}^{r=n} \frac{e^{-(x-rk\sigma_s)^2/2(\sigma_k^2+\sigma_s^2)} \, dx}{(2n+1)\sqrt{\{2\pi(\sigma_k^2+\sigma_s^2)\}}}$$ 4.61(1)

Putting

$$\frac{(x - rk\sigma_s)}{(\sigma_k^2 + \sigma_s^2)} = y$$

gives

$$dx = \sqrt{(\sigma_k^2 + \sigma_s^2)} \, dy$$

The new limits are

$$y = \frac{-(p\sigma + rk\sigma_s)}{\sqrt{(\sigma_k^2 + \sigma_s^2)}} \qquad \text{to} \qquad \frac{p\sigma - rk\sigma_s}{\sqrt{(\sigma_k^2 + \sigma_s^2)}}$$

for each term of the series. Thus

$$P = \frac{1}{(2n+1)\sqrt{(2\pi)}} \sum_{r=-n}^{r=n} \int_{\frac{-(p\sigma+rk\sigma_s)}{(\sigma_k^2+\sigma_s^2)}}^{\frac{p\sigma-rk\sigma_s}{\sqrt{\sigma_k^2+\sigma_s^2}}} e^{-y^2/2} \, dy$$ 4.61(2)

Putting $r = q$ and $r = -q$, we have the two contributions to the complete integral as

$$\int_{\frac{-(p\sigma+qk\sigma_s)}{\sqrt{(\sigma_k^2+\sigma_s^2)}}}^{\frac{p\sigma-qk\sigma_s}{\sqrt{(\sigma_k^2+\sigma_s^2)}}} e^{-y^2/2} \, dy + \int_{\frac{-(p\sigma+qk\sigma_s)}{\sqrt{(\sigma_k^2+\sigma_s^2)}}}^{\frac{p\sigma+qk\sigma_s}{\sqrt{(\sigma_k^2+\sigma_s^2)}}} e^{-y^2/2} \, dy$$

This is equivalent to

$$\int_{-a}^{b} e^{-y^2/2} \, dy + \int_{-b}^{a} e^{-y^2/2} \, dy$$ 4.61(3)

Put $y = -x$ in the second integral of 4.61(3). Thus $dy = -dx$ and the limits become b to $-a$. The second integral thus becomes

$$-\int_{b}^{-a} e^{-x^2/2} \, dx = \int_{-a}^{b} e^{-x^2/2} \, dx$$

and 4.61(3) becomes

$$2 \int_{-a}^{b} e^{-y^2/2} \, dy$$

Now

$$\int_{-a}^{b} e^{-y^2/2} \, dy = \int_{-a}^{0} e^{-y^2/2} \, dy + \int_{0}^{b} e^{-y^2/2} \, dy$$

$$= -\int_{0}^{-a} e^{-y^2/2} \, dy + \int_{0}^{b} e^{-y^2/2} \, dy$$

$$= \int_{0}^{a} e^{-y^2/2} \, dy + \int_{0}^{b} e^{-y^2/2} \, dy$$

$$= \frac{1}{2} \left(\int_{-a}^{a} e^{-y^2/2} \, dy + \int_{-b}^{b} e^{-y^2/2} \, dy \right)$$

because $e^{-y^2/2}$ is symmetrical about $y = 0$. Thus P can be transformed into

$$P = \frac{1}{(2n+1)\sqrt{(2\pi)}} \left\{ \sum_{r=1}^{r=n} \left(\int_{-(p\sigma+rk\sigma_s)/\sqrt{(\sigma_k^2+\sigma_s^2)}}^{p\sigma+rk\sigma_s/\sqrt{(\sigma_k^2+\sigma_s^2)}} e^{-y^2/2} \, dy + \int_{-(p\sigma-rk\sigma_s)/\sqrt{(\sigma_k^2+\sigma_s^2)}}^{p\sigma-rk\sigma_s/\sqrt{(\sigma_k^2+\sigma_s^2)}} e^{-y^2/2} \, dy \right) \right.$$

$$\left. + \underbrace{\int_{-p\sigma/\sqrt{(\sigma_k^2+\sigma_s^2)}}^{p\sigma/\sqrt{(\sigma_k^{4^{28}}+{}^8\sigma_s^{4^2})}} e^{-y^2/2} \, dy}_{\text{this term corresponds to } r = 0 \text{ in } \sum_{r=-n}^{r=n}} \right\} \qquad 4.61(4)$$

Referring to Table II, Appendix I, P can thus finally be written as

$$P = \frac{1}{(2n+1)} \left\{ \sum_{r=1}^{r=n} \left(P_{-\alpha_r \text{ to } \alpha_r} + P_{-\beta_r \text{ to } \beta_r} \right) + P_{-\gamma \text{ to } \gamma} \right\} \qquad 4.61(5)$$

where

$$\alpha_r = \frac{p\sigma + rk\sigma_s}{\sqrt{(\sigma_k^2 + \sigma_s^2)}} \qquad 4.61(6)$$

$$\beta_r = \frac{p\sigma - rk\sigma_s}{\sqrt{(\sigma_k^2 + \sigma_s^2)}} \qquad 4.61(7)$$

$$\gamma = \frac{p\sigma}{\sqrt{(\sigma_k^2 + \sigma_s^2)}} \qquad 4.61(8)$$

and where

$$\sigma = \sqrt{\left[\sigma_k^2 + \sigma_s^2 \left(1 + \frac{k^2 n(n+1)}{3} \right) \right]} \qquad 4.61(9)$$

from 4.60(2), p is an integer and $k = 2.10$. If $p\sigma$ is smaller than $rk\sigma_s$, then β is negative. Thus a term $\underset{-\beta\,\text{to}\,\beta}{P}$ becomes $\underset{|\beta|\,\text{to}\,-|\beta|}{P}$ which is equal to $\underset{-|\beta|\,\text{to}\,|\beta|}{-P}$ where $|\beta|$ means the magnitude or positive value of β. Thus if β is negative, the contribution to the integral from this term is negative and equal to

$$-\underset{-|\beta|\,\text{to}\,|\beta|}{P}$$

Numerical Computation

4.62 We will now consider an example, utilizing the formulae we have developed for an approximation to a rectangular distribution, and for the combination of this approximate distribution with a Gaussian distribution. Let us first consider the approximation P_{σ_s} to a rectangular distribution. Now P_{σ_s} is given by

$$P_{\sigma_s} = \sum_{r=-n}^{r=n} \frac{e^{-(x-rk\sigma_s)^2/2\sigma_s^2}}{(2n+1)\sigma_s\sqrt{(2\pi)}} \qquad 4.55(1)$$

where σ_s is the standard deviation of each of the $2n + 1$ components of the function. We have also shown that the standard deviation of this function is given by σ' where

$$\sigma'^2 = \sigma_s^2\left\{1 + \frac{k^2 n(n+1)}{3}\right\} \qquad 4.56(3)$$

Rectangular distribution: probability of an uncertainty in the range $-p\sigma'$ to $p\sigma'$

The probability of an error between the limits $-p\sigma'$ to $p\sigma'$ for P_{σ_s} of 4.55(1), where p is $\geqslant 0$, is thus given by

$$P = \int_{-p\sigma'}^{p\sigma'} \sum_{r=-n}^{r=n} \frac{e^{-(x-rk\sigma_s)^2/2\sigma_s^2}}{(2n+1)\sigma_s\sqrt{(2\pi)}} \qquad 4.62(1)$$

Comparing this integral with 4.61(1) and with the result of integrating 4.61(1), that is 4.61(4), we see that the integral of 4.62(1) is given by

$$P = \frac{1}{(2n+1)\sqrt{(2\pi)}}\left\{\sum_{r=1}^{r=n}\left(\int_{-(p\sigma'+rk\sigma_s)/\sigma_s}^{(p\sigma'+rk\sigma_s)/\sigma_s} e^{-y^2/2}\,dy + \int_{-(p\sigma'-rk\sigma_s)/\sigma_s}^{(p\sigma'-rk\sigma_s)/\sigma_s} e^{-y^2/2}\,dy\right)\right.$$
$$\left. + \int_{-p\sigma'/\sigma_s}^{p\sigma'/\sigma_s} e^{-y^2/2}\,dy\right\} \qquad 4.62(2)$$

4.63 This expression can be transformed and written in just the same form as 4.61(5), that is

$$P = \frac{1}{(2n+1)} \left\{ \sum_{r=1}^{r=n} \left(\underset{-\alpha_r \text{ to } \alpha_r}{P} + \underset{-\beta_r \text{ to } \beta_r}{P} \right) + \underset{-\gamma \text{ to } \gamma}{P} \right\} \qquad 4.61(5)$$

but where α_r, β_r and γ are now given by

$$\alpha_r = \frac{p\sigma' + rk\sigma_s}{\sigma_s} \qquad 4.63(1)$$

$$\beta_r = \frac{p\sigma' - rk\sigma_s}{\sigma_s} \qquad 4.63(2)$$

$$\gamma = p\sigma'/\sigma_s \qquad 4.63(3)$$

and σ' is given by 4.56(3). As before, if β is negative, its modulus is taken and the value of $\underset{|\beta| \text{ to } |\beta|}{P}$ taken as negative.

4.64 α_r, β_r and γ can be rewritten as

$$\alpha_r = p \left\{ 1 + \frac{k^2 n(n+1)}{3} \right\}^{1/2} + rk \qquad 4.64(1)$$

$$\beta_r = p \left\{ 1 + \frac{k^2 n(n+1)}{3} \right\}^{1/2} - rk \qquad 4.64(2)$$

$$\gamma = p \left\{ 1 + \frac{k^2 n(n+1)}{3} \right\}^{1/2} \qquad 4.64(3)$$

by substituting for σ' from 4.56(3). Thus putting $n = 10$, $k = 2.10$ and $p = 1$ and evaluating 4.61(3) using Table II of Appendix I, we find that the probability of an uncertainty between plus and minus one standard deviation for the approximation to a rectangular distribution is equal to 0.577, in excellent agreement with the value for a true rectangular distribution, i.e.

$$\frac{1}{\sqrt{3}} = 0.577.$$

Combination of a Rectangular Distribution with a Gaussian Distribution: Probability of an Uncertainty in the Range -2σ to 2σ

4.65 We will now calculate the probability of an uncertainty between plus and minus two standard deviations for the combination of a rectangular distribution with a Gaussian distribution of equal standard deviation. Before proceeding, we will modify equations 4.61(5) to 4.61(9) so that they are more immediately applicable to any combination of a rectangular distribution with a Gaussian one. σ_k = standard deviation of the Gaussian distribution and

$$\sigma' = \sigma_s \left\{ 1 + \frac{k^2 n(n+1)}{3} \right\}^{1/2} \qquad 4.56(3)$$

the standard deviation of the approximation to a rectangular distribution. Let $\sigma' = \sigma_s \tau$ where

$$\tau = \left\{ 1 + \frac{k^2 n(n+1)}{3} \right\}^{1/2} \qquad 4.65(1)$$

Let us also put

$$\sigma' = m\sigma_k \qquad 4.65(2)$$

where m is a positive number. Now

$$\sigma = \sqrt{\left\{ \sigma_k^2 + \sigma_s^2 \left(1 + \frac{kn(n+1)}{3} \right) \right\}} \qquad 4.61(9)$$

$$= (\sigma_k^2 + \sigma'^2)^{1/2}$$

$$= \sigma_k (1 + m^2)^{1/2} \qquad 4.65(3)$$

using 4.65(2). Also

$$\sigma_s = \frac{\sigma'}{\tau} = \frac{m}{\tau} \sigma_k \qquad 4.65(4)$$

and thus

$$(\sigma_s^2 + \sigma_k^2)^{1/2} = \sigma_k \left(1 + \frac{m^2}{\tau^2} \right)^{1/2} \qquad 4.65(5)$$

Thus substituting in 4.61(6), 4.61(7) and 4.61(8) we have

$$\alpha_r = \frac{p\tau(1 + m^2)^{1/2} + rkm}{(m^2 + \tau^2)^{1/2}} \qquad 4.65(6)$$

$$\beta_r = \frac{p\tau(1 + m^2)^{1/2} - rkm}{(m^2 + \tau^2)^{1/2}} \qquad 4.65(7)$$

$$\gamma = \frac{p\tau(1 + m^2)^{1/2}}{(m^2 + \tau^2)^{1/2}} \qquad 4.65(8)$$

which are independent of σ_s, σ_k and σ'.

4.66 In order to calculate the required probability we again use 4.61(5), giving the probability of an uncertainty between -2σ to 2σ as

$$P = \frac{1}{(2n+1)}\left\{\sum_{r=1}^{r=n}\left(\underset{-\alpha_r \text{to}\alpha_r}{P} + \underset{-\beta_r \text{to}\beta_r}{P}\right) + \underset{-\gamma \text{to}\gamma}{P}\right\} \qquad 4.61(5)$$

where $k = 2.10$, $m = 1$, $n = 10$ and $p = 2$ and where α_r, β_r and γ are obtained from 4.65(6), (7) and (8) and $\underset{-\alpha_r \text{to}\alpha_r}{P}$ etc. of 4.61(5) from Table II of Appendix I. The value of P obtained is 0.9599, or the probability of an uncertainty lying outside plus or minus two standard deviations is 0.0401 as indicated in Figure 4.53.

4.67 Figure 4.67 shows another example of the combination of a rectangular distribution with a Gaussian one, in which the standard deviation of the former distribution is five times that of the latter. The probability of an error lying outside -2σ to 2σ for the combination is equal to 0.0036, much less than the corresponding Gaussian figure of 0.0455. Table XXIIIa of Appendix I has been prepared giving values for the tolerance factor k for indexed values of the probability P and of the ratio $\eta = \sigma_G/\sigma_R$, where $\sigma_G \equiv$ standard deviation of the Gaussian distribution and $\sigma_R \equiv$ standard deviation of the rectangular distribution. Table XXIIIb gives values for the probability P between the limits $\mu - k\sigma$ to $\mu + k\sigma$ for indexed values of k and of η, where μ is the mean value of the combination.

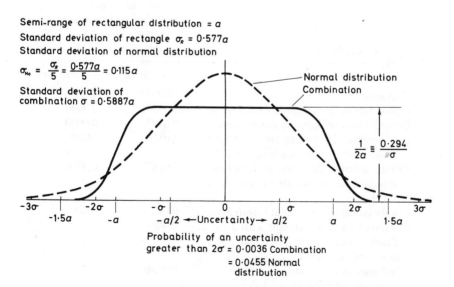

Figure 4.67 Combination of a normal distribution with a rectangular distribution having a standard deviation five times as large

Calculation of Approximate Probabilities for the Combination of Rectangular Distributions with Themselves and with a Gaussian Distribution

4.68   Table 4.68 summarizes the results we have obtained by combining rectangular distributions with themselves and with Gaussian distributions. The most important point to notice about the results is that the probabilities of an uncertainty outside either $\pm 2\sigma$ or $\pm 3\sigma$ are greater for a Gaussian distribution than for any of the combinations. This is a highly important point since it means that if we combine rectangular distributions both with themselves or with Gaussian distributions, we can always be sure that the probability of an uncertainty greater than $\pm 2\sigma$, $\pm 3\sigma$ etc. is always less than in the corresponding Gaussian case, and that the true probability will in general not be much less than the Gaussian one.

4.69   Let us look at this another way round. If we know the standard deviations of the Gaussian uncertainties of an item of calibration, and thus the total Gaussian standard deviation, and if we assess the maximum value of each of the estimated uncertainties, and assume that these have a

Table 4.68

Type of distribution	Probability of an error lying outside the following multiples of the standard deviation for each combination	
	-2σ to 2σ	-3σ to 3σ
Rectangular	0.0000	0.0000
Combination of two similar rectangles	0.0337	0.0000
Combination of three similar rectangles	0.0417	0.0000
Combination of three rectangles whose standard deviations are in ratio 3, 2, 1	0.0329	0.0000
Combination of three rectangles whose standard deviations are in ratio 3, 1.5, 1	0.0287	0.0000
Combination of four equal rectangles	0.0420	0.0007
Combination of a rectangle with a Gaussian distribution of equal standard deviation	0.0401	
Combination of a rectangle with a Gaussian distribution of $\frac{1}{5}$ the standard deviation of the rectangle	0.0036	
Gaussian	0.0455	0.0027

rectangular distribution, we can at once find the total standard deviation of the measurements. We can then state that the probability of an uncertainty lying outside $\pm 2\sigma$ is less than 0.0455 or the probability of an uncertainty lying outside $\pm 3\sigma$ is less than 0.0027, without having to go to all the trouble of calculating the combined distribution and calculating the required probabilities. The correct probabilities will always be *less* than the corresponding Gaussian ones having the same tolerance limits and standard deviation, but usually by only a small amount.

Examples on Chapter 4

Example 4.1   The semi-ranges of three rectangular uncertainty distributions are 0.0012 in, 0.0037 in and 0.005 in. If one part from each of these distributions is chosen at random and placed one against the other, find the probability of the sum of the dimensions of these three parts exceeding the sum of the three means by more than two standard deviations.

Answer: The standard deviation of the combination is equal to

$$\sigma = \frac{1}{\sqrt{3}} [0.0012^2 + 0.0037^2 + 0.005^2]^{1/2}$$

$$= 0.003\,657$$

Thus

$$2\sigma = 0.007\,315$$

Now, since $0.005 > 0.0012 + 0.0037$, the appropriate distribution is that given by equations 4.37(1) to 4.37(7). Since we require the probability of the mean being exceeded we need only consider equations 4.37(4) to 4.37(7). Put

$$j = 0.005, \qquad k = 0.0037 \qquad \text{and} \qquad h = 0.0012$$

then

$$4.37(4) \text{ is valid from } -0.0001 \text{ to } 0.0001$$

$$4.37(5) \text{ is valid from } \quad 0.0001 \text{ to } 0.0025$$

$$4.37(6) \text{ is valid from } \quad 0.0025 \text{ to } 0.0075$$

and

$$4.37(7) \text{ is valid from } \quad 0.0075 \text{ to } 0.0099$$

Thus in order to find the probability of an uncertainty greater than two standard deviations we must integrate 4.37(6) from 0.007 315 to 0.0075, and

4.37(7) from 0.0075 to 0.0099. 4.37(6) when the values of j, k and h are substituted becomes

$$P'_6 = 1351.513(0.0087 - \mu)$$

whilst 4.37(7) gives

$$P'_7 = 2\,815\,315.315(0.0099 - \mu)^2$$

The required probability is thus

$$P = 13\,513.513 \int_{0.007\,315}^{0.0075} (0.0087 - \mu)\, d\mu$$

$$+ \int_{0.0075}^{0.0099} 2\,815\,315.515(0.0099 - \mu)^2\, d\mu$$

$$= \left\{ -13\,513.513\, \frac{(0.0087 - \mu)^2}{2} \right\}_{0.007\,315}^{0.0075}$$

$$- \left\{ 2\,815\,315.315\, \frac{(0.0099 - \mu)^3}{3} \right\}_{0.0075}^{0.0099}$$

$$= \frac{13\,513.513}{2} \{(0.0012)^2 - (0.001\,385)^2\} - \frac{2\,815\,315.315}{3} \{0 - (0.0024)^3\}$$

$$= 0.0162$$

This value is a little less than the value which would have been obtained had the distributions been Gaussian, namely, 0.0227.

Example 4.2   Three rectangular estimated uncertainties have semi-ranges of 0.02, 0.06 and 0.05 units. Find the standard deviation of the combination and the probability of an uncertainty outside two standard deviations and two-and-a-half standard deviations.

Answer: Standard deviation = 0.046 55 units. Probability of an uncertainty greater than 2σ is 0.0349, greater than 2.5σ is 0.001 757.

Example 4.3   Ten 1 lb weights each have fractional uncertainties in their weights of up to 0.001. If these uncertainties are assumed to be rectangular, find the fractional standard deviation for combinations of 4, 6, 8 and 10 weights, and also state the approximate probability of the fractional uncertainty exceeding two standard deviations in each case.

Answer: = 0.000 2887, 0.000 2357, 0.000 2041, 0.000 1826.
Probability is less than 0.0455.

Example 4.4   Find the probability of an uncertainty exceeding two standard deviations for the combination of a rectangular distribution with a Gaussian distribution having a standard deviation twice the value of a rectangular one.

Answer: 0.0449.

Example 4.5 Find an approximate expression for the density function for the combination of two rectangular distributions, with semi-ranges h and k, $k > h$, and a Gaussian distribution with standard deviation of σ_k.

Hint: Use as a starting point the three functions already obtained for the combination of two rectangles, that is 40.8(2), (3) and (4), and convert the integrals between the finite limits obtained to integrals between $-\infty$ to $+\infty$ by multiplying each integrand by a unit multiplier, which is unity over the range of integration and zero elsewhere. An approximate multiplier is given by

$$Q_y = \sum_{r=-n}^{r=n} \frac{e^{-(y-y_0-rk\sigma_s)^2/2\sigma_s^2}}{\sqrt{(2\pi)}(0.476\,15)}$$

If $2h$ is the required range

$$2h = \sqrt{3}\sigma_s\left\{1 + \frac{k'^2 n(n+1)}{3}\right\}^{1/2}$$

whence σ_s. Take the value of $k' = 2.1$. The mean or centre of the range of $2h$ is y_0.

Answer:

$$P(\mu) = \frac{\sigma_k}{4hk\sqrt{(2\pi)}}\left\{e^{-(\mu-k-h)^2/2\sigma_k^2} + e^{-(\mu+k+h)^2/2\sigma_k^2} - e^{-(\mu-k+h)^2/2\sigma_k^2}\right.$$

$$\left. - e^{-(\mu+k-h)^2/2\sigma_k^2}\right\} + \frac{2\sigma_s \sum_{r=-n}^{r=n} e^{-(\mu-rk_2'\sigma_s)^2/2(\sigma_k^2+{}_2\sigma_s^2)}}{0.9523k\sqrt{(2\pi)}(\sigma_k^2+{}_2\sigma_s^2)^{1/2}}$$

$$+ \frac{{}_1\sigma_s}{1.9046\sqrt{(2\pi)}hk(\sigma_k^2+{}_1\sigma_s^2)^{1/2}}\left\{(h+k-\mu)\sum_{r=-n}^{r=n} e^{-(\mu-k-rk_1'\sigma_s)^2/2(\sigma_k^2+{}_1\sigma_s^2)}\right.$$

$$\left. + (h+k+\mu)\sum_{r=-n}^{r=n} e^{-(\mu+k-rk_1'\sigma_s)^2/2(\sigma_k^2+{}_1\sigma_s^2)}\right\}$$

where

$$_1\sigma_s = \frac{2h}{\sqrt{3}}\bigg/\left\{1 + \frac{k'n(n+1)}{3}\right\}^{1/2}$$

$$_2\sigma_s = \frac{2(k-h)}{\sqrt{3}}\bigg/\left\{1 + \frac{k'n(n+1)}{3}\right\}^{1/2}$$

and

$$k' = 2.1$$

For numerical work n can be taken as any integer greater than 5, the larger n is, the steeper the sides of the unit function. If the value of n is taken as 10 and substituted in the unit function, then a good approximation to a true rectangle of unit height is obtained.

5

Applications

5.01 In this chapter we shall look at several problems that are often encountered in industry. These are concerned with the effect of measurements made using an imperfect measuring machine, i.e. one that has a finite uncertainty of measurement, and on the reliance which can be placed on the sizes of the parts accepted as lying within chosen tolerance limits.

Sizing Machine with Two Measuring Heads

5.02 The first application deals with a sizing machine which will accept parts lying between two tolerance limits and will reject those lying outside the tolerance range. The machine will be assumed to have two measuring heads, one of which will accept parts equal to or below the highest limit, but reject parts above this limit, whilst the other will accept parts equal to or above the lowest limit, but will reject parts which are less than this limit.

5.03 Let us assume that the sizing machine is set to accept parts with sizes between a and b, $b > a$, and that the standard deviation of the parts is σ_T, with a mean of $\bar{x}_T$. We assume also that each limit point has a probability function associated with it, such that the probability of the part being accepted varies from 0 to 1 over a small range of sizes. We will also need to assume a suitable form for this function and to derive expressions for the fraction of parts the machine will accept which are (1) the correct size, (2) undersize and (3) oversize. Also derive expressions for (4) the fraction of correct parts rejected, (5) the fraction of oversize parts rejected and (6) the fraction of undersize parts rejected.

5.04 Let P_P be the density function of the manufactured parts, which we will assume to be Gaussian, that is

$$P_P = \frac{e^{-(x - \bar{x}_T)^2 / 2\sigma_T}}{\sigma_T \sqrt{(2\pi)}} \qquad 5.04(1)$$

(See Figure 5.04.)

150

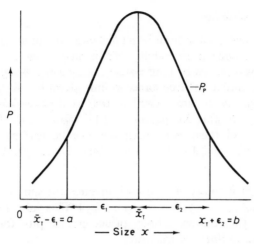

Figure 5.04

Let the sizing machine be set to accept parts between the sizes a and b, $b > a$. If the machine were perfect then the fraction of undersize parts rejected would be $\int_{-\infty}^{a} P_P \, dx$ whilst the fraction of oversize parts rejected would be $\int_{b}^{\infty} P_P \, dx$. **5.05** However, since no sizing machine is perfect, some oversize parts will be accepted as will some undersize parts, and correspondingly some correct parts will be rejected either as oversize or undersize. Now with each limit point a or b at which the sizing machine has been set, there will be an associated distribution function. Let this function be denoted by $f(x)$. At the lower limit point a we write $f(x)$ as $f_1(x - a)$, whilst at b we write it as $f_2(x - b)$. The meaning of this function is as follows. Consider $f_1(x - a)$; if we put $x = a$, then $f(0)$ is the probability that a part of size a would be accepted. As x is increased we should expect the probability of a part being accepted to increase, that is $f_1(x - a) > f(0)$ for $x > a$. Correspondingly if $x < a$, then we should expect $f_1(x - a) < f(0)$. Similarly with $f_2(x - b)$. When $x > b$ we should expect $f_2(x - b) < f(0)$ and vice versa. The probability of rejection at the point x for f_1 is $\{1 - f_1(x - a)\}$ whilst the probability of rejection by f_2 is $\{1 - f_2(x - b)\}$. The probability to be associated with $x = a$ or b will be discussed later in paragraph 5.07 et seq. In order to size components, parts must first pass through the *upper* sizing limit b and then proceed to the lower sizing limit a. The probability of a part of size x being accepted is given as follows. The probability of *acceptance* at limit b, which means passing *through* the limit point, is $f_2(x - b)$, whilst the probability of *acceptance* at limit a, which means *not* passing through the limit point, is given by $f_1(x - a)$, giving the total probability of acceptance as the product of the probabilities of these two separate events, that is $f_2(x - b)f_1(x - a)$. The probability of rejection is found as follows. There are now two possibilities.

(i) The part is *rejected* at limit point b, that is, it will *not* pass through the limit point, and the probability of this happening is $[1 - f_2(x - b)]$.

(ii) The part is *accepted* by limit point b, and passes *through* to limit point a. The probability of acceptance at limit point b is $f_2(x - b)$. The part is now *rejected* at limit point a, that is it passes *through*, and the probability of this happening is $[1 - f_1(x - a)]$. The compound probability of these two successive events, leading to a rejection, is thus $f_2(x - b)[1 - f_1(x - a)]$, that is the product of the two probabilities.

5.06 The total probability of *rejection* is given as the sum of the probabilities of (i) and (ii), that is $[1 - f_1(x - a)f_2(x - b)]$. This value also follows from the relationship that the probability of acceptance plus the probability of rejection is equal to unity. A fraction

$$\frac{e^{-(x - \bar{x}_T)^2/2\sigma_T^2} \, dx}{\sigma_T \sqrt{(2\pi)}} \qquad 5.06(1)$$

of the manufactured parts lies between x and $x + dx$. Thus the fraction dP_A of these accepted by the sizing machine will be

$$dP_A = \frac{e^{-(x - \bar{x}_T)^2/2\sigma_T^2}}{\sigma_T \sqrt{(2\pi)}} f_1(x - a)f_2(x - b)dx \qquad 5.06(2)$$

Correspondingly the fraction dP_R rejected will be

$$dP_R = \frac{e^{-(x - \bar{x}_T)^2/2\sigma_T^2}}{\sigma_T \sqrt{(2\pi)}} \{1 - f_1(x - a)f_2(x - b)\}dx \qquad 5.06(3)$$

Thus the fraction of parts whose sizes lie between a and b, that is correct size parts which are accepted, will be

$$C_A = \int_a^b dP_A \qquad 5.06(4)$$

The fraction of oversize parts accepted will be

$$O_A = \int_b^\infty dP_A \qquad 5.06(5)$$

whilst the fraction of undersize parts accepted will be

$$U_A = \int_{-\infty}^a dP_A \qquad 5.06(6)$$

Similarly the fraction of correct size parts rejected will be

$$C_R = \int_a^b dP_R \qquad \text{5.06(7)}$$

and the fraction of oversize parts rejected will be

$$O_R = \int_b^\infty dP_R \qquad \text{5.06(8)}$$

whilst the fraction of undersize parts rejected will be

$$U_R = \int_{-\infty}^a dP_R \qquad \text{5.06(9)}$$

Note that

$$C_A + U_A + O_A + C_R + U_R + O_R = 1 \qquad \text{5.06(10)}$$

The form of f(x)

5.07 In order to proceed we must now make some assumptions about the form of $f(x)$. If we consider the lower limit of a sizing machine, we require a function which is small when x is much less than a and which approaches 1 when x is much larger than a. Also when $x = a$ there is a case for assuming that $f(x)$ is 0.5, that is that acceptance or rejection is equi-probable. Further there is some case for considering $f(x)$ to be anti-symmetrical about the limit point, that is, the probability of acceptance is increased when a part is a given amount larger than the value a, by the same amount as its probability of acceptance is decreased, when its size is the same amount smaller than the value a. Also the slope of the function should be linear in the region of the set position of the limit point a, where the probability of acceptance is 0.5. A simple function fulfilling these requirements is

$$f_1(x - a) = \frac{e^{-(x - a - \rho)^2/2\rho^2}}{\tau} \qquad \text{5.07(1)}$$

valid for $-\infty \leqslant x \leqslant a$ and

$$f_1(x - a) = 1 - \frac{e^{-(x - a + \rho)^2/2\rho^2}}{\tau} \qquad \text{5.07(2)}$$

valid for $a \leqslant x \leqslant \infty$ where

$$\tau = 1.213\,0613 \qquad \text{5.07(3)}$$

(See Figure 5.07(1).)

When $x = a$ this function has a value 0.50, when $x = a + \rho$ its value is 0.8885, and when $x = a - \rho$ its value is 0.1115. These latter two values are

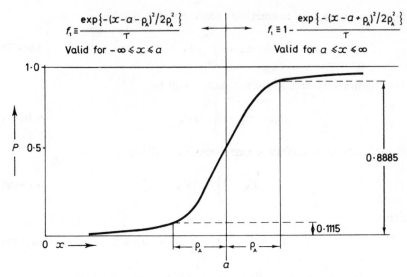

Figure 5.07(1)   Appearance of $f_1(x - a)$ distribution function

thus the probabilities of acceptance at plus and minus one ρ from the limit point. Note in this instance ρ is not a true standard deviation in the defined sense of the word, and it is for this reason that the symbol $\rho\dagger$ has been used rather than σ. For the upper limit, the two functions will be

$$f_2(x - b) = \left\{ 1 - \frac{e^{-(x - b - \rho)^2/2\rho^2}}{\tau} \right\} \qquad 5.07(4)$$

† The distribution function $f_1(x - a)$ has a density function given by $d\{f_1(x - a)\}/dx$, which has a standard deviation given by $\sigma = 0.8298\rho$. A similar relation holds for $f_2(x - b)$. The distribution functions $f_1(x - a)$ and $f_2(x - b)$ are reasonable approximations to the normal distribution as the following table shows:

	P		$1 - P$	
$k\sigma$	Normal	$f(x)$	Normal	$f(x)$
σ	0.6827	0.6909	0.3173	0.3090
2σ	0.9545	0.9517	0.0455	0.0483
3σ	0.9973	0.9963	0.0027	0.0037

where $P \equiv$ Probability that $k\sigma \geqslant x - a \geqslant -k\sigma$
$1 - P \equiv$ Probability that $x - a \geqslant k\sigma$ and $x - a \leqslant -k\sigma$
where $f(x) = f_2\{(b - k\sigma) - b\} - f_2\{(b + k\sigma) - b\} = f_2(-k\sigma) - f_2(k\sigma)$
$= P, k\sigma \geqslant x - b \geqslant -k\sigma$
and $f(x) = f_1\{(a - k\sigma) - a\} - f_1\{(a - k\sigma) - a\} = f_1(k\sigma) - f_1(-k\sigma)$
$= P, k\sigma \geqslant x - a \geqslant -k\sigma$

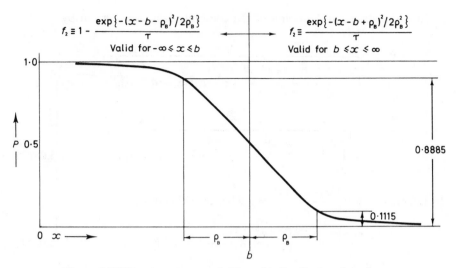

Figure 5.07(2)   Appearance of $f_2(x - b)$ distribution function

valid for $-\infty \leqslant x \leqslant b$ and

$$f_2(x - b) = \frac{e^{-(x - b + \rho)^2/2\rho^2}}{\tau} \qquad \qquad 5.07(5)$$

valid for $b \leqslant x \leqslant \infty$. (See Figure 5.07(2).)

5.08   The derivation of the standard deviation of f_1 is obtained as follows:

$$f_1(x - a) = \frac{e^{-(x - a - \rho)^2/2\rho^2}}{\tau} \qquad \qquad 5.07(1)$$

valid for $-\infty \leqslant x \leqslant a$ and

$$f_1(x - a) = 1 - \frac{e^{-(x - a + \rho)^2/2\rho^2}}{\tau} \qquad \qquad 5.07(2)$$

valid for $a \leqslant x \leqslant \infty$.

Since $f_1(x - a)$ is a distribution function, its density function is found by differentiation. Whence

$$\frac{df_1}{dx} = \frac{-e^{-(x - a - \rho)^2/2\rho^2}}{\tau\rho^2} \cdot (x - a - \rho) \qquad \qquad 5.08(1)$$

valid for $-\infty \leqslant x \leqslant a$ and

$$\frac{df_1}{dx} = \frac{e^{-(x - a + \rho)^2/2\rho^2}}{\tau\rho^2} \cdot (x - a + \rho) \qquad \qquad 5.08(2)$$

valid for $a \leqslant x \leqslant \infty$.

Thus the standard deviation of the complete function is given by

$$
\sigma^2 = \overbrace{\int_{-\infty}^{a} \frac{-e^{-(x-a-p)^2/2\rho^2}}{\tau\rho^2} \cdot (x-a)^2(x-a-\rho)\,dx}^{\textcircled{1}}
$$

$$
+ \overbrace{\int_{a}^{\infty} \frac{e^{-(x-a+\rho)^2/2\rho^2}}{\tau\rho^2} \cdot (x-a)^2(x-a+\rho)\,dx}^{\textcircled{2}} \qquad 5.08(3)
$$

In integral ① put $\dfrac{(x-a-\rho)}{\rho} = y$ whence ① becomes

$$
-\int_{-\infty}^{-1} \frac{e^{-y^2/2}}{\tau}(y+1)^2 y\rho^2\,dy \qquad 5.08(4)
$$

In integral ② put $\dfrac{(x-a+\rho)}{\rho} = z$ whence ② becomes

$$
\int_{1}^{\infty} e^{-z^2/2}(z-1)^2 z\rho^2\,dz \qquad 5.08(5)
$$

If we now put $y = -V$ in ①, i.e. 5.08(4), we get

$$
-\int_{\infty}^{1} \frac{e^{-V^2/2}}{\tau}(V-1)^2 V\rho^2\,dV = \int_{1}^{\infty} \frac{e^{-V^2/2}}{\tau}(V-1)^2 V\rho^2\,dV \quad 5.08(6)
$$

The total integral is thus

$$
\frac{2}{\tau}\int_{1}^{\infty} e^{-V^2/2}(V^2-1)V\rho^2\,dV = I = \sigma^2 \qquad 5.08(7)
$$

Integrating by parts we have

$$
I = \frac{-2\rho^2}{\tau}\int_{1}^{\infty}(V^2-1)d(e^{-V^2/2})
$$

$$
= \frac{-2\rho^2}{\tau}\left[(V-1)^2 e^{-V^2/2}\right]_{1}^{\infty} + \frac{4\rho^2}{\tau}\int_{1}^{\infty} e^{-V^2/2}(V-1)\,dV
$$

$$
= \frac{-2\rho^2}{\tau}\left[(V-1)^2 e^{-V^2/2}\right]_{1}^{\infty} - \frac{4\rho^2}{\tau}\int_{1}^{\infty} d(e^{-V^2/2}) - \frac{4\rho^2}{\tau}\int_{1}^{\infty} e^{-V^2/2}\,dV
$$

$$
= \frac{-2\rho^2}{\tau}\left[(V-1)^2 e^{-V^2/2}\right]_{1}^{\infty} - \frac{4\rho^2}{\tau}\left[e^{-V^2/2}\right]_{1}^{\infty} - \frac{4\rho^2}{\tau}\int_{1}^{\infty} e^{-V^2/2}\,dV
$$

$$
\qquad\qquad \downarrow
$$

$$
= \qquad 0 \qquad\qquad + \frac{4\rho^2}{\tau}e^{-1/2} \qquad - \frac{4\rho^2}{\tau}\int_{1}^{\infty} e^{-V^2/2}\,dV
$$

Considering the last integral we have

$$\int_1^\infty = \frac{1}{2}\left[\int_1^\infty + \int_{-\infty}^{-1}\right] = \frac{1}{2}\left[\int_1^\infty - \int_{-1}^{-\infty}\right]$$

since $e^{-V^2/2}$ is a symmetric function and

$$\int_{-\infty}^{-1} + \int_{-1}^{+1} + \int_{+1}^\infty = \int_{-\infty}^\infty$$

whence

$$\int_1^\infty = \frac{1}{2}\left[\int_{-\infty}^\infty - \int_{-1}^{+1}\right]$$

Thus $\int_1^\infty e^{-V^2/2} dV$ can be found from Table II of Appendix I.
Evaluating I gives

$$\sigma = 0.8298\rho \qquad\qquad 5.08(8)$$

or

$$\rho = 1.205\,08\sigma \qquad\qquad 5.08(9)$$

Using f_2 gives a similar value for σ.

δ_1 and δ_2 positive. Lower limit measuring system

5.09 We now have the problem of finding ρ for each limit point, whilst a and b are the sizes between which the machine is required to accept parts. ρ is found as follows. Take a measured part of known size, say a, and set the sizing machine to reject parts less than a using only the lower limit measuring system. But suppose that the setting dial of the lower limit measuring system reads in error, and that when it is set to the size a it is really set to reject parts less than $a' = a + \delta_1$, where δ_1 is a small positive number. Since a is less than a', function 5.07(1) is appropriate, and its index is now $x - a' - \rho$. Thus when $x = a$, the index becomes $-\delta_1 - \rho$, and the fraction of parts of size a accepted will be less than 0.5. The part of known size a is now put through the machine a large number of times, say 100, using only the lower limit measuring system and a note made of the number of rejections and acceptances. The machine is now set to reject parts of size $a + \delta_2$ by adjusting the dial of the lower limit measuring system. The machine will now in fact be set to reject parts less than $a + \delta_1 + \delta_2$, and when the part of size a is put through the machine again, the index of the probability function for the part will now be $-\delta_1 - \delta_2 - \rho$, and the number of parts accepted will be less than that for the first set of measurements. Let N_1 be the number of parts *accepted* for the first throughput of the part and N_2 the number of parts *accepted* for the second throughput of the part, where the part is put through the machine N times in each case, thus giving

$$\frac{e^{-(-\delta_1 - \rho)^2/2\rho^2}}{\tau} = \frac{N_1}{N} \qquad\qquad 5.09(1)$$

for the first throughput of the part, and

$$\frac{e^{-(-\delta_1 - \delta_2 - \rho)^2/2\rho^2}}{\tau} = \frac{N_2}{N} \qquad\qquad 5.09(2)$$

for the second throughput of the part. Taking logarithms of both sides to the base e, we have

$$\delta_1 + \rho = + \left\{ 2 \log_e \frac{N}{\tau N_1} \right\}^{1/2} \qquad\qquad 5.09(3)$$

and

$$\delta_1 + \delta_2 + \rho = + \left\{ 2 \log_e \frac{N}{\tau N_2} \right\}^{1/2} \qquad\qquad 5.09(4)$$

and substituting for $\delta_1 + \rho$ from the first equation into the second one, we obtain ρ as

$$\rho = \delta_2 \bigg/ \left[\left\{ 2 \log_e \frac{N}{\tau N_2} \right\}^{1/2} - \left\{ 2 \log_e \frac{N}{\tau N_1} \right\}^{1/2} \right] \qquad\qquad 5.09(5)$$

δ_1 if required is given by

$$\delta_1 = \rho \left[\left\{ 2 \log_e \frac{N}{\tau N_1} \right\}^{1/2} - 1 \right] \qquad\qquad 5.09(6)$$

from 5.09(3) and where ρ is given by 5.08(8). Note that the positive root has been taken in equations 5.09(3) and 5.09(4) because δ_1 and δ_2 have been assumed to be positive, as is ρ, and that δ_2 is known, i.e. set into the machine.

δ_1 and δ_2 negative. Lower limit measuring system

5.10   If N_1/N is less than 0.5 then it means that δ_1 is positive (as assumed in paragraph 5.09), which thus means that the fraction of parts of size a accepted will be less than 0.5. If however N_1/N is greater than 0.5, then it means that δ_1 is negative and that the fraction of parts of size a accepted will be greater than 0.5. In this case a is greater than a' and we write $a' = a - |\delta_1|$. The appropriate function will now be 5.07(2). The index of the function will thus be $x - a' + \rho$, and substituting for a' we have the index as $x - a + |\delta_1| + \rho$. When $x = a$, the index becomes $|\delta_1| + \rho$, where $|\delta_1|$ is the modulus of δ_1. After putting the part of size a through the machine N times and using 5.07(2), we have

$$1 - \frac{e^{-(|\delta_1| + \rho)^2/2\rho^2}}{\tau} = \frac{N_1}{N} \qquad\qquad 5.10(1)$$

The lower limit measuring system is now set to accept parts δ_2 less than a, i.e. δ_2 is negative. The machine is now set to accept parts of size

$$a' = a - |\delta_1| - |\delta_2|$$

The index is now $x - a + |\delta_1| + |\delta_2| + \rho$. Thus when $x = a$ the index becomes $|\delta_1| + |\delta_2| + \rho$ and the fraction of parts accepted becomes

$$1 - \frac{e^{-(|\delta_1| + |\delta_2| + \rho)^2/2\rho^2}}{\tau} = \frac{N_2}{N} \qquad 5.10(2)$$

Thus we have

$$|\delta_1| + \rho = \rho\left[2 \log_e\left(\frac{1}{1 - \tau\, N_1/N}\right)\right]^{1/2} \qquad 5.10(3)$$

and

$$|\delta_1| + |\delta_2| + \rho = \rho\left[2 \log_e\left(\frac{1}{1 - \tau\, N_2/N}\right)\right]^{1/2} \qquad 5.10(4)$$

Subtracting 5.10(3) from 5.10(4) gives

$$\rho = \delta_2 \Big/ \left[\left(2 \log_e\left(\frac{1}{1 - \tau\, N_2/N}\right)\right)^{1/2} - \left(2 \log_e\left(\frac{1}{1 - \tau\, N_1/N}\right)\right)^{1/2}\right]$$

$$5.10(5)$$

5.11   Obtaining δ_1 enables the machine to be calibrated, i.e. if δ_1 is positive, it means that the machine is set to accept parts of size $a + \delta_1$. Thus to accept parts of size a, the setting device should be made to read δ_1 less. Conversely if δ_1 is negative it means the machine is set to accept parts of size $a - |\delta_1|$. Thus to accept parts of size a the setting device should be made to read $|\delta_1|$ more. A similar procedure will find ρ and δ_1 for the upper limit measuring system.

δ_1 and δ_2 positive. Upper limit measuring system

5.12   Dealing now with the second or upper limit measuring system, we see that the distribution function for accepted parts is the mirror image of the first system. As before we assume an error in the setting of the measuring system, so that when set to reject parts greater than b it is really set to reject parts less than $b' = b + \delta_1$, where δ_1 is a small positive number. Since b is less than b', the appropriate function is 5.07(4). The index is now $x - b' - \rho = x - b - \delta_1 - \rho$. Thus when $x = b$ the index becomes $-\delta_1 - \rho$ and the fraction of parts accepted will be *greater* than 0.5. The part of size b is now put through the upper limit measuring system, say one hundred times, and a note made of the number of rejections and acceptances.

5.13   The machine is now set to reject parts of size $b + \delta_2$ by adjusting the dial of the upper limit measuring system. The machine will now in fact be set to reject parts greater than $b + \delta_1 + \delta_2 = b'$. When the part of size b is put through the machine the index will be $x - b' - \rho = x - b - \delta_1 - \delta_2 - \rho$. Thus when $x = b$ the index becomes $-\delta_1 - \delta_2 - \rho$ and the number of parts of size b will show an increase on the first set of measurements. As for the lower limit measuring system, let N_1 be the number of parts accepted for the first throughput, N_2 the number of parts accepted for the second throughput and N the total number of parts for each throughput of parts. For the first throughput of parts, using 5.07(4), we have

$$1 - \frac{e^{-(-\delta_1 - \rho)^2/2\rho^2}}{\tau} = \frac{N_1}{N} \qquad\qquad 5.13(1)$$

For the second throughput of parts, again using 5.07(4), we have

$$1 - \frac{e^{-(-\delta_1 - \delta_2 - \rho)^2/2\rho^2}}{\tau} = \frac{N_2}{N} \qquad\qquad 5.13(2)$$

Rearranging and taking logarithms of both sides to the base e, we have

$$\delta_1 + \rho = \rho\left[2\log_e\left(\frac{1}{1 - \tau N_1/N}\right)\right]^{1/2} \qquad\qquad 5.13(3)$$

and

$$\delta_1 + \delta_2 + \rho = \rho\left[2\log_e\left(\frac{1}{1 - \tau N_2/N}\right)\right]^{1/2} \qquad\qquad 5.13(4)$$

Thus subtracting 5.13(3) from 5.13(4) and transposing we have

$$\rho = \delta_2 \bigg/ \left[\left(2\log_e\left(\frac{1}{1 - \tau N_2/N_1}\right)\right)^{1/2} - \left(2\log_e\left(\frac{1}{1 - \tau N_1/N}\right)\right)^{1/2}\right]$$
$$5.13(5)$$

δ_1 if required is given by

$$\delta_1 = \rho\left(\left[2\log_e\left(\frac{1}{1 - \tau N_1/N}\right)\right]^{1/2} - 1\right) \qquad\qquad 5.13(6)$$

δ_1 and δ_2 negative. Upper limit measuring system

5.14   If N_1/N is greater than 0.5, then δ_1 is positive as assumed in paragraph 5.13. If however N_1/N is less than 0.5, then it means that δ_1 is negative. Thus when the machine is set to reject parts greater than b it is really set to reject parts greater than $b - |\delta_1| = b'$. Since b is greater than b', the appropriate function is 5.07(5).

The index is now $x - b' + \rho = x - b + |\delta_1| + \rho$. Thus when $x = b$, the index becomes $|\delta_1| + \rho$ and the fraction of parts accepted will be *less* than 0.5. The part of size b is now put through the upper limit measuring system, say one hundred times, and a note made of the number of rejections and acceptances.

5.15 The machine is now set to reject parts of size $b - |\delta_2|$ where δ_2 is negative, by adjusting the dial of the upper limit measuring system. The machine will now in fact be set to reject parts greater than $b - |\delta_1| - |\delta_2| = b'$. Since b is greater than b', the appropriate function is again 5.07(5). The index is now $x - b' + \rho = x - b + |\delta_1| + |\delta_2| + \rho$. Thus when $x = b$, the index becomes $|\delta_1| + |\delta_2| + \rho$ and the fraction of parts accepted will be even less than that given in paragraph 5.14 and less than 0.5. The part b is now put through the upper limit measuring system, say one hundred times, and a note made of the number of rejections and acceptances. Once again let N_1 be the number of parts accepted for the first throughput, N_2 the number of parts for the second and N the number of parts put through for each throughput. For the first throughput of parts using 5.07(5) we have

$$\frac{e^{-(|\delta_1| + \rho)^2/2\rho^2}}{\tau} = \frac{N_1}{N} \qquad 5.15(1)$$

For the second throughput of parts

$$\frac{e^{-(|\delta_1| + |\delta_2| + \rho)^2/2\rho^2}}{\tau} = \frac{N_2}{N} \qquad 5.15(2)$$

Rearranging and taking logarithms of both sides to the base e, we have

$$|\delta_1| + \rho = \rho \left[2 \log_e \frac{N}{\tau N_1} \right]^{1/2} \qquad 5.15(3)$$

and

$$|\delta_1| + |\delta_2| + \rho = \rho \left[2 \log_e \frac{N}{\tau N_2} \right]^{1/2} \qquad 5.15(4)$$

Subtracting 5.15(3) from 5.15(4) we have

$$\rho = \delta_2 \left/ \left\{ \left(2 \log_e \frac{N}{\tau N_2} \right)^{1/2} - \left(2 \log_e \frac{N}{\tau N_1} \right)^{1/2} \right\} \right. \qquad 5.15(5)$$

δ_1 if required is given by

$$|\delta_1| = \rho \left[\left(2 \log_e \frac{N}{N_1} \right)^{1/2} - 1 \right] \qquad 5.15(6)$$

Once again δ_1 can be used to calibrate the machine.

5.16 It should be noted that for the lower limit measuring system if N_1/N is less than 0.5, then δ_1 is positive. If N_1/N is greater than 0.5 then δ_1 is negative. For the upper limit measuring system, if N_1/N is greater than 0.5 then δ_1 is positive. If however N_1/N is less than 0.5, then δ_1 is negative. The appropriate formulae should be used according to the sign of δ_1.

5.17 Returning to equation 5.06(2) and inserting f_1 and f_2† into it we have the fraction of parts accepted as the integrals of the following:

$$dU_A = \frac{e^{-(x-\bar{x}_T)^2/2\sigma_T^2}}{\sigma_T\sqrt{(2\pi)}} \cdot \frac{e^{-(x-a-\rho_A)^2/2\rho_A^2}}{\tau}$$

$$\cdot \left\{1 - e^{-(x-b-\rho_B)^2/2\rho_B^2}\right\} \qquad 5.17(1)$$

valid for $-\infty \leqslant x \leqslant a$, using 5.06(2), 5.07(1) and 5.07(4)

$$dC_A = \frac{e^{-(x-\bar{x}_T)^2/2\sigma_T^2}}{\sigma_T\sqrt{(2\pi)}} \left\{1 - \frac{e^{-(x-a+\rho_A)^2/2\rho_A^2}}{\tau}\right\} \left\{1 - \frac{e^{-(x-b-\rho_B)^2/2\rho_B^2}}{\tau}\right\}dx$$

$$5.17(2)$$

valid for $a \leqslant x \leqslant b$ using 5.06(2), 5.07(2) and 5.07(4) and

$$dO_A = \frac{e^{-(x-\bar{x}_T)^2/2\sigma_T^2}}{\sigma_T\sqrt{(2\pi)}} \left\{1 - \frac{e^{-(x-a+\rho_A)^2/2\rho_A^2}}{\tau}\right\} \frac{e^{-(x-b+\rho_B)^2/2\rho_B^2}}{\tau} dx \quad 5.17(3)$$

valid for $b \leqslant x \leqslant \infty$ using 5.06(2), 5.07(2) and 5.07(5).

The fraction of parts rejected is obtained as the integral of the expressions for dP_R which are obtained by substituting for f_1 and f_2 in 5.06(3), and which are valid for the same ranges as the expressions for dP_A above

$$dU_R = \frac{e^{-(x-\bar{x}_T)^2/2\sigma_T^2}}{\sigma_T\sqrt{(2\pi)}} \left\{1 - \frac{e^{-(x-a-\rho_A)^2/2\rho_A^2}}{\tau}\left(1 - \frac{e^{-(x-b-\rho_B)^2/2\rho_B^2}}{\tau}\right)\right\}dx$$

$$5.17(4)$$

valid for $-\infty \leqslant x \leqslant a$ using 5.06(3), 5.07(1) and 5.07(4)

$$dC_R = \frac{e^{-(x-\bar{x}_T)^2/2\sigma_T^2}}{\sigma_T\sqrt{(2\pi)}} \left\{1 - \left(1 - \frac{e^{-(x-a+\rho_A)^2/2\rho_A^2}}{\tau}\right)\left(1 - \frac{e^{-(x-b-\rho_B)^2/2\rho_B^2}}{\tau}\right)\right\}dx$$

$$5.17(5)$$

valid for $a \leqslant x \leqslant b$ using 5.06(3), 5.07(2) and 5.07(4) and

$$dO_R = \frac{e^{-(x-\bar{x}_T)^2/2\sigma_T^2}}{\sigma_T\sqrt{(2\pi)}} \left\{1 - \left(1 - \frac{e^{-(x-a+\rho_A)^2/2\rho_A^2}}{\tau}\right)\frac{e^{-(x-b+\rho_B)^2/2\rho_B^2}}{\tau}\right\}dx$$

$$5.17(6)$$

valid for $b \leqslant x \leqslant \infty$ using 5.06(3), 5.07(2) and 5.07(5).

† The ρ of f_1 and f_2 have been given suffixes because they may not be equal. The suffix for f_1 is A and that for f_2 is B.

The number of correct parts accepted is given by equation 5.06(4), taken between the limits a and b. This is the integral of equation 5.17(2). Thus

$$C_A = \int_a^b dP_A = \int_a^b \frac{e^{-(x-\bar{x}_T)^2/2\sigma_T^2}}{\sigma_T\sqrt{(2\pi)}}\left\{1 - \frac{e^{-(x-a+\rho_A)^2/2\rho_A^2}}{\tau}\right\}\left\{1 - \frac{e^{-(x-b-\rho_B)^2/2\rho_B^2}}{\tau}\right\}dx$$

5.17(7)

This integral contains four parts and can be written

$$\int_a^b \frac{e^{-\alpha_0(x-\bar{x}_T)^2}dx}{\sigma_T\sqrt{(2\pi)}} + e^{-\gamma_3^{-+}}\int_a^b \frac{e^{-\alpha_3(x-\beta_3^{-+})^2}dx}{\sigma_T\tau^2\sqrt{(2\pi)}}$$

$$- e^{-\gamma_1^-}\int_a^b \frac{e^{-\alpha_1(x-\beta_1^-)^2}dx}{\sigma_T\tau\sqrt{(2\pi)}} - e^{-\gamma_2^+}\int_a^b \frac{e^{-\alpha_2(x-\beta_2^+)^2}dx}{\sigma_T\tau\sqrt{(2\pi)}}$$

$$= \frac{1}{\sigma_T\sqrt{(2\alpha_0)}}\int_{\sqrt{(2\alpha_0)}(a-\bar{x}_T)}^{\sqrt{(2\alpha_0)}(b-\bar{x}_T)} \frac{e^{-c^2/2}dc}{\sqrt{(2\pi)}} + \frac{e^{-\gamma_3^{-+}}}{\sigma_T\tau^2\sqrt{(2\alpha_3)}}\int_{\sqrt{(2\alpha_3)}(a-\beta_3^{-+})}^{\sqrt{(2\alpha_3)}(b-\beta_3^{-+})} \frac{e^{-c^2/2}dc}{\sqrt{(2\pi)}}$$

$$- \frac{e^{-\gamma_1^-}}{\sigma_T\tau\sqrt{(2\alpha_1)}}\int_{\sqrt{(2\alpha_1)}(a-\beta_1^-)}^{\sqrt{(2\alpha_1)}(b-\beta_1^-)} \frac{e^{-c^2/2}dc}{\sqrt{(2\pi)}}$$

$$- \frac{e^{-\gamma_2^+}}{\sigma_T\tau\sqrt{(2\alpha_2)}}\int_{\sqrt{(2\alpha_2)}(a-\beta_2^+)}^{\sqrt{(2\alpha_2)}(b-\beta_2^+)} \frac{e^{-c^2/2}dc}{\sqrt{(2\pi)}}$$

5.17(8)

where the new symbols used are given at the end of this problem. Bearing in mind the terminology used for Table II of Appendix I and introducing new symbols for the limits (see the end of this problem) we have

$$C_A = \frac{1}{2}\left\{P_{-_bK_0\,\text{to}\,_bK_0} - P_{-_aK_0\,\text{to}\,_aK_0}\right\}$$

$$+ \frac{1}{2}\frac{e^{-\gamma_3^{-+}}}{\sigma_T\tau^2\sqrt{(2\alpha_3)}}\left\{P_{-_bK_3^{-+}\,\text{to}\,_bK_3^{-+}} - P_{-_aK_3^{-+}\,\text{to}\,_aK_3^{-+}}\right\}$$

$$- \frac{-e^{-\gamma_1^-}}{2\sigma_T\tau\sqrt{(2\alpha_1)}}\left\{P_{-_bK_1^-\,\text{to}\,_bK_1^-} - P_{-_aK_1^-\,\text{to}\,_aK_1^-}\right\}$$

$$- \frac{e^{-\gamma_2^+}}{2\sigma_T\tau\sqrt{(2\alpha_2)}}\left\{P_{-_bK_2^+\,\text{to}\,_bK_2^+} - P_{-_aK_2^+\,\text{to}\,_aK_2^+}\right\}$$

5.17(9)

where the P are obtained from Table II of Appendix I. Similar expressions can be obtained for the other fractions, and simplifying a little further, we have:

5.18 1 C_A, the fraction of correct size parts with dimensions between a and b which are accepted, is given by

$$C_A = \frac{1}{2}\left\{ \underset{-_bK_0 \text{ to } _bK_0}{P} - \underset{-_aK_0 \text{ to } _aK_0}{P} \right\}$$

$$+ A_3^{-+}\left\{ \underset{-_bK_3^{-+} \text{ to } _bK_3^{-+}}{P} - \underset{-_aK_3^{-+} \text{ to } _aK_3^{-+}}{P} \right\}$$

$$- A_1^{-}\left\{ \underset{-_bK_1^{-} \text{ to } _bK_1^{-}}{P} - \underset{-_aK_1^{-} \text{ to } _aK_1^{-}}{P} \right\}$$

$$- A_2^{+}\left\{ \underset{-_bK_2^{+} \text{ to } _bK_2^{+}}{P} - \underset{-_aK_2^{+} \text{ to } _aK_2^{+}}{P} \right\} \qquad 5.18(1)$$

2 U_A, the fraction of undersize parts accepted, that is parts less than a, is given by

$$U_A = A_1^{+}\left\{ \underset{-_aK_1^{+} \text{ to } _aK_1^{+}}{P} + 1 \right\} - A_3^{++}\left\{ \underset{-_aK_3^{+} \text{ to } _aK_3^{+}}{P} + 1 \right\} \qquad 5.18(2)$$

3 O_A, the fraction of oversize parts accepted, that is parts greater than b, is given by

$$O_A = A_2^{-}\left\{ 1 - \underset{-_bK_2^{-} \text{ to } _bK_2^{-}}{P} \right\} - A_3^{--}\left\{ 1 - \underset{-_bK_3^{--} \text{ to } _bK_3^{--}}{P} \right\} \qquad 5.18(3)$$

4 C_R, the fraction of correct size parts rejected, is given by

$$C_R = \frac{1}{2}\left\{ \underset{-_bK_0 \text{ to } _bK_0}{P} - \underset{-_aK_0 \text{ to } _aK_0}{P} \right\} - C_A\dagger$$

$$= A_1^{-}\left\{ \underset{-_bK_1^{-} \text{ to } _bK_1^{-}}{P} - \underset{-_aK_1^{-} \text{ to } _aK_1^{-}}{P} \right\}$$

$$+ A_2^{+}\left\{ \underset{-_bK_2^{+} \text{ to } _bK_2^{+}}{P} - \underset{-_aK_2^{+} \text{ to } _aK_2^{+}}{P} \right\}$$

$$- A_3^{-+}\left\{ \underset{-_bK_3^{-+} \text{ to } _bK_3^{-+}}{P} - \underset{-_aK_3^{-+} \text{ to } _aK_3^{-+}}{P} \right\} \qquad 5.18(4)$$

† Since adding 5.06(2) and 5.06(3) and integrating between a and b gives the first term of the expression, i.e. the total of correct parts

$$C_A + C_R = \int_a^b \frac{e^{-(x-x_T)^2/2\sigma_T^2} dx}{\sigma_T\sqrt{(2\pi)}} = \frac{1}{2}\left[\int_{-b}^b - \int_{-a}^a \right] = \frac{1}{2}\left[\underset{-_bK_0 \text{ to } _bK_0}{P} - \underset{-_aK_0 \text{ to } _aK_0}{P} \right]$$

5 U_R, the fraction of undersize parts rejected, is given by

$$U_R = \frac{1}{2}\left\{1 + \underset{-_aK_0 \text{ to } _aK_0}{P}\right\} - U_A$$

$$= \frac{1}{2}\left\{1 + \underset{-_aK_0 \text{ to } _aK_0}{P}\right\} - A_1^+\left\{\underset{-_aK_1^+ \text{ to } _aK_1^+}{P} + 1\right\}$$

$$+ A_3^{++}\left\{\underset{-_aK_3^{++} \text{ to } _aK_3^{++}}{P} + 1\right\} \qquad 5.18(5)$$

Since $U_R + U_A = U$ and

$$U = \int_{-\infty}^{a} e^{-(x - x_T)^2/2\sigma_T^2}/\sigma_T\sqrt{(2\pi)}$$

$$U = \int_{-\infty}^{a} = \int_{-\infty}^{0} + \int_{0}^{a} = \frac{1}{2}\left[\int_{-\infty}^{\infty} + \int_{-a}^{a}\right]$$

since the integrand is an even function and thus

$$U = \frac{1}{2}\left[1 + \underset{-_aK_0 \text{ to } _aK_0}{P}\right]$$

whence U_R using U_A and U is obtained by adding 5.06(2) and 5.06(3) and integrating between $-\infty$ and a.

6 O_R, the fraction of oversize parts rejected, is given by

$$O_R = \frac{1}{2}\left\{1 - \underset{-_bK_0 \text{ to } _bK_0}{P}\right\} - O_A$$

$$= \frac{1}{2}\left\{1 - \underset{-_bK_0 \text{ to } _bK_0}{P}\right\} - A_2^-\left\{1 - \underset{-_bK_2^- \text{ to } _bK_2^-}{P}\right\}$$

$$+ A_3^{--}\left\{1 - \underset{-_bK_3^{--} \text{ to } _bK_3^{--}}{P}\right\} \qquad 5.18(6)$$

obtained using $O_R + O_A = O$ where

$$O = \int_{b}^{\infty} e^{-(x - x_T)^2/2\sigma_T^2}/\sigma_T\sqrt{(2\pi)}$$

$$= \frac{1}{2}\left[1 - \underset{-_bK_0 \text{ to } _bK_0}{P}\right].$$

The values of P can be found from Table II of Appendix 1. The constants involved are given below

Explanation of contants used in foregoing equations

5.19

$$_aK_0 = \frac{a - \bar{x}_T}{\sigma_T} \qquad\qquad _bK_0 = \frac{b - \bar{x}_T}{\sigma_T}\dagger$$

$$_aK_1^+ = \sqrt{(2\alpha_1)}(a - \beta_1^+) \qquad _aK_1^- = \sqrt{(2\alpha_1)}(a - \beta_1^-)$$

$$_bK_1^+ = \sqrt{(2\alpha_1)}(b - \beta_1^+) \qquad _bK_1^- = \sqrt{(2\alpha_1)}(b - \beta_1^-)$$

$$_aK_2^+ = \sqrt{(2\alpha_2)}(a - \beta_2^+) \qquad _aK_2^- = \sqrt{(2\alpha_2)}(a - \beta_2^-)$$

$$_bK_2^+ = \sqrt{(2\alpha_2)}(b - \beta_2^+) \qquad _bK_2^- = \sqrt{(2\alpha_2)}(b - \beta_2^-)$$

$$_aK_3^{++} = \sqrt{(2\alpha_3)}(a - \beta_3^{++}) \quad _aK_3^{--} = \sqrt{(2\alpha_3)}(a - \beta_3^{--})$$

$$_aK_3^{-+} = \sqrt{(2\alpha_3)}(a - \beta_3^{-+})$$

$$_bK_3^{++} = \sqrt{(2\alpha_3)}(b - \beta_3^{++}) \quad _bK_3^{--} = \sqrt{(2\alpha_3)}(b - \beta_3^{--})$$

$$_bK_3^{-+} = \sqrt{(2\alpha_3)}(b - \beta_3^{-+})$$

$$\alpha_0 = \frac{1}{2\sigma_T^2} \qquad\qquad \alpha_1 = \frac{1}{2\sigma_T^2} + \frac{1}{2\rho_A^2}$$

$$\alpha_2 = \frac{1}{2\sigma_T^2} + \frac{1}{2\rho_B^2} \qquad \alpha_3 = \frac{1}{2\sigma_T^2} + \frac{1}{2\rho_A^2} + \frac{1}{2\rho_B^2}$$

$$\beta_1^+ = \left(\frac{\bar{x}_T}{2\sigma_T^2} + \frac{a + \rho_A}{2\rho_A^2}\right)\Big/ \alpha_1 \quad \beta_1^- = \left(\frac{\bar{x}_T}{2\sigma_T^2} + \frac{a - \rho_A}{2\rho_A^2}\right)\Big/ \alpha_1$$

$$\beta_2^+ = \left(\frac{\bar{x}_T}{2\sigma_T^2} + \frac{b + \rho_B}{2\rho_B^2}\right)\Big/ \alpha_2 \quad \beta_2^- = \left(\frac{\bar{x}_T}{2\sigma_T^2} + \frac{b - \rho_B}{2\rho_B^2}\right)\Big/ \alpha_2$$

$$\beta_3^{++} = \left(\frac{\bar{x}_T}{2\sigma_T^2} + \frac{a + \rho_A}{2\rho_A^2} + \frac{b + \rho_B}{2\rho_B^2}\right)\Big/ \alpha_3$$

$$\beta_3^{--} = \left(\frac{\bar{x}_T}{2\sigma_T^2} + \frac{a - \rho_A}{2\rho_A^2} + \frac{b - \rho_B}{2\rho_B^2}\right)\Big/ \alpha_3$$

$$\beta_3^{-+} = \left(\frac{\bar{x}_T}{2\sigma_T^2} + \frac{a - \rho_A}{2\rho_A^2} + \frac{b + \rho_B}{2\rho_B^2}\right)\Big/ \alpha_3$$

$$\gamma_1^+ = \left(\frac{(a + \rho_A - \bar{x}_T)^2}{4\sigma_T^2 \rho_A^2 \alpha_1}\right) \qquad \gamma_1^- = \frac{(a - \rho_A - \bar{x}_T)^2}{4\sigma_T^2 \rho_A^2 \alpha_1}$$

$$\gamma_2^+ = \frac{(b + \rho_B - \bar{x}_T)^2}{4\sigma_T^2 \rho_B^2 \alpha_2} \qquad \gamma_2^- = \frac{(b - \rho_B - \bar{x}_T)^2}{4\sigma_T^2 \rho_B^2 \alpha_2}$$

† The K are the limit points for the probability P between $-K$ to $+K$.

$$\gamma_3^{++} = \left(\frac{(a + \rho_A - \bar{x}_T)^2}{4\sigma_T^2 \rho_A^2} + \frac{(b + \rho_B - \bar{x}_T)^2}{4\sigma_T^2 \rho_B^2} + \frac{\{a + \rho_A - (b + \rho_B)\}^2}{4\rho_A^2 \rho_B^2} \right) \Bigg/ \alpha_3$$

$$\gamma_3^{--} = \left(\frac{(a - \rho_A - \bar{x}_T)^2}{4\sigma_T^2 \rho_A^2} + \frac{(b - \rho_B - \bar{x}_T)^2}{4\sigma_T^2 \rho_B^2} + \frac{\{a - \rho_A - (b + \rho_B)\}^2}{4\rho_A^2 \rho_B^2} \right) \Bigg/ \alpha_3$$

$$\gamma_3^{-+} = \left(\frac{(a - \rho_A - \bar{x}_T)^2}{4\sigma_T^2 \rho_A^2} + \frac{(b + \rho_B - \bar{x}_T)^2}{4\sigma_T^2 \rho_B^2} + \frac{\{a - \rho_A - (b + \rho_B)\}^2}{4\rho_A^2 \rho_B^2} \right) \Bigg/ \alpha_3$$

$$A_1^+ = \frac{e^{-\gamma_1^+}}{2\sqrt{(2\alpha_1)}\sigma_T \tau} \qquad A_1^- = \frac{e^{-\gamma_1^-}}{2\sqrt{(2\alpha_1)}\sigma_T \tau}$$

$$A_2^+ = \frac{e^{-\gamma_2^+}}{2\sqrt{(2\alpha_2)}\sigma_T \tau} \qquad A_2^- = \frac{e^{-\gamma_2^-}}{2\sqrt{(2\alpha_2)}\sigma_T \tau}$$

$$A_3^{++} = \frac{e^{-\gamma_3^{++}}}{2\sqrt{(2\alpha_3)}\sigma_T \tau^2} \qquad A_3^{--} = \frac{e^{-\gamma_3^{--}}}{2\sqrt{(2\alpha_3)}\sigma_T \tau^2}$$

$$A_3^{-+} = \frac{e^{-\gamma_3^{-+}}}{2\sqrt{(2\alpha_3)}\sigma_T \tau^2}$$

Worked Example

5.20 The standard deviation of some manufactured parts is 0.001 in and the limit points of a sizing machine have been set at 1.998 in and 2.002 in, and the mean of the manufactured parts is 2.00 in. If the ρ_A and ρ_B of the limit points of the sizing machine are 0.0002 in, find the ratio of the fraction of the correct parts accepted to the fraction of correct parts which would have been accepted had the sizing machine been perfect, that is $\rho_A = \rho_B = 0$.

5.21 The fraction of correct parts accepted by the sizing machine is given by 5.18(1). It is also to be noted that the first term of this gives the fraction of parts accepted by a perfect sizing machine. Since the mean of the density distribution of the parts is midway between the tolerances, we have

$$- {}_aK_0 = {}_bK_0 = \frac{0.002}{0.001} = 2.000$$

Thus, using Table II of Appendix I, we have
First term of 5.18(1)

$$\frac{1}{2}\left(\underset{-{}_bK_0 \text{ to } {}_bK_0}{P} - \underset{-{}_aK_0 \text{ to } {}_aK_0}{P} \right) = 0.954\,50$$

Note that the second term in the bracket is negative since ${}_aK_0$ is negative, and thus the second term is added to the first term.

Second term of 5.18(1)
Now

$$_bK_3^{-+} = \sqrt{(2\alpha_3)}(b - \beta_3^{-+})$$

$a = 1.998$, $b = 2.002$.

$$\alpha_3 = \frac{1}{2\sigma_T^2} + \frac{1}{2\rho_A^2} + \frac{1}{2\rho_B^2}$$

therefore $\sqrt{(2\alpha_3)} = 7.141 \times 10^3$. Thus $_bK_3^{-+} = 7.141 \times 10^3(b - \beta_3^{-+}) = 7.141 \times 10^3(2 \times 10^{-3}) = 14.2828$. Likewise $_aK_3^{-+}$ is a number of comparable size but *negative*, and so

$$\left(\frac{P}{_{-bK_3^{-+} \text{ to } _bK_3^{-+}}} - \frac{P}{_{-aK_3^{-+} \text{ to } _aK_3^{-+}}} \right)$$

is approximately equal to 2.0000.

$$A_3^{-+} = \frac{e^{-\gamma_3^{-+}}}{2\sigma_T\tau^2\sqrt{(2\alpha_3)}} = \frac{e^{-121.00}}{2 \times 7.141 \times 10^3 \times 10^{-3} \times (1.213\,06)^2}$$

which is utterly negligible.
Third term of 5.18(1)
The third term is

$$-A_1^- \left(\frac{P}{_{-bK_1^- \text{ to } _bK_1^-}} - \frac{P}{_{-aK_1^- \text{ to } _aK_1^-}} \right)$$

$_bK_1^- = \sqrt{(2\alpha_1)}(b - \beta_1^-)$ and $_aK_1^- = \sqrt{(2\alpha_1)}(a - \beta_1^-)$ giving $_bK_1^- = 20.9845$ and $_aK_1^- = 0.588\,35$.

$$A_1^- = \frac{e^{-\gamma_1^-}}{2\sigma_T\tau\sqrt{(2\alpha_1)}}$$

$$\gamma_1^- = \frac{(a - \rho_A - \bar{x}_T)^2}{4\sigma_T\rho_A^2\alpha_1} = 2.3269$$

$$2\sigma_T\tau\sqrt{(2\alpha_1)} = 12.3708$$

$$A_1^- = \frac{e^{-2.3269}}{12.3708} = \frac{0.097\,59}{12.3708} = 0.007\,88$$

Thus

$$\left(\frac{P}{_{-bK_1^- \text{ to } _bK_1^-}} - \frac{P}{_{-aK_1^- \text{ to } _aK_1^-}} \right) = (1 - 0.4437) = 0.5563$$

Therefore the third term $= -0.004\,383$.

Fourth term of 5.18 (1)

The fourth term is

$$-A_2^+ \left[\underset{-_bK_2^+ \text{ to } _bK_2^+}{P} - \underset{-_aK_2^+ \text{ to } _aK_2^+}{P} \right]$$

Now $_bK_2^+ = \sqrt{(2\alpha_2)}(b - \beta_2^+) = -0.588\,54$ and $_aK_2^+ = \sqrt{(2\alpha_2)}(a - \beta_2^+) = -20.9845$. Therefore

$$\left[\underset{-_bK_2^+ \text{ to } _bK_2^+}{P} - \underset{-_aK_2^+ \text{ to } _aK_2^+}{P} \right] = [-0.4437 + 1] = 0.5563$$

$$A_2^+ = \frac{e^{-\gamma_2^+}}{2\sigma_T \tau \sqrt{(2\alpha_2)}}$$

$$\gamma_2^+ = \frac{(b + \rho_R - \bar{x}_T)^2}{4\sigma_T^2 \rho_B^2 \alpha_2} = 2.3269$$

$$2\sigma_T \tau \sqrt{(2\alpha_2)} = 12.3708$$

Therefore the fourth term $= -0.004\,383$ and the required fraction

$$= 1 - \frac{2 \times 0.004\,383}{0.9545}$$

$$= 0.9908$$

Thus approximately 1 per cent of correct sized parts are rejected or wasted.

5.22 If $\sigma_T \gg \rho_A$ or ρ_B it can be shown that an approximate relationship for C_A, the expression 5.18(1) for the number of correct parts accepted, is given by

$$C_A = \frac{1}{2}\left(\underset{-_bK_0 \text{ to } _bK_0}{P} - \underset{-_aK_0 \text{ to } _aK_0}{P} \right) - A_1^-\left(1 - \underset{-_aK_1^- \text{ to } _aK_1^-}{P} \right)$$

$$- A_2^+\left(\underset{-_bK_2^+ \text{ to } _bK_2^+}{P} + 1 \right)$$

where

$$_bK_0 = \frac{(b - \bar{x}_T)}{\sigma_T} = \frac{\varepsilon_2}{\sigma_T}$$

$$_aK_0 = \frac{(a - \bar{x}_T)}{\sigma_T} = \frac{\varepsilon_1}{\sigma_T}$$

that is $b = \bar{x}_T + \varepsilon_2$ and $a = \bar{x}_T + \varepsilon_1$, and

$$A_1^- = \frac{e^{-\varepsilon_1^2/2\sigma_T^2}}{2\tau(1 + \sigma_T/\rho_A^2)^{1/2}} \qquad A_2^+ = \frac{e^{-\varepsilon_2^2/\sigma_T^2}}{2\tau(1 + \sigma_T^2/\rho_B^2)^{1/2}}$$

whilst

$$_aK_1^- = 1 + \frac{\rho_A \varepsilon_1}{\sigma_T^2} \quad _bK_2^+ = \frac{\rho_B \varepsilon_2}{\sigma_T} - 1$$

Note that usually ε_1 is *negative*.

5.23 It is interesting to note that the fraction of correct size parts rejected can be decreased by setting the sizing machine to the limit points $a' = a - q\rho_A$ and $b' = b + q\rho_B$ where q is a positive number. a and b are still the limiting sizes of the workpiece, and the integrals are integrated between a and b. The constants α, β, and γ now contain the new values of the constants, that is a' and b'. The first term in the second bracket of the Ks is still either a or b because the limits of integration are still the same. The new Ks are denoted by a bar down the left-hand side—thus $|K$, whilst the new As and fractions of parts accepted or rejected are shown primed. The new fractional sizes are given by

$$C_A' = \frac{1}{2}\left(\underset{-_bK_0 \text{ to }_bK_0}{P} - \underset{-_aK_0 \text{ to }_aK_0}{P} \right)$$

$$+ A_3'^{-+}\left(\underset{-|_bK_3^{-+} \text{ to }|_bK_3^{-+}}{P} - \underset{-|_aK_3^{-+} \text{ to }|_aK_3^{-+}}{P} \right)$$

$$- A_1'^{-}\left(\underset{-|_bK_1^- \text{ to }|_aK_1^-}{P} - \underset{-|_aK_1^- \text{ to }|_aK_1^-}{P} \right)$$

$$- A_2'^{+}\left(\underset{-|_bK_2^+ \text{ to }|_bK_2^+}{P} - \underset{-|_aK_2^+ \text{ to }|_aK_2^+}{P} \right) \qquad \text{5.23(1)}$$

$$U_A' = A_1'^{+}\left(\underset{-|_{a'}K_1^+ \text{ to }|_{a'}K_1^+}{P} + 1 \right)$$

$$- A_3'^{++}\left(\underset{-|_{a'}K_3^{++} \text{ to }|_{a'}K_3^{++}}{P} + 1 \right)$$

$$+ \frac{1}{2}\left(\underset{-_aK_0 \text{ to }_aK_0}{P} - \underset{-_{a'}K_0 \text{ to }_{a'}K_0}{P} \right)$$

$$+ A_3'^{-+}\left(\underset{-|_aK_3^{-+} \text{ to }|_aK_3^{-+}}{P} - \underset{-|_{a'}K_3^{-+} \text{ to }|_{a'}K_3^{-+}}{P} \right)$$

$$- A_1'^{-}\left(\underset{-|_aK_1^- \text{ to }|_aK_1^-}{P} - \underset{-|_{a'}K_1^- \text{ to }|_{a'}K_1^-}{P} \right)$$

$$- A_2'^{+}\left(\underset{-|_aK_3^+ \text{ to }|_aK_3^+}{P} - \underset{-|_{a'}K_2 \text{ to }|_{a'}K_2^+}{P} \right) \qquad \text{5.23(2)}$$

$$O'_A = A'^{+}_1 \left(\underset{-|_a K^{+}_1 \text{ to } |_a K^{+}_1}{P} + 1 \right) - A'^{+}_3{}^{+} \left(\underset{-|_a K^{+}_3{}^{+} \text{ to } |_a K^{+}_3{}^{+}}{P} + 1 \right)$$

$$+ \frac{1}{2} \left(\underset{-_{b'} K_0 \text{ to } _{b'} K_0}{P} - \underset{-_b K_0 \text{ to } _b K_0}{P} \right)$$

$$+ A'^{-}_3{}^{+} \left(\underset{-|_{b'} K^{-}_3{}^{+} \text{ to } |_{b'} K^{-}_3{}^{+}}{P} - \underset{-|_b K^{-}_3{}^{+} \text{ to } _b K^{-}_3{}^{+}}{P} \right)$$

$$- A'^{-}_1 \left(\underset{-|_{b'} K^{-}_1 \text{ to } |_{b'} K^{-}_1}{P} - \underset{-|_b K^{-}_1 \text{ to } |_b K^{-}_1}{P} \right)$$

$$- A'^{+}_2 \left(\underset{-|_{b'} K^{+}_2 \text{ to } |_{b'} K^{+}_2}{P} - \underset{-|_b K^{+}_2 \text{ to } |_b K^{+}_2}{P} \right) \qquad \text{5.23(3)}$$

$$C'_R = \frac{1}{2} \left\{ \underset{-_b K_0 \text{ to } _b K_0}{P} - \underset{-_a K_0 \text{ to } _a K_0}{P} \right\} - C'_A \qquad \text{5.23(4)}$$

$$U'_R = \frac{1}{2} \left\{ 1 + \underset{-_a K_0 \text{ to } _a K_0}{P} \right\} - U'_A \qquad \text{5.23(5)}$$

$$O'_R = \frac{1}{2} \left\{ 1 - \underset{-_b K_0 \text{ to } _b K_0}{P} \right\} - O'_A \qquad \text{5.23(6)}$$

5.24 Whilst the fraction of correct size parts rejected is decreased by this procedure, the fraction of oversize and undersize parts accepted is however increased. The fraction of undersize and oversize parts rejected also decreases. Thus it is a 'swings and roundabouts' problem. The fraction of undersize and oversize parts accepted can be reduced at the expense of increasing the fraction of correct size parts rejected. This can be done by setting the sizing machine to the limits $a' = a + q\rho_A$ and $b' = b - q\rho_B$. A new set of equations for the C_A, U_A, O_A, C_R, U_R and O_R is required and this can be obtained in a similar manner to the last set by taking into account the discontinuity of the limit point density function which is discontinuous at a' and b', not at a and b.

5.25 If the distribution function of the limit points of the sizing machine had been Gaussian, then it would have been given by a function of the form

$$F(x) = \int_{-\infty}^{x} \frac{e^{-y^2/2\sigma_L^2} \, dy}{\sigma_L \sqrt{(2\pi)}} \qquad \text{5.25(1)}$$

where σ_L is the standard deviation of the probability distribution of each limit point. The lower limit point distribution function is thus given by $F_1(x - a)$ where

$$F_1(x - a) = \int_{-\infty}^{x-a} \frac{e^{-y^2/2\sigma_L^2} \, dy}{\sigma_L \sqrt{(2\pi)}} \qquad \text{5.25(2)}$$

The upper limit point distribution function $F_2(x - b)$ is given by

$$F_2(x - b) = \int_{x-b}^{\infty} \frac{e^{-z^2/2\sigma_L^2}\, dz}{\sigma_L\sqrt{(2\pi)}} \dagger \qquad 5.25(3)$$

As before the various fractions of parts accepted and rejected will be given by equations 5.06(4) to 5.06(9). For example the number of correct parts accepted between the limits a and b is given by C_A where

$$C_A = \int_a^b \frac{e^{-(x - \bar{x}_T)^2/2\sigma_T^2}\, dx}{\sigma_T\sqrt{(2\pi)}} \int_{-\infty}^{x-a} \frac{e^{-y^2/2\sigma_L^2}\, dy}{\sigma_L\sqrt{(2\pi)}} \int_{x-b}^{\infty} \frac{e^{-z^2/2\sigma_L^2}\, dz}{\sigma_L\sqrt{(2\pi)}} \qquad 5.25(4)$$

The above case when the distribution function of the limit points is Gaussian was not considered in the numerical example because of the difficulty of calculating the above integrals, for which there is no simple analytical solution. However the distribution function used is a very good approximation to a Gaussian function and therefore the answers obtained using it will be a good approximation to that obtained if a Gaussian function were used.

5.26   As before, the number of incorrect parts accepted or the number of correct parts rejected can be reduced by writing in 5.25(4) respectively the upper limit of the y integral as $x - (a + \delta_1)$ and the lower limit of the z integral as $x - (b - \delta_2)$, or writing the upper limit of the y integral as $x - (a - \delta_1)$ and the lower limit of the z integral as $x - (b + \delta_2)$. Usually δ_1 will be taken equal to δ_2, and δ_1, and δ_2 will be of the order of σ_L, their exact magnitudes depending on the results required. If the number of incorrect parts accepted is reduced, then as before the number of correct parts rejected will be increased and vice versa.

† Since the probability F_1 increases as x increases, whilst the probability F_2 increases as x decreases, i.e. F_2 is the mirror image of F_1 reflected in the ordinate axis. Therefore

$$F_2(x - b) = 1 - F_1(x - b)$$

Let

$$\frac{e^{-y^2/2\sigma_L^2}}{\sigma_L\sqrt{(2\pi)}} = f_1$$

then

$$F_2(x - b) = 1 - \int_{-\infty}^{x-b} f_1\, dy = \int_{-\infty}^{+\infty} f_1\, dy - \int_{-\infty}^{x-b} f_1\, dy$$

$$= \int_{-\infty}^{+\infty} f_1\, dy + \int_{x-b}^{-\infty} f_1\, dy$$

$$= \int_{x-b}^{\infty} f_1\, dy$$

as given by 5.25(3).

Sizing Machine with One Measuring Head with Known Uncertainty of Measurement

5.27 The foregoing problem dealt with the question of a sizing machine set to accept parts lying between set limits and to reject parts lying outside these limits. We saw that because the limit points defined by the measuring machine were uncertain within a small range, that this led to the acceptance of oversize and undersize parts and also to the rejection of correct parts, the proportion of which parts was estimated. We shall now consider a very similar problem, where we start with a distribution of manufactured parts which have been measured by a single measuring device which has a known uncertainty of measurement. The difference between the two problems is one of procedure. The first one dealt with used two measuring heads each set to accept or reject parts of a given size. This lends itself to automatic processing of the parts, since the heads will either pass a part or not allow it to pass. The present method uses only one measuring head, the size of each part is measured, and the selection or rejection of each part will generally be done by manual selection. The measuring head could of course be connected electronically to reject or pass gates which could sort the parts as they came off the manufacturing machine. The problem we shall tackle in this example is to estimate the proportion of parts lying between the limits c_1 and c_2 which are accepted and rejected and to estimate the proportion of parts lying in the ranges $-\infty$ to c_1 and c_2 to ∞ which are accepted and rejected, when the acceptance limits are taken as b_1 and b_2. The acceptance limits b_1 and b_2 are usually taken between the required limits c_1 and c_2 in order to allow for the uncertainty in the measuring equipment. Consider the diagram shown (Figure 5.27), where the density function of the manufactured parts is given by $f_p(x - a)$, where a is the mean value, c_1 and c_2 the required limits and b_1 and b_2 the so-called acceptance limits. Let $f_m(z - x)$ be the density function of the measuring or sizing equipment. Thus the fraction of parts lying between

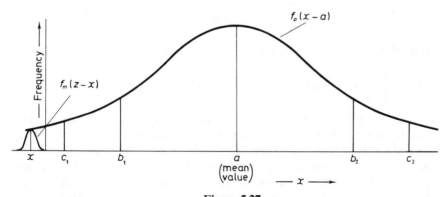

Figure 5.27

x and $x + dx$ is equal to $f_p(x - a)dx$. The fraction of these parts when measured by the sizing machine which appear to lie between b_1 and b_2 is thus equal to

$$f_p(x - a)dx \int_{b_1}^{b_2} f_m(z - x)dz \qquad 5.27(1)$$

Thus the fraction of correct parts lying between c_1 and c_2, which are accepted, when the acceptance limits are set at b_1 and b_2 is equal to

$$C_A = \int_{c_1}^{c_2} f_p(x - a)dx \int_{b_1}^{b_2} f_m(z - x)dz \qquad 5.27(2)$$

Correspondingly the fraction of correct parts lying between c_1 and c_2 which are rejected when the acceptance limits are set at b_1 and b_2 is equal to

$$C_R = \int_{c_1}^{c_2} f_p(x - a)dx \left[\int_{-\infty}^{b_1} f_m(z - x)dz + \int_{b_2}^{\infty} f_m(z - x)dz \right] \qquad 5.27(3)$$

In a similar manner, the fraction of oversize parts, that is parts lying between c_2 and ∞ which are accepted when the acceptance limits are set between b_1 and b_2, is equal to

$$O_A = \int_{c_2}^{\infty} f_p(x - a)dx \int_{b_1}^{b_2} f_m(z - x)dz \qquad 5.27(4)$$

Correspondingly the fraction of oversize parts rejected when the acceptance limits are set between b_1 and b_2 is equal to

$$O_R = \int_{c_2}^{\infty} f_p(x - a)dx \left[\int_{-\infty}^{b_1} f_m(z - x)dz + \int_{b_2}^{\infty} f_m(z - x)dz \right] \qquad 5.27(5)$$

Finally the fraction of undersize parts, that is parts whose size lies between $-\infty$ and c_1 which are accepted when the acceptance limits are set between b_1 and b_2, will be equal to

$$U_A = \int_{-\infty}^{c_1} f_p(x - a)dx \int_{b_2}^{b_2} f_m(z - x)dz \qquad 5.27(6)$$

Similarly the fraction of undersize parts rejected is given by

$$U_R = \int_{-\infty}^{c_1} f_p(x - a)dx \left[\int_{-\infty}^{b_1} f_m(z - x)dz + \int_{b_2}^{\infty} f_m(z - x)dz \right] \qquad 5.27(7)$$

It is fairly obvious that if the number of undersize and oversize parts accepted is to be reduced, then the acceptance limits b_1 and b_2 should lie within the required limits c_1 and c_2 in order to provide a margin of safety to cover the uncertainty of the measuring equipment, that is $|c_2 - a| > |b_2 - a|$ and $|c_1 - a| > |b_1 - a|$. Correspondingly the fraction of correct-size parts rejected will increase as the margin of safety is increased. In fact, in order to reduce the number of correct-size parts rejected we require $|b_2 - a| > |c_2 - a|$ and $|b_1 - a| > |c_1 - a|$. This in turn would increase the fraction of *undersize* and *oversize* parts *accepted*.

5.28 Thus to summarize, if O_A and U_A *are required to be small*, then $|c_2 - a| > |b_2 - a|$ and $|c_1 - a| > |b_1 - a|$. These conditions lead to an increased value of C_R. If C_R *is required to be small*, then $|b_2 - a| > |c_2 - a|$ and $|b_1 - a| > |c_1 - a|$. These conditions lead inevitably to an increase in value of O_A and U_A. The following relationships exist between the designated fractional parts

$$C_A + C_R = \int_{c_1}^{c_2} f_p(x - a)dx \equiv C_p \equiv \text{fraction of correct parts} \qquad 5.28(1)$$

$$O_A + O_R = \int_{c_2}^{\infty} f_p(x - a)dx \equiv O_p \equiv \text{fraction of oversize parts} \qquad 5.28(2)$$

$$U_A + U_R = \int_{-\infty}^{c_1} f_p(x - a)dx \equiv U_p \equiv \text{fraction of undersize parts} \qquad 5.28(3)$$

Also

$$C_p + O_p + U_p = 1 \equiv C_A + C_R + O_A + O_R + U_A + U_R \qquad 5.28(4)$$

The number of incorrect parts accepted is equal to

$$O_A + U_A \equiv W_A \qquad 5.28(5)$$

The number of incorrect parts rejected is equal to

$$O_R + U_R \equiv W_R \qquad 5.28(6)$$

Thus

$$1 = C_A + C_R + W_A + W_R \qquad 5.28(7)$$

Let us now assume that the density function of the manufactured parts $f_p(x - a)$ is a Gaussian function with standard deviation equal to σ_p. The density function can thus be written as

$$\frac{e^{-(x - a)2\sigma_p^2}}{\sigma_p \sqrt{(2\pi)}}$$

Similarly, let us assume that the density function of the measuring equipment is also a Gaussian function, which can thus be written as

$$\frac{e^{-(z - x)^2/2\sigma_m^2}}{\sigma_m \sqrt{(2\pi)}} \qquad 5.28(8)$$

Now we have seen from 5.07(1) and 5.07(2) that the distribution function for the density function

$$\frac{e^{-(z-x)^2/2\sigma_m^2}}{\sigma_m\sqrt{(2\pi)}} \qquad 5.28(9)$$

can be written as

$$F_{Ap1}(z_1 - x) = \int_{-\infty}^{z_1} \frac{e^{-(z-x)^2/2\sigma_m^2}dz}{\sigma_m\sqrt{(2\pi)}} \simeq \frac{e^{-(z_1-x-\rho)^2/2\rho^2}}{\tau} \qquad 5.28(10)$$

for $x \geqslant z_1 \geqslant -\infty$ and

$$F_{Ap2}(z_1 - x) = \int_{-\infty}^{z_1} \frac{e^{-(z-x)^2/2\sigma_m^2}dz}{\sigma_m\sqrt{(2\pi)}} \simeq \left[1 - \frac{e^{-(z_1-x+\rho)^2/2\rho^2}}{\tau}\right] \qquad 5.28(11)$$

for $\infty > z_1 \geqslant x$ where $\tau = 1.213\,0613$ and $\rho = 1.205\,0795\sigma_m = \eta\sigma_m$ or $\sigma_m = 0.8298\rho$ (see page 154 under Figure 5.07(1) where σ_m is the standard deviation of the distribution. Since our aim is to reduce the fraction of incorrect parts accepted we require $c_2 > b_2 > b_1 > c_1$.

Parts With Gaussian Distribution

5.29 Let us first consider

$$C_A = \int_{c_1}^{c_2} f_p(x - a)dx \int_{b_1}^{b_2} f_m(z - x)dz \qquad 5.27(2)$$

This can be written as

$$C_A = \int_{c_1}^{b_1} f_p dx \int_{b_1}^{b_2} f_m dz + \int_{b_1}^{b_2} f_p dx \int_{b_1}^{b_2} f_m dz + \int_{b_2}^{c_2} f_p dx \int_{b_1}^{b_2} f_m dz \qquad 5.29(1)$$
$$\qquad\qquad (1) \qquad\qquad\qquad\qquad (2) \qquad\qquad\qquad\qquad (3)$$

Now

$$\int_{b_1}^{b_2} f_m dz = \int_{-\infty}^{b_2} f_m dz + \int_{b_1}^{-\infty} f_m dz$$

$$= \int_{-\infty}^{b_2} f_m dz - \int_{-\infty}^{b_1} f_m dz \qquad 5.29(2)$$

The first double integral of C_A can thus be written as

$$\int_{c}^{b_1} f_p dx \left[\int_{-\infty}^{b_2} f_m dz - \int_{-\infty}^{b_1} f_m dz\right] \qquad 5.29(3)$$

Let us consider $\int_{-\infty}^{b_2} f_m dz$ and compare it with either F_{Ap1} or F_{Ap2}. Now $z_1 \equiv b_2$, and the range of values of x goes from c_1 to b_1, that is the limits of the second integration. Thus, since $b_2 > b_1 > c_1$, $z_1 > x$ and so F_{Ap2} is the appropriate form to use for $\int_{-\infty}^{b_2} f_m dz$. In the case of $\int_{-\infty}^{b_1} f_m dz$, $z_1 \equiv b_1$ and the range of values of x once again goes from c_1 to b_1, and so in this case $z_1 \geqslant x$ and it is again appropriate to use F_{Ap2}. We therefore write

$$\int_{-\infty}^{b_2} f_m dz = 1 - \frac{e^{-(b_2 - x + \rho)^2/2\rho^2}}{\tau} \qquad 5.29(4)$$

and

$$\int_{-\infty}^{b_1} f_m dz = 1 - \frac{e^{-(b_1 - x + \rho)^2/2\rho^2}}{\tau} \qquad 5.29(5)$$

The second double integral of C_A can be written similarly to the first one as

$$\int_{b_1}^{b_2} f_p dx \int_{b_1}^{b_2} f_m dz = \int_{b_1}^{b_2} f_p dx \left[\int_{-\infty}^{b_2} f_m dz - \int_{-\infty}^{b_1} dz \right] \qquad 5.29(6)$$

Comparing $\int_{-\infty}^{b_2} f_m dz$ with F_{Ap1} or F_{Ap2} we have $z_1 \equiv b_2$ and the range of values of x goes from b_1 to b_2, and so $z_1 \geqslant x$ and the form F_{Ap2} is appropriate (5.28(11)). Dealing with $\int_{-\infty}^{b_1} f_m dz$, $z_1 \equiv b_1$, but the range of values of x goes from b_1 to b_2, and so $x \geqslant z_1$ and the form F_{Ap1} is appropriate (5.28(10)).

Lastly, the third double integral of C_A can be written as

$$\int_{b_2}^{c_2} f_p dx \left[\int_{-\infty}^{b_2} f_m dz - \int_{-\infty}^{b_1} f_m dz \right] \qquad 5.29(7)$$

Dealing with $\int_{-\infty}^{b_2} f_m dz$ and comparing with F_{Ap1} or F_{Ap2} we have $z_1 \equiv b_2$ and the range x is from b_2 to c_2, and so $x \geqslant z_1$ and the form F_{Ap1} is appropriate. Similarly in the case of $\int_{-\infty}^{b_1} f_m dz$, $z_1 \equiv b_1$ and the range of x is once again from b_2 to c_2, and so $x > z_1$ and once again F_{Ap1} is appropriate. Finally if we replace $f_p dx$ by

$$\frac{e^{-(x - a)^2/2\sigma_p^2}}{\sigma_p \sqrt{(2\pi)}} \qquad 5.29(8)$$

and the various integrals of f_m by their approximations F_{Ap1} or F_{Ap2} with $z_1 \equiv b_1$ or b_2 as appropriate, then C_A can be written as

$$
\begin{aligned}
C_A = & \int_{c_1}^{b_1} \frac{e^{-(x - a)^2/2\sigma_p^2}}{\sigma_p \sqrt{(2\pi)}} \left\{ \frac{e^{-(b_1 - x + \rho)^2/2\rho^2} - e^{-(b_2 - x + \rho)^2/2\rho^2}}{\tau} \right\} dx \\
& + \int_{b_1}^{b_2} \frac{e^{-(x - a)^2/2\sigma_p^2}}{\sigma_p \sqrt{(2\pi)}} \left\{ 1 - \frac{e^{-(b_2 - x + \rho)^2/2\rho^2}}{\tau} - \frac{e^{-(b_1 - x - \rho)^2/2\rho^2}}{\tau} \right\} dx \\
& + \int_{b_2}^{c_2} \frac{e^{-(x - a)^2/2\sigma_p^2}}{\sigma_p \sqrt{(2\pi)}} \left\{ \frac{e^{-(b_2 - x - \rho)^2/2\rho^2} - e^{-(b_1 - x - \rho)^2/2\rho^2}}{\tau} \right\} dx \qquad 5.29(9)
\end{aligned}
$$

Note that if $c_1 = b_1$ and $b_2 = c_2$ the first and third integrals disappear and we are left with the case when the acceptance limits are equal to the required limits.

5.30 Dealing now with the fraction of correct parts rejected C_R, we can write the expression for this quantity, that is

$$C_R = \int_{c_1}^{c_2} f_p(x-a)dx\left[\int_{-\infty}^{b_1} f_m(z-x)dz + \int_{b_2}^{\infty} f_m(z-x)dz\right] \quad 5.27(3)$$

as

$$= \int_{c_1}^{b_1} f_p dx \int_{-\infty}^{b_1} f_m dz + \int_{b_1}^{c_2} f_p dx \int_{-\infty}^{b_1} f_m dz + \int_{c_1}^{b_2} f_p dx \int_{b_2}^{\infty} f_m dz$$

$$(1) \qquad\qquad (2) \qquad\qquad (3)$$

$$+ \int_{b_2}^{c_2} f_p dx \int_{b_2}^{\infty} f_m dz$$

$$(4)$$

We now replace the integrals with respect to z with their approximations F_{Ap1} or F_{Ap2}, where the range of x is given in each case by the limits of the integral of f_p and z_1 is put equal to the upper limit of the integral of f_m. Integral (1)

$$\int_{-\infty}^{b_1} f_m dz, \quad z_1 \equiv b_1$$

and range of x is from c_1 to b_1. Therefore $z_1 \geqslant x$ and we write

$$\int_{-\infty}^{b_1} f_m dz = \left\{1 - \frac{e^{-(b_1 - x + \rho)^2/2\rho}}{\tau}\right\} \quad 5.30(1)$$

Integral (2)

$$\int_{-\infty}^{b_1} f_m dz, \quad z_1 \equiv b_1$$

and range of x is from b_1 to c_2. Therefore $x \geqslant z_1$ and we write

$$\int_{-\infty}^{b_1} f_m dz = \frac{e^{-(b_1 - x - \rho)^2/2\rho}}{\tau} \quad 5.30(2)$$

Integral (3)

$$\int_{b_2}^{\infty} f_m dz = \int_{b_2}^{-\infty} f_m dz + \int_{-\infty}^{\infty} f_m dz = 1 - \int_{-\infty}^{b_2} f_m dz, \quad z_1 \equiv b_2$$

$$5.30(3)$$

and range of x is from c_1 to b_2. Therefore $z_1 \geqslant x$ and we write the integral using the form F_{Ap2} and

$$
\int_{b_2}^{\infty} f_m dz = 1 - \left[1 - \frac{e^{-(b_2 - x + \rho)^2/2\rho^2}}{\tau} \right]
$$

$$
= \frac{e^{-(b_2 - x + \rho)^2/2\rho^2}}{\tau} \qquad 5.30(4)
$$

Integral (4), as in integral (3)

$$
\int_{b_2}^{\infty} f_m dz = 1 - \int_{-\infty}^{b_2} f_m dz, \quad z_1 \equiv b_2 \qquad 5.30(5)
$$

and the range of x is from b_2 to c_2. Therefore $x \geqslant z_1$ amd we use the form F_{Ap1} giving

$$
\int_{b_2}^{\infty} f_m dz = 1 - \frac{e^{-(b_2 - x - \rho)^2/2\rho^2}}{\tau} \qquad 5.30(6)
$$

Replacing f_p by the Gaussian density function

$$
\frac{e^{-(x - a)^2/2\sigma_p^2}}{\sigma_p \sqrt{(2\pi)}} \qquad 5.30(7)
$$

and using the values just found above for the various integrals of f_m, we have that

$$
C_R = \int_{c_1}^{b_1} \frac{e^{-(x - a)^2/2\sigma_p^2}}{\sigma_p \sqrt{(2\pi)}} \left\{ 1 - \frac{e^{-(b_1 - x + \rho)^2/2\rho^2}}{\tau} \right\} dx
$$

$$
+ \int_{b_1}^{c_2} \frac{e^{-(x - a)^2/2\sigma_p^2}}{\sigma_p \sqrt{(2\pi)}} \cdot \frac{e^{-(b_1 - x - \rho)^2/2\rho^2}}{\tau} dx
$$

$$
+ \int_{c_1}^{b_2} \frac{e^{-(x - a)^2/2\sigma_p^2}}{\sigma_p \sqrt{(2\pi)}} \cdot \frac{e^{-(b_2 - x + \rho)^2/2\rho^2}}{\tau} dx
$$

$$
+ \int_{b_2}^{c_2} \frac{e^{-(x - a)^2/2\sigma_p^2}}{\sigma_p \sqrt{(2\pi)}} \left\{ 1 - \frac{e^{-(b_2 - x - \rho)^2/2\rho^2}}{\tau} \right\} dx \qquad 5.30(8)
$$

5.31 Similarly

$$
O_A = \int_{c_2}^{\infty} f_p(x - a) dx \int_{b_1}^{b_2} f_m(z - x) dz \qquad 5.31(1)
$$

and can be written

$$
= \int_{c_2}^{\infty} f_p(x - a) dx \left[\int_{-\infty}^{b_2} f_m dz - \int_{-\infty}^{b_1} f_m dz \right] \qquad 5.31(2)
$$

$$
(1) \qquad\qquad (2)
$$

In the first integral (1), comparing with F_{Ap1} or F_{Ap2}, $z_1 \equiv b_2$ and the range of x is from c_2 to ∞. Therefore $x > z_1$ and we write

$$\int_{-\infty}^{b_2} f_m dz = F_{Ap1} = \frac{e^{-(b_2 - x - \rho)^2/2\rho^2}}{\tau} \qquad 5.31(3)$$

In the second integral (2), $z_1 \equiv b_1$ and the range of x is from c_2 to ∞. Therefore $x > z_1$ and we write

$$\int_{-\infty}^{b_1} f_m dz = F_{Ap1} = \frac{e^{-(b_1 - x - \rho)^2/2\rho^2}}{\tau} \qquad 5.31(4)$$

Thus substituting in O_A for f_p and $\int f_m dz$ we have

$$O_A = \int_{c_2}^{\infty} \frac{e^{-(x - a)^2/2\sigma_p^2}}{\sigma_p \sqrt{(2\pi)}} \left\{ \frac{e^{-(b_2 - x - \rho)^2/2\rho^2} - e^{-(b_1 - x - \rho)^2/2\rho^2}}{\tau} \right\} dx \qquad 5.31(5)$$

5.32 Dealing in a similar way with

$$U_A = \int_{-\infty}^{c_1} f_p(x - a)\, dx \int_{b_1}^{b_2} f_m(z - x) dz \qquad 5.32(1)$$

we obtain

$$U_A = \int_{-\infty}^{c_1} \frac{e^{-(x - a)^2/2\sigma_p^2}}{\sigma_p \sqrt{(2\pi)}} \left\{ \frac{e^{-(b_1 - x + \rho)^2/2\rho^2} - e^{-(b_2 - x + \rho)^2/2\rho^2}}{\tau} \right\} dx \qquad 5.32(2)$$

Thus

$$W_A = O_A + U_A$$

$$= \int_{c_2}^{\infty} \frac{e^{-(x - a)^2/2\sigma_p^2}}{\sigma_p \sqrt{(2\pi)}} \left\{ \frac{e^{-(b_2 - x - \rho)^2/2\rho^2} - e^{-(b_1 - x - \rho)^2/2\rho^2}}{\tau} \right\} dx$$

$$+ \int_{-\infty}^{c_1} \frac{e^{-(x - a)^2/2\sigma_p^2}}{\sigma_p \sqrt{(2\pi)}} \left\{ \frac{e^{-(b_1 - x - \rho)^2/2\rho^2} - e^{-(b_2 - x - \rho)^2/2\rho^2}}{\tau} \right\} dx$$

$$5.32(3)$$

Similarly it can be shown that

$$O_R = \int_{c_2}^{\infty} \frac{e^{-(x - a)^2/2\sigma_p^2}}{\sigma_p \sqrt{(2\pi)}} \left\{ \frac{e^{-(b_1 - x - \rho)^2/2\rho^2}}{\tau} + 1 - \frac{e^{-(b_2 - x - \rho)^2/2\rho^2}}{\tau} \right\} dx$$

$$5.32(4)$$

and that

$$U_R = \int_{-\infty}^{c_1} \frac{e^{-(x - a)^2/2\sigma_p^2}}{\sigma_p \sqrt{(2\pi)}} \left\{ 1 - \frac{e^{-(b_1 - x + \rho)^2/2\rho^2}}{\tau} + \frac{e^{-(b_2 - x + \rho)^2/2\rho^2}}{\tau} \right\} dx$$

$$5.32(5)$$

W_R, the fraction of incorrect parts rejected, is of course equal to the sum of $O_R + U_R \equiv W_R$.

Solution of the integrals just derived

5.33   These are all of the form

$$I = \int_{y_1}^{y_2} \frac{e^{-(x-a)^2/2\sigma_p^2}}{\sigma_p\sqrt{(2\pi)}} \cdot \frac{e^{-(b-x+\rho)^2/2\rho^2}}{\tau} \qquad 5.33(1)$$

where ρ can be negative.

The index of e is equal to

$$-\left\{\frac{x^2 + a^2 - 2xa}{2\sigma_p^2} + \frac{x^2 + (b+\rho)^2 - 2x(b+\rho)}{2\rho^2}\right\}$$

$$= -\left\{x^2\left(\frac{1}{2\sigma_p^2} + \frac{1}{2\rho^2}\right) - x\left(\frac{a}{\sigma_p^2} + \frac{b+\rho}{\rho^2}\right) + \frac{a^2}{2\sigma_p^2} + \frac{(b+\rho)^2}{2\rho^2}\right\}$$

$$= -(Ax^2 - Bx + C)$$

$$= -\left\{A\left(x - \frac{B}{2A}\right)^2 + C - \frac{B^2}{4A}\right\} \qquad 5.33(2)$$

Therefore

$$I = \frac{1}{\tau\sigma_p\sqrt{(2\pi)}} \int_{y_1}^{y_2} e^{-\left\{A\left(x - \frac{B}{2A}\right)^2 + C - \frac{B^2}{4A}\right\}} dx$$

$$= \frac{e^{B^2/4A - C}}{\tau\sigma_p\sqrt{(2\pi)}} \int_{y_1}^{y_2} e^{-A\left(x - \frac{B}{2A}\right)^2} dx \qquad 5.33(3)$$

Now put $\sqrt{A}(x - B/2A) = z/\sqrt{2}$. Then $dx = dz/\sqrt{(2A)}$.

The new limits are when

$$x = y_1, \quad \text{then} \quad z_1 = \sqrt{(2A)}\left(y_1 - \frac{B}{2A}\right) \qquad 5.33(4)$$

and when

$$x = y_2, \quad \text{then} \quad z_2 = \sqrt{(2A)}\left(y_2 - \frac{B}{2A}\right) \qquad 5.33(5)$$

and so

$$I = \frac{e^{(B^2/4A) - C}}{\tau\sigma_p\sqrt{(4\pi A)}} \int_{\sqrt{(2A)}\left(y_1 - \frac{B}{2A}\right)}^{\sqrt{(2A)}\left(y_2 - \frac{B}{2A}\right)} e^{-z^2/2} dz \qquad 5.33(6)$$

Now

$$\int_{z_1}^{z_2} \frac{e^{-z^2/2}\, dz}{\sqrt{(2\pi)}} = \int_{z_1}^{0} \frac{e^{-z^2/2}\, dz}{\sqrt{(2\pi)}} + \int_{0}^{z_2} \frac{e^{-z^2/2}\, dz}{\sqrt{(2\pi)}}$$

$$= \int_{0}^{z_2} \frac{e^{-z^2/2}\, dz}{\sqrt{(2\pi)}} - \int_{0}^{z_1} \frac{e^{-z^2/2}}{\sqrt{(2\pi)}}$$

$$= \frac{1}{2}\left(\underset{-z_2\ \text{to}\ z_2}{P} - \underset{-z_1\ \text{to}\ z_1}{P} \right) \qquad \text{5.33(7)}$$

Therefore

$$I = \frac{e^{(B^2/4A) - C}}{2\tau\sigma_p\sqrt{(2A)}}\left(\underset{-k_2\ \text{to}\ k_2}{P} - \underset{-k_1\ \text{to}\ k_1}{P} \right) \qquad \text{5.33(8)}$$

where

$$k_2 = \sqrt{(2A)}(y_2 - B/2A) \qquad \text{5.33(9)}$$

and

$$k_1 = \sqrt{(2A)}(y_1 - B/2A) \qquad \text{5.33(10)}$$

and $\underset{-k\ \text{to}\ k}{P}$ is the probability of occurrence of an event between the limits $-k$ to k for a Gaussian distribution of unit standard deviation. $\underset{-k\ \text{to}\ k}{P}$ is given in Table II of Appendix I.

5.34 In the *special case* when the b and ρ are absent from the integral, put $b = -\rho$ and $\rho = \infty$, whence $(B^2/4A) - C = 0$. Also put $\tau\sigma_p\sqrt{(2A)} = 1$ in this case. Finally, k_1 and k_2 reduce to $k_1 = (y_1 - a)/\sigma_p$ and $k_2 = (y_2 - a)/\sigma_p$ for this special case. Each integral can be divided into several separate integrals; considering C_A we see that this consists of seven separate integrals. Thus as an example of the nomenclature adopted here, the B constant of the fourth integral of C_A is designated as B_{CA4}, with other terms similarly designated. The following list gives the constants of the integrals for C_A, C_R, O_A, U_A, O_R and U_R. The A and B used in the ks are of course the appropriate values of A and B with suffixes, which are omitted in the interests of economy. For computational purposes it is convenient to express the limits c_1, c_2, b_1 and b_2, and the constants ρ and σ_p, in terms of the total tolerance range T. ρ is found from the expression $\rho = 1.205\,08\ \sigma_m$ where σ_m is the standard deviation of the measuring equipment and σ_p is the standard deviation of the parts to be measured. The constant $\tau = 1.213\,06$.

Since $B^2/4A - C$ is dimensionless, as are k_1 and k_2, only the coefficient of T the tolerance range need be entered for each expression.

It should be noted that if k_1 or k_2 are found to be negative, then the corresponding P should also be treated as negative. The reason for this is easily seen as follows. All the integrals are of the form

$$\int_{z_1}^{z_2} e^{-z^2/2} \, dz = \int_{z_1}^{0} + \int_{0}^{z_2} = \int_{0}^{z_2} - \int_{0}^{z_1} \qquad 5.34(1)$$

Now if either z_1 and z_2 are negative, i.e. for z_2 negative, we have

$$\int_{0}^{-z_2} = -\int_{-z_2}^{0} \qquad 5.34(2)$$

Now since $e^{-z^2/2}$ is a symmetrical function

$$\int_{-z_0}^{0} = \int_{0}^{z_2} = \frac{1}{2} \int_{-z_2}^{z_2} \qquad 5.34(3)$$

Therefore

$$\int_{0}^{-z_2} = -\frac{1}{2} \int_{-z_2}^{z_2} \qquad 5.34(4)$$

A similar argument applies for z_1.

$$C_A$$

5.35

$$A_{CA1} = \frac{1}{2\sigma_p^2} + \frac{1}{2\rho^2}, \quad B_{CA1} = \frac{a}{\sigma_p^2} + \frac{b_1 + \rho}{\rho^2}, \quad C_{CA1} = \frac{a^2}{2\sigma_p^2} + \frac{(b_1 + \rho)^2}{2\rho^2}$$

$$5.35(1)$$

$$k_1 = \sqrt{(2A)} \left(c_1 - \frac{B}{2A} \right), \quad k_2 = \sqrt{(2A)} \left(b_1 - \frac{B}{2A} \right) \qquad 5.35(2)$$

$$A_{CA2} = \frac{1}{2\sigma_p^2} + \frac{1}{2\rho^2}, \quad B_{CA2} = \frac{a}{\sigma_p^2} + \frac{b_2 + \rho}{\rho^2}, \quad C_{CA2} = \frac{a^2}{2\sigma_p^2} + \frac{(b_2 + \rho)^2}{2\rho^2}$$

$$5.35(3)$$

$$k_1 = \sqrt{(2A)} \left(c_1 - \frac{B}{2A} \right), \quad k_2 = \sqrt{(2A)} \left(b_1 - \frac{B}{2A} \right) \qquad 5.35(4)$$

$$A_{CA3} = \frac{1}{2\sigma_p^2}, \quad B_{CA3} = \frac{a}{\sigma_p^2}, \quad C_{CA3} = \frac{a^2}{2\sigma_p^2} \qquad 5.35(5)$$

$$\frac{B^2}{4A} - C = 0, \quad k_1 = \frac{b_1 - a}{\sigma_p}, \quad k_2 = \frac{b_2 - a}{\sigma_p}, \quad \text{and put } \tau\sigma_p\sqrt{(2A)} = 1$$

$$5.35(6)$$

$$A_{CA4} = \frac{1}{2\sigma_p^2} + \frac{1}{2\rho^2}, \quad B_{CA4} = \frac{a}{\sigma_p^2} + \frac{b_2 + \rho}{\rho^2}, \quad C_{CA4} = \frac{a^2}{2\sigma_p^2} + \frac{(b_2 + \rho)^2}{2\rho^2}$$

$$5.35(7)$$

$$k_1 = \sqrt{(2A)}\left(b_1 - \frac{B}{2A}\right), \quad k_2 = \sqrt{(2A)}\left(b_2 - \frac{B}{2A}\right) \qquad 5.35(8)$$

$$A_{CA5} = \frac{1}{2\sigma_p^2} + \frac{1}{2\rho^2}, \quad B_{CA5} = \frac{a}{\sigma_p^2} + \frac{b_1 - \rho}{\rho^2}, \quad C_{CA5} = \frac{a^2}{2\sigma_p^2} + \frac{(b_1 - \rho)^2}{2\rho^2}$$

$$5.35(9)$$

$$k_1 = \sqrt{(2A)}\left(b_1 - \frac{B}{2A}\right), \quad k_2 = \sqrt{(2A)}\left(b_2 - \frac{B}{2A}\right) \qquad 5.35(10)$$

$$A_{CA6} = \frac{1}{2\sigma_p^2} + \frac{1}{2\rho^2}, \quad B_{CA6} = \frac{a}{\sigma_p^2} + \frac{b_2 - \rho}{\rho^2}, \quad C_{CA6} = \frac{a^2}{2\sigma_p^2} + \frac{(b_2 - \rho)^2}{2\rho^2}$$

$$5.35(11)$$

$$k_1 = \sqrt{(2A)}\left(b_2 - \frac{B}{2A}\right), \quad k_2 = \sqrt{(2A)}\left(c_2 - \frac{B}{2A}\right) \qquad 5.35(12)$$

$$A_{CA7} = \frac{1}{2\sigma_p^2} + \frac{1}{2\rho^2}, \quad B_{CA7} = \frac{a}{\sigma_p^2} + \frac{b_1 - \rho}{\rho^2}, \quad C_{CA7} = \frac{a^2}{2\sigma_p^2} + \frac{(b_1 - \rho)^2}{2\rho^2}$$

$$5.35(13)$$

$$k_1 = \sqrt{(2A)}\left(b_2 - \frac{B}{2A}\right), \quad k_2 = \sqrt{(2A)}\left(c_2 - \frac{B}{2A}\right) \qquad 5.35(14)$$

whence

$$C_A = \left[\frac{e^{(B^2/4A) - C}}{2\tau\sigma_p\sqrt{(2A)}}\left(\underset{-k_2 \text{ to } k_2}{P} - \underset{-k_1 \text{ to } k_1}{P}\right)\right]_1 - [\quad]_2$$

$$+ [\quad]_3 - [\quad]_4 - [\quad]_5 + [\quad]_6 - [\quad]_7 \qquad 5.35(15)$$

where the suffix of each bracket refers to the constants with that suffix for C_A.

$$C_R$$

5.36

$$A_{CR1} = \frac{1}{2\sigma_p^2}, \quad B_{CR1} = \frac{a}{\sigma_p^2}, \quad C_{CR1} = \frac{a^2}{2\sigma_p^2} \qquad 5.36(1)$$

$$\frac{B^2}{4A} - C = 0, \quad k_1 = \frac{c_1 - a}{\sigma_p}, \quad k_2 = \frac{b_1 - a}{\sigma_p} \quad \text{and put } \tau\sigma_p\sqrt{(2A)} = 1$$

$$5.36(2)$$

$$A_{CR2} = \frac{1}{2\sigma_p^2} + \frac{1}{2\rho^2}, \quad B_{CR2} = \frac{a}{\sigma_p^2} + \frac{b_1 + \rho}{\rho^2}, \quad C_{CR2} = \frac{a^2}{2\sigma_p^2} + \frac{(b_1 + \rho)^2}{2\rho^2}$$

5.36(3)

$$k_1 = \sqrt{(2A)}\left(c_1 - \frac{B}{2A}\right), \quad k_2 = \sqrt{(2A)}\left(b_1 - \frac{B}{2A}\right)$$

5.36(4)

$$A_{CR3} = \frac{1}{2\sigma_p^2} + \frac{1}{2\rho^2}, \quad B_{CR3} = \frac{a}{\sigma_p^2} + \frac{b_1 - \rho}{\rho^2}, \quad C_{CR3} = \frac{a^2}{2\sigma_p^2} + \frac{(b_1 - \rho)^2}{2\rho^2}$$

5.36(5)

$$k_1 = \sqrt{(2A)}\left(b_1 - \frac{B}{2A}\right), \quad k_2 = \sqrt{(2A)}\left(c_2 - \frac{B}{2A}\right)$$

5.36(6)

$$A_{CR4} = \frac{1}{2\sigma_p^2} + \frac{1}{2\rho^2}, \quad B_{CR4} = \frac{a}{\sigma_p^2} + \frac{b_2 + \rho}{\rho^2}, \quad C_{CR4} = \frac{a^2}{2\sigma_p^2} + \frac{(b_2 + \rho)^2}{2\rho^2}$$

5.36(7)

$$k_1 = \sqrt{(2A)}\left(c_1 - \frac{B}{2A}\right), \quad k_2 = \sqrt{(2A)}\left(b_2 - \frac{B}{2A}\right)$$

5.36(8)

$$A_{CR5} = \frac{1}{2\sigma_p^2}, \quad B_{CR5} = \frac{a}{\sigma_p^2}, \quad C_{CR5} = \frac{a^2}{2\sigma_p^2}$$

5.36(9)

$$\frac{B^2}{4A} - C = 0, \quad k_1 = \frac{b_2 - a}{\sigma_p}, \quad k_2 = \frac{c_2 - a}{\sigma_p} \quad \text{and put } \tau\sigma_p\sqrt{(2A)} = 1$$

5.36(10)

$$A_{CR6} = \frac{1}{2\sigma_p^2} + \frac{1}{2\rho^2}, \quad B_{CR6} = \frac{a}{\sigma_p^2} + \frac{b_2 - \rho}{\rho^2}, \quad C_{CR6} = \frac{a^2}{2\sigma_p^2} + \frac{(b_2 - \rho)^2}{2\rho^2}$$

5.36(11)

$$k_1 = \sqrt{(2A)}\left(b_2 - \frac{B}{2A}\right), \quad k_2 = \sqrt{(2A)}\left(c_2 - \frac{B}{2A}\right)$$

whence

$$C_R = \left[\frac{e^{(B^2/4A) - C}}{2\tau\sigma_p\sqrt{(2A)}}\left(\mathop{P}_{-k_2 \text{ to } k_2} - \mathop{P}_{-k_1 \text{ to } k_1}\right)\right]_1 - [\quad]_2 + [\quad]_3$$
$$+ [\quad]_4 + [\quad]_5 - [\quad]_6$$

5.36(12)

$$O_A$$

5.37

$$A_{OA1} = \frac{1}{2\sigma_p^2} + \frac{1}{2\rho^2}, \quad B_{OA1} = \frac{a}{\sigma_p^2} + \frac{b_2 - \rho}{\rho^2}, \quad C_{OA1} = \frac{a^2}{2\sigma_p^2} + \frac{(b_2 - \rho)^2}{2\rho^2}$$

$$5.37(1)$$

$$k_1 = \sqrt{(2A)}\left(c_2 - \frac{B}{2A}\right), \quad k_2 = \infty, \quad \text{that is}\left(\underset{-k_2 \text{ to } k_2}{P} = 1\right) \qquad 5.37(2)$$

$$A_{OA2} = \frac{1}{2\sigma_p^2} + \frac{1}{2\rho^2}, \quad B_{OA2} = \frac{a}{\sigma_p^2} + \frac{b_1 - \rho}{\rho^2}, \quad C_{OA2} = \frac{a^2}{2\sigma_p^2} + \frac{(b_1 - \rho)^2}{2\rho^2}$$

$$5.37(3)$$

$$k_1 = \sqrt{(2A)}\left(c_2 - \frac{B}{2A}\right), \quad k_2 = \infty, \quad \text{that is}\left(\underset{-k_2 \text{ to } k_2}{P} = 1\right) \qquad 5.37(4)$$

whence

$$O_A = \left[\frac{e^{(B^2/4A) - C}}{2\tau\sigma_p\sqrt{(2A)}}\left(1 - \underset{-k_1 \text{ to } k_1}{P}\right)\right]_1 - \left[\frac{e^{(B^2/4A) - C}}{2\tau\sigma_p\sqrt{(2A)}}\left(1 - \underset{-k_1 \text{ to } k_1}{P}\right)\right]_2$$

$$5.37(5)$$

$$U_A$$

5.38

$$A_{UA1} = \frac{1}{2\sigma_p^2} + \frac{1}{2\rho^2}, \quad B_{UA1} = \frac{a}{\sigma_p^2} + \frac{b_1 + \rho}{\rho^2}, \quad C_{UA1} = \frac{a^2}{2\sigma_p^2} + \frac{(b_1 + \rho)^2}{\rho^2}$$

$$5.38(1)$$

$$k_1 = -\infty, \quad \text{that is}\left(\underset{-k_1 \text{ to } k_1}{P} = -1\right), \quad k_2 = \sqrt{(2A)}\left(c_1 - \frac{B}{2A}\right)$$

$$5.38(2)$$

$$A_{UA2} = \frac{1}{2\sigma_p^2} + \frac{1}{2\rho^2}, \quad B_{UA2} = \frac{a}{\sigma_p^2} + \frac{b_2 + \rho}{\rho^2}, \quad C_{UA2} = \frac{a^2}{2\sigma_p^2} + \frac{(b_2 + \rho)^2}{\rho^2}$$

$$5.38(3)$$

$$k_1 = -\infty, \quad \text{that is}\left(\underset{-k_1 \text{ to } k_1}{P} = -1\right), \quad k_2 = \sqrt{(2A)}\left(c_1 - \frac{B}{2A}\right)$$

$$5.38(4)$$

whence

$$U_A = \left[\frac{e^{(B^2/4A)-C}}{2\tau\sigma_p\sqrt{(2A)}} \left(\underset{-k_2 \text{ to } k_2}{P} + 1 \right) \right]_1 - \left[\frac{e^{(B^2/4A)-C}}{2\tau\sigma_p\sqrt{(2A)}} \left(\underset{-k_2 \text{ to } k_2}{P} + 1 \right) \right]_2$$

5.38(5)

$$O_R$$

5.39

$$A_{OR1} = \frac{1}{2\sigma_p^2} + \frac{1}{2\rho^2}, \quad B_{OR1} = \frac{a}{\sigma_p^2} + \frac{b_1 - \rho}{\rho^2}, \quad C_{OR1} = \frac{a^2}{2\sigma_p^2} + \frac{(b_1 - \rho)^2}{2\rho^2}$$

5.39(1)

$$k_1 = \sqrt{(2A)} \left(c_2 - \frac{B}{2A} \right), \quad k_2 = \infty, \quad \text{that is} \left(\underset{-k_2 \text{ to } k_2}{P} = 1 \right) \quad 5.39(2)$$

$$A_{OR2} = \frac{1}{2\sigma_p^2}, \quad B_{OR2} = \frac{a}{\sigma_p^2}, \quad C_{OR2} = \frac{a^2}{2\sigma_p^2}$$

5.39(3)

$$\frac{B^2}{4A} - C = 0, \quad k_1 = \frac{c_2 - a}{\sigma_p}, \quad k_2 = \infty, \quad \text{that is} \left(\underset{-k_2 \text{ to } k_2}{P} = 1 \right)$$

and put $\tau\sigma_p\sqrt{(2A)} = 1$ 5.39(4)

$$A_{OR3} = \frac{1}{2\sigma_p^2} + \frac{1}{2\rho^2}, \quad B_{OR3} = \frac{a}{\sigma_p^2} + \frac{b_2 - \rho}{\rho^2}, \quad C_{OR3} = \frac{a^2}{2\sigma_p^2} + \frac{(b_2 - \rho)^2}{2\rho^2}$$

5.39(5)

$$k_1 = \sqrt{(2A)} \left(c_2 - \frac{B}{2A} \right), \quad k_2 = \infty, \quad \text{that is} \left(\underset{-k_2 \text{ to } k_2}{P} = 1 \right) \quad 5.39(6)$$

whence

$$O_R = \left[\frac{e^{(B^2/4A)-C}}{2\tau\sigma_p\sqrt{(2A)}} \left(1 - \underset{-k_1 \text{ to } k_1}{P} \right) \right]_1 + \left[\frac{1}{2} \left(1 - \underset{-k_1 \text{ to } k_1}{P} \right) \right]_2$$

$$- \left[\frac{e^{(B^2/4A)-C}}{2\tau\sigma_p\sqrt{(2A)}} \left(1 - \underset{-k_1 \text{ to } k_1}{P} \right) \right]_3$$

5.39(7)

$$U_R$$

5.40

$$A_{UR1} = \frac{1}{2\sigma_p^2}, \quad B_{UR1} = \frac{a}{\sigma_p^2}, \quad C_{UR1} = \frac{a^2}{2\sigma_p^2} \qquad 5.40(1)$$

$$\frac{B^2}{4A} - C = 0, \quad k_1 = -\infty, \quad \text{that is} \left(\underset{-k_1 \text{ to } k_1}{P} = -1 \right),$$

$$k_2 = \frac{c_1 - a}{\sigma_p} \quad \text{and put } \tau\sigma_p\sqrt{(2A)} = 1 \qquad 5.40(2)$$

$$A_{UR2} = \frac{1}{2\sigma_p^2} + \frac{1}{2\rho^2}, \quad B_{UR2} = \frac{a}{\sigma_p^2} + \frac{b_1 + \rho}{\rho^2}, \quad C_{UR2} = \frac{a^2}{2\sigma_p^2} + \frac{(b_1 + \rho)^2}{2\rho^2}$$

$$5.40(3)$$

$$k_1 = -\infty, \quad \text{that is} \left(\underset{-k_1 \text{ to } k_1}{P} = -1 \right), \quad k_2 = \sqrt{(2A)} \left(c_1 - \frac{B}{2A} \right)$$

$$5.40(4)$$

$$A_{UR3} = \frac{1}{2\sigma_p^2} + \frac{1}{2\rho^2}, \quad B_{UR3} = \frac{a}{\sigma_p^2} + \frac{b_2 + \rho}{\rho^2}, \quad C_{UR3} = \frac{a^2}{2\sigma_p^2} + \frac{(b_2 + \rho)^2}{2\rho^2}$$

$$5.40(5)$$

$$k_1 = -\infty, \quad \text{that is} \left(\underset{-k_1 \text{ to } k_1}{P} = -1 \right), \quad k_2 = \sqrt{(2A)} \left(c_1 - \frac{B}{2A} \right)$$

$$5.40(6)$$

whence

$$U_R = \left[\frac{1}{2} \left(\underset{-k_2 \text{ to } k_2}{P} + 1 \right) \right]_1 - \left[\frac{e^{(B^2/4A) - C}}{2\tau\sigma_p\sqrt{(2A)}} \left(\underset{-k_2 \text{ to } k_2}{P} + 1 \right) \right]_2 \quad 5.40(7)$$

$$+ \left[\frac{e^{(B^2/4A) - C}}{2\tau\sigma_p\sqrt{(2A)}} \left(\underset{-k_2 \text{ to } k_2}{P} + 1 \right) \right]_3$$

It is to be noted that

$$C_A + C_R = \int_{c_1}^{c_2} \frac{e^{-(x - a)^2/2\sigma_p^2}}{\sigma_p\sqrt{(2\pi)}} = C_p \qquad 5.40(8)$$

$$U_A + U_R = \int_{-\infty}^{c_1} \frac{e^{-(x - a)^2/2\sigma_p^2}}{\sigma_p\sqrt{(2\pi)}} = U_p \qquad 5.40(9)$$

and that

$$O_A + O_R = \int_{c_2}^{\infty} \frac{e^{-(x - a)^2/2\sigma_p^2}}{\sigma_p\sqrt{(2\pi)}} = O_p \qquad 5.40(10)$$

as they should, as has been seen from the generalized expressions for these quantities. (See 5.28(1) to 5.28(3).)

Modified expressions for C_A, C_R, etc. when $b_2 > c_2$ and $b_1 < c_1$

5.41 If $b_2 > c_2$ and $b_1 < c_1$, the expressions for C_A, C_R, O_A etc. need to be modified because of the changed ranges of x in F_1 and F_2.

It is easily shown that the new values of the above quantities are

$$C_A = \int_{c_1}^{c_2} \frac{e^{-(x-a)^2/2\sigma_p^2}}{\sigma_p\sqrt{(2\pi)}} \left\{ 1 - \frac{e^{-(b_2-x+\rho)^2/2\rho^2}}{\tau} - \frac{e^{-(b_1-x-\rho)^2/2\rho^2}}{\tau} \right\} dx \quad 5.41(1)$$

$$C_R = \int_{c_1}^{c_2} \frac{e^{-(x-a)^2/2\sigma_p^2}}{\sigma_p\sqrt{(2\pi)}} \left\{ \frac{e^{-(b_1-x-\rho)^2/2\rho^2}}{\tau} + \frac{e^{-(b_2-x+\rho)^2/2\rho^2}}{\tau} \right\} dx \quad 5.41(2)$$

$$C = C_A + C_R \quad 5.41(3)$$

$$O_A = \int_{c_2}^{b_2} \frac{e^{-(x-a)^2/2\sigma_p^2}}{\sigma_p\sqrt{(2\pi)}} \left\{ 1 - \frac{e^{-(b_2-x+\rho)^2/2\rho^2}}{\tau} \right\} dx$$

$$+ \int_{b_2}^{\infty} \frac{e^{-(x-a)^2/2\sigma_p^2}}{\sigma_p\sqrt{(2\pi)}} \left\{ \frac{e^{-(b_2-x-\rho)^2/2\rho^2}}{\tau} \right\} dx$$

$$- \int_{c_2}^{\infty} \frac{e^{-(x-a)^2/2\sigma_p^2}}{\sigma_p\sqrt{(2\pi)}} \left\{ \frac{e^{-(b_1-x-\rho)^2/2\rho^2}}{\tau} \right\} dx \quad 5.41(4)$$

$$O_R = \int_{c_2}^{\infty} \frac{e^{-(x-a)^2/2\sigma_p^2}}{\sigma_p\sqrt{(2\pi)}} \left\{ \frac{e^{-(b_1-x-\rho)^2/2\rho^2}}{\tau} \right\} dx$$

$$+ \int_{c_2}^{b_2} \frac{e^{-(x-a)^2/2\sigma_p^2}}{\sigma_p\sqrt{(2\pi)}} \left\{ \frac{e^{-(b_2-x+\rho)^2/2\rho^2}}{\tau} \right\} dx$$

$$+ \int_{b_2}^{\infty} \frac{e^{-(x-a)^2/2\sigma_p^2}}{\sigma_p\sqrt{(2\pi)}} \left\{ 1 - \frac{e^{-(b_2-x-\rho)^2/2\rho^2}}{\tau} \right\} dx \quad 5.41(5)$$

$$O = O_A + O_R \quad 5.41(6)$$

$$U_A = \int_{-\infty}^{c_1} \frac{e^{-(x-a)^2/2\sigma_p^2}}{\sigma_p\sqrt{(2\pi)}} \left\{ 1 - \frac{e^{-(b_2-x+\rho)^2/2\rho^2}}{\tau} \right\} dx$$

$$- \int_{-\infty}^{b_1} \frac{e^{-(x-a)^2/2\sigma_p^2}}{\sigma_p\sqrt{(2\pi)}} \left\{ 1 - \frac{e^{-(b_1-x+\rho)^2/2\rho^2}}{\tau} \right\} dx$$

$$- \int_{b_1}^{c_1} \frac{e^{-(x-a)^2/2\sigma_p^2}}{\sigma_p\sqrt{(2\pi)}} \left\{ \frac{e^{-(b_1-x-\rho)^2/2\rho^2}}{\tau} \right\} dx \quad 5.41(7)$$

$$U_R = \int_{-\infty}^{b_1} \frac{e^{-(x-a)^2/2\sigma_p^2}}{\sigma_p \sqrt{(2\pi)}} \left\{ 1 - \frac{e^{-(b_1 - x + \rho)^2/2\rho^2}}{\tau} \right\} dx$$

$$+ \int_{b_1}^{c_1} \frac{e^{-(x-a)^2/2\sigma_p^2}}{\sigma_p \sqrt{(2\pi)}} \left\{ \frac{e^{-(b_1 - x - \rho)^2/2\rho^2}}{\tau} \right\} dx$$

$$+ \int_{-\infty}^{c_1} \frac{e^{-(x-a)^2/2\sigma_p^2}}{\sigma_p \sqrt{(2\pi)}} \left\{ \frac{e^{-(b_2 - x + \rho)^2/2\rho^2}}{\tau} \right\} dx \qquad 5.41(8)$$

$$U = U_A + U_R \qquad\qquad\qquad\qquad 5.41(9)$$

The various constants A, B and C can be seen by comparison with the previous sets.

Worked Example

5.42 The standard deviation of parts manufactured on a certain machine is known to be σ_p, and parts lying in the range $\pm 3\sigma_p$ about the mean value can be accepted as correct parts. These parts are measured by means of a sizing device which gives the individual size of each piece measured, unlike the sizing machine of paragraph 5.02 et seq. which either accepted parts lying in a given tolerance range or rejected them if they lay outside this range. The present sizing device has an uncertainty of measurement such that its standard deviation σ_m is equal to the total tolerance range of manufacture, that is $6\sigma_p$ divided by 20. Let the tolerance range be denoted by T, and in order to minimize the fraction of incorrect parts accepted, let the acceptance limits be set within the tolerance limits by $2\sigma_m$ each side of the tolerance zone. Find the fraction of incorrect parts accepted, the fraction of correct parts rejected, the fraction of incorrect parts rejected and the fraction of correct parts accepted on the assumption that the distribution functions involved are Gaussian, and that the mean value of the manufactured parts is mid-way between the tolerance limits.

If we express the various constants in terms of the total tolerance T we have

$$\sigma_p = T/6 = 0.166'T$$

$$\sigma_m = T/20 = 0.05T$$

The acceptance limits will thus be set in by $2\sigma_m = 0.1T$ from the tolerance limits. It is easily verified that the limits k_1 and k_2 for each integral, and the exponent $(B^2/4A) - C$ are all independent of a. This can be shown by typically writing $c_1 = a - \gamma_1$, $b_1 = a - \beta_1$, $b_2 = a + \beta_2$ and $c_2 = a + \gamma_2$ and substituting in a typical expression for k_1, k_2 or $(B^2/4A) - C$. In effect this means that we can treat a as zero in our expressions for the constants, and b and c are now measured from a, that is zero, and typically $c_1 = -\gamma$,

$b_1 = -\beta_1$, $b_2 = \beta_2$ and $c_2 = \gamma_2$ if we use the expressions given for b and c above. Now we were given that the tolerance limits were $\pm 3\sigma_p = \pm 0.5T$. The acceptance limits are thus

$$\pm(+3\sigma_p - 2\sigma_m) = \pm 0.4T$$

With $a = 0$, we thus have

$$c_1 = -0.5T, \quad b_1 = -0.4T$$

$$c_2 = +0.5T, \quad b_2 = +0.4T$$

since the mean of the manufactured parts is mid-way between the tolerance limits. From the expression 5.08(9) $\rho = 1.205\,08$ $\sigma_m = 0.060\,254\,T$, and from expression 5.07(3) $2\tau = 2.426\,123$.

C_R

5.43 Let us first calculate the fraction of correct parts rejected, namely C_R.

First Integral. Referring to the constants for C_R given in paragraph 5.36 we have

$$(B^2/4A) - C = 0, \quad \tau\sigma_p\sqrt{(2A)} = 1$$

$$k_1 = -3.0, \quad k_2 = -2.4$$

whence the first integral becomes

$$\tfrac{1}{2}(-0.983\,60 + 0.997\,30) \equiv 0.685\% \equiv 0.006\,85$$

as a fraction.

Second Integral

$$(B^2/4A) - C = -1.837\,53$$

$$k_1 = -3.521\,17, \quad k_2 = -1.756\,40$$

and

$$2\tau\sigma_p\sqrt{(2A)} = 7.135\,91$$

whence the second integral yields $0.175\% \equiv 0.001\,75$ as a fraction.

Third Integral

$$(B^2/4A) - C = -3.372\,26$$

$$k_1 = 0.124\,46, \quad k_2 = 16.0074$$

and

$$2\tau\sigma_p\sqrt{(2A)} = 7.135\,91$$

yielding the third integral as $0.433\% \equiv 0.004\,33$ as a fraction.

From an inspection of the integrals for C_R it is at once apparent that if the limits are symmetrical about the mean value a, then integral (1) is equal to integral (5), integral (2) is equal to integral (6) and integral (3) is equal to integral (4). Thus the total integral for C_R is equal to twice the sum of the first three integrals†, whence $C_R = 1.886\% \equiv 0.018\ 86$ as a fraction.

$$W_A$$

5.44   We now calculate the fraction or percentage of incorrect parts accepted, where $W_A = U_A + O_A$. First we calculate U_A (see paragraph 5.38).

First Integral. For the first integral

$$(B^2/4A) - C = -1.8375$$

$$k_1 = -\infty, \quad k_2 = -3.521\ 17$$

$$2\tau\sigma_p\sqrt{(2A)} = 7.135\ 91$$

yielding the first integral as $0.000\ 964\% \equiv 0.000\ 009\ 64$ as a fraction.

Second Integral

$$(B^2/4A) - C = -3.3722$$

$$k_1 = -\infty, \quad k_2 = -16.0074$$

$$2\tau\sigma_p\sqrt{(2A)} = 7.135\ 91$$

yielding integral (2) equal to zero.

5.45   From a consideration of the integrals for W_A, in view of the symmetry of the limits, it is apparent that integral (1) equals integral (3) and integral (2) equals integral (4), and so W_A is equal to twice the sum of the first two integrals, that is $W_A = 0.001\ 928\% \equiv 0.000\ 019\ 28$ as a fraction. The percentage of incorrect workpieces made, since the tolerance limits were fixed at $\pm 3\sigma_p$, is equal to 0.27%. Thus, using the relation $W_A + W_R = W_p$, the fraction of incorrect workpieces made, we have

$$W_R = W_p - W_A = 0.2700 - 0.001\ 928 = 0.268\ 072\%$$

Similarly, since $C_A + C_R = C_p$, the fraction of correct workpieces made; we have

$$C_A = 99.73 - 1.886 = 97.844\%$$

Note that whilst the percentage of incorrect workpieces accepted has been reduced to a very small percentage, $0.001\ 928\%$, the percentage of correctly made workpieces rejected, C_R, is equal to 1.886% and is a much larger percentage than that actually made incorrectly, that is 0.27%.

† Due account must be taken of signs. See expression for C_R.

Parts with Rectangular Distribution

5.46 The third application we shall deal with is an extreme one, because it deals with the case when the distribution of manufactured parts is rectangular. It is interesting because it allows one to investigate a number of special cases. Figure 5.46 gives the layout of the case to be considered.

As before a is the offset of the mean of the parts distribution with respect to the centre of the tolerance range. $c_1 b_1$ and $b_2 c_2$ are amounts the tolerance limits are inset in order to reduce the acceptance of oversize and undersize parts. The range of the rectangular distribution is $2f$.

5.47 We obtain the required expressions by replacing the Gaussian function

$$\frac{e^{-(x-a)^2/2\sigma_p^2}}{\sigma_p\sqrt{(2\pi)}}$$

in equations 5.29(9), 5.30(8), 5.31(5), 5.32(2), 5.32(4) and 5.32(5) by the rectangular expression $1/2f$. Also where $-\infty$ and $+\infty$ limits were involved, these must be replaced by the finite limits of the rectangular distribution, namely $-f + a$ and $f + a$ respectively.

5.48 Further, if $-f + a$ is greater than either c_1 or b_1 then these limits must be replaced by $-f + a$. Similarly if $f + a$ is less than either c_2 or b_2 then limits involving these quantities must be replaced by $f + a$.

5.49 Making the substitution mentioned above, we derive the required expressions as

$$
\begin{aligned}
C_A = & \int_{c_1}^{b_1} \frac{dx}{2f}\left\{\frac{e^{-(b_1-x+\rho)^2/2\rho^2}}{\tau} - \frac{e^{-(b_2-x+\rho)^2/2\rho^2}}{\tau}\right\} \\
& + \int_{b_1}^{b_2} \frac{dx}{2f}\left\{1 - \frac{e^{-(b_2-x+\rho)^2/2\rho^2}}{\tau} - \frac{e^{-(b_1-x-\rho)^2/2\rho^2}}{\tau}\right\} \\
& + \int_{b_2}^{c_2} \frac{dx}{2f}\left\{\frac{e^{-(b_2-x-\rho)^2/2\rho^2}}{\tau} - \frac{e^{-(b_1-x-\rho)^2/2\rho^2}}{\tau}\right\}
\end{aligned}
\qquad 5.49(1)
$$

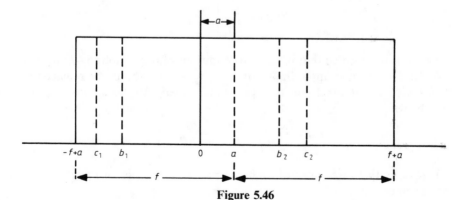

Figure 5.46

All integrals are of the form

$$\int_{y_1}^{y_2} e^{-(b-x+\rho)^2/2\rho^2} \, dx = I \qquad\qquad 5.49(2)$$

Putting $(b - x + \rho)/\sqrt{2}\rho = z/\sqrt{2}$, we have on differentiating that

$$-dx = dz\rho$$

The new limits are thus

$$z_1 = (b - y_1 + \rho)/\rho \qquad\qquad 5.49(3)$$

and

$$z_2 = (b - y_2 + \rho)/\rho \qquad\qquad 5.49(4)$$

Thus

$$I = \frac{-\rho}{2f\tau} \int_{(b - y_1 + \rho)/\rho}^{(b - y_2 + \rho)/\rho} e^{-z^2/2} \, dz$$

$$= \frac{\rho}{2f\tau} \int_{(b - y_2 + \rho)/\rho}^{(b - y_1 + \rho)/\rho} e^{-z^2/2} \, dz \qquad\qquad 5.49(5)$$

If 5.49(1) is now transformed as above, we have that

$$C_A = \frac{\rho}{2f\tau} \underbrace{\int_{(b_1 - b_1 + \rho)/\rho}^{(b_1 - c_1 + \rho)/\rho}}_{1} e^{-z^2/2} \, dz - \frac{\rho}{2f\tau} \int_{(b_2 - b_1 + \rho)/\rho}^{(b_2 - c_1 + \rho)/\rho} e^{-z^2/2} \, dz$$

$$+ \frac{(b_2 - b_1)}{2f} - \frac{\rho}{2f\tau} \underbrace{\int_{(b_2 - b_2 + \rho)/\rho}^{(b_2 - b_1 + \rho)/\rho}}_{1} e^{-z^2/2} \, dz - \frac{\rho}{2f\tau} \overbrace{\int_{(b_1 - b_2 - \rho)/\rho}^{(b_1 - b_1 - \rho)/\rho}}^{-1} e^{-z^2/2} \, dz$$

$$+ \frac{\rho}{2f\tau} \overbrace{\int_{(b_2 - c_2 - \rho)/\rho}^{(b_2 - b_2 - \rho)/\rho}}^{-1} e^{-z^2/2} \, dz - \frac{\rho}{2f\tau} \int_{(b_1 - c_2 - \rho)/\rho}^{(b_1 - b_2 - \rho)/\rho} e^{-z^2/2} \, dz \qquad 5.49(6)$$

5.50 It is to be noted that where the conditions of paragraphs 5.47 and 5.48 prevail, the lower or upper limits, or both, y_1 and y_2, should be replaced by $-f + a$ or $f + a$ whichever is appropriate. Finally, the components of C_A can be written as

$$\frac{\rho\sqrt{(2\pi)}}{4f\tau} \left[\underset{-k_2 \, \text{to} \, k_2}{P} - \underset{-k_1 \, \text{to} \, k_1}{P} \right]$$

where k_1 is the lower limit of each component and k_2 is an upper limit of each component.

$P_{-k\,\text{to}\,k}$ is the Gaussian probability between the limits $-k\sigma$ to $k\sigma$. For an explanation of the transformation of

$$\int_{k_1}^{k_2} e^{-z^2/2}\, dz \quad \text{to} \quad \frac{\sqrt{(2\pi)}}{2}\left[P_{-k_2\,\text{to}\,k_2} - P_{-k_1\,\text{to}\,k_1} \right]$$

see equation 5.33(7).

It is to be noted that if either k_1 or k_2 is negative that the sign of P for that term should be changed to negative.

5.51 The components of C_A can now be written as

$$C_{A1} = \frac{\rho\sqrt{(2\pi)}}{4f\tau}\left[P_{-k_2\,\text{to}\,k_2} - P_{-k_1\,\text{to}\,k_1} \right] \qquad 5.51(1)$$

where

if $-f + a \leqslant c_1$ then $k_2 = (b_1 - c_1 + \rho)/\rho$
    and $k_1 = (b_1 - b_1 + \rho)/\rho = 1$
if $-f + a > c_1$ and $-f + a \leqslant b_1$ then $k_2 = (b_1 + f - a + \rho)/\rho$
    and $k_1 = (b_1 - b_1 + \rho)/\rho = 1$
if $-f + a > b_1$ then $k_2 = (b_1 + f - a + \rho)/\rho$
    and $k_1 = (b_1 + f - a + \rho)/\rho = 1$, which means that the integral is zero, i.e. equal limits.

$$C_{A2} = \frac{\rho\sqrt{(2\pi)}}{4f\tau}\left[P_{-k_2\,\text{to}\,k_2} - P_{-k_1\,\text{to}\,k_1} \right] \qquad 5.51(2)$$

where

if $-f + a \leqslant c_1$ then $k_2 = (b_2 - c_1 + \rho)/\rho$
    and $k_1 = (b_2 - b_1 + \rho)/\rho$
if $-f + a > c_1$ and $-f + a \leqslant b_1$ then $k_2 = (b_2 + f - a + \rho)/\rho$
    and $k_1 = (b_2 - b_1 + \rho)/\rho$
if $-f + a > b_1$ then $k_2 = (b_2 + f - a + \rho)/\rho$
    and $k_1 = (b_2 + f - a + \rho)$, i.e. integral $= 0$ (equal limits).

If $-f + a \geqslant c_1$ and $-f + a \leqslant b_1$ then
$$C_{A3} = (b_2 - b_1)/2f$$
if $-f + a > b_1$ and $f + a \geqslant b_2$ then
$$C_{A3} = (b_2 + f - a)/2f \qquad 5.51(3)$$
if $-f + a > b_1$ and $f + a < b_2$ then
$$C_{A3} = (f + a + f - a)/2f = 1$$

$$C_{A4} = \frac{\rho\sqrt{(2\pi)}}{4f\tau}\left[P_{-k_2\,\text{to}\,k_2} - P_{-k_1\,\text{and}\,k_1} \right] \qquad 5.51(4)$$

where

if $-f + a \leqslant c_1$ then $k_2 = (b_2 - b_1 + \rho)/\rho$
    and $k_1 = (b_2 - b_2 + \rho)/\rho = 1$
if $-f + a > c_1$ and $-f + a \leqslant b_1$ then $k_2 = (b_2 - b_1 + \rho)/\rho$
    and $k_1 = (b_2 - b_2 + \rho)/\rho = 1$
if $-f + a > b_1$ and $f + a \geqslant b_2$ then $k_2 = (b_2 + f - a + \rho)/\rho$
    and $k_1 = (b_2 - b_2 + \rho)/\rho = 1$
if $-f + a > b_1$ and $f + a < b_2$ then $k_2 = (b_2 + f - a + \rho)/\rho$
    and $k_1 = (b_2 - f - a + \rho)/\rho$

$$C_{A5} = \frac{\rho\sqrt{(2\pi)}}{4f\tau}\left[\underset{-k_2 \text{ to } k_2}{P} - \underset{-k_1 \text{ to } k_1}{P}\right] \qquad 5.51(5)$$

where

if $-f + a \leqslant c_1$ then $k_2 = (b_1 - b_1 - \rho)/\rho = -1$
    and $k_1 = (b_1 - b_2 - \rho)/\rho$
if $-f + a > c_1$ and $-f + a \leqslant b_1$ then $k_2 = (b_1 - b_1 - \rho)/\rho = -1$
    and $k_1 = (b_1 - b_2 - \rho)/\rho$
if $-f + a > b_1$ and $f + a \geqslant b_2$ then $k_2 = (b_1 + f - a - \rho)/\rho$
    and $k_1 = (b_1 - b_2 - \rho)/\rho$
if $-f + a > b_1$ and $f + a < b_2$ then $k_2 = (b_1 + f - a - \rho)/\rho$
    and $k_1 = (b_1 - f - a - \rho)/\rho$

$$C_{A6} = \frac{\rho\sqrt{(2\pi)}}{4f\tau}\left[\underset{-k_2 \text{ to } k_2}{P} - \underset{-k_1 \text{ to } k_1}{P}\right] \qquad 5.51(6)$$

where

if $f + a \geqslant c_2$ then $k_2 = (b_2 - b_2 - \rho)/\rho = -1$
    and $k_1 = (b_2 - c_2 - \rho)/\rho$
if $f + a \geqslant b_2$ and $f + a < c_2$ then $k_2 = (b_2 - b_2 - \rho)/\rho = -1$
    and $k_1 = (b_2 - f - a - \rho)/\rho$
if $f + a < b_2$ then $k_2 = (b_2 - f - a - \rho)/\rho$
    and $k_1 = (b_2 - f - a - \rho)/\rho$, i.e. integral is zero.

$$C_{A7} = \frac{\rho\sqrt{(2\pi)}}{4f\tau}\left[\underset{-k_2 \text{ to } k_2}{P} - \underset{-k_1 \text{ to } k_1}{P}\right] \qquad 5.51(7)$$

where

if $f + a \geqslant c_2$ then $k_2 = (b_1 - b_2 - \rho)/\rho$ and $k_1 = (b_1 - c_2 - \rho)/\rho$
if $f + a \geqslant b_2$ and $f + a < c_2$ then $k_2 = (b_1 - b_2 - \rho)/\rho$
    and $k_1 = (b_1 - f - a - \rho)/\rho$
if $f + a < b_2$ then $k_2 = (b_1 - f - a - \rho)/\rho$
    and $k_1 = (b_1 - f - a - \rho)/\rho$, i.e. integral is zero.

Finally

$$C_A = C_{A1} - C_{A2} + C_{A3} - C_{A4} - C_{A5} + C_{A6} - C_{A7} \qquad 5.51(8)$$

5.52 The required expression for C_R after substitution is

$$C_R = \int_{c_1}^{b_1} \frac{dx}{2f} \left\{ 1 - \frac{e^{-(b_1 - x + \rho)^2/2\rho^2}}{\tau} \right\} + \int_{b_1}^{c_2} \frac{dx}{2f} \left\{ \frac{e^{-(b_1 - x - \rho)^2/2\rho^2}}{\tau} \right\}$$

$$+ \int_{c_1}^{b_2} \frac{dx}{2f} \left\{ \frac{e^{-(b_2 - x + \rho)^2/2\rho^2}}{\tau} \right\} + \int_{b_2}^{c_2} \frac{dx}{2f} \left\{ 1 - \frac{e^{-(b_2 - x - \rho)^2/2\rho^2}}{\tau} \right\} \qquad 5.52(1)$$

As before each integral reduces to the form

$$I = \frac{\rho}{2f\tau} \int_{(b - y_2 + \rho)/\rho}^{(b - y_1 + \rho)/\rho} e^{-z^2/2} \qquad 5.52(2)$$

where y_1 is the lower limit and y_2 is the upper limit in the appropriate term in 5.52(1). Thus

$$C_R = \left[\frac{b_1 - c_1}{2f} \right] - \underbrace{\int_{(b_1 - b_1 + \rho)/\rho}^{(b_1 - c_1 + \rho)/\rho} e^{-z^2/2}\, dz}_{1}$$

$$+ \frac{\rho}{2f\tau} \int_{(b_1 - c_2 - \rho)/\rho}^{\overbrace{(b_1 - b_1 - \rho)/\rho}^{-1}} e^{-z^2/2}\, dz$$

$$+ \frac{\rho}{2f\tau} \underbrace{\int_{(b_2 - b_2 + \rho)/\rho}^{(b_2 - c_1 + \rho)/\rho} e^{-z^2/2}\, dz}_{1}$$

$$+ \left[\frac{c_2 - b_2}{2f} \right] - \frac{\rho}{2f\tau} \int_{(b_2 - c_2 - \rho)/\rho}^{\overbrace{(b_2 - b_2 - \rho)/\rho}^{-1}} e^{-z^2/2}\, dz \qquad 5.52(3)$$

5.53 Proceeding as before the components of C_R can now be written as

$$\left.\begin{array}{l} \text{if } -f + a \leqslant c_1 \text{ then} \\ \quad C_{R1} = (b_1 - c_1)/2f \\ \text{if } -f + a > c_1 \text{ and } -f + a \leqslant b_1 \text{ then} \\ \quad C_{R1} = (b_1 + f - a)/2f \\ \text{if } -f + a > b_1 \text{ then} \\ \quad C_{R1} = 0 \end{array}\right\} \qquad 5.53(1)$$

$$C_{R2} = \frac{\rho\sqrt{(2\pi)}}{4f\tau} \left[P_{-k_2 \text{ to } k_2} - P_{-k_1 \text{ to } k_1} \right] \qquad 5.53(2)$$

where

if $-f + a \leqslant c_1$ then $k_2 = (b_1 - c_1 + \rho)/\rho$
  and $k_1 = (b_1 - b_1 + \rho)/\rho = 1$
if $-f + a > c_1$ and $-f + a \leqslant b_1$ then $k_2 = (b_1 + f - a + \rho)/\rho$
  and $k_1 = (b_1 - b_1 + \rho)/\rho = 1$
if $-f + a > b_1$ then $k_2 = (b_1 + f - a + \rho)/\rho$
  and $k_1 = (b_1 + f - a + \rho)/\rho$, i.e. limits are equal and so $C_{R2} = 0$
in this case.

$$C_{R3} = \frac{\rho \sqrt{(2\pi)}}{4f\tau} \left[\underset{-k_2 \text{ to } k_2}{P} - \underset{-k_1 \text{ to } k_1}{P} \right] \qquad 5.53(3)$$

where

if $-f + a \leqslant c_1$ then $k_2 = (b_1 - b_1 - \rho)/\rho = -1$
  and $k_1 = (b_1 - c_2 - \rho)/\rho$
if $-f + a > c_1$ and $-f + a \leqslant b_1$ and $f + a > c_2$ then
  $k_2 = (b_1 - b_1 - \rho)/\rho = -1$ and $k_1 = (b_1 - c_2 - \rho)/\rho$
if $-f + a > c_1$ and $-f + a \leqslant b_1$ and $f + a \leqslant c_2$ then
  $k_2 = (b_1 - b_1 - \rho)/\rho = -1$ and $k_1 = (b_1 - f - a - \rho)/\rho$
if $-f + a > b_1$ and $f + a \leqslant c_2$ then $k_2 = (b_1 + f - a - \rho)/\rho$
  and $k_1 = (b_1 - f - a - \rho)/\rho$
if $-f + a > b_1$ and $f + a > c_2$ then $k_2 = (b_1 + f - a - \rho)/\rho$
  and $k_1 = (b_1 - c_2 - \rho)/\rho$

$$C_{R4} = \frac{\rho \sqrt{(2\pi)}}{4f\tau} \left[\underset{-k_2 \text{ to } k_2}{P} - \underset{-k_1 \text{ to } k_1}{P} \right] \qquad 5.53(4)$$

where

if $-f + a \leqslant c_1$ then $k_2 = (b_2 - c_1 + \rho)/\rho$
  and $k_1 = (b_2 - b_2 + \rho)/\rho = 1$
if $-f + a > c_1$ and $-f + a \leqslant b_1$ and $f + a > b_2$ then
  $k_2 = (b_2 + f - a + \rho)/\rho$ and $k_1 = (b_2 - b_2 + \rho)/\rho = 1$
if $-f + a > b_1$ and $f + a \geqslant b_2$ then $k_2 = (b_2 + f - a + \rho)/\rho$
  and $k_1 = (b_2 - b_2 + \rho)/\rho = 1$
if $-f + a > b_1$ and $f + a < b_2$ then
  $k_2 = (b_2 + f - a + \rho)/\rho$ and $k_1 = (b_2 - f - a + \rho)/\rho$

$$\left. \begin{array}{l} \text{If } f + a > c_2 \text{ then} \\[6pt] \qquad C_{R5} = (c_2 - b_2)/2f \\[6pt] \text{if } f + a \geqslant b_2 \text{ and } f + a \leqslant c_2 \text{ then} \\[6pt] \qquad C_{R5} = (f + a - b_2)/2f \\[6pt] \text{if } f + a < b_2 \text{ then} \\[6pt] \qquad C_{R5} = 0 \end{array} \right\} \qquad 5.53(5)$$

$$C_{R6} = \frac{\rho \sqrt{(2\pi)}}{4f\tau} \left[\underset{-k_2 \text{ to } k_2}{P} - \underset{-k_1 \text{ to } k_1}{P} \right] \qquad 5.53(6)$$

where

if $f + a > c_2$ then $k_2 = (b_2 - b_2 - \rho)/\rho = -1$
 and $k_1 = (b_2 - c_2 - \rho)/\rho$
if $f + a \geqslant b_2$ and $f + a \leqslant c_2$ then
  $k_2 = (b_2 - b_2 - \rho)/\rho = -1$ and $k_1 = (b_2 - f - a - \rho)/\rho$
if $f + a < b_2$ then $k_2 = (b_2 - f - a - \rho)/\rho$
 and $k_1 = (b_2 - f - a - \rho)/\rho$, i.e. in this case of equal limits $C_{R6} = 0$.

Finally

$$C_R = C_{R1} - C_{R2} + C_{R3} + C_{R4} + C_{R5} - C_{R6} \qquad 5.53(7)$$

5.54 The required expression for O_A after substitution is

$$O_A = \int_{c_2}^{\infty} \frac{dx}{2f} \left\{ \frac{e^{-(b_2 - x - \rho)^2/2\rho^2}}{\tau} - \frac{e^{-(b_1 - x - \rho)^2/2\rho^2}}{\tau} \right\} \qquad 5.54(1)$$

As before each integral can be transformed to the form

$$I = \frac{\rho}{2f\tau} \int_{(b - y_2 + \rho)/\rho}^{(b - y_1 + \rho)/\rho} e^{-z^2/2} \, dz \qquad 5.54(2)$$

If 5.54(1) is now transferred as above, we have that

$$O_A = \frac{\rho}{2f\tau} \int_{(b_2 - f - a - \rho)/\rho}^{(b_2 - c_2 - \rho)/\rho} e^{-z^2/2} \, dz - \frac{\rho}{2f\tau} \int_{(b_1 - f - a - \rho)/\rho}^{(b_1 - c_2 - \rho)/\rho} e^{-z^2/2} \, dz$$

5.55 The components of O_A can now be written as

$$O_{A1} = \frac{\rho\sqrt{(2\pi)}}{4f\tau} \left[\underset{-k_2 \text{ to } k_2}{P} - \underset{-k_1 \text{ to } k_1}{P} \right] \qquad 5.55(1)$$

where

if $f + a \geqslant c_2$ then $k_2 = (b_2 - c_2 - \rho)/\rho$
 and $k_1 = (b_2 - f - a - \rho)/\rho$
if $f + a < c_2$ then $k_2 = (b_2 - f - a - \rho)/\rho$
 and $k_1 = (b_2 - f - a - \rho)/\rho$, i.e. in this case of equal limits $O_{A1} = 0$.

$$O_{A2} = \frac{\rho\sqrt{(2\pi)}}{4f\tau} \left[\underset{-k_2 \text{ to } k_2}{P} - \underset{-k_1 \text{ to } k_1}{P} \right] \qquad 5.55(2)$$

where

if $f + a \geqslant c_2$ then $k_2 = (b_1 - c_2 - \rho)/\rho$
 and $k_1 = (b_1 - f - a - \rho)/\rho$
if $f + a < c_2$ then $k_2 = (b_1 - f - a - \rho)/\rho$
 and $k_1 = (b_1 - f - a - \rho)/\rho$, i.e. in this case of equal limits $O_{A2} = 0$.
Finally

$$O_A = O_{A1} - O_{A2} \qquad 5.55(3)$$

5.56   The expression for U_A after substitution of the rectangular distribution is given by

$$U_A = \int_{-\infty}^{c_1} \frac{dx}{2f} \left\{ \frac{e^{-(b_1-x+\rho)^2/2\rho^2}}{\tau} - \frac{e^{-(b_2-x+\rho)^2/2\rho^2}}{\tau} \right\} \qquad 5.56(1)$$

Transforming each integral to the form

$$I = \frac{\rho}{2f\tau} \int_{(b-y_2+\rho)/\rho}^{(b-y_1+\rho)/\rho} e^{-z^2/2} \, dz \qquad 5.56(2)$$

as before we have

$$U_A = \frac{\rho}{2f\tau} \int_{(b_1-c_1+\rho)/\rho}^{(b_1+f-a+\rho)/\rho} e^{-z^2/2} \, dz$$

$$- \frac{\rho}{2f\tau} \int_{(b_2-c_1+\rho)/\rho}^{(b_2+f-a+\rho)/\rho} e^{-z^2/2} \, dz \qquad 5.56(3)$$

5.57   The components of U_A can now be written as

$$U_{A1} = \frac{\rho\sqrt{(2\pi)}}{4f\tau} \left[\underset{-k_2 \text{ to } k_2}{P} - \underset{-k_1 \text{ to } k_1}{P} \right] \qquad 5.57(1)$$

where

if $-f + a \leqslant c_1$ then $k_2 = (b_1 + f - a + \rho)/\rho$
        and $k_1 = (b_1 - c_1 + \rho)/\rho$
if $-f + a > c_1$ then $k_2 = (b_1 + f - a + \rho)/\rho$
        and $k_1 = (b_1 + f - a + \rho)/\rho$, i.e. in this case of equal limits $U_{A1} = 0$.

$$U_{A2} = \frac{\rho\sqrt{(2\pi)}}{4f\tau} \left[\underset{-k_2 \text{ to } k_2}{P} - \underset{-k_1 \text{ to } k_1}{P} \right] \qquad 5.57(2)$$

where

if $-f + a \leqslant c_1$ then $k_2 = (b_2 + f - a + \rho)/\rho$
        and $k_1 = (b_2 - c_1 + \rho)/\rho$
if $-f + a > c_1$ then $k_2 = (b_2 + f - a + \rho)/\rho$
        and $k_1 = (b_2 + f - a + \rho)/\rho$, i.e. in this case $U_{A2} = 0$.
Finally

$$U_A = U_{A1} - U_{A2} \qquad 5.57(3)$$

5.58   The expression for O_R after substitutiong of the rectangular distribution is given by

$$O_R = \int_{c_2}^{\infty} \frac{dx}{2f} \left\{ \frac{e^{-(b_1-x-\rho)^2/2\rho^2}}{\tau} + 1 - \frac{e^{-(b_2-x-\rho)^2/2\rho^2}}{\tau} \right\} \qquad 5.58(1)$$

Transforming each integral, as before, to the form

$$I = \frac{\rho}{2f\tau} \int_{(b-y_2+\rho)/\rho}^{(b-y_1+\rho)/\rho} e^{-z^2/2}\, dz \qquad 5.58(2)$$

we have

$$O_R = \int_{(b_1-f-a-\rho)/\rho}^{(b_1-c_2-\rho)/\rho} e^{-z^2/2}\, dz + \left[\frac{f+a-c_2}{2f}\right]$$
$$- \int_{(b_2-f-a-\rho)/\rho}^{(b_2-c_2-\rho)/\rho} e^{-z^2/2}\, dz \qquad 5.58(3)$$

5.59 The components of O_R are given as

$$O_{R1} = \frac{\rho\sqrt{(2\pi)}}{4f\tau}\left[\underset{-k_2 \text{ to } k_2}{P} - \underset{-k_1 \text{ to } k_1}{P} \right] \qquad 5.59(1)$$

where

if $f + a > c_2$ then $k_2 = (b_1 - c_2 - \rho)/\rho$
  and $k_1 = (b_1 - f - a - \rho)/\rho$
if $f + a \leqslant c_2$ then $k_2 = (b_1 - f - a - \rho)/\rho$
  and $k_1 = (b_1 - f - a - \rho)/\rho$, i.e. in this case of equal limits $O_{R1} = 0$.

$$\left.\begin{array}{l} \text{If } f + a > c_2 \text{ then} \\[4pt] \qquad O_{R2} = (f + a - c_2)/2f \\[4pt] \text{if } f + a \leqslant c_2 \text{ then} \\[4pt] \qquad O_{R2} = 0 \end{array}\right\} \qquad 5.59(2)$$

$$O_{R3} = \frac{\rho\sqrt{(2\pi)}}{4f\tau}\left[\underset{-k_2 \text{ to } k_2}{P} - \underset{-k_1 \text{ to } k_1}{P} \right] \qquad 5.59(3)$$

where

if $f + a > c_2$ then $k_2 = (b_2 - c_2 - \rho)/\rho$
  and $k_1 = (b_2 - f - a - \rho)/\rho$
if $f + a \leqslant c_2$ then $k_2 = (b_2 - f - a - \rho)/\rho$
  and $k_1 = (b_2 - f - a - \rho)/\rho$, i.e. in this case of equal limits $O_{R3} = 0$.
Finally

$$O_R = O_{R1} + O_{R2} - O_{R3} \qquad 5.59(4)$$

5.60 Lastly the expression for U_R after substitution of the rectangular distribution is given by

$$U_R = \int_{-\infty}^{c_1} \frac{dx}{2f}\left\{ 1 - \frac{e^{-(b_1-x+\rho)^2/2\rho^2}}{\tau} + \frac{e^{-(b_2-x+\rho)^2/2\rho^2}}{\tau} \right\} \qquad 5.60(1)$$

Transforming each integral, as before, to the form

$$I = \frac{\rho}{2f\tau} \int_{(b - y_2 + \rho)/\rho}^{(b - y_1 + \rho)/\rho} e^{-z^2/2} \, dz \qquad 5.60(2)$$

we have

$$U_R = \left[\frac{c + f - a}{2f} \right] - \frac{\rho}{2f\tau} \int_{(b_1 - c_1 + \rho)/\rho}^{(b_1 + f - a + \rho)/\rho} e^{-z^2/2}$$

$$+ \frac{\rho}{2f\tau} \int_{(b_2 - c_1 + \rho)/\rho}^{(b_2 + f - a + \rho)/\rho} e^{-z^2/2} \, dz \qquad 5.60(3)$$

5.61   The components of U_R are given as

$$\left.\begin{array}{l} \text{if } -f + a \leqslant c_1 \text{ then} \\[4pt] \quad U_{R1} = (c_1 + f - a)/2f \\[4pt] \text{if } -f + a > c_1 \text{ then} \\[4pt] \quad U_{R1} = 0 \end{array}\right\} \qquad 5.61(1)$$

$$U_{R2} = \frac{\rho\sqrt{(2\pi)}}{4f\tau} \left[\underset{-k_2 \text{ to } k_2}{P} - \underset{-k_1 \text{ to } k_1}{P} \right] \qquad 5.61(2)$$

where

if $-f + a \leqslant c_1$ then $k_2 = (b_1 + f - a + \rho)/\rho$
  and $k_1 = (b_1 - c_1 + \rho)/\rho$
if $-f + a > c_1$ then $k_2 = (b_1 + f - a + \rho)/\rho$
  and $k_1 = (b_1 + f - a + \rho)/\rho$, i.e. in this case of equal limits $U_{R2} = 0$.

$$U_{R3} = \frac{\rho\sqrt{(2\pi)}}{4f\tau} \left[\underset{-k_2 \text{ to } k_2}{P} - \underset{-k_1 \text{ to } k_1}{P} \right] \qquad 5.61(3)$$

where

if $-f + a \leqslant c_1$ then $k_2 = (b_2 + f - a + \rho)/\rho$
  and $k_1 = (b_2 - c_1 + \rho)/\rho$
if $-f + a > c_1$ then $k_2 = (b_2 + f - a + \rho)/\rho$
  and $k_1 = (b_2 + f - a + \rho)/\rho$, i.e. in this case of equal limits $U_{R3} = 0$.
Finally

$$U_R = U_{R1} - U_{R2} + U_{R3} \qquad 5.61(4)$$

5.62   The formulae of application two have been used to prepare a set of tables so that the effect of varying the standard deviation of the manufactured parts σ_p and that of the measuring machine σ_m, together with the offset of the mean a of the manufactured parts and the inset of the tolerance limits

$\pm T_p/2$ by $i = k_m \sigma_m$ can be seen. The table concerned is numbered XXII, dealing with the Gaussian distribution of manufactured parts.

Worked Example

5.63 The following worked example will show how Table XXII may be used. A manufacturer is required to make parts which have a diameter which must be maintained within a tolerance range of -0.0003 to $+0.0003$ in for a given size. He has manufacturing equipment which can produce parts such that their standard deviation from the mean varies by ± 0.00015 in. His measuring equipment has a standard deviation of 0.000030 in and the manufacturing equipment can produce parts such that the mean of the distribution of the manufactured parts lies within 0.0001 in of the required value for the parts. It is required to find the measurement conditions necessary for the proportion of incorrect parts accepted not to exceed one per cent of the total parts made and for the proportion of correct parts rejected to be kept to a minimum.

5.64 Firstly we convert all the required variables to a percentage of the total tolerance range, i.e. 0.0006 in. 'a' the offset of the mean as a percentage is $0.0001/0.0006 \times 100\%$, i.e. 16.67%, $\sigma_p = 0.00015/0.0006 \times 100 = 25\%$ and $\sigma_m = 0.00003/0.0006 \times 100 = 5.0\%$. k_p the Gaussian tolerance factor for the manufactured parts is thus given by $2k_p\sigma_p = T_p$, i.e. $2k_p \times 25 = 100$ or $k_p = 2$. Consulting Table XXII we look at the section having $a = 15$, $\sigma_p = 25$ and $\sigma_m = 5$. There are three lines which meet these conditions, namely with $k_m = 0$, 1 and 2. *Because of the offset of the mean of the manufactured parts far more oversize parts are made than undersize parts.* (See Figure 5.64.)

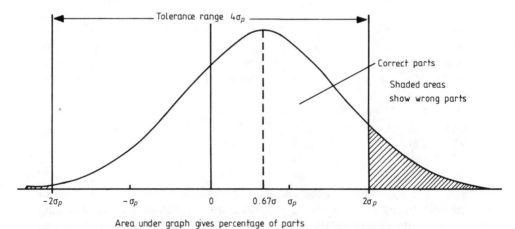

Figure 5.64 Area under graph gives percentage of parts

Uncertainty, Calibration and Probability

Looking up the three lines mentioned in the tables, we find on tabulating the results, the following values

a	i	k_m	s_m	k_p	s_p	C	C_A	C_R	U	U_A	U_R
15	0	0	5	2	25	91.46	89.90	1.56	0.47	0.08	0.39
15	5	1	5	2	25	91.46	86.87	4.59	0.47	0.02	0.45
15	10	2	5	2	25	91.46	82.08	9.38	0.47	0.00	0.46

O	O_A	O_R	W	W_A	W_R
8.08	0.99	7.08	8.54	1.07	7.47
8.08	0.22	7.86	8.54	0.24	8.3
8.08	0.03	8.05	8.54	0.03	8.51

We now look at the tables again, but with 'a' changed to 20, and with i, k_m, s_m, k_p and s_p the same. This gives the following results

a	i	k_m	s_m	k_p	s_p	C	C_A	C_R	U	U_A	U_R
20	0	0	5	2	25	88.24	86.37	1.86	0.26	0.05	0.21
20	5	1	5	2	25	88.24	82.82	5.42	0.26	0.01	0.25
20	10	2	5	2	25	88.24	77.40	10.84	0.26	0.00	0.25

O	O_A	O_R	W	W_A	W_R
11.51	1.3	10.19	11.76	1.36	10.40
11.51	0.29	11.22	11.76	0.30	11.46
11.51	0.03	11.47	11.76	0.04	11.73

We now interpolate (linear interpolation will be sufficient) for 'a' = 16.67 between corresponding lines of the three groups of three lines, giving

a	i	k_m	s_m	k_p	s_p	C	C_A	C_R	U	U_A	U_R
16.67	0	0	5	2	25	90.38	88.72	1.66	0.40	0.07	0.33
16.67	5	1	5	2	25	90.38	88.52	4.87	0.40	0.02	0.38
16.67	10	2	5	2	25	90.38	80.52	9.87	0.40	0.00	0.39

O	O_A	O_R	W	W_A	W_R
9.23	1.10	8.12	9.62	1.17	8.45
9.23	0.24	8.98	9.62	0.26	9.36
9.23	0.03	9.19	9.62	0.03	9.55

5.65 The parameters in the last three lines are now the same as those of the problem, i.e. $a = 16.67\%$, $s_m = 5.0\%$ and $s_p = 25\%$ with $k_p = 2$. Looking at the results we see that $W_A = 1.17$ is just larger than the required 1%, so we interpolate between lines 1 and 2 to find the value of k_m which will give 1%.

The final set of results is as follows:

a	i	k_m	s_m	k_p	s_p	C	C_A	C_R	U	U_A	U_R
16.67	0.95	0.19	5	2	25	90.38	88.12	2.27	0.40	0.06	0.34

O	O_A	O_R	W	W_A	W_R
9.23	0.94	8.28	9.62	1.00	8.62

We see that the number of correct parts rejected is 2.27%. If W_A is reduced below 1%, then the percentage of correct parts rejected rises rapidly. It is interesting to note that our criterion of 1% of wrong parts accepted has been attained with an inset of only $0.19 \times 5 = 0.95\%$ of the tolerance range.

Interpolation Procedure

5.66 If the values of k_p, s_p, s_m and a do not occur in the tables, the required result can be obtained by means of interpolation. k_p will of course be determined by the value of s_p and so will not be part of the interpolation. This leaves three variables s_p, s_m and a. Suppose that s_p were 17.8%, s_m were 6.2% and 'a' were 11.6% and it is required that the percentage of wrong parts accepted should not be more than 1%.

Firstly we look for two lines where tabulated values of s_p straddle the value of s_p required and where the values of s_m and a are the nearest to the required values both either above or below these values and where the value of k_m is the lowest consistent with satisfying the W_A requirement. Since the W_A required is 1% we can choose the two lines

a	i	k_m	s_m	k_p	s_p	C_R	W_A	
10	0	0	5	3	16.67	0.45	0.18	5.66(1)
10	0	0	5	2.5	20.00	0.81	0.42	5.66(2)

We also choose the two lines

10	5	1	5	3	16.67	1.5	0.04	5.66(3)
10	5	1	5	2.5	20.00	2.5	0.10	5.66(4)

From lines 5.66(1) and 5.66(2) we get

10	0	0	5	2.81	17.8	5.66(5)

and from lines 5.66(3) and 5.66(4) we get

10	5	1	5	2.81	17.8	5.66(6)

Now choose pairs of lines where $s_m > 6.2$, i.e. 7.5 say, and we get

a	i	k_m	s_m	k_p	s_p	C_R	W_A	
10	0	0	7.5	3	16.67			5.66(7)

and

10	0	0	7.5	2.5	20.00			5.66(8)

and

10	7.5	1	7.5	3	16.67			5.66(9)

and

10	7.5	1	7.5	2.5	20.00			5.66(10)

Lines 5.66(7) and 5.66(8) yield

10	0	0	7.5	2.81	17.8			5.66(11)

and lines 5.66(9) and 5.66(10) give

10	7.5	1	7.5	2.81	17.8			5.66(12)

From 5.66(5) and 5.66(11) interpolating between 5 and 7.5 for $s_m = 6.2$ we get

10	0	0	6.2	2.81	17.8			5.66(13)

Interpolating between 5.66(6) and 5.66(12) for s_m again equal to 6.2 we get

10	6.2	1	6.2	2.81	17.8			5.66(14)

We now repeat the above process for $a = 15$ with the other variables the same. This will yield

15	0	0	6.2	2.81	17.8			5.66(15)
15	6.2	1	6.2	2.81	17.8			5.66(16)

Interpolating now to obtain $a = 11.6\%$ we get using 5.66(13) and 5.66(15)

11.6	0	0	6.2	2.81	17.8			5.66(17)

and from 5.66(14) and 5.66(16) we get

11.6	6.2	1	6.2	2.81	17.8			5.66(18)

We can now interpolate between 5.66(17) and 5.66(18) to obtain the required value for W_A using k_m as the variable. Other problems may need the pairing of lines containing $k_m = 1$ and 2, 2 and 3, or intermediate values as required.

Examples on Chapter 5

Examples 5.1   Parts are manufactured on a machine, and these are acceptable provided they lie within the range $\pm T/2$ on either side of the mean. The standard deviation of the manufactured parts is equal to $\sigma_p \equiv T/6$. If the standard deviation of the equipment used to measure the manufactured parts is $\sigma_m \equiv T/10$, find the percentage W_A of wrong parts accepted and the percentage C_R of correct parts wrongly rejected where the acceptance limits are taken σ_m inside the tolerance limits $\pm T/2$ on either side, and the mean of the manufactured parts is in the middle of the tolerance range.

Answer: $W_A = 0.023\%$, $C_R = 3.768\%$.

Example 5.2   In the previous problem, if $\sigma_p \equiv T/4$ and $\sigma_m \equiv T/20$ and the acceptance limits are set in from the tolerance limits by $2\sigma_m \equiv T/10$, find W_A and C_R.

Answer: $W_A = 0.018\%$; $C_R = 7.134\%$.

Example 5.3   If $\sigma_p \equiv T/6$ and $\sigma_m = T/20$ and the acceptance limits are the same as the tolerance limits, find W_A and C_R. The mean of the manufactured parts is in the middle of the tolerance range.

Answer: $W_A = 0.064\%$; $C_R = 0.202\%$.

Example 5.4   Show that approximately

$$W_A = \frac{1}{T\sigma_p\sqrt{(2\pi)}}(e^{\varphi_1 - \psi_1} - e^{\varphi_2 - \psi_2} + e^{\varphi_3 - \psi_3} - e^{\varphi_4 - \psi_4})$$

where

$$\varphi_1 = [B^2/4A - C]_{OA1}, \qquad \varphi_2 = [B^2/4A - C]_{OA2}$$
$$\varphi_3 = [B^2/4A - C]_{UA1}, \qquad \varphi_4 = [B^2/4A - C]_{UA2}$$

$$\psi_1 = \left[\left\{\sqrt{(2A)}\left(c_2 - \frac{B}{2A}\right)\right\}_{UA1} + \rho\right]^2 \Bigg/ 2\rho^2$$

$$\psi_2 = \left[\left\{\sqrt{(2A)}\left(c_2 - \frac{B}{2A}\right)\right\}_{UA2} + \rho\right]^2 \Bigg/ 2\rho^2$$

$$\psi_3 = \left[\left\{\sqrt{(2A)}\left(c_1 - \frac{B}{2A}\right)\right\}_{UA1} - \rho\right]^2 \Bigg/ 2\rho^2$$

$$\psi_4 = \left[\left\{\sqrt{(2A)}\left(c_1 - \frac{B}{2A}\right)\right\}_{UA2} + \rho\right]^2 \Bigg/ 2\rho^2$$

and where the A, B, and C terms are those given in paragraphs 5.37 and 5.38.

Example 5.5   A firm is asked to make parts within ± 0.0002 in of a nominal size. It can produce parts with a distribution whose standard deviation is 0.0001 in. The centre of this distribution can be maintained within ± 0.0001 in of the nominal value. The measuring equipment has a standard deviation of 0.000 05 in. Find the percentage of correct parts rejected (as a percentage of the total parts made) if the percentage of wrong parts accepted must not exceed 1%. State also the required inset i to produce the required result, i.e. k_m, i as a percentage of T_p and i in inches, and also the percentage of correct parts produced.

Answer: $C_R = 17.9\%$, $k_m = 0.92$, $i = 11.5\%$, $i = 0.000\,046$ in, $C = 84\%$.

Example 5.6   A manufacturer is asked to produce parts within a tolerance $T_p/2$ of ± 0.0002 in of a nominal size. He is told that he must not produce more than 5% scrap because of the value of the work already done on the parts. Further, not more than 1% of incorrect parts may be accepted. If the firm can produce parts with a standard deviation of 16.67% of the total tolerance T_p with a mean lying within $\pm 15\%$ of the centre of T_p find the maximum value of the standard deviation of the measuring equipment that will meet the specification, together with the required inset and the percentage of correct parts rejected.

Answer: $\sigma_m = 16.2\%$, $k_m = 0.54$, $i = -8.37\%$, i.e. $\sigma_m = 0.000\,065$ in, $i = -0.000\,035$ in, $W_A = 1.00\%$, $(C_R + W_R) = 5.00\%$, $C_R = 4.26\%$.

Note: that i the inset is negative, i.e. is an offset, and also that most of the 5% scrap is made up of correct parts rejected, i.e. 4.26%. This value could of course be reduced by making σ_m smaller.

Example 5.7   A customer requires parts to be made to lie between the tolerance limits ± 0.0001 in of a nominal size. He specifies that no incorrect parts will be accepted. The manufacturer has production equipment which can produce parts with a spread of $\pm 3\sigma = \pm 0.0001$ in and with the mean of this spread lying within 0.000 03 in of the centre of the tolerance limits. His measuring equipment has a standard deviation of 0.000 01 in. Find the necessary inset of the measuring limits in order to achieve the necessary 'no incorrect parts to be accepted' proviso. Find also the percentage of incorrect parts produced W, the percentage of correct parts rejected C_R, and thus the total of parts rejected.

Answer: Inset $= 15\% \equiv 0.000\,03$ in. Percentage of incorrect parts produced $= 1.79$. Percentage of correct parts rejected $= 10.93$. Total percentage of parts rejected $= 12.72$.

Note: the high percentage of correct parts rejected. This is mainly because of the non-centrality of the mean of the manufactured parts with the centre of the tolerance limits. Some improvement can of course be attained by reducing σ_m, the standard deviation of the measuring equipment.

6

Distributions Ancillary to the Gaussian

6.01 In our consideration of frequency or probability distributions hitherto, we have assumed that the constants of these distributions, that is the standard deviation σ and the mean value μ, have been known exactly. In practice these constants are derived from a finite number of readings, and so their values will have a degree of uncertainty associated with them which will diminish with the increase in the number of readings taken. In this chapter we shall deal briefly with methods and ancillary distributions which allow the uncertainty in the basic constants to be allowed for.

The Student or t Distribution[1]

6.02 The first of these special distributions is known as the Student (pseudonym for W S Gosset) or t distribution. It will be recalled that Table II of Appendix I gives values of

$$\frac{1}{\sqrt{(2\pi)}} \int_{-k_1}^{k_1} e^{-c^2/2}\, dc = \underset{-k_1 \text{ to } k_1}{P} \qquad\qquad 6.02(1)$$

where

$$c = \frac{x - \mu}{\sigma} \qquad \text{and} \qquad -k_1 \leqslant c \leqslant k_1 \qquad\qquad 6.02(2)$$

Thus $\underset{-k_1 \text{ to } k_1}{P}$ gives the probability for x lying between the limits

$$\mu - k_1\sigma \leqslant x \leqslant \mu + k_1\sigma \qquad\qquad 6.02(3)$$

where σ is the standard deviation of the distribution and μ is the mean value.

6.03 Let us replace σ in 6.02(2) by the estimate s_v of the standard deviation, derived from n readings and given by

$$s_v = \sqrt{\left\{\sum_1^n \frac{(x_r - \bar{x})^2}{n-1}\right\}} \qquad 2.24(9)$$

where $v \equiv$ the number of degrees of freedom. In the usual case, where a single variable is considered, $v = n - 1$. Thus 6.02(2) becomes

$$t_v = \frac{x - \mu}{s_v} \qquad 6.03(1)$$

on making the substitution stated above, and replacing c by t_v.

6.04 If we now wish to calculate the probability of $-k_2 \leqslant t \leqslant k_2$ or its alternative expression, the probability of

$$\mu - k_2 s_v \leqslant x \leqslant \mu + k_2 s_v \qquad 6.04(1)$$

then the required probability is no longer expressed by 6.02(1), but by the integral of the Student distrubution, that is

$$\int_{-k_2}^{k_2} f(t_v)\, dt_v = \frac{\Gamma((v+1)/2)}{\Gamma(v/2)\sqrt{(v\pi)}} \int_{-k_2}^{k_2} \left(1 + \frac{t_v^2}{v}\right)^{-(v+1)/2} dt_v \qquad 6.04(2)$$

where $\Gamma(\alpha)$ is the gamma function. When α is even

$$\Gamma(\alpha/2) = (\alpha/2 - 1)(\alpha/2 - 2)\ldots 3.2.1 \qquad 6.04(3)$$

valid for $\alpha \geqslant 4$. If $\alpha = 2$, $\Gamma(1) = 1$.

When α is odd

$$\Gamma(\alpha/2) = (\alpha/2 - 1)(\alpha/2 - 2)\ldots 5/2.3/2.1/2.\sqrt{\pi} \qquad 6.04(4)$$

valid for $\alpha \geqslant 3$. If $\alpha = 1$, $\Gamma(\tfrac{1}{2}) = \sqrt{\pi}$.

6.05 Figure 6.05 shows a plot of $f(t_v)$, for $v = n - 1 = 4$, with a plot of a normal distribution of equal standard deviation superposed. It is to be noted

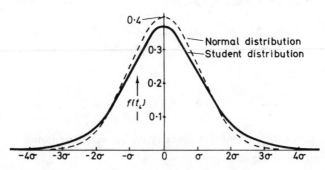

Figure 6.05 Graph of Student distribution $f(t_v)$ with $n = 5$ (that is four degrees of freedom) with a normal distribution of the same standard deviation superposed (broken curve) ($n =$ number of observations, $v = n - 1$)

that the Student distribution, in contrast to the normal distribution, has more probability concentrated in its tails. Like the normal distribution it is symmetrical, continuous and bell shaped, and with increasing v or n rapidly converges to the normal or Gaussian distribution. It should also be noticed that $f(t_v)$ is independent of both μ and σ.

6.06 Whilst the t distribution can be used to calculate the probability of the inequality expressed by 6.04(1), it is not usually used for this purpose, because in most practical cases μ, the mean of a very large number of readings, is not known. For a limited number of readings, only $\bar{x}$ and s_v can be calculated, and in this case it is the probability of the inequality $\bar{x} - ks_v \leqslant x \leqslant \bar{x} + ks_v$ that is required, and the t distribution does not provide this.

Probability of the Mean μ Living Within a Given Range

6.07 Let us consider the variable

$$\frac{\bar{x} - \mu}{\sigma_\mu} = c \qquad \qquad 6.07(1)$$

The probability of this variable is given by Table II of Appendix I and if $-k_1 \leqslant c \leqslant k_1$ then we can calculate the probability of

$$\bar{x} - k_1 \sigma_\mu \leqslant \mu \leqslant \bar{x} + k_1 \sigma_\mu$$

where σ_μ is the standard error or standard deviation of the mean μ.

6.08 Now if σ_μ in 6.07(1) is not known, then we can replace it by its estimate $_\mu s_v$, in which case the right-hand side of 6.07(1) becomes equal to t_v, where v is equal to the number of degrees of freedom, that is $n - 1$ in the case of a single variable. Thus we have

$$\frac{\bar{x} - \mu}{_\mu s_v} = t_v \qquad \qquad 6.08(1)$$

from 6.03(1). Now

$$_\mu s_v = s_v / \sqrt{n} \qquad \qquad 6.08(2)$$

(see 2.30(1)). Thus 6.08(1) becomes

$$\frac{(\bar{x} - \mu)\sqrt{n}}{s_v} = t_v \qquad \qquad 6.08(3)$$

6.09 We can now use 6.04(2) to calculate the probability of $-k_2 \leqslant t_v \leqslant k_2$, or its alternative mode of expression

$$\bar{x} - \frac{k_2 s_v}{\sqrt{n}} \leqslant \mu \leqslant \bar{x} + \frac{k_2 s_v}{\sqrt{n}} \qquad \qquad 6.09(1)$$

thus giving the probability of the mean μ between chosen limits.

6.10   Table IV of Appendix I gives values of k_2 against values of v ($= n - 1$ for a single variable) for six values of the integral, that is 0.90, 0.95, 0.98, 0.99, 0.995 and 0.999, so that when using this table, the upper and lower limits of the position of μ can be found for these six tolerance probabilities, the range being equally disposed about $\bar{x}$. Table V, Appendix I, gives values of k_3, where

$$k_3 = k_2/\sqrt{n} \qquad\qquad 6.10(1)$$

and gives the probability

$$\beta_{cp} \text{ of } \bar{x} - k_3 s_v \leqslant \mu \leqslant \bar{x} + k_3 s_v \qquad\qquad 6.10(2)$$

for two confidence probabilities, that is 0.95 and 0.99, where $\beta_{cp} \equiv$ confidence probability.

6.11   The t distribution thus enables us to calculate the range in which the mean μ is likely to lie, for a chosen probability. As with Table II a probability is often associated with a range by stating the probability of the range being exceeded; in the cases cited the associated probabilities are 0.05, 0.01, 0.005 and 0.001. A little study of Table IV will show that there is a considerable difference in the k_1 of Table II and the k_2 of Table IV for small values of n. For instance for a probability of 0.99, k_1 of Table II is 2.576, whilst k_2 of Table IV for the same probability and $n = 10$ is equal to 3.2498. Thus for small n, the limits for μ for the same probability and the same number of readings are considerably smaller if σ is known accurately.

6.12   The t distribution has other uses and these will be described in Chapter 9 dealing with consistency or significance tests.

The Chi-square (χ^2) Distribution[2]

6.13   Consider the variable

$$c_i^2 = \left(\frac{x_i - \mu}{\sigma} \right)^2 \qquad\qquad 6.13(1)$$

where μ is the mean and σ the standard deviation of a normal or Gaussian distribution. Then

$$\chi^2 = \sum_1^v \left(\frac{x_i - \mu}{\sigma} \right)^2 = \sum_1^v c_i^2 \qquad\qquad 6.13(2)$$

where v is the number of degrees of freedom of χ^2. χ^2 has a probability density distribution given by

$$f(\chi_v^2) = \frac{e^{-\chi^2/2}(\chi^2)^{v/2 - 1}}{2^{v/2}\Gamma(v/2)} \qquad\qquad 6.13(3)$$

valid for $0 \leqslant \chi^2 \leqslant \infty$. $\Gamma(v/2)$ is given as before by 6.04(3) and 6.04(4).

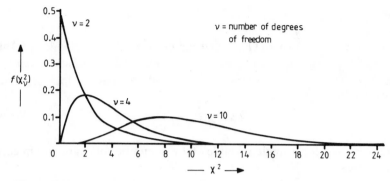

Figure 6.14 Graphs of the density function of χ^2 for three values of v

6.14 Figure 6.14 gives three graphs of $f(\chi_v^2)$ for $v = 2$, 4 and 10 against χ^2 as abscissa. It will be seen that the χ^2 distribution is continuous and asymmetrical. With increasing number of degrees of freedom it slowly converges to a Gaussian distribution. Like the Student t distribution its form is dependent on the number of degrees of freedom, and it should be noted that it is also independent of μ and σ.

Parameters of the χ^2 Distribution

6.15 The mode or maximum of the function is given when $\chi^2 = v - 2$, whilst the mean is given when $\chi^2 = v$, and the standard deviation by $\sigma = 2v$. Table VI of Appendix I gives the probability of χ^2 exceeding the values of χ^2 plotted against v (columns 2 to 5) and also the probability of χ^2 being less than the values of χ^2 plotted against v (columns 6 to 9). In each case four values of probability are used, that is 0.05, 0.025, 0.01 and 0.005.

Use of χ^2 Distribution to find a Maximum and a Minimum Value of σ, for a Given Probability, where the Given Probability is that of the Maximum Value of σ being Exceeded and also that of σ being less than the Minimum Value of σ

6.16 If in 6.13(2) we replace μ by $\bar{x}$ we have

$$\sum_{1}^{n}(x - \bar{x})^2 = \chi_v^2 \sigma^2 \qquad 6.16(1)$$

Also

$$\sum_1^n (x - \bar{x})^2 = s_v^2(n - 1) = s_v^2 v \qquad 6.16(2)$$

Thus

$$\chi_v^2 = \frac{v s_v^2}{\sigma^2} \qquad 6.16(3)$$

where $v = n - 1$ for a single variable and $v =$ number of degrees of freedom. 6.16(3) gives

$$\sigma = s_v \sqrt{(v/\chi^2)} \qquad 6.16(4)$$

If a value of χ_v^2 is chosen then 6.16(4) can be used to find the corresponding value of σ. From

$$\int_{\chi_v^2}^{\infty} f(\chi_v^2) \, d\chi_v^2 \qquad \text{or} \qquad \int_0^{\chi_v^2} f(\chi_v^2) \, d\chi_v^2$$

the probability of σ respectively exceeding or being less than the value found from 6.16(4) can be found. Table VII of Appendix I gives four values for the confidence factor

$$\left\{ 1 - \int_{_1\chi_v^2}^{\infty} f(\chi_v^2) \, d\chi_v^2 - \int_0^{_2\chi_v^2} f(\chi_v^2) \, d\chi_v^2 \right\} \qquad 6.16(5)$$

$$(1) \qquad\qquad\qquad (2)$$

where integral (1) is equal to integral (2) for each value of v, the sum being equal to 0.10, 0.05, 0.02 and 0.01 respectively, giving four probabilities, that is 0.90, 0.95, 0.98 and 0.99. Under each probability is given the corresponding maximum and minimum value of the constant k_9 where

$$_{min}k_9 s_v \leqslant \sigma \leqslant {}_{max}k_9 s_v \qquad 6.16(6)$$

and

$$_{max}k_9 = \sqrt{(v/_1\chi_v^2)}, \qquad {}_{min}k_9 = \sqrt{(v/_2\chi_v^2)} \qquad 6.16(7)$$

and where k_9 is derived from 6.16(4), that is

$$\sigma = s_v \sqrt{(v/\chi_v^2)} = s_v k_9$$

The figures in brackets under the confidence factors in Table VII are the probabilities of k_9 not exceeding $_{max}k_9$ and of not being less than $_{min}k_9$.

F Distribution[3]

6.17 This distribution, which was discovered by R A Fisher, is related to the χ^2 distribution. Let s_1^2 and s_2^2 be two independent estimates of the variance σ^2 of a normally distributed population. Thus following 6.16(3) we can write

$$s_1^2 = \sigma^2 \chi_1^2 / v_1 \qquad\qquad 6.17(1)$$

and

$$s_2^2 = \sigma^2 \chi_2^2 / v_2 \qquad\qquad 6.17(2)$$

F is then defined as

$$F = s_1^2/s_2^2 = \frac{\chi_1^2/v_1}{\chi_2^2/v_2} \qquad\qquad 6.17(3)$$

where $0 \leqslant F \leqslant \infty$. v_1 and v_2 are the number of degrees of freedom of s_1 and s_2 respectively. Fisher showed that F had a probability density function given by

$$f(F) = \frac{\Gamma((v_1 + v_2)/2)}{\Gamma(v_1/2)\Gamma(v_2/2)} v_1^{v_1/2} v_2^{v_2/2} \frac{F^{v_1/2-1}}{(v_2 + v_1 F)^{(v_1+v_2)/2}} \qquad 6.17(4)$$

6.18 Like the χ^2 distribution, the F distribution is a continuous asymmetrical distribution with a range from zero to infinity. Figure 6.18 shows some graphs of the F distribution for several degrees of freedom (v_1, v_2).

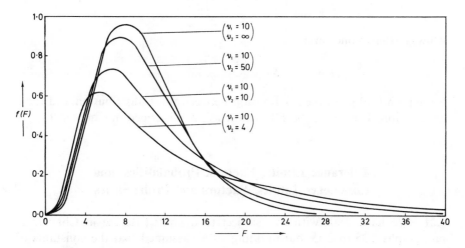

Figure 6.18 Graphs of the density function of the F distribution for several values of the degrees of freedom v_1 and v_2

Parameters of the *F* Distribution

6.19 The mean value of the distribution is given when

$$F = \mu = v_2/(v_2 - 2) \qquad\qquad 6.19(1)$$

where $v_2 > 2$. The variance σ^2 is given by

$$2\left(\frac{v_2}{v_2 - 2}\right)^2 \cdot \left\{\frac{v_1 + v_2 - 2}{v_1(v_2 - 4)}\right\} \qquad\qquad 6.19(2)$$

where $v_2 > 4$.

6.20 Tables VIII and IX of Appendix I give values of *F* tabulated against v_1 and v_2. Table VIII is given for a probability of 0.05 and Table IX for a probability of 0.01, and these two probabilities represent respectively the probability of the tabulated values of *F* in each table being exceeded, that is

$$\text{probability } P \text{ of each table} = \int_{F_0}^{\infty} f(F)\, dF$$

$$= \text{probability of } F > F_0$$

A property of the *F* distribution which is sometimes useful is that on interchanging v_1 and v_2, we have the relationship

$$F_1(p; v_1; v_2) = 1/F_2(1 - p; v_2; v_1)\dagger \qquad\qquad 6.20(1)$$

where *p* in the left-hand term is the probability of *F* being either larger or smaller than F_1, whilst $1 - p$ in the right-hand term is the probability of *F* being either larger or smaller than F_2. In the first term

$$F_1 = \frac{s_1^2}{s_2^2} = \frac{\chi_1^2/v_1}{\chi_2^2/v_2}$$

whilst in the second term

$$F_2 = \frac{s_2^2}{s_1^2} = \frac{\chi_2^2/v_2}{\chi_1^2/v_1}$$

Paragraph 9.10 gives the relationship between the *F* distribution and the *t* distribution. The use of the *F* function will be described in Chapter 9.

Tolerance Limits, Factors, Probabilities, and Confidence Limits, Factors and Probabilities

6.21 We have dealt with this subject to a limited degree in Chapter 2, paragraphs 2.35 to 2.53, but in doing so we assumed that the constants of

---

† See Appendix II for proof.

the distributions considered, that is the standard deviation σ and the mean μ, were known. In many if not most practical cases, σ and μ are derived from a fairly limited number of observations, usually of the order of ten or twenty, and so the values of σ and μ obtained are themselves random variables.

6.22 Before proceeding with how to deal with this dilemma we shall define some terms.

1a *Tolerance Limits or Interval*

This is a range or interval defined by limit points between which a given fraction or percentage of observations or readings is expected to lie.

1b *Tolerance Factor*

This is a quality which when multiplied by the estimated standard deviation or standard deviation gives the semi-range over which a fraction of observations (readings) is expected to occur.

1c *Tolerance Probability β_{tp}*

A tolerance probability is the fraction or percentage of observations expected to be within a tolerance range or interval.

2a *Confidence Limits or Interval*

This is the range or interval defined by limit points between which the *mean* value μ of a quantity is expected to lie with a probability expressed by the associated confidence probability.

2b *Confidence Factor*

This is a quantity which when multiplied by the estimated standard deviation or standard deviation gives the semi-range over which the quantity is expected to occur with a given probability.

2c *Confidence Probability β_{cp}*

This is the probability associated with a given confidence range, when the range is associated with the occurrence of the *mean* value of a quantity. Alternatively, if a confidence probability is associated with the occurrence of a *single* reading or observation, then it is defined as follows. If a large number of groups of readings have been taken, each containing the same number of readings, then the confidence probability β_{cp} is equal to the proportion of groups in which at least a certain proportion of readings (tolerance probability) lies between specified limits (tolerance limits). In practice the confidence probability to be associated with a single group or sample can be found without taking any further readings (see paragraphs

6.31 to 6.32). A confidence probability and a tolerance probability taken together thus deal with the occurrence of a *single* reading, whilst a confidence probability by itself deals with the occurrence of a *mean* value. As the number of observations n tends to infinity, a tolerance limit tends to a constant, whilst under the same circumstances a confidence limit tends to zero.

The Uncertainty of a Single Reading, and the Probability of it Occurring over a Given Range: Without Confidence Probabilities.† Known Parameters μ and σ

6.23   This case is one that we have already dealt with in paragraphs 2.35 to 2.46, but we will restate very briefly how the probability of an error lying in a given range is found or alternatively how to find the range associated with a given probability.

Given the range of uncertainty x_1 to x_2, to find the probability of occurrence of an uncertainty in this range

6.24   Calculate $|\mu - x_1|$ and $|\mu - x_2|$, where the straight lines round each difference mean the modulus or positive value of the difference, and divide each modulus by σ the standard deviation. Let

$$\frac{|\mu - x_1|}{\sigma} = {}_1k_1 \qquad\qquad 6.24(1)$$

and

$$\frac{|\mu - x_2|}{\sigma} = {}_2k_1 \qquad\qquad 6.24(2)$$

Consult Table II, Appendix I, and look up the probability associated with each of the values of k; let these be P_1 and P_2 respectively. There are four possible answers, depending on the signs of $\mu - x_1$ and $\mu - x_2$, thus:

1   $(\mu - x_1)$ *and* $(\mu - x_2)$ *both positive*

Take the larger probability, subtract the smaller and divide by two, then this is the required probability, that is

$$\text{probability required} = P = \frac{P_2 - P_1}{2}$$

---

† This means that in fact the confidence probability is 0.50 for this case.

2   $(\mu - x_1)$ *negative*, $(\mu - x_2)$ *positive*

Add P_1 to P_2 and divide by two, that is

$$P = \frac{P_1 + P_2}{2}$$

3   $(\mu - x_1)$ *and* $(\mu - x_2)$ *both negative*

Subtract smaller probability from larger and divide by two, that is

$$P = \frac{P_2 - P_1}{2}$$

4   *Usual case, when* $(\mu - x_1) = -(\mu - x_2)$, *that is* $\mu = (x_1 + x_2)/2$ *or as it is more usually written, the probability of an uncertainty in the range* $\mu - \delta$ *to* $\mu + \delta$

Divide δ by σ and look up this value under k_1 of Table II. The corresponding probability gives the required probability for the range.

Given a probability for a range, to find the associated range

6.25   We will deal here only with ranges symmetrically distributed about the mean. Suppose the probability of a range is given as P, then Table III of Appendix I, should be consulted and the appropriate probability located from the columns carrying the integral sign. Let k_1 be the corresponding value from the k column. Then the range required is from $\mu - k_1\sigma$ to $\mu + k_1\sigma$.

Uncertainty of a Single Reading: Without Confidence Probabilities. Known Parameters $\bar{x}$ and σ

6.26   This is a case that occurs fairly frequently. For instance, if an instrument such as a measuring microscope is used to make repeat readings of the position of a line, then a very large number of readings can be made and a reasonably accurate figure found for σ, the standard deviation of the spread of readings. If the microscope is now used to locate the position of a workpiece from a comparatively small number of readings, then the range of a single uncertainty for a chosen probability can be found as follows.

Calculation of the range of a single reading for a selected probability σ known

6.27   Let n readings be taken and the mean $\bar{x}$ found, then the tolerance range for a given tolerance probability for a single reading is given by consulting Table X of Appendix I and choosing the value of k_4 from the appropriate probability column and from the row corresponding to n the number of readings. The required range for x is then given by $\bar{x} - k_4\sigma$ to $\bar{x} + k_4\sigma$. It is to be noticed that if the estimate of σ had been found from n readings, then the range would have been much larger (see known parameters $\bar{x}$ and s, paragraph 6.29). The value of the coefficient k_4 of σ is given by

$$|c_\alpha| \sqrt{\left(1 + \frac{1}{n}\right)} = k_4$$

where c_α is the value of c which satisfies the equation

$$\frac{1}{\sqrt{(2\pi)}} \int_{c_\alpha}^{\infty} e^{-c^2/2} \, dc = \alpha$$

Alternatively

$$1 - 2\alpha = \frac{1}{\sqrt{(2\pi)}} \int_{-c_\alpha}^{c_\alpha} e^{-c^2/2} \, dc$$

(see Table II), and thus

$$k_4 = k_1 \sqrt{\left(1 + \frac{1}{n}\right)} \qquad\qquad 6.27(1)$$

Derivation of k_4

6.28   If $\bar{x}$ were the true mean then the uncertainty in x would be given by $k\sigma$, but $\bar{x}$ itself is a variable with a standard deviation of $\sigma/\sqrt{n}$, and so the uncertainty in x will be obtained by combining the distribution for x with that of $\bar{x}$, to give a distribution centred on $\bar{x}$, where the latter is the mean of x found from n readings. The standard deviation of the combination is given by

$$\sigma\left(1 + \frac{1}{n}\right)^{1/2}$$

(see 3.41(1)) and since the distribution of x and $\bar{x}$ are normal, so is that of the combination, whose variable we will call z. Thus z may be looked on as the total uncertainty due to the variation of $\bar{x}$ and x. An uncertainty of given probability is thus given by

$$k_1\sigma\left(1 + \frac{1}{n}\right)^{1/2}, \qquad \text{giving} \qquad k_4 = k_1\left(1 + \frac{1}{n}\right)^{1/2} \qquad 6.28(1)$$

Uncertainty of a Single Reading: Without Confidence Probabilities. Known Parameters $\bar{x}$ and s

6.29   This is a very common case, and occurs when $\bar{x}$ and s are found from n readings.

$$\bar{x} = \sum_1^n x_r/n \qquad\qquad 2.22(1)$$

and

$$s_v^2 = \sum_1^n \frac{(x_r - \bar{x})^2}{n - 1} \qquad\qquad 2.24(8)$$

where $v = n - 1$ the number of degrees of freedom of a single variable. Since $\bar{x}$ and s_v are both approximations to μ and σ, respectively, obtained from a limited number of readings, then the possibility of the limiting values of $\bar{x}$ and s_v being different from those found must be allowed for.

6.30   Table XI of Appendix I gives two values of the tolerance probability β_{tp}, that is 0.95 and 0.99, and for either of these probabilities enables the range associated with n readings to be found in terms of $\bar{x}$ and s_v. The range of x is given by

$$\bar{x} - k_5 s_v \qquad \text{to} \qquad \bar{x} + k_5 s_v$$

where k_5 is chosen from the appropriate probability column β_{tp} and row n. k_5 is given by the expression

$$k_5 = k_{n-1,2\alpha} \sqrt{\left(1 + \frac{1}{n}\right)} \qquad\qquad 6.30(1)$$

where $k_{n-1,2\alpha}$ is derived from the equation

$$\beta_{tp} = 1 - 2\alpha = \int_{-t_{v,2\alpha}}^{t_{v,2\alpha}} f(t_v)\,dt_v \qquad\qquad 6.30(2)$$

(see Table IV). As $t_{v,2\alpha} = k_2$; $v = n - 1$, thus

$$k_5 = k_2 \left(1 + \frac{1}{n}\right)^{1/2} \qquad\qquad 6.30(3)$$

Derivation of k_5

An indication of how k_5 is obtained is as follows. Both x and $\bar{x}$ are obtained from n values of x and thus both have a t_v distribution. As in the previous case, since x and $\bar{x}$ are both variables, in order to find the total uncertainty, we combine the two distributions, which together have a combined estimated standard deviation of $s_v(1 + 1/n)^{1/2}$. If we call the variable of the combination

z, then we can write

$$\frac{z - \mu}{s_v(1 + 1/n)^{1/2}} = t_v \qquad 6.30(4)$$

Whence it follows that the uncertainty of z is given by

$$\bar{x} - k_2\left(1 + \frac{1}{n}\right)^{1/2} s_v \qquad \text{to} \qquad \bar{x} + k_2\left(1 + \frac{1}{n}\right)^{1/2}$$

and thus

$$k_5 = k_2\left(1 + \frac{1}{n}\right)^{1/2} \qquad 6.30(3)$$

Uncertainty of a Single Reading: Without Confidence Probabilities. Known Parameters μ and s

6.31 We have already mentioned this case, as it represents the Student 't' distribution; s_v is found from 2.24(8), that is

$$s_v^2 = \sum_1^n \frac{(x_r - \bar{x})^2}{n - 1} \qquad 2.24(8)$$

The range for a selected probability is obtained as

$$\mu - k_2 s_v \leqslant x \leqslant \mu + k_2 s_v$$

Table IV, Appendix I, gives k_2 for four probabilities.

$$\int_{-t_v}^{t_v} f(t_v)\, dt_v = \beta_{tp} = 1 - 2\alpha$$

where

$$\alpha = \int_{t_v}^{\infty} f(t_v)\, dt_v$$

and

$$t_v = k_2$$

See 6.04(2) for $f(t_v)$.

Uncertainty of a Single Reading: With Confidence Probabilities. Known Parameters $\bar{x}$ and σ

6.32 Let $\bar{x}$ be derived from n readings and let σ be known. Now $\bar{x}$ is not the limiting mean and is itself a variable. Since σ is known, the standard deviation of $\bar{x}$ is $\sigma/\sqrt{n}$. Let us decide that the tolerance probability required is β_{tp} and since $\bar{x}$ is a variable, the confidence probability will be proportional to the position of $\bar{x}$. Let us choose a confidence probability for $\bar{x}$, say β_{cp}, then

$$\beta_{cp} = \int_{-k_1}^{k_1} f(c)\,dc = \frac{1}{\sqrt{(2\pi)}} \int_{-k_1}^{k_1} e^{-c^2/2}\,dc \qquad 6.32(1)$$

Since the standard deviation of $\bar{x}$ is $\sigma/\sqrt{n}$ the range of $\bar{x}$ is

$$\bar{x}_0 - k_1 \frac{\sigma}{\sqrt{n}} \leqslant \bar{x} \leqslant \bar{x}_0 + k_1 \frac{\sigma}{\sqrt{n}} \qquad 6.32(2)$$

where $\bar{x}_0$ is the value of the mean from n readings. Consider Figure 6.32. The normal distribution curve shown has a standard deviation of unity, and its mean is displaced by $k_1/\sqrt{n}$ from the zero position. This value is obtained by putting $\sigma = 1$ in 6.32(2). The value of k_6, such that the probability of an error is equal to β_{tp}, is given by

$$\beta_{tp} = \frac{1}{\sqrt{(2\pi)}} \int_{-k_6}^{k_6} e^{-(c - k_1/\sqrt{n})^2/2}\,dc \qquad 6.32(3)$$

Writing

$$z = k_1/\sqrt{n} - c \qquad 6.32(4)$$

6.32(3) becomes

$$\frac{1}{\sqrt{(2\pi)}} \int_{-k_6 + k_1/\sqrt{n}}^{k_6 + k_1/\sqrt{n}} e^{-z^2/2}\,dz = \beta_{tp} \qquad 6.32(5)$$

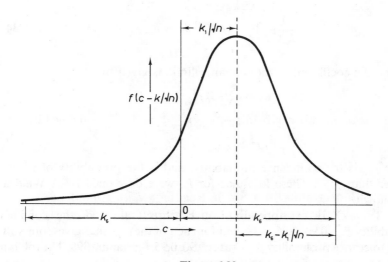

Figure 6.32

6.33 It is at once obvious that as the mean of the curve in Figure 6.32 moves to the left the proportion of readings between the limits $-k_6$ to k_6 increases until the mean reaches the position 0 and then decreases again until the position $-k_1/\sqrt{n}$ on the other side of the ordinate axis is reached, when the proportion is the same as for $z = k_1/\sqrt{n}$. Thus in a proportion β_{cp} of sample groups of n readings, each group will contain at least a proportion β_{tp} of readings lying between the limits $-k_6$ to k_6. Table XII of Appendix I gives two values of tolerance probability for each of the two values of the confidence probability given. Having selected a tolerance probability and a confidence probability, k_6 is chosen from the appropriate column and the row corresponding to the number of readings. The required range is then given by $\bar{x} - k_6\sigma$ to $\bar{x} + k_6\sigma$.

Uncertainty of a Single Reading: With Confidence Probabilities. Known Parameters $\bar{x}$ and s

6.34 This is the most usual case, that is when $\bar{x}$ and s are obtained from n readings. As before

$$\bar{x} = \sum_{r=1}^{r=n} \frac{x_r}{n} \qquad \qquad 2.22(1)$$

and

$$s_v^2 = \sum_{r=1}^{r=n} \frac{(x_r - \bar{x})^2}{n-1} \qquad \qquad 2.24(8)$$

The tolerance probability β_{tp} is derived from

$$\frac{1}{\sqrt{(2\pi)}} \int_{\frac{1}{\sqrt{n}}-r}^{\frac{1}{\sqrt{n}}+r} e^{-c^2/2} \, dc = \beta_{tp} \qquad \qquad 6.34(1)$$

and k_7, the coefficient of the tolerance limit, is given by

$$k_7 = r\sqrt{(n/\chi^2_{v,\beta_{cp}})} \qquad \qquad 6.34(2)$$

where n is the number of readings, $v = n - 1$, and $\chi^2_{v,\beta_{cp}}$ is defined by

$$P(\chi^2 > \chi^2_{v,\beta_{cp}}) = \beta_{cp} \qquad \qquad 6.34(3)$$

where β_{cp} is the confidence probability, and is the probability of χ^2 being greater than $\chi^2_{v,\beta_{cp}}$. These formulae for k_6 were developed by A Wald and J Wolfowitz.

6.35 Table XIII of Appendix I gives three values of the confidence probability β_{cp}, 0.90, 0.95, 0.99, and for each of these values gives four values of the tolerance probability β_{tp}, that is 0.90, 0.95, 0.99 and 0.999. The tolerance limits are given as $\bar{x} - k_7 s$ to $\bar{x} + k_7 s$.

Uncertainty of a Single Reading: With Confidence Probability. Known Parameters μ and s

6.36 The required tolerance limits are obtained by consulting Table XIV of Appendix I and the required tolerance limits are $\mu - k_8 s$ to $\mu + k_8 s$, where s is found from n readings using 2.24(8). Two values of the confidence probability β_{cp} are given, and for each of these, two values of the tolerance probability β_{tp} are given.

6.37 The coefficient k_8 is derived as follows. The value of $\chi^2_{v,(1-\beta_{cp})}$ is obtained from tables of chi-square, that is

$$\text{Prob}(\chi^2 > \chi^2_{v,(1-\beta_{cp})}) = 1 - \beta_{cp} \qquad 6.37(1)$$

where β_{cp} is the confidence probability. Alternatively

$$\text{Prob}(\chi^2 < \chi^2_{v,\beta_{cp}}) = \beta_{cp} \qquad 6.37(2)$$

Since from 6.16(4)

$$\frac{\sigma^2}{s_v^2} = \frac{v}{\chi^2_{v,(1-\beta_{cp})}} \qquad 6.37(3)$$

the maximum value of σ is given by

$$\sigma_{max} = s_v \sqrt{\left(\frac{v}{\chi^2_{v,(1-\beta_{cp})}}\right)} \qquad 6.37(4)$$

Now consider Figure 6.37. If the tolerance probability is β_{tp} and we consider the curve of σ_{max} shown in the figure, it is at once apparent that k_1 is given by

$$\beta_{tp} = \frac{1}{\sigma_{max}\sqrt{(2\pi)}} \int_{-k_1\sigma_{max}}^{k_1\sigma_{max}} e^{-x^2/2\sigma_{max}^2} \, dx \qquad 6.37(5)$$

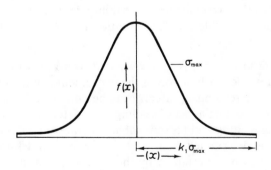

Figure 6.37

Writing $x/\sigma_{max} = c$, this reduces to

$$\beta_{tp} = \frac{1}{\sqrt{(2\pi)}} \int_{-k_1}^{k_1} e^{-c^2/2} \, dc \qquad 6.37(6)$$

Table III enables k_1 to be found.

6.38 Now $k_1 \sigma_{max}$ is the required range since the proportion β_{cp} of distributions with σs smaller than σ_{max} all have at least a proportion β_{tp} of their readings lying in the range $k_1 \sigma_{max}$.

Now

$$k_1 \sigma_{max} = k_1 s_v \sqrt{\left(\frac{v}{\chi^2_{v,(1-\beta_{cp})}} \right)} \qquad 6.38(1)$$

Thus

$$k_8 = k_1 \sqrt{\left(\frac{v}{\chi^2_{v,(1-\beta_{cp})}} \right)} \qquad 6.38(2)$$

where $v = n - 1$ and n is the number of readings.

Probability of an Uncertainty in the Mean: Confidence Probability and Confidence Limits

1 Known parameters $\bar{x}$ and σ

6.39 If $\bar{x}$ has been found from n readings the mean is itself variable, with a standard deviation of $\sigma/\sqrt{n}$, where σ is the known standard deviation of the variable x. The required confidence limits for a range of the mean are thus given by

$$\bar{x} - k_1 \frac{\sigma}{\sqrt{n}} \qquad \text{to} \qquad x + k_1 \frac{\sigma}{\sqrt{n}} \qquad 6.39(1)$$

where k_1 is the tolerance factor obtained from Table III and corresponds to a chosen probability (confidence probability in this case) selected from columns 1, 4 or 7 of Table III. This situation is often met with and usually occurs in two ways.

a   A measuring instrument is used over and over again for measurements and the σ for a particular function of the instrument can thus be found from a large number of readings and can thus be considered a constant of the instrument. For a particular set of readings, $\bar{x}$ is determined from a relatively small number of readings, and is thus subject to uncertainty. Since σ is known the range can be determined from 6.39(1). If σ were not known, the range would have had to be determined by the method given in the next section, leading to a much larger range.

b   A particular type of instrument is manufactured, and each new instrument is calibrated. Since all the instruments are basically similar, the σ found for one instrument can be used for the others of the same type.

2   *Known parameters $\bar{x}$ and s*

6.40   This case has already been dealt with in paragraphs 6.07 to 6.11. Tables IV or V can be used to calculate the confidence limits, but Table V is to be preferred because less work is involved. The confidence range using Table IV is given by

$$\bar{x} - k_2 \frac{s_v}{\sqrt{n}} \leqslant \mu \leqslant x + k_2 \frac{s_v}{\sqrt{n}} \qquad\qquad 6.09(1)$$

or, using Table V it is given by

$$\bar{x} - k_3 s_v \leqslant \mu \leqslant \bar{x} + k_3 s_v \qquad\qquad 6.10(2)$$

The probability associated with either k_2 or k_3 is known as the confidence factor.

Probability of an Uncertainty in σ

6.41   The probability of an uncertainty in σ is obtained from the χ^2 chi-square distribution. (See paragraphs 6.15 to 6.16.) Table VI enables the probability of χ^2 being larger than or less than the values to be tabulated, and whence from 6.16(4), that is $\sigma = s_v \sqrt{(v/\chi_v^2)}$, the probability of σ exceeding or being less than the values given by 6.16(4), that is if the probability of χ_v^2 exceeding g_1 say is η_1, then the probability of σ exceeding $s_v \sqrt{(v/g_1)}$ is $1 - \eta_1$. Similarly, if the probability of χ^2 being less than g_2 is η_2, then the probability of σ being less than $s_v \sqrt{(v/g_2)}$ is $1 - \eta_2$.

6.42   Table VII gives ranges of σ against $v = n - 1$ for four confidence probabilities. This table can be derived from Table VI as indicated above, where n is the number of readings.

Index to Uncertainties (Gaussian)

6.43   The following index to uncertainties summarizes the methods that have just been discussed and provides a quick reference to the appropriate methods to be employed in order to find the probability of an uncertainty of given size or to find the size of error associated with a given probability, depending on the constants that are known.

1. The uncertainty of a single reading

a   *Without confidence probabilities*

Known parameters μ and σ. See paragraphs 2.35 to 2.46 and 6.23 to 6.25. For the calculation of *tolerance probability* $P_{-k_1 \text{ to } k_1}$ for a given range, $\mu - k_1\sigma$ to $\mu + k_1\sigma$, see Table II for values of $P_{-k_1 \text{ to } k_1}$, given the value of k_1. For the calculation of *tolerance range* (limits) for a given tolerance probability $P_{-k_1 \text{ to } k_1}$, see Table III for the value of k_1, given the value of $P_{-k_1 \text{ to } k_1}$. Tolerance range $= \mu - k_1\sigma$ to $\mu + k_1\sigma$.

Known parameters $\bar{x}$ and σ. See paragraphs 6.26 to 6.28. For the calculation of *tolerance range* $\bar{x} - k_4\sigma$ to $\bar{x} + k_4\sigma$ for two values of tolerance probability, 0.95 and 0.99, see Table X for values of k_4.

Known parameters $\bar{x}$ and s. See paragraphs 6.29 to 6.30. For the calculation of *tolerance range* $\bar{x} - k_5 s$ to $\bar{x} + k_5 s$ for two tolerance probabilities, 0.95 and 0.99, see Table XI for values of k_5.

Known parameters μ and s. See paragraph 6.31. For the calculation of *tolerance range* $\mu - k_2 s$ to $\mu + k_2 s$ for six tolerance probabilities, 0.90, 0.95, 0.98, 0.99, 0.995 and 0.999, see Table IV for values of k_2.

b   *With confidence probabilities*

Known parameters $\bar{x}$ and σ. See paragraphs 6.32 to 6.33. For the calculation of *tolerance range* $\bar{x} - k_6\sigma$ to $\bar{x} + k_6\sigma$ for a confidence probability β_{cp}, 0.95, and two tolerance probabilities β_{tp}, 0.90 and 0.95, and for a confidence probability 0.99 and two tolerance probabilities, 0.95 and 0.99, see Table XII for values of k_6.

Known parameters $\bar{x}$ and s. See paragraphs 6.34 to 6.35. For the calculation of *tolerance range* $\bar{x} - k_7 s$ to $\bar{x} + k_7 s$ for three confidence probabilities β_{cp}, 0.90, 0.95 and 0.99, each having four associated tolerance probabilities β_{tp}, 0.90, 0.95, 0.99 and 0.999, see Table XIII for values of k_7.

Known parameters μ and s. See paragraphs 6.36 to 6.38. For the calculation of *tolerance range* $\mu - k_8 s$ to $\mu + k_8 s$ for a confidence probability β_{cp}, 0.95, and two tolerance probabilities β_{tp}, 0.90 and 0.95, and for a confidence probability β_{cp}, 0.99, and two tolerance probabilities β_{tp}, 0.95 and 0.99, see Table XIV for values of k_8.

2. Probability of an uncertainty in the mean: confidence probability and confidence limits

Known parameters $\bar{x}$ and σ. See paragraph 6.39. For the calculation of *confidence range* $\bar{x} - k_1\sigma/\sqrt{n}$ to $\bar{x} + k_1\sigma/\sqrt{n}$ for a given confidence probability β_{cp}, $P_{-k_1 \text{ to } k_1}$, see Table III for values of k_1. For the calculation of *confidence probability* β_{cp} for a given confidence range $\bar{x} - k_1\sigma/\sqrt{n}$ to $\bar{x} + k_1\sigma/\sqrt{n}$ find $\underset{-k_1 \text{ to } k_1}{P}$ from Table II using a value of k_1 from the range.

Known parameters $\bar{x}$ and s. See paragraphs 6.40 and 6.07 to 6.11. For the calculation of the *confidence range* $\bar{x} - k_2s_v/\sqrt{n}$ to $\bar{x} + k_2s_v/\sqrt{n}$ for four values of the confidence probability β_{cp}, 0.95, 0.99, 0.995 and 0.999, see Table IV which gives values of k_2. Alternatively Table V gives values of k_3 for two confidence probabilities, 0.95 and 0.99, allowing the *confidence range* to be found as $\bar{x} - k_3s_v$ to $\bar{x} + k_3s_v$. Note: $k_3 = k_2/\sqrt{n}$.

3. Given the probability of an uncertainty in the mean (confidence probability) to find the associated uncertainty range from a small number of readings using a precalibrated measuring instrument with known standard deviations

a *Expected or most probable uncertainty range for mean value*
Let s_n be the standard deviation of a piece of measuring equipment found from n readings, $n \geqslant 20$. Further, let p be the number of observations made on a component using the measuring equipment, $n \gg p$. p will normally be in the range 3 to 7. The required uncertainty range for the mean $\bar{x}_p$ of p readings is then given by

$$\bar{x}_p - k_1 s_n/\sqrt{p} < \mu < \bar{x}_p + k_1 s_n/\sqrt{p} \qquad 6.43(1)$$

where μ is the true mean and k_1 is the tolerance factor corresponding to a chosen probability P. (See Table III.)

b *Uncertainty range for mean with stated frequency of it being exceeded*
If more security is required for the uncertainty range of the mean given by 6.43(1), then Table VII may be used to obtain a maximum value for the standard deviation s_n for a given probability. The value of $_{max}k_q$ should be selected for the appropriate value of n and the chosen bracketed confidence probability β_{cp} used. Note that $(1 - \beta_{cp})$ gives the probability that σ will exceed $_{max}k_q s_n$. The uncertainty in the mean is now given by

$$\bar{x}_p - k_1 \cdot {}_{max}k_q s_n/\sqrt{p} < \mu < \bar{x}_p + k_1 \cdot {}_{max}k_q s_n/\sqrt{p} \qquad (6.43(2))$$

where k_1 is the normal tolerance factor corresponding to a chosen probability P. k_1 is obtained from Table III.

The meaning to be attached to 6.43(2) is that the mean value of p readings of x will lie within the uncertainty range given, with a probability P corresponding to the value of k_1 derived from Table III. Further if

100 sets of p readings of x were taken, then in $100.\beta_{cp}$ cases the uncertainty range found would be expected to be within that given by 6.43(2). In the case of 6.43(1) out of 100 sets of p readings, only 50 would be expected to lie within the range given by 6.43(1), i.e. 6.43(1) gives the most probable result.

4.   *Probability of an uncertainty in σ*

Known parameter s. See paragraphs 6.41 to 6.42. For a *confidence range* of σ for four confidence probabilities β_{cp}, 0.90, 0.95, 0.98, 0.99, see Table VII

$$\sigma_{min} = s_{min}k_9$$

$$\sigma_{max} = s_{max}k_9$$

For the calculation of maximum or minimum values of σ, see Table VI, giving values of χ_v^2. The maximum or minimum value of $\sigma = s_v\sqrt{(v/\chi_v^2)}$, $v = n - 1$, where n = number of readings.

Derivation of Standard Deviation σ from Extreme Range

6.44   It is possible to obtain the standard deviation of a set of observations with reasonable accuracy without having recourse to finding the sum of the squares of the residuals, provided the number of observations lie approximately between five and ten. If, from a number of observations of a variable, x_n is the largest value and x_1 the smallest value, then $x_n - x_1$ is known as the extreme range w_n of the sample. The standardized extreme range of a sample of size n, taken from a population whose standard deviation is σ, is given by

$$W_n = \frac{w_n}{\sigma} = \frac{x_n - x_1}{\sigma}$$

6.44(1)

6.45   The mean extreme range of m samples, each of size n, obtained from measurements on the same variable is given by

$$\bar{w}_{m,n} = \sum_1^m \frac{w_{q,n}}{m} = \sum_1^m \frac{(x_{qn} - x_{q1})}{m}$$

6.45(1)

where w_{qn} is the extreme range of the qth sample which contains n observations, and x_{qn} is the maximum value of x in the qth sample, whilst x_{q1} is the minimum value of x in the qth sample. The mean standardized extreme range of the n groups of observations is given by

$$\bar{W}_{m,n} = \frac{\bar{w}_{m,n}}{\sigma} = \sum_{q=1}^{q=m} \frac{w_{qn}}{m\sigma} = \sum_{q=1}^{q=m} \frac{(x_{qn} - x_{q1})}{m\sigma}$$

6.45(2)

When $n \to \infty$, $\overline{W}_{m,n} \to \overline{W}_n$, which can be calculated. Thus

$$\frac{\bar{w}_{m,n}}{\overline{W}_n} \simeq \sigma \qquad 6.45(3)$$

or

$$\sigma \simeq \sum_{q=1}^{q=m} \frac{(x_{qn} - x_{q1})}{m\overline{W}_n} \qquad 6.45(4)$$

6.46 Table XVI of Appendix I gives values of $1/W_n$. The method gives an unbiased estimate of σ, which improves with increasing m, but worsens rapidly as n is increased beyond about ten. The best values for n are from five to ten. The estimate obtained is greater than that obtained by using the root mean square of the sum of the residuals. Whilst it is best to use a fair number m of sets of observations, quite reasonable results can be obtained even with $m = 1$, provided that n lies between five and ten.

Derivation of Standard Error $\sigma_{\bar{x}}$ (Standard Deviation of Mean) from Extreme Range

6.47 Since the standard deviation of the mean is given by $\sigma/\sqrt{N}$ we have, using 6.45(3)

$$\sigma_{\bar{x}} = \frac{\bar{w}_{m,n}}{\overline{W}_n \sqrt{(mn)}} \qquad 6.47(1)$$

where $mn = N$, the total number of readings. Table XVII of Appendix I gives values of

$$\frac{1}{\overline{W}_n \sqrt{(mn)}}$$

for different values of m and n. As before $\bar{w}_{m,n}$ is given by 6.45(1). The same remarks apply to the values of n and m as for the case of σ (see paragraph 6.46).

References

For proofs of the t, χ^2 and F distributions respectively, the following references should be consulted

1. [**6.02**] See B W Lindgren *Statistical Theory* p 379
    and H Cramér *Mathematical Methods of Statistics* p 237
2. [**6.13**] See B W Lindgren *Statistical Theory* p 373
    and H Cramér *Mathematical Methods of Statistics* p 233

3.   [**6.17**] See B W Lindgren *Statistical Theory* p 379
     and H Cramér *Mathematical Methods of Statistics* p 241

   A simplified proof of the *t* distribution is given in
     H Levy and L Roth *Elements of Probability* pp 184–9
and a simplified proof of the χ^2 distribution is given in
     L G Parratt *Probability and Experimental Errors in Science* pp 136–7
   The figures in square brackets give the paragraph numbers where the
references are cited in the present text.

Examples on Chapter 6

Example 6.1   A set of ten readings has been taken to determine the position
of a line, and the mean value found. If a confidence probability of 0.99 is
required for the position of the mean, find the confidence limits, assuming
first that the distribution has a Gaussian, and then that it has a Student
distribution. Express the answer in terms of the value of the confidence factors
involved, that is the number by which the estimated standard deviation must
be multiplied.

Answer: 0.8146 and 1.0277.

Example 6.2   A workpiece is measured for length, and the mean value found
from ten readings. What is the confidence factor for a confidence probability
for the mean of 0.999?

Answer: 1.513.

If the distribution is assumed Gaussian, what is the corresponding confidence
factor?

Answer: 1.041.

   It is interesting to note the relative increase in the confidence factor for
the Student distribution relative to that for a Gaussian distribution as the
confidence probability is increased.

Example 6.3   An estimated standard deviation s_v has been found from ten
readings. Find the approximate probability that s_v shall be twice this value
or half this value.

Answer:
From 6.16(3)

$$\frac{\sigma^2}{s_v^2} = \frac{v}{\chi^2} = k^2$$

Putting $k = 2$ we have

$$\chi^2 = \frac{v}{k^2} = \frac{9}{4} = 2.25$$

For σ/s_v to be greater than 2, χ^2 must be less than 2.25. Thus we consult Table VI with $v = n - 1 = 9$ to find the probability of χ^2 being less than 2.25. It is seen that the required probability lies between 0.01 and 0.025. Putting $k = 0.5$ we have $\chi^2 = 36.0$. For σ/s_v to be less than 0.5, χ^2 must be greater than 36.0. Once again, consulting Table VI, for $v = 9$, it is seen that the required probability is less than 0.005.

Example 6.4 If an estimated standard deviation s_v is found from twenty readings, find the approximate probability of its being twice or half the value found.

Answer: The probability of σ being greater than $2s_v$ is less than 0.005. The probability of σ being less than 0.5 is less than 0.005.

Example 6.5 Two estimated standard deviations have been found, one from twenty readings, and the other from thirty readings. If $s_{20} = 3.1$ units, and $s_{30} = 4.2$ units, find if it is likely that these two values are consistent.

Answer:

$$F = \frac{s_1^2}{s_2^2} = \left(\frac{4.2}{3.1}\right)^2 = 1.835$$

Consulting Table VIII, we see that the probability of F exceeding this value is somewhat greater than 0.05, and so the two values are reasonably consistent.

Example 6.6 If the ratio of two estimated standard deviations is 2.7, the larger of which was obtained from fifteen readings and the smaller from ten readings, find the approximate probability of this ratio.

Answer: The probability of the ratio is less than 0.01.

Example 6.7 If the ratio of two estimated standard deviations s_1/s_2 is 1.533, for which v_1 is ten and v_2 is twenty, find the probability of s_1^2/s_2^2 being exceeded. Find also the probability of s_1^2/s_2^2 being less than $1/1.533$.

Answer: 0.05 in each case.

Example 6.8 Ten readings of the size of a workpiece have been taken, yielding (in inches) 1.2484, 1.2483, 1.2481, 1.2486, 1.2487, 1.2488, 1.2489, 1.2486, 1.2485, 1.2486. Find the mean value $\bar{x}$ and the estimated standard deviation s of any reading. Assuming that s is the correct value of σ, find the tolerance range for a single reading for a tolerance probability of 0.95.

If $\bar{x}$ and s are assumed to be only estimates, find the range of a single reading for the same tolerance probability. (Hint: use Table XI.)

Answer: $\bar{x} = 1.248\,55$, $s = 0.000\,2236$, first range $\equiv \bar{x} \pm 0.000\,438$, second range $\equiv \pm 0.000\,530$.

Example 6.9 In the previous example, find the tolerance range for the same tolerance probability of 0.95, but also for confidence probabilities of 0.90, 0.95 and 0.99.

Answer: Ranges $\equiv \pm 0.000\,675$, $\pm 0.000\,746$, and $\pm 0.000\,954$.

Example 6.10 An instrument, whose standard deviation is known, is used to find the size of a workpiece. If ten readings are taken, find the tolerance factor for a tolerance probability of 0.95.

Answer: 2.0556.

Example 6.11 The following two sets of readings were obtained for the determination of the angle of a workpiece. *Set* (1): 0°1′22.4″, 0°1′22.8″, 0°1′24.3″, 0°1′23.6″, 0°1′22.6″, 0°1′22.9″, 0°1′24.1″, 0°1′24.4″, 0°1′23.3″, 0°1′23.6″. *Set* (2): 45°1′28.3″, 45°1′27.4″, 45°1′27.4″, 45°1′26.5″, 45°1′29.2″, 45°1′27.6″, 45°1′28.2″, 45°1′28.7″, 45°1′29.2″, 45°1′28.3″, 45°1′27.9″.

Find the mean value of the angle of the workpiece and the standard error in the mean. Find also (1) the uncertainty in the mean for a confidence probability of 0.95, assuming that the estimated value of the standard error is the correct one; (2) the uncertainty in the mean for the same confidence probability, but assume that the value obtained for the standard error in the mean is only an estimate.

Answer: Mean angle $= 45°0′4.73″$.
     Standard error in mean angle $= 0.347″$
(1) Uncertainty in mean angle $= 0.680″$
(2) Uncertainty in mean angle $= 0.785″$.

Example 6.12 The following ten readings have been taken: 0.225, 0.226, 0.228, 0.229, 0.227, 0.226, 0.229, 0.227, 0.226, 0.224. Find the standard deviation using the root mean square of the deviations and also using the extreme range formula of Table XVI.

Answer: Root mean square $= 0.001\,63$
     Extreme range $= 0.001\,62$.

Example 6.13 In Example 6.8 find an estimate of the standard deviation of the mean, and find the uncertainty in the mean for a confidence probability of 0.95, assuming that the estimated σ is the correct value. If the value of σ

is treated as an estimate only, find the uncertainty in the mean for the same confidence probability.

Answer: Estimate of standard error of mean = 0.000 070 70 in
Uncertainty = 0.000 138 in assuming σ is correct
Uncertainty = 0.000 160 in assuming σ is only an estimate.

7

A General Theory of Uncertainty

Introduction

7.01 The subject of the estimation of uncertainty of measurement, especially the linking of a probability with a given range of uncertainty has been fraught with controversy. The combination of random uncertainty distributions having a Gaussian or near Gaussian form has long been accepted, but the convolution (to use the technical term for the special process needed to combine distributions) of non-Gaussian distributions has led to much misunderstanding and confusion amongst scientists, especially when these distributions have been associated with systematic errors. The convolution of random uncertainties with systematic uncertainties has only led to further controversy.

7.02 It may be useful to start with a definition of random and systematic uncertainties so that their meaning will be quite clear in what follows.

Random Distributions

7.03 A random uncertainty occurs when the repeated measurement of a physical quantity leads to a variation in the measured size of that quantity. Whether the physical quantity is itself really varying is a moot point. By this it is meant that the quantity itself may be constant but the measuring system and environment may be introducing random fluctuations in its value. Alternatively the quantity itself may be varying. Which of these suppositions is true will depend on independent investigation. In any case the value of the mean of the quantity obtained is usually taken to be of significance, either as the true value or just as the value to be accepted from that particular measuring system. The value obtained for the mean may of course not be the true value because of systematic errors produced by the environment and measuring system.

7.04 The important thing about a random variation is that it is impossible to predict the size of each succeeding variation with respect to the previous one. If enough results are taken some pattern of the size of the variations with respect to their frequency will emerge. The mathematical function describing the form of this frequency distribution is known as the density distribution of the uncertainty concerned. The most usually met with form of a random distribution is the so-called normal or Gaussian distribution, with its characteristic bell-shaped envelope. A random distribution may however take other forms. The only predictability associated with random events is that it is possible to predict the probability of an event lying between chosen limits, or the proportion of events lying between chosen limits. If we consider a rectangular frequency distribution then we have the very common case of a random number occurring over a given range. The frequency distribution being rectangular means that a number is equally likely to occur over the whole range.

7.05 A random distribution can also take the form of a stochastic distribution, which means that an event has a probability of occurring only at specified points. A very simple example of a stochastic distribution is given by $\pm a$, which means that there is a probability of 0.5 that an event will occur at the position of either $+a$ or $-a$.

Systematic Uncertainties

7.06 Systematic uncertainties are not essentially different from random ones, although there are several important differences. The essential difference is that when a set of readings is obtained, then if the readings are repeated, we generally find that the mean value of the random uncertainties has not changed significantly. This leads to the supposition that when measurements are repeated, the systematic uncertainties take on the same values each time. This means that simply repeating a set of measurements will not reveal the presence of systematic uncertainties. When systematic uncertainties are estimated, a reasonable deduction is that their combined value ought to influence the measured result, either in the form of a greatly increased standard deviation for the measured results or by a detectable shift in the mean. The fact that this shift does not happen thus leads to the supposition of the systematics repeating their previous values, and thus leads to a constant but unknown shift of the measured mean from the true value. The problem we are now faced with is how to estimate and to combine the systematic uncertainties with the random uncertainties.

7.07 Let us first try to define a systematic uncertainty. When a set of measurements is made on a physical quantity, then apart from the obvious random uncertainties which we observe, it is reasonable to assume that the result may be being influenced by a constant amount by environmental conditions, or by the uncertainty of the calibration of the measuring equipment. The equipment measuring temperature may be in error, the voltmeter or ammeter may be in error, standard resistances may be in error, connectors may introduce spurious resistances, lack of screening against draughts or heat sources may produce current changes. Each set of measurements requires a careful analysis of the influences likely to affect the outcome of the measurements, and an assessment made as to the maximum amount that the result of the measurements are likely to be influenced by each possible source of uncertainty. Since we cannot know, except in exceptional circumstances, whether the mean of a set of measurements has been increased or decreased by each systematic error we have to assume that a systematic error can produce a shift of the mean value by up to a maximum of say $\pm a$ units. What else can we say about such an error? It would be unwise to assume that the constant value taken by the systematic error is always $\pm a$; it is prudent therefore to assume that the constant value which it does take lies anywhere between $\pm a$. There is still one more assumption to be made, and this concerns the density function of such a distribution. For almost all random distributions met with, the density function is Gaussian or near Gaussian, i.e. the probability of an error diminishes with an increase in its size. The worst case we can assume is therefore that the systematic error has an equal chance of occurrence over the whole of its assumed range, i.e. that the density function is a constant and thus the distribution function is rectangular. Thus we end up with the assumption that a systematic uncertainty will have a finite plus or minus range about the true mean value, together with a continuous rectangular distribution function. This does not of course mean that any systematic uncertainty has a rectangular distribution. What it means is that given the maximum limits, the assumed rectangular distribution will give the largest standard deviation and thus the most *pessimistic* value for the uncertainty.

7.08 The systematic errors can thus be treated as a number of variables, all having a rectangular distribution. We can only assume that each variable can take up any value in its designated range. The fact that each time the variables are combined they take up the same value, leading to a constant shift of the mean, is of no statistical significance. Each variable is assumed independent of any other variable, and so we can combine them using standard convolution theory, to predict the probability of an error lying between given limits. Since we cannot know what value the systematic errors are going to assume, we can only assume that it is any value over their combined range. The random uncertainties which we detect can be thought of as interacting with the systematic errors to produce the result obtained. The systematic errors can thus be combined with the random ones to determine the total uncertainty. The fact that the systematic errors each assume the same value each time is irrelevant, since this value is unknown to the observer and may be any of the possible values that the systematic errors could assume.

7.09 Some people have tried to evoke Bayes theorem on the grounds that the systematic errors always produce the same value and that therefore the final answers should be found in terms of conditional probabilities. Unfortunately although Bayes theorem deals with conditional probabilities it also requires knowing the results of one or more events. In the case of systematic errors we know that an event has occurred, and that the same event will be repeated, but we do not know the value of that event, and so Bayes theorem is inapplicable, and as far as the observer goes the result is unknown.

7.10 The next problem is to estimate the distribution function obtained by convoluting the rectangular distributions of the systematic errors. To convolute a number of rectangular distributions is a very time consuming exercise, and as we shall show is not strictly necessary, it being possible to obtain a very close approximation to the correct answer, which errs on the safe side; i.e. the value of the probability between limits for the approximation will be shown to be slightly less than the true value for the convoluted rectangles.

Central Limit Theorem

7.11 Before proceeding further it is appropriate to mention a very important statistical theorem known as the central limit theorem†. This states that if a number of random variables with similar distributions are convoluted, then the resulting distribution function tends towards the normal Gaussian distribution as the number of convoluted distributions is increased. The more nearly normal the convoluted distributions, the fewer distributions are needed to achieve near normality. Further, the convergence towards the normal distribution is greater, the smaller the spread of standard deviations amongst the components.

7.12 We have seen from Chapter 4 in Table 4.68 remarkable evidence of the truth of the central limit theorem. All the convoluted distributions have probabilities between limits which are slightly greater than for a Gaussian distribution of the same standard deviation. This is equivalent to saying that the probability of an uncertainty outside stated limits is greater for a Gaussian distribution than for a convoluted distribution of the same standard deviation.

7.13 Thus calculating the total standard deviation of a group of systematic uncertainties, combined with the accompanying random uncertainty, and

† Cramér H *Mathematical Methods of Statistics* pages 213, 316.
  Lindgren B W *Statistical Theory* pages 143–5.

treating the resulting distribution as Gaussian, will result in the probability of an uncertainty between limits being slightly less than the correct value had the systematic and random uncertainties been combined by convolution. The uncertainty can then be stated in the form 'The uncertainty between stated limits has a probability of at least the Gaussian value'.

7.14 The uncertainty of a system can thus be expressed as follows. Let $U_R \equiv$ random uncertainty, where $U_R = ks_R$ and where $k \equiv$ Gaussian tolerance probability and s_R is the estimated random standard deviation. Also let $U_S \equiv$ systematic uncertainty where $U_S = k\sigma_S = k[\sum_{m=1}^{m=n} {}_m\sigma_S^2]^{1/2}$, where σ_S is the total systematic standard deviation and ${}_m\sigma_S$ is the standard deviation of the mth systematic component. The overall uncertainty is thus

$$U = [U_R^2 + U_S^2]^{1/2} = k[s_R^2 + \sigma_S^2]^{1/2} \qquad 7.14(1)$$

Alternatively this can be expressed as

$$U = k\left[s_R^2 + \sum_{m=1}^{m=n} a_m^2/3 \right]^{1/2} \qquad 7.14(2)$$

where a_m is the semi-range of the mth systematic component. s_R is usually taken as the standard deviation of the mean of the random component, since the mean is generally assumed to be the best approximation to the correct value. If s_R is the estimated standard deviation of the random component, then the standard deviation of the mean is given by $s_R\sqrt{q}$ where 'q' is the number of observations made of the random variable. U can then be expressed as

$$U = k\left[s_R^2/q + \sum_{m=1}^{m=n} a_m^2/3 \right]^{1/2} \qquad 7.14(3)$$

7.15 If the number of random readings is small, then the value of s_R derived may be inaccurate, and the distribution for the random component is then best represented by a Student 't' distribution. However replacing k by t and writing

$$U = \left[t^2 s_R^2/q + k^2 \sum_{m=1}^{m=n} a_m^2/3 \right]^{1/2} \qquad 7.15(1)$$

where q is the number of random observations would lead to an overestimate of the uncertainty probability, especially if q were small and s_R and $[\sum_{m=1}^{m=n} a_m^2/3]^{1/2}$ were comparable in size. The problem can be resolved by using an approximate formula derived by Welch† which enables the effective number of degrees of freedom of combined Student 't' distributions or combined Student 't' and Gaussian distributions to be found and treating the resultant distribution as a 't' distribution with the effective number of

† Welch B L *Biometrika* **36** 290 (1949) and *Biometrika* **34** 28 (1947).

degrees of freedom found. The formula below gives the effective number of degrees of freedom of the combined distributions as

$$1/v_{\text{eff}} = \sum_{i=1}^{i=n} s_i^4/v_i \bigg/ \left(\sum_{i=1}^{i=n} s_i^2 \right)^2 \qquad 7.15(2)$$

where $s_i \equiv$ estimated standard deviation of the 'i' component derived from $n_i - 1 = v_i$ degrees of freedom. Now the standard deviation of the systematic uncertainties is assumed as known, and so each degree of freedom for a systematic uncertainty can be put equal to ∞. $7.15(2)$ reduces in this case to

$$v_{\text{eff}} = \left[\sum_{i=1}^{i=n} a_m^2/3 + s_R^2/q \right]^2 v_R/s_R^4/q^2 \qquad 7.15(3)$$

where s_R is the standard deviation of the random variable derived from q readings.

This equation reduces to

$$v_{\text{eff}} = [q\sigma_S^2/s_R^2 + 1]^2 v_R \qquad 7.15(3A)$$

$$= (\alpha^2 + 1)^2 v_R \qquad 7.15(3B)$$

where $\sum_{i=1}^{i=n} a_m^2/3 = \sigma_S^2$, the square of the standard deviation of the systematic uncertainties and $\sqrt{(q)}\sigma_S/s_R = \alpha$. In general v_{eff} yields a non-integer number which should be rounded down to the next integral number. The 't' corresponding to v_{eff} degrees of freedom is written t_{eff} and substituted for the Gaussian k of $7.14(2)$, giving

$$U = t_{\text{eff}} \left[s_R^2/q + \sum_{i=1}^{i=n} a_m^2/3 \right] \qquad 7.15(4)$$

If several random components are involved

$$v_{\text{eff}} = \left[\sum_{R=1}^{R=U} s_R^2 + \sum_{i=1}^{i=n} a_m^2/3 \right]^2 \bigg/ \left(\sum_{R=1}^{R=U} s_R^4/v_R \right) \qquad 7.15(5)$$

where s_R is either the standard deviation or the standard error in the mean of each variable. If some systematic errors have estimated standard deviations then these should also be included as $\sum_{p=1}^{p=1} \sigma_p^2$ and added inside the brackets of $7.15(3)$ and $7.15(4)$. 't' is then found as before from 't' tables to give t_{eff}.

7.16 It is instructive to find the difference in value for the uncertainty range obtained using a number of numerical values for the ratio $[\sum_{i=1}^{i=n} a_R^2/3]^{1/2}/s_R/\sqrt{q}$ and for v the number of degrees of freedom of s_R. We can compare $7.14(2)$ and $7.15(4)$ directly by comparing t_{eff} of $7.15(4)$ with the k of $7.14(2)$. We can include $7.15(1)$ by writing it as

$$U = \frac{[t^2 s_R^2/q + k^2 \sum_{m=1}^{m=n} a_m^2/3]^{1/2}}{[s_R^2/q + \sum_{m=1}^{m=n} a_m^2/3]^{1/2}} \times \left[s_R^2/q + \sum_{m=1}^{m=n} a_m^2/3 \right]^{1/2}$$

$$= t' \left[s_R^2/q + \sum_{m=1}^{m=n} a_m^2/3 \right]^{1/2}$$

where

$$t' = \frac{[t^2 s_R^2/q + k^2 \sum_{m=1}^{m=n} a_m^2/3]^{1/2}}{[s_R^2/q + \sum_{m=1}^{m=n} a_m^2/3]^{1/2}} \qquad 7.16(1)$$

Writing $\sum_{m=1}^{m=n} a_m^2/3 = \sigma_S^2$ where σ_S = systematic standard deviation we have

$$t' = [t^2 + k^2 q \sigma_S^2/s_R^2]^{1/2} \Big/ \left[1 + \frac{q\sigma_S^2}{s_R^2} \right]^{1/2} \qquad 7.16(2)$$

The ranges of 7.14(2), 7.15(1) and 7.15(4) can now be compared by comparing k, t' and t_{eff} (see Figure 7.16). Table 7.16 gives values of t', t_{eff} and k, and the ratio t'/t_{eff} and k/t_{eff} for a number of values of $\sqrt{(q)}\sigma_S/s_R$ and q the number of observations for s_R for probabilities of 0.95 and 0.995. It is seen that if the ratio of the standard deviation of the systematic errors to that of the random error is 5 or greater and q the number of readings for the random component is 5 or greater, then there is not more than 5% difference between any of the three methods of estimating the total uncertainty. Only if q is less than 5 and the ratio of standard deviations less than 5 need the method of 7.15(4) be used. The t' approximation always gives a safe if sometimes too large uncertainty.

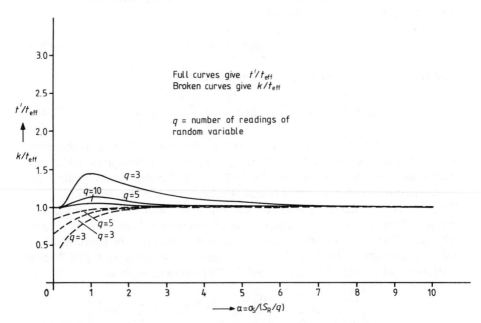

Figure 7.16 Graph comparing t', t_{eff} and k for $P = 0.95$. (See paragraphs 7.15 and 7.16 and Table 7.16.)

7.17 To summarize, the Central Limit Theorem gives a firm foundation for the convolution of non-Gaussian distributions, predicting an asymptotic convergence to the Gaussian form as the number of independent systematic contributions is increased. As we have seen in Chapter 4, and in particular from Table 4.68, the convolution of rectangular distributions leads to a very rapid convergence to the Gaussian form. Further, for all finite range distributions, from a rectangular form down to a near Gaussian one, the probability between limits, for convolutions of these distributions for k greater than 2, is greater than for the corresponding Gaussian distribution of equal standard deviation. The difference in probability is small, one or two per cent, and diminishes with increasing number of convoluted components. Thus assuming that the resulting distribution of systematic uncertainties is Gaussian gives a good approximation to the correct value, which errs on the conservative side for the probability between the chosen limits.

7.18 The assumption that the systematic uncertainties are independent of one another would appear to be perfectly reasonable. However if two systematic uncertainties are known to be highly correlated, then their contribution to the systematic standard deviation can be written as $(a_m + a_{m+1})/\sqrt{3}$ rather than $(a_m^2 + a_{m+1}^2)^{1/2}/\sqrt{3}$ for the uncorrelated case.

7.19 If each systematic component were assumed to take up its maximum value, i.e. $\pm a_m$, then each would become a stochastic distribution of two components. Since each distribution would now have a standard deviation of a_m (see Chapter 3), the total systematic standard deviation becomes $[\sum_{m=1}^{m=n} a_m^2]^{1/2}$ leading to a systematic uncertainty of

$$k \left[\sum_{m=1}^{m=n} a_m^2 \right]^{1/2}. \qquad 7.19(1)$$

Comparing with the continuous rectangular distribution case we see that the systematic uncertainty is increased by a factor of $\sqrt{3}$. Assuming the maximum values only for the systematic uncertainties would in practice be extremely unreasonable.

7.20 One of the drawbacks of the method of writing the uncertainty of a system as $k s_R / \sqrt{n} + \sum_{m=1}^{m=n} |a_m|$ is that no probability can be assigned to the result, other than a probability of unity for the systematic component. One has no means of calculating the probability between diminished limits, and thus no means of estimating how much over-safety has been built into the result. A few simple examples will give a feel for the amount of safety that has been built up, if one uses the method given here to calculate the uncertainty and also to calculate the uncertainty probability given by adding the systematic contribution. Let us take ten equal systematic uncertainties with semi-ranges equal to say 1 unit. Using the proposed method

$$\sigma_S = \left[\sum_{m=1}^{m=10} 1^2 \right]^{1/2} \sqrt{3} = 1.826$$

Table 7.16 The table gives relative values of uncertainty which are proportional to t', t_{eff} and k.

P	q	$\sigma_S/(s_R/\sqrt{q})=\alpha$ 0.2 t'	t'/t_{eff}	t_{eff}	k	k/t_{eff}	1 t'	t'/t_{eff}	t_{eff}	k	k/t_{eff}	5 t'	t'/t_{eff}	t_{eff}	k	k/t_{eff}	10 t'	t'/t_{eff}	t_{eff}	k	k/t_{eff}
$P=0.90$	3	2.88	0.99	2.92	1.645	0.56	2.37	1.27	1.86	1.645	0.88	1.71	1.04	1.645	1.645	1.00	1.66	1.01	1.645	1.645	1.00
	5	2.12	1.00	2.13	1.645	0.77	1.90	1.09	1.75	1.645	0.94	1.67	1.02	1.645	1.645	1.00	1.65	1.00	1.645	1.645	1.00
	10	1.83	1.00	1.83	1.645	0.90	1.74	1.02	1.70	1.645	0.97	1.65	1.00	1.645	1.645	1.00	1.65	1.00	1.645	1.645	1.00
	20	1.73	1.01	1.72	1.645	0.96	1.69	1.01	1.67	1.645	0.98	1.65	1.00	1.645	1.645	1.00	1.65	1.00	1.645	1.645	1.00
$P=0.95$	3	4.24	0.99	4.3	1.96	0.46	3.34	1.45	2.31	1.96	0.85	2.10	1.07	1.96	1.96	1.00	2.00	1.02	1.96	1.96	1.00
	5	2.75	0.99	2.78	1.96	0.71	2.40	1.13	2.12	1.96	0.92	2.00	1.02	1.96	1.96	1.00	1.97	1.00	1.96	1.96	1.00
	10	2.25	1.00	2.26	1.96	0.87	2.12	1.04	2.03	1.96	0.97	1.97	1.00	1.96	1.96	1.00	1.96	1.00	1.96	1.96	1.00
	20	2.09	1.00	2.09	1.96	0.94	2.03	1.02	1.99	1.96	0.98	1.97	1.00	1.96	1.96	1.00	1.96	1.00	1.96	1.96	1.00

P	q												
0.995	3	13.83	14.09	2.81	10.16	3.83	2.81	3.90	2.81	2.81	3.13	2.81	2.81
		0.98	0.20		2.65	0.73		1.39		1.00	1.11		1.00
	5	5.52	5.60	2.81	4.43	3.25	2.81	2.96	2.81	2.81	2.85	2.81	2.81
		0.99	0.50		1.36	0.86		1.05		1.00	1.01		1.00
	10	3.66	3.69	2.81	3.28	2.99	2.81	2.84	2.81	2.81	2.82	2.81	2.81
		0.99	0.76		1.10	0.94		1.01		1.00	1.00		1.00
	20	3.16	3.15	2.81	3.00	2.89	2.81	2.82	2.81	2.81	2.81	2.81	2.81
		1.00	0.89		1.04	0.97		1.00		1.00	1.00		1.00

t' is a pseudo value of the tolerance coefficient when the approximation 7.15(1) is used, t_{eff} is the tolerance coefficient when the Welch approximation for combining 't' distributions is used and 'k' is the Gaussian distribution coefficient. t_{eff} is the preferred coefficient, q is the number of observations made for each set of measurements. $\sigma_{S}/(s_{R}\sqrt{q}) \equiv \alpha$ is the ratio of the standard deviation of the systematic uncertainties to the standard deviation of the mean of the random uncertainty.

Assuming a Gaussian distribution, and taking $k = 2.0$, giving a minimum probability of 0.954, we have an uncertainty of ± 3.652 units. Adding the semi-ranges leads to an uncertainty of 10 units. Dividing this figure by the standard deviation of 1.826 we obtain $k = 5.476$, leading to a probability of unity. As most commercial laboratories doing calibration work use a tolerance probability of 0.95, we see that adding the systematic uncertainties leads to a result approximately three times the value given for the accepted 0.95 probability. The other model of assessing systematic uncertainties, which is sometimes used, sums the uncertainties as $[\sum_{m=1}^{m=n} a_m^2]^{1/2}$. Comparing this with the method described here we see that the ratio of the two methods gives the value

$$R = \left[\sum_{m=1}^{m=n} a_m^2\right]^{1/2} \Bigg/ k\left[\sum_{m=1}^{m=n} a_m^2/3\right]^{1/2} = \sqrt{3}/k$$

If $k = 2$ this gives $R = 0.866$. If $k = 3$ the result is even worse, i.e. $R = 0.577$.

Thus this method gives too small a value of the uncertainty and like the adding method of paragraph 7.19 gives no accompanying probability.

7.21   One final point on the suggested method, the only time that assuming a Gaussian distribution will lead to a significant difference between the probability for the convoluted systematics and an assumed Gaussian distribution is when one of the systematic uncertainties has a range much larger than the others. The best way to deal with this is to calculate the total standard deviation comprising the random component and all the systematics excluding the large one, i.e.

$$\left[s_R^2 + \sum_{m=1}^{m=n-1} a_m^2/3\right]^{1/2}$$

If a_n is the large systematic, then form the ratio

$$\left[s_R^2 + \sum_{m=1}^{m=n-1} a_m^2/3\right]^{1/2} \Bigg/ a_n/\sqrt{3} \equiv \eta$$

If the value of the above ratio is less than about 0.8, then the true probability of the combined systematics and random component is likely to be at least 1% higher than the 95.4% for $k = 2$. For $k = 3$ the difference falls to about 0.18%. If the ratio is less than 0.5, the corresponding percentages are about 2.2 and 0.26.

7.22   In order to deal with this special case, Table XXIII has been produced, which gives values of k for given values of the probability against the ratio η and also values of the probability P for given values of k against the ratio. η is defined as the ratio of σ_G/σ_R where σ_G = Gaussian standard deviation and σ_R = standard deviation of a rectangular distribution. The Table gives the method for dealing with the case of a large rectangular distribution to be combined with a Gaussian distribution or a Gaussian distribution and smaller rectangular distributions.

Conclusion

7.23 Overall the method for the combination of all the uncertainties is simple and has the merit of giving a minimum uncertainty to be associated with a given tolerance range. The assumption of a rectangular distribution for a given semi-range also maximizes the associated standard deviation. The method also has the merit of associating a probability with a given tolerance range.

8

The Estimation of Calibration Uncertainties

8.01 In this chapter we shall deal with the application of the methods for calculating the various types of uncertainties which have been discussed in previous chapters. Before this is done, however, it is necessary to consider functional relationships, and the magnitude of the uncertainties that they produce.

Standard Deviation of a Function of n Variables

8.02 If F is a function of $x_1, x_2, \ldots, x_n$, and in a particular measurement situation, suppose we are interested in the value of F for which the values of the independent variables are $_ix_1, _ix_2, _ix_3, _ix_4, \ldots, _ix_n$. Thus, providing F is continuous and has continuous derivatives in the region near $_ix_1, _ix_2, \ldots, _ix_n$, we can expand F in a Taylor series. Thus

$$F = \left\{ \sum_{r=0}^{r=\infty} \left(\sum_{i=1}^{i=n} \delta x_i \frac{\partial}{\partial x_i} \right)^r \right\} F(_ix_1, _ix_2, \ldots, _ix_n) \qquad 8.02(1)$$

If we neglect all terms above the first order, this reduces to

$$F = F_1 + \frac{\partial F_1}{\partial x_1} \delta x_1 + \frac{\partial F_1}{\partial x_2} \delta x_2 + \cdots + \frac{\partial F_1}{\partial x_n} \delta x_n \qquad 8.02(2)$$

where

$$F_1 = F(_1x_1, _1x_2, _1x_3, \ldots, _1x_n) \qquad 8.02(3)$$

and

$$\frac{\partial F_1}{\partial x_1} = \frac{\partial F}{\partial x_1} \qquad \text{with} \qquad x_1 = {_1x_1}, x_2 = {_1x_2}, \ldots, x_n = {_1x_n} \qquad 8.02(4)$$

248

and similarly with

$$\frac{\partial F_1}{\partial x_2} \qquad \text{to} \qquad \frac{\partial F_1}{\partial x_n}$$

We can neglect the terms above the first order because we have assumed that the

$$\frac{\partial^r F}{\partial x_i^r}$$

are continuous and finite, and since the δx_i are small, terms involving powers of the δx_i above one with coefficients involving the

$$\frac{\partial^r F}{\partial x_i^r}$$

can be neglected. If we write $F - F_1 = \delta F$ then we can write 8.02(2) as

$$\delta F = \sum_{i=1}^{i=n} \delta x_i \frac{\partial F}{\partial x_i} \qquad\qquad 8.02(5)$$

8.03 This expression gives the error in F from F_1 caused by a change or error in the variables $_1x_1, {}_1x_2, \ldots, {}_1x_n$ to $_1x_1 + \delta x_1, {}_1x_2 + \delta x_2, \ldots, {}_1x_n + \delta x_n$. The δxs can of course be plus or minus. If we now square the expression for δF, we have

$$\delta F^2 = \sum_{i=1}^{i=n} \delta x_i^2 \left(\frac{\partial F}{\partial x_i}\right)^2 + \sum_{p=1}^{p=n} \sum_{q=1}^{q=n} \left(\frac{\partial F}{\partial x_p}\right)\left(\frac{\partial F}{\partial x_q}\right)_{p \neq q} \delta x_p \delta x_q \quad 8.03(1)$$

Suppose many measurements are made of δF, each leading to a new set of δx_i. We will distinguish a particular set of measurements by adding the suffix r to the variables δx and δF. Thus the rth set of measurements, using 8.03(1), becomes

$$(\delta F)_r^2 = \sum_{i=1}^{i=n} \delta x_{ir} \left(\frac{\partial F}{\partial x_i}\right)^2 + \sum_{p=1}^{p=n} \sum_{q=1}^{q=n} \left(\frac{\partial F}{\partial x_p}\right)\left(\frac{\partial F}{\partial x_q}\right)\delta x_{pr} \delta x_{qr} \quad 8.03(2)$$

If t sets of measurements are taken, and the sum of 8.03(2) is found for all r from 1 to t we have the sum

$$\sum_{r=1}^{r=t} (\delta F_r)^2$$

If we divide this by t we have the square of the standard deviation of F, that is

$$\sigma_F^2 = \sum_{r=1}^{r=t} \frac{(\delta F_r)^2}{t} = \sum_{r=1}^{r=t} \sum_{i=1}^{i=n} \delta x_{ir}^2 \left(\frac{\partial F}{\partial x_i}\right)^2 / t$$

$$+ \sum_{r=1}^{r=t} \sum_{p=1}^{p=n} \sum_{q=1}^{q=n} \left(\frac{\partial F}{\partial x_p}\right)_{p \neq q} \left(\frac{\partial F}{\partial x_q}\right)\delta x_{pr} \delta x_{qr} / t \qquad\qquad 8.03(3)$$

8.04 Now for any p and q the sum

$$\sum_{r=1}^{r=t} \left(\frac{\partial F}{\partial x_p}\right)\left(\frac{\partial F}{\partial x_q}\right) \delta x_{pr} \delta x_{qr}/t$$

will tend to zero as t gets large, since any product $\delta x_{pr} \delta x_{qr}$ is as likely to be negative as positive. Thus in the limit when $t \rightarrow \infty$

$$\sigma_F^2 = \sum_{r=1}^{r=t} \sum_{i=1}^{i=n} \delta x_{ir}^2 \left(\frac{\partial F}{\partial x_i}\right)^2 / t \qquad 8.04(1)$$

Now

$$\sum_{r=1}^{r=t} \frac{\delta x_{ir}^2}{t} = \sigma_{x_i}^2 \qquad 8.04(2)$$

the variance or standard deviation squared, of the ith variable of x. Thus

$$\sigma_F^2 = \sum_{i=1}^{i=n} \sigma_{x_i}^2 \left(\frac{\partial F}{\partial x_i}\right)^2 \qquad 8.04(3)$$

If the σs are derived from m readings where m is not very large, then the σs should be replaced by ss where s is the best estimation of the standard deviation for a small number of readings. Therefore

$$s_F^2 = \sum_{i=1}^{i=n} s_{x_i}^2 \left(\frac{\partial F}{\partial x_i}\right)^2 \qquad 8.04(4)$$

and

$$\sigma^2 = s^2 \left(\frac{m-1}{m}\right) \qquad 2.24(5)$$

Writing

$$\left(\frac{\partial F}{\partial x_i}\right) = A_i \qquad 8.04(5)$$

a constant, we have

$$\sigma_F^2 = \sum_{i=1}^{i=n} \sigma_{x_i}^2 A_i^2 \qquad 8.04(6)$$

or

$$s_F^2 = \sum_{i=1}^{i=n} s_{x_i}^2 A_i^2 \qquad 8.04(7)$$

8.05 This formula can be derived alternatively as follows:

$$F = F_1 + \sum_{i=1}^{i=n} \left(\frac{\partial F}{\partial x_i}\right)\delta x_i + \text{(neglected higher terms)}$$

$$= F_1 + \sum_{i=1}^{i=n} A_i \delta x_i \qquad\qquad 8.05(1)$$

Comparing with 3.40(4) we see that since 8.05(1) is a linear function of x_i in the region of $x_i = {}_1x_i$, that is $\delta x_i = (x_i - {}_1x_i)$, then the standard deviation of F, by comparison with 3.40(5), is given by

$$\sigma_F^2 = \sum_{i=1}^{i=n} \sigma_{x_i}^2 A_i^2 \qquad\qquad 8.05(2)$$

or

$$s_F^2 = \sum_{i=1}^{i=n} s_{x_i}^2 A_i^2 \qquad\qquad 8.05(3)$$

Standard Deviation of the Mean of a Function of n Variables

8.06 If each of the variables x_i has been measured t times, then from 3.42(3) the standard deviation of $\bar{x}_i$ is equal to

$$\frac{\sigma_{x_i}}{\sqrt{t}} = \sigma_{\bar{x}_i} \qquad\qquad 8.06(1)$$

Now

$$\sigma_F^2/t = \sum_{i=1}^{i=n} \frac{\sigma_{x_i}^2 A_i^2}{t} \qquad\qquad 8.06(2)$$

Thus writing

$$\sigma_F^2/t = \sigma_{\bar{F}}^2 \qquad\qquad 8.06(3)$$

where $\sigma_{\bar{F}}$ is defined as the standard error in the mean of F, we have, using 8.06(1), that

$$\sigma_{\bar{F}}^2 = \sum_{i=1}^{i=n} \sigma_{\bar{x}_i}^2 A_i^2 \qquad\qquad 8.06(4)$$

or

$$s_{\bar{F}}^2 = \sum_{i=1}^{i=n} s_{\bar{x}_i}^2 A_i^2 \qquad\qquad 8.06(5)$$

Generalized Standard Deviation in the Mean

8.07   If the standard deviation σ_i for each variable x_i is found from a different number of readings m_i, then the standard deviation in the mean for each variable is

$$\sigma_{\bar{x}_i} = \sigma_{x_i}/\sqrt{m_i} \qquad\qquad 8.07(1)$$

Now if we consider the variation in the mean of each variable, that is of $\bar{x}_i$, then we can write 8.02(5) as

$$\delta F = \sum_{i=1}^{i=n} \delta \bar{x}_i \, \frac{\partial F}{\partial x_i} \qquad\qquad 8.07(2)$$

where

$$\delta \bar{x}_i = \bar{x}_i - {}_1 x_i \qquad\qquad 8.07(3)$$

Now 8.07(2) is a linear function in $\bar{x}_i$ and thus the standard deviation of 8.07(2) is given by

$$\sigma_{\bar{F}G}^2 = \sum_{i=1}^{i=n} \sigma_{\bar{x}_i}^2 A_i^2 \qquad\qquad 8.06(4)$$

where σ_{FG} is the generalized standard deviation of F when the standard errors for each variable are derived from different numbers of observations m_i. When the m_is are all equal, then of course

$$\sigma_{F\bar{G}} = \sigma_F \qquad\qquad 8.07(4)$$

or

$$s_{F\bar{G}} = s_F \qquad\qquad 8.07(5)$$

Estimation of Component Uncertainties

8.08   When an instrument or piece of equipment is calibrated it is important to state the uncertainty of the measurements, and in order to do this, it is necessary to take into account every possible factor likely to cause a variation in the value of the parameter being measured.

Measurable Uncertainties

8.09   The easiest uncertainties to calculate are those whose distribution constants can be determined from repeated readings. These uncertainties will be designated *measurable uncertainties*, since the associated constants of the distributions involved can be obtained with increasing accuracy as the number of readings is increased. The larger the number of readings the smaller will be the uncertainty in the calculated value of the standard deviation and of the mean. In the great majority of cases the distributions obtained will be of the Gaussian form. An important point to be considered when measurable uncertainties are to be combined with non-measurable uncertainties is to be clear whether it is the uncertainty in the mean, or the uncertainty of a single reading for the measurable uncertainty, that is to be combined. If a measuring instrument is being used to find the magnitude of a parameter or of a workpiece, then it is the uncertainty in the mean that should be used. If, on the other hand, one is expressing the uncertainty to be expected from a single reading, then it is the deviation of a single reading from the mean that should be combined.

8.10 When Gaussian uncertainties are calculated it is necessary to make up one's mind whether confidence probabilities are required. In general the distribution constants are found from a limited number of readings, usually well under a hundred and often under twenty, and so only the estimates for μ and σ are known, that is $\bar{x}$ and s respectively. Under these circumstances the calculated tolerance range for a given tolerance probability is likely to be too small, whilst correspondingly the calculated tolerance probability for a selected tolerance range is likely to be too large, when the calculations are made assuming a Gaussian distribution and using Tables II and III of Appendix I. A better approximation to the correct tolerance probability or tolerance range is obtained by using one or other of the tables allied to the Gaussian distribution.

Guidance on Which Tables to Use

8.11 The information given under the heading 'Choice of Table' in the section devoted to tables (see Appendix I) sets out in tabular form the range that can be derived from each table, together with the table number and appropriate constant that each table gives. The first column of each set of information gives the parameters that have been found or are known, and it is these in conjunction with column 2 that determine which table should be used.

8.12 Section (i) enables the uncertainty range associated with a given probability to be found (a) without confidence probabilities and (b) with confidence probabilities. Section (ii) enables the probability associated with a given uncertainty range to be found. This section is however restricted to large values of n, the number of readings, although its results are approximately true for smaller values of n. Section (iii) enables the uncertainty range for the mean to be found when the associated probability is given.

Selection of Tolerance Probability

8.13 It is important to note that whatever value is selected for the total tolerance probability, the same value must be selected for each component uncertainty.

Estimated Uncertainties

8.14 Under this heading is included all uncertainties which cannot be directly measured, and since we have to rely on previous experience or on calculation, we will designate this class of uncertainties as *estimated uncertainties*.

Rectangular Uncertainties

8.15 We now come to that class of uncertainties which are very difficult to calculate. In many cases all that can be said is that a particular variable can lie between two limit points, usually plus and minus some value on either side of the mean. It is necessary of course to find the functional relationship between the variable in question and the parameter which is the subject of measurement. This should be done when finding the appropriate constant of proportionality between the measured variable and the variable whose variation we are considering. Often there is a known mathematical relationship and, if so, then $\partial F/\partial x$ is the required coefficient, where F is the measured parameter and x the variable whose plus and minus range we are considering. If $\partial F/\partial x$ is not known, then it should be found by measuring the values of F for two discrete values of x, whence

$$\frac{\partial F}{\partial x} \simeq \frac{F_1 - F_2}{x_1 - x_2} \qquad\qquad 8.15(1)$$

8.16 Since nothing is known about the distribution of the type of variables we are considering they should be assumed to be rectangular. Thus, if the assumed range is $\pm a$, about the mean, the standard deviation is, by paragraph 4.06, equal to $\pm a/\sqrt{3}$.

Maximum Value Only Uncertainties

8.17 There are some cases where the uncertainty distribution for a particular variable consists of just two values, usually equally distributed about the mean. If these two values are $\pm b$, then the standard deviation is given by b (see paragraphs 3.54 to 3.57).

Gaussian Estimated Uncertainties

8.18   This is a very difficult case to estimate, since it is usually much easier to say that a given variable always lies within some tolerance range than to say that the probability of a variable lying within some stated range is such-and-such a figure. One procedure commonly adopted is to choose for the uncertainty range selected the same tolerance probability as that selected for the total uncertainty. For instance, if the estimated tolerance range is $\pm j$ and the tolerance coefficient found from Table III for a chosen probability P is k_1, then the equivalent standard deviation is j/k_1. When the total uncertainty is found and the combined standard deviation is multiplied by k_1, this effectively means that the contribution to the root mean square uncertainty for this particular uncertainty is j. Thus the doubt in this instance is really centred on the probability to be associated with j; it might be too low or too high.

8.19   If estimated uncertainties of the type we are considering are known to be Gaussian it is usually not very difficult to estimate a maximum value, but it is extremely difficult to estimate the associated probability. In general it is much safer to estimate the probability on the low side, that is to 0.954 or even 0.917, corresponding to 2.00 and 1.732 standard deviations respectively. In this case if the uncertainty range is $\pm j$ as before, the equivalent standard deviation is j/k_1 where k_1 is found from Table III for the chosen value of probability. If now the tolerance probability for the total uncertainty is greater than the probability selected for the uncertainty j, and the total tolerance coefficient is k'_1, then the effective contribution to the total uncertainty by the Gaussian-estimated uncertainties is represented by the term $k'j/k_1$ which is larger than j and thus gives some margin of safety. It is not recommended that this type of uncertainty be used. It is better to use a rectangular distribution.

Combination of Uncertainties to Give Total Uncertainty

Gaussian-type measurable uncertainties U_{GM} (random uncertainties)

8.20   Take each Gaussian-type standard deviation σ or estimated standard deviation s_v and multiply it by its appropriate tolerance factor (see paragraphs 8.09 to 8.13 and particularly paragraphs 8.11 to 8.13). Each of these products should now be squared and the results added together.

8.21   If each Gaussian-type measurable standard deviation is s_r and $n \geqslant r \geqslant 1$, where n is equal to the number of uncertainties of this type, and each Gaussian-type tolerance coefficient is $_rk_i$, each corresponding to the probability for which the total uncertainty is required, then the effective uncertainty contribution from the Gaussian-type errors U_{GM} is given by

$$U_{GM} = \left\{ \sum_{r=1}^{r=n} (_r k_i s_r)^2 \right\}^{1/2} \qquad 8.21(1)$$

$_r k_i$ is the 'r' component uncertainty and i is the particular number chosen, designating which tolerance coefficient has been chosen, i.e. k_1, k_2, k_3, k_4, k_5, k_6 and k_7. The values of the $_r k_i$s will depend on the tables used and of course on the number of readings taken. If a contribution $_r k_i s_r$ represents the uncertainty in the *mean*, a little care must be exercised over the relation between the $_r k_i$ and the s_r. If the s_r is the standard deviation of a variable, and Table III is used the $_r k_i$ is $k_1 \sqrt{n}$ where k_1 is the k selected from Table III and n is the number of readings. If Table IV is used the $_r k_i$ is equal to $k_2/\sqrt{n}$. If, however, Table V is used the $_r k_i$ is equal to k_3. If the standard error is found, that is $s_r/\sqrt{n}$, and this is used in place of s_r, then $_r k_i$ when Table III is used is equal to k_1, and when Table IV is used $_r k_i$ is equal to k_2. If the *standard error* is used for s_r then Table V should not be used, but Table IV used instead.

8.22 Some caution needs to be observed here in the use of equation 8.21(1). If any of the tolerance coefficients chosen are k_2, k_3 or k_5, from Tables IV, V and XI respectively, then it is necessary to find v_{eff} as described in paragraph 7.15 when, of course, *all* standard deviations contributing to the uncertainty must be used to find v_{eff}. U_{GM} should now be written as

$$U'_{GM} = k_{2_{veff}} \left[\sum_{r=1}^{r=n} \left\{ \frac{_r k_i s_r}{_r k_{1or2}} \right\}^{2} \right]^{1/2} \qquad 8.22(1)$$

If $i = 2$, 3 or 5 then the k in the denominator for each term r should be k_2 for the probability chosen for the number of readings made to obtain s_r. Correspondingly if i is *not* 2, 3 or 5, then the k in the denominator for each term r should be k_1 from Table III for the chosen probability.

8.23 If k_2, k_3 or k_5 have not been used in preparing U_{GM}, then U_{GM} remains unaltered as in 8.21(1).

Combination of estimated systematic rectangular system uncertainties (see paragraphs 8.15 and 8.16) U_{RE}

8.24 If the range of each rectangular error is a_r, then the standard deviation of each is $a_r/\sqrt{3}$ and the total standard deviation is

$$\left(\sum_{r=1}^{r=n} \frac{a_r^2}{3} \right)^{1/2} = \sigma_{RE} \qquad 8.24(1)$$

for n such distributions. The probability for which the total uncertainty is required is now used to find the tolerance coefficient k_1 which is found from Table III. The effective uncertainty contribution from the rectangular errors

U_{RE} to the total uncertainty is thus given by

$$U_{RE} = k_1\left(\sum_1^n \frac{a_r^2}{3}\right)^{1/2} \qquad 8.24(2)$$

This procedure is justified by the results obtained in Chapter 4 and summarized in paragraphs 4.67 and 4.68 and Table VI. The true probability of an error occurring in the range

$$\bar{x} - k_1\sigma_{RE} < x < \bar{x} + k_1\sigma_{RE} \qquad 8.24(3)$$

is always greater than that given by the procedure of this paragraph, but in general the value found is very close to the correct value. Alternatively it means that the probability of an uncertainty occurring outside the range 8.24(3) is slightly less than that calculated. If U'_{GM} has been calculated, i.e. k_2, k_3 and k_5 have been used, then U_{RE} should be replaced by

$$U'_{RE} = k_{2_{\text{eff}}}\left(\sum \frac{a_r^2}{3}\right)^{1/2} \qquad 8.24(4)$$

i.e. the k_2 corresponding to v_{eff} obtained from the t distribution.

Combination of maximum value only systematic uncertainties (see paragraph 8.17) U_{MU}

8.25 If the value of each of these uncertainties is b_r, the total standard deviation should be calculated as

$$\left(\sum_{r=1}^{r=n} b_r^2\right)^{1/2} \qquad 8.25(1)$$

The effective uncertainty contribution U_{MU} from the maximum value only of systematic uncertainties to the total uncertainty is thus given by

$$U_{MU} = k_1\left(\sum_{r=1}^{r=n} b_r^2\right)^{1/2} \qquad 8.25(2)$$

or

$$U'_{MU} = k_{2_{\text{eff}}}\left(\sum_{r=1}^{r=n} b_r^2\right)^{1/2} \qquad 8.25(3)$$

where k_1 is the tolerance coefficient obtained from Table III corresponding to the required probability $\underset{-k_1 \text{ to } k_1}{P}$ for the total uncertainty. The justification for the mode of calculation of this contribution to the total uncertainty is given in the section beginning at paragraph 3.48 where it is shown that the probability associated with 8.25(2) is slightly less than the correct value. U'_{MU} is used if v_{eff} has been used.

Combination of Gaussian systematic estimated uncertainties U_{GE}

8.26 If j_r is the assumed magnitude of a Gaussian estimated uncertainty of this type for a given probability, then the effective standard deviation is $j_r/_r k_1$, where $_r k_1$ is the tolerance coefficient obtained from Table III for the assumed probability of occurrence of the uncertainty. The resulting effective standard deviation is thus

$$\left\{ \sum_{r=1}^{r=n} (j_r/_r k_1)^2 \right\}^{1/2}$$
8.26(1)

and if k_1 is the tolerance coefficient for the total uncertainty, the effective contribution to the total uncertainty from the systematic Gaussian estimated uncertainties U_{GU} is given by

$$U_{GE} = k_1 \left\{ \sum_{r=1}^{r=n} (j_r/_r k_1)^2 \right\}^{1/2}$$
8.26(2)

If k_2, k_3 or k_5 have been used in U_{GM}, then as before U_{GE} should be replaced by

$$U'_{GE} = k_{2_{eff}} \left(\sum_{r=1}^{r=n} (j_r/_r k_1)^2 \right\}^{1/2}$$
8.26(3)

Total uncertainty of measurement

8.27 The total uncertainty of measurement U_T is given by the square root of the sum of the squares of the effective component uncertainties given in paragraphs 8.09 to 8.26 and thus

$$U_T = (U_{GM}^2 + U_{RE}^2 + U_{MU}^2 + U_{GE}^2)^{1/2}$$
8.27(1)

a U_{GM} = total of Gaussian-type measurable uncertainties for which observations have been made (see paragraphs 8.09 to 8.13 and 8.20 to 8.25).

b U_{RE} = total of rectangular-type systematic uncertainties (see paragraphs 8.15, 8.16 and 8.26).

c U_{MU} = total of maximum value only systematic uncertainties (see paragraphs 8.17 and 8.27).

d U_{GE} = total of Gaussian-type estimated uncertainties (see paragraphs 8.18, 8.19 and 8.28).

It is to be noted that the appropriate tolerance coefficient by which the total standard deviation is multiplied is either the Gaussian k_1 or the effective 't' coefficient k_2 derived from equations 7.15(2), 7.15(3) or 7.15(5).

Summary of Terms

a

$$U_{GM} = \left\{ \sum_{r=1}^{r=n} (_r k_i s_r)^2 \right\}^{1/2}$$

8.21(1)

or

$$U'_{GM} = k_{2_{eff}} \left\{ \sum_{r=1}^{r=n} \left\{ \frac{_r k_i s_r}{_r k_{1 \, or \, 2}} \right\}^2 \right\}^{1/2}$$

8.22(1)

where: n = number of Gaussian-type distributions.

$_r k_i$ = the appropriate tolerance coefficient of the rth variable, selected from an appropriate table (see paragraph 8.11 et seq.) for the tolerance probability chosen for the total uncertainty. The value of i will depend on the table used and corresponds to the suffix given to k in a particular table.

s_r = the estimated standard deviation of the rth variable, and is given by

$$s_r = \left\{ \sum_{i=1}^{i=n} \frac{(x_i - \bar{x})^2}{n-1} \right\}^{1/2}$$

2.24(8)

where n = number of observations made of the rth variable and

$$\bar{x} = \sum_{i=1}^{i=n} \frac{x_i}{n}$$

2.22(1)

b

$$U_{RE} = k_1 \left\{ \sum_{r=1}^{r=n} (a_r/3)^2 \right\}^{1/2}$$

8.24(2)

or

$$U'_{RE} = k_{2_{eff}} \left\{ \sum_{r=1}^{r=n} (a_r/3)^2 \right\}^{1/2} \qquad \text{if } v_{eff} \text{ is used}$$

8.24(3)

where: n = number of rectangular distributions.

k_1 = the tolerance coefficient corresponding to the chosen tolerance probability for the total uncertainty $P_{-k_1 \, to \, k_1}$. k_1 is selected from Table III from the value of $P_{-k_1 \, to \, k_1}$. $k_{2_{eff}}$ is found from v_{eff} using Table IV.

a_r = semi-range of rth rectangular uncertainty.

c

$$U_{MU} = k_1 \left\{ \sum_{r=1}^{r=n} (b_r)^2 \right\}^{1/2}$$

8.25(2)

where: k_1 = the same as that for U_{RE}.

b_r = the magnitude of the rth maximum value only systematic uncertainty.

d

$$U_{GE} = k_1 \left\{ \sum_{r=1}^{r=n} (j_r/_r k_1)^2 \right\}^{1/2}$$

8.26(2)

or

$$U'_{GE} = k_{2_{eff}} \left\{ \sum_{r=1}^{r=n} (j_r /_r k_1)^2 \right\}^{1/2}$$

8.26(3)

where: k_1 = the same as that for U_{RE} and U_{ME}.

j_r = estimated magnitude of rth Gaussian estimated uncertainty for estimate probability P_r.

$_r k_1$ = tolerance coefficient for the above variable corresponding to the estimated probability $P_r ._r k_1$ is obtained from Table III, $k_1 \geqslant _r k_1$.

$k_{2_{eff}}$ = tolerance coefficient if v_{eff} has been used.

8.28 Consideration of the contributions from the component uncertainties may reveal what appear to be anomalies. For instance, suppose that the rectangular contribution is made up of two uncertainty distributions of equal standard deviation $a/\sqrt{3}$, then the combined standard deviation is $\sqrt{\frac{2}{3}}a$. Thus if k_1 is the tolerance coefficient corresponding to a tolerance probability of 0.997, then $k_1 = 3$ and the rectangular contribution is $k_1 \sigma = 3 \times \sqrt{\frac{2}{3}}a = 2.4489$. But the total range of the combined rectangular distributions is only $2a$, and they thus appear to be contributing an amount greater than their combined range.

Examples on Chapter 8

Example 8.1 The area A of a triangle is given by $\frac{1}{2}ab \sin \theta$ where a and b are adjacent sides and θ is the included angle. If s_a is the estimated standard deviation of the side a, s_b that of side b, and s_θ the estimated standard deviation of the included angle θ, show that the estimated standard deviation s_A of the area of the triangle is given by

$$s_A^2 = A^2 \left(\frac{s_a^2}{a^2} + \frac{s_b^2}{b^2} + \frac{s_\theta^2}{\tan^2 \theta} \right)$$

Answer: Now

$$A = \frac{ab \sin \theta}{2}$$

and so

$$\log A = \log a + \log b + \log(\sin \theta) + \log \tfrac{1}{2}$$

Differentiating we have

$$\frac{dA}{A} = \frac{da}{a} + \frac{db}{b} + \cot \theta \, d\theta$$

and thus, using 8.04(4), we have

$$s_A^2 = A^2\left(\frac{s_a^2}{a^2} + \frac{s_b^2}{b^2} + \frac{s_\theta^2}{\tan^2\theta}\right)$$

Example 8.2   Ohm's law states that current, voltage and resistance are related by the expression $E/R = I$, where E is the voltage across the resistance R, and I is the resulting current. If the standard deviations of these quantitites are respectively σ_E, σ_R and σ_I, find an expression for σ_I in terms of σ_E and σ_R.

Answer:

$$\sigma_I^2 = I^2\left(\frac{\sigma_E^2}{E^2} + \frac{\sigma_R^2}{R^2}\right)$$

where I, E and R are the nominal or mean values of these quantities.

Example 8.3   Charles' Law states that pressure P, volume V, and absolute temperature T are related by the expression $PV = RT$ where R is the gas constant for one gramme molecule of gas. Find an expression for the standard deviation σ_P of the pressure in terms of the standard deviations of the volume σ_V and temperature σ_T.

Answer:

$$\sigma_P^2 = P^2\left(\frac{\sigma_V^2}{V^2} + \frac{\sigma_T^2}{T^2}\right)$$

where P, V and T are the nominal or mean values of these quantities in the case considered.

Example 8.4   The wattage W of an alternating electric current is given by the expression $W = EI\cos\Phi$, where $E \equiv$ voltage, $I \equiv$ current, and Φ is the phase angle between E and I; $\cos\Phi$ is the power factor. Derive an expression for the standard deviation of the wattage σ_W in terms of the standard deviation of the voltage σ_E, the current σ_I and the phase angle σ_Φ.

Answer:

$$\sigma_W^2 = W_I^2\left(\frac{\sigma_I^2}{I^2} + \frac{\sigma_E^2}{E^2} + \sigma_\Phi^2\tan^2\Phi\right)$$

Note that σ_Φ is measured in radians.

Example 8.5   The refractive index of a prism is given by

$$\mu = \frac{\sin(A + D)/2}{\sin A/2}$$

where A is the refracting angle of the prism, and D is the deviation of the incident beam of light when the incident and emergent beams make equal angles with the prism surfaces (condition of minimum deviation). Find an expression for the standard deviation of the refractive index σ_μ in terms of the standard deviation σ_D of the ray deviation and σ_A the standard deviation of the prism angle.

Answer:

$$\sigma_\mu^2 = \frac{\sigma_A^2 \sin^2 D/2}{4 \sin^4 A/2} + \frac{\sigma_D^2 \mu^2}{4 \tan^2 (D + A)/2}$$

Example 8.6 Measurements are taken on the size of a certain dimension of a workpiece; the standard deviation of the measurements made with the measuring machine is known to be 0.000 02 in, and six measurements are made at each end of the workpiece. The uncertainty in the mean measured length due to inaccuracies in the measuring machine is known to have an upper limit of 0.000 05 in. During the measurements the temperature varied over a range of $\pm 0.4°C$. Further, the temperature indicator had a possible uncertainty of $\pm 0.3°C$ from the mean indicated temperature. The coefficient of expansion of the workpiece is given as $1.7 \times 10^{-5} °C^{-1}$, whilst that of the measuring equipment can be taken as $1.1 \times 10^{-5} °C^{-1}$. Find the uncertainty in the measured dimension of the workpiece for a tolerance probability of 0.95, given that the *nominal size* of the workpiece is 3.00 in.

Answer: The standard deviation of the measurements is given as 0.000 02 in, and thus the standard error in the mean at each end of the workpiece is equal to $0.000 02/\sqrt{6}$. Since the required tolerance probability is 0.95, the appropriate tolerance factor k from Table III is equal to 1.96. The contribution to the total uncertainty from the observational uncertainties is thus

$$1.96 \left\{ 2 \times \frac{(0.000 02)^2}{6} \right\}^{1/2} = 0.000 0266 \text{ in}$$

The uncertainty due to the measuring machine itself can have an upper limit of 0.000 05 in, so we assume this uncertainty to have a rectangular distribution, and so its standard deviation is $0.000 05/\sqrt{3}$, and its effective contribution to the total uncertainty is thus

$$\frac{1.96 \times 0.000 05}{\sqrt{3}} = 0.000 0566$$

The temperature uncertainties should be treated as two separate rectangular contributions. The uncertainty in the size of the workpiece caused by a change of temperature is equal to the size of the workpiece multiplied by the difference in the expansion coefficients of the workpiece and the measuring machine times the change in temperature. Thus the two maximum uncertainties from this cause are $0.4 \times 0.6 \times 10^{-5} \times 3$ and $0.3 \times 0.6 \times 10^{-5} \times 3$. The standard deviations of each contribution are obtained by dividing each by $\sqrt{3}$ and the contribution of each to the total uncertainty is obtained by multiplying each standard deviation by 1.96. The two effective temperature contributions are thus

$$\frac{0.4 \times 0.6 \times 10^{-5} \times 3 \times 1.96}{\sqrt{3}} \quad \text{and} \quad \frac{0.3 \times 0.6 \times 10^{-5} \times 3 \times 1.96}{\sqrt{3}}$$

or

$$0.000\,0081 \quad \text{and} \quad 0.000\,0061$$

The total uncertainty is thus

$$(0.000\,0226^2 + 0.000\,0566^2 + 0.000\,0081^2 + 0.000\,0061^2)^{1/2} = 0.000\,062 \text{ in}$$

Note how the largest uncertainty, the instrumental error of $0.000\,0566$, dominates the final uncertainty.

Example 8.7 The readings of a flow-rate meter are checked by collecting a hundred gallons of water which has passed through the flowmeter in one minute. The uncertainty in the time measurement can be up to 0.02 s, and the uncertainty in the mean volume measurement can be up to 0.2%. The standard deviation of the readings indicated by the flowmeter is 0.1% about a mean value. Give in percentage form the uncertainty in the mean flow rate for a tolerance probability of 0.95.

Hint: treat volume and time uncertainties as rectangular.

Answer: 0.30%.

Example 8.8 The light intensity of a tungsten filament lamp when run at a given voltage is to be measured by comparison with a standard lamp, also to be run at a stated voltage. The voltage to the standard and test lamps can each independently fluctuate by up to 0.1%. From twenty measurements made it was found that the standard deviation of the position of the photometer head was 0.3% of the mean distance d_S of the standard lamp from the photometer head, whilst the mean distance d_T of the test lamp from the photometer head was $0.7\,d_S$. Given that a fluctuation of 1% in the voltage of a lamp can cause a 3.5% fluctuation in the light intensity, calculate the percentage tolerance limits for the uncertainty in the measured test lamp intensity relative to the standard lamp for a tolerance probability of 0.95. The test lamp intensity I_T is related to the standard lamp intensity I_S by the relation

$$\frac{I_T}{I_S} = \frac{d_T^2}{d_S^2}$$

Hint: treat uncertainties caused by voltage fluctuations as rectangular.

Answer: 0.69%

Example 8.9 The relative refractive index μ of glass, with respect to air at a given temperature, when measured by the minimum deviation method, is a function of the prism angle A, the minimum deviation D and the temperature T of the sample. Let $\mu = f(A, D, T)$. Therefore

$$\delta\mu = \frac{\partial\mu}{\partial A} \cdot \delta A + \frac{\partial\mu}{\partial D} \cdot \delta D + \frac{\partial\mu}{\partial T} \cdot \delta T$$

and so by 8.04(4)

$$\sigma_\mu^2 = \sigma_A^2 \left(\frac{\partial\mu}{\partial A}\right)^2 + \sigma_D^2 \left(\frac{\partial\mu}{\partial D}\right)^2 + \sigma_T \left(\frac{\partial\mu}{\partial T}\right)^2$$

If

$$\mu = \frac{\sin(D + A)/2}{\sin A/2}$$

at temperature T then by Example 8.5 or by differentiating the expression for μ, we have

$$\sigma_\mu^2 = \sigma_A^2 \frac{\sin^2 D/2}{4\sin^4 A/2} + \frac{\sigma_D^2 \mu^2}{4\tan^2(D + A)/2} + \left(\frac{\partial\mu}{\partial T}\right)^2 \sigma_T^2$$

Thus if $\sigma_A = \sigma_D = 2.6$ s, $A = 60°$, $\mu = 1.62$ approximately,

$$\left(\frac{\partial\mu}{\partial T}\right) = 5 \times 10^{-6}\,°C^{-1}$$

and $\delta T_{max} = \pm 1°C$, find the uncertainty in μ for a tolerance probability of 0.95. Treat the temperature variation as rectangular.

Answer: Tolerance limits for $\mu = \pm 2.62 \times 10^{-5}$.

Example 8.10 Nominally $1000\,\Omega$ resistances have been graded into groups, where the groups contain resistances varying between $1000 \pm \frac{1}{2}\Omega$, $1001 \pm \frac{1}{2}\Omega$, $1002 \pm \frac{1}{2}\Omega, \ldots$, and $999 \pm \frac{1}{2}\Omega$, $998 \pm \frac{1}{2}\Omega$, $997 \pm \frac{1}{2}\Omega$, etc. The standard deviation about the mean of $1000\,\Omega$ is found to be $3.5\,\Omega$. If twenty resistances are picked at random and connected in series what is the likely variation of the total resistance from $20\,000\,\Omega$ for a probability of 0.95? If the resistances are connected in parallel, what is the likely variation of the resistance from $50\,\Omega$ to the same probability? Assume first of all a Gaussian distribution for the chosen resistors, then a 't' distribution.

Answer: 30.7 Ω, 1.53 Ω. Gaussian
32.7 Ω, 1.63 Ω. 't' distribution

Note:
$$R_{\text{series}} = \sum_{1}^{n} r_i \qquad \text{and} \qquad \frac{1}{R_{\text{parallel}}} = \sum_{1}^{n} \frac{1}{r_i}$$

where R_{parallel} = total resistance if the resistors are connected in parallel and r_i is the resistance of each component.

9

Consistency and Significance Tests

t Test. Comparison of Two Means when Both Sets of Observations Belong to the Same Population, and so $\sigma_1 = \sigma_2$, that is Both Sets have Equal Variances

9.01 The first test we shall deal with is known as the t test and it is based on the Student t distribution. Suppose we have two sets of observations for a variable x, for each of which an estimate of the standard deviation and of the mean value has been made. If the two means and standard deviations differ it is useful to be able to decide if the differences are significant or whether they are to be expected, and are consistent with statistical fluctuations caused by a relatively small number of readings having been used.

9.02 Let the two sets of observations be distinguished by the suffixes 1 and 2. Thus

$$\bar{x}_1 = \frac{\sum_{r=1}^{r=n_1} x_r}{n_1} \qquad\qquad 9.02(1)$$

$$\bar{x}_2 = \frac{\sum_{r=1}^{r=n_2} x_r}{n_2} \qquad\qquad 9.02(2)$$

where the bar denotes the mean value of the variable x and n_1 and n_2 are the numbers of readings taken for the two sets 1 and 2. Let

$$S_1^2 = \sum_{r=1}^{r=n_1} (x_r - \bar{x}_1)^2 = s_1^2 v_1 \qquad\qquad 9.02(3)$$

and

$$S_2^2 = \sum_{r=1}^{r=n_2} (x_r - \bar{x}_2)^2 = s_2^2 v_2 \qquad\qquad 9.02(4)$$

where s_1 and s_2 are the estimated values of the variances of x using the two sets of observations. $v_1 = n_1 = 1$ and $v_2 = n_2 - 1$ where v_1 and v_2 are the number of degrees of freedom for each set of observations.

9.03 Now both sets of observations have been made on the same variable under similar circumstances, and thus should have a common standard deviation σ. The best estimate of this standard deviation using the combined data of both sets of observations is given by

$$s_v = \sqrt{\left(\frac{S_1^2 + S_2^2}{n_1 + n_2 - 2}\right)}^{\dagger} \qquad 9.03(1)$$

$$= \sqrt{\left(\frac{s_1^2 v_1 + s_2^2 v_2}{v_1 + v_2}\right)} \qquad 9.03(1a)$$

that is s_v is the weighted root mean square value of s_1 and s_2. The estimate of the standard deviation of each of these two means is

$$\frac{s_v}{\sqrt{n_1}} \qquad \text{and} \qquad \frac{s_v}{\sqrt{n_2}}$$

Thus the estimate of the standard deviation of the difference of these two means is

$$s_{v_d} = s_v \sqrt{\left(\frac{1}{n_1} + \frac{1}{n_2}\right)} \qquad 9.03(2)$$

This is thus the standard deviation of $\bar{x}_1 - \bar{x}_2$.

9.04 If we normalize $\bar{x}_1 - \bar{x}_2$ by dividing by s_{v_d} we have

$$t_{v_d} = \frac{\bar{x}_1 - \bar{x}_2}{s_{v_d}} = \frac{(\bar{x}_1 - \bar{x}_2)}{s_v} \sqrt{\left(\frac{n_1 n_2}{n_1 + n_2}\right)} \qquad 9.04(1)$$

Note that $v_d = n_1 + n_2 - 2 = v_1 + v_2$. The expression given by the right-hand side of 9.04(1) is now found from the given data, whilst s_v is given by 9.03(1a) and $\bar{x}_1$ and $\bar{x}_2$ by 9.02(1) and 9.02(2) respectively. If $\bar{x}_1$ is assumed to be the correct value of the mean, that is μ, then it is seen that

$$\frac{\bar{x}_1 - \bar{x}_2}{s_{v_d}} \equiv \frac{\mu - \bar{x}_2}{s_{v_d}}$$

is equal to the variable t_{v_d}, that is the Student t variable. Note that the value v_d to be used in Table IV of Appendix I is in this case equal to

$$v_1 + v_2 = n_1 + n_2 - 2$$

$\dagger$ The denominator is $n_1 + n_2 - 2$ because two means are involved and the number of degrees of freedom is thus two less than the total number of readings.

9.05   As a test of the value of t_{v_d} obtained from the data, two courses of action are open: (1) the t distribution can be used to find the probability P_t of the value of t_{v_d}, found by calculation from the data, being exceeded, or (2) a probability P_t can be selected, and the corresponding t_{v_d} found from Table IV of the Student distribution, and its value compared with the t_{v_d} calculated from the data. If the value of t_{v_d} found from the data exceeds that for the selected probability, then the difference between the two means is considered to be too large and the two means are said to be inconsistent; one of the means should be considered to be suspect. Method (2) is usually adopted for the simple reason that tables of t_{v_d} usually give P_t for a limited number of values only. The probability P_t is equal to $(1 - \beta_{tp})/2$ where β_{tp} is the probability obtained from Table IV and is equal to $\int_{-t_v}^{t_v} f(t_v)\, dt_v$. Table IV gives six values of β_{tp}, that is 0.90, 0.95, 0.98, 0.99, 0.995 and 0.999. The corresponding values of P_t are thus 0.05, 0.025, 0.01, 0.005, 0.0025 and 0.0005.

9.06   Having found the value of t_{v_d} from the two sets of data it is suggested that this value of t_{v_d} is compared successively with the values of t_{v_d} from Table IV, for which the values of P_t are successively 0.05, 0.025, 0.01, 0.005, 0.0025 and 0.0005. If the value of t_{v_d} from the data is smaller than that for the t_{v_d} from Table IV corresponding to $P_t = 0.05$, then the two means are consistent. But if the value of t_{v_d} from the data lies between the values of t_{v_d} from Table IV, having probabilities of P_t equal to 0.05 and 0.01, then the two means should be treated with some caution. If the data for t_{v_d} is larger than the t_{v_d} from Table IV corresponding to $P_t = 0.01$, 0.005, 0.0025 or even 0.0005, then it must be suspected that one of the means is incorrect; but it is not possible to deduce which one is at fault.

Combined Mean and Standard Deviation of Two Consistent Means

9.07   If the two means are considered to be consistent then $\bar{x}$ and s are given by

$$\bar{x} = \frac{n_1 \bar{x}_1 + n_2 \bar{x}_2}{n_1 + n_2} \qquad 9.07(1)$$

and

$$s^2 = \left\{ v_1 s_1^2 + v_2 s_2^2 + \frac{n_1 n_2 (\bar{x}_1 - \bar{x}_2)^2}{n_1 + n_2} \right\} / (n_1 + n_2 - 1) \qquad 9.07(2)$$

These two expressions are obtained from the expressions for $\bar{x}_1$, $\bar{x}_2$, s_1^2 and s_2^2 as follows:

$$\bar{x} = \sum_{r=1}^{r = n_1 + n_2} \frac{x_r}{n_1 + n_2} \qquad 9.07(3)$$

Now

$$\sum_{r=1}^{r=n} x_r = n_1 \bar{x}_1$$

from 9.02(1) and

$$\sum_{r=1}^{r=n_2} x_r = n_2 \bar{x}_2$$

from 9.02(2). Therefore, adding the two above expressions, we have

$$\sum_{r=1}^{r=n_1} x_r + \sum_{r=1}^{r=n_2} x_r = \sum_{r=1}^{r=n_1+n_2} x_r = n_1 \bar{x}_1 + n_2 \bar{x}_2$$

therefore

$$\bar{x} = \frac{\sum_{r=1}^{r=n_1+n_2} x_r}{n_1 + n_2} = \frac{n_1 \bar{x}_1 + n_2 \bar{x}_2}{n_1 + n_2} \qquad 9.07(1)$$

Now

$$s^2 = \sum_{1}^{n_1+n_2} \frac{(x_r - \bar{x})^2}{n_1 + n_2 - 1}$$

$$= \left(\sum_{r=1}^{r=n_1+n_2} x_r^2 + \sum_{r=1}^{r=n_1+n_2} \bar{x}^2 - 2\bar{x} \sum_{r=1}^{r=n_1+n_2} x_r \right) \Big/ (n_1 + n_2 - 1)$$

$$= \left(\sum_{r=1}^{r=n_1+n_2} x_r^2 + (n_1 + n_2)\bar{x}^2 - 2\bar{x}(n_1 + n_2)\bar{x} \right) \Big/ (n_1 + n_2 - 1) \qquad 9.07(4)$$

(from 9.07(3))

$$= \sum_{r=1}^{r=n_1+n_2} \frac{x_r^2 - (n_1 + n_2)\bar{x}^2}{n_1 + n_2 - 1} \qquad 9.07(5)$$

Now

$$s_1^2 v = \sum_{r=1}^{r=n_1} (x_r - \bar{x}_1)^2 = \sum_{r=1}^{r=n_1} x_r^2 - n_1 \bar{x}_1^2$$

and

$$s_2^2 v_2 = \sum_{r=1}^{r=n_2} (x_r - \bar{x}_2)^2 = \sum_{r=1}^{r=n_2} x_r^2 - n_2 \bar{x}_2^2$$

Adding we have

$$s_1^2 v_1 + s_2^2 v_2 = \sum_{r=1}^{r=n_1} x_r^2 + \sum_{r=1}^{r=n_2} x_r^2 - n_1 \bar{x}_1^2 - n_2 \bar{x}_2^2$$

Therefore

$$\sum_{r=1}^{r=n_1+n_2} x_r^2 = s_1^2 v_1 + s_2^2 v_2 + n_1 \bar{x}_1^2 + n_2 \bar{x}_2^2 \qquad 9.07(6)$$

Thus, substituting in 9.07(5) for $\sum_{r=1}^{r=n_1+n_2} x_r^2$ and using 9.07(1) to express $\bar{x}$ in terms of $\bar{x}_1$ and $\bar{x}_2$ we have after a little reduction

$$s^2 = \left\{ s_1^2 v_1 + s_2^2 v_2 + \frac{n_1 n_2 (\bar{x}_1 - \bar{x}_2)^2}{n_1 + n_2} \right\} \Big/ (n_1 + n_2 - 1) \qquad 9.07(2)$$

It should be noted that the s of 9.07(2) is not identical with the weighted mean s_v obtained by using s_1^2 and s_2^2, and given by 9.03(1a). If n_1 and n_2 are large, however, and $\bar{x}_1 \rightarrow \bar{x}_2$, as would be the case if the two means were consistent, then the two variances will tend to equality.

t Test. Comparison of Two Means which do not come from the Same Population, and so $\sigma_1 \neq \sigma_2$, that is their Variances are not Equal (i.e. inconsistent means)

9.08   If it is suspected that for various reasons $\sigma_1 \neq \sigma_2$, then the estimate of the standard deviation of the difference of the two means should be taken as

$$s_{v_d} = \left(\frac{s_1^2}{n_1} + \frac{s_2^2}{n_2} \right)^{1/2} \qquad 9.08(1)$$

As before

$$t_{v_d} = \frac{\bar{x}_1 - \bar{x}_2}{s_{v_d}} = (\bar{x}_1 - \bar{x}_2) \Big/ \left(\frac{s_1^2}{n_1} + \frac{s_2^2}{n_2} \right)^{1/2} \qquad 9.08(2)$$

but v_d is not now given by $v_1 + v_2 = n_1 + n_2 - 2$, but by

$$v_d = (n_2 s_1^2 + n_1 s_2^2)^2 \Big/ \left(\frac{n_2^2 s_1^4}{v_1} + \frac{n_1^2 s_2^4}{v_2} \right)^{\dagger} \qquad 9.08(3)$$

† 7.15(2) due to Welch B L gives

$$\frac{1}{v_{\text{eff}}} = \sum_{i=1}^{i=n} \frac{s_i^4}{v_i} \Big/ \left(\sum_{i=1}^{i=n} s_i^2 \right)^2$$

for the effective number of degrees of freedom for the convolution of n Student t distributions. In 9.08(1) we are dealing with the standard deviation of the difference of two means. Thus we replace s_i by s_i/n_i in 7.15(2) and putting $n = 2$ we have

$$\frac{1}{v_{\text{eff}}} = \left(\frac{s_1^4}{n_1^2 v_1} + \frac{s_2^2}{n_2 v_2} \right) \Big/ \left(\frac{s_1^2}{n_1} + \frac{s_2^2}{n_2} \right)^2$$

Inverting and multiplying through top and bottom by $n_1^2 n_2^2$ we get $v_{\text{eff}} \equiv v_d = 9.08(3)$.

If the two means are found to be consistent, then the best value of the mean is given by

$$\bar{x} = \left(\frac{\bar{x}_1}{s_1^2/n_1} + \frac{\bar{x}_2}{s_2^2/n_2}\right) \Big/ \left(\frac{1}{s_1^2/n_1} + \frac{1}{s_2^2/n_2}\right)$$

$$= \frac{n_1 s_2^2 \bar{x}_1 + n_2 s_1^2 \bar{x}_2}{n_1^2 s_2^2 + n_2^2 s_1^2} \qquad 9.08(4)$$

The estimate of the weighted mean standard deviation is given by

$$\frac{1}{s^2} = \frac{n_1}{s_1^2} + \frac{n_2}{s_2^2} \qquad 9.08(5)$$

F Test. Comparison of Two Standard Deviations when Both Samples Come from the Same Population

9.09   If s_1 and s_2 are the estimated standard deviations of two samples, where

$$s_1 = \sqrt{\left\{\frac{\sum (_r x_1 - \bar{x}_1)^2}{n_1 - 1}\right\}} \qquad 9.09(1)$$

and

$$s_2 = \sqrt{\left\{\frac{\sum (_r x_2 - \bar{x}_2)^2}{n_2 - 1}\right\}} \qquad 9.09(2)$$

then

$$F = s_1^2/s_2^2 \qquad 9.09(3)$$

In tables giving values of F, n is usually replaced by v, that is $n_1 - 1 = v_1$ and $n_2 - 1 = v_2$ where v_1 and v_2 are the number of degrees of freedom of the variable x_r in each case. For the use of Tables VIII or IX, Appendix I, F should always be larger than unity, and so it may be necessary to interchange the values of s_1 and s_2. Note that the value of v_1 in the tables always applies to the numerator s_1 of the expression for F. Having calculated F, the values of F are found from Tables VIII and IX corresponding to the number of degrees of freedom of s_1 and s_2. If the calculated F is smaller than that found from Table VIII then it means that the probability of occurrence of the square of the ratio of the two estimated standard deviations having the calculated value is greater than 0.05. If the calculated F is greater than the value found from Table VIII but smaller than the value found from Table IX, then the probability of occurrence of this value lies between 0.05 and 0.01. If F calculated is greater than the value found from Table IX then the probability of occurrence of a ratio of this magnitude for F is less than 0.01.

In this case one would strongly suspect one of the standard deviations of being in error.

Relationship of F Distribution to Other Distributions

Relationship to Student t distribution

9.10 The probability $F > F_0$, where $v_1 = 1$ and v_2 is $\geqslant 1$ is equal to the probability of $(t^2 > t_0^2)$ for v_2 which in turn is equal to twice the probability of $(t > t_0)$ for v_2. In shortened form this is written

$$\text{Prob.}(F > F_0 | 1 : v_2) = \text{Prob.}(t^2 > t_0^2 | v_2)$$

$$= 2 \,\text{Prob.}(t_2 > t_0 | v_2) \qquad 9.10(1)$$

Table XV of Appendix I gives some tabulated values of t^2 for $v = 1$ to 200 and for four probabilities of t^2 being exceeded. The proof of 9.10(1) is as follows. If we take the expression for the density function for F, that is

$$f(F) = \frac{\Gamma((v_1 + v_2)/2)v_1^{v_1/2} \cdot v_2^{v_2/2} F^{v_1/2 - 1}}{\Gamma(v_1/2)\Gamma(v_2/2)(v_2 + v_1 F)^{(v_1 + v_2)/2}}$$

where $0 \leqslant F \leqslant \infty$ and put $v_1 = 1$ and $v_2 = v$, we obtain

$$f(F | 1 : v) = \frac{\Gamma((1 + v)/2)v^{v/2}F^{-1/2}}{\Gamma(v/2)\Gamma(1/2)(v + F)^{(1 + v)/2}} \qquad 9.10(2)$$

Now

$$\Gamma(\tfrac{1}{2}) = \sqrt{\pi}$$

Thus with a little rearrangement and manipulation

$$f(F | 1 : v) = \frac{\Gamma((1 + v)/2)(1 + F/v)^{-(v + 1)/2}F^{-1/2}}{\Gamma(v/2)\sqrt{(v\pi)}} \qquad 9.10(3)$$

and

$$\int_{F_0}^{\infty} f(F | 1 : v) = \text{Prob.}\ F > F_0 \qquad \text{for} \qquad v_1 = 1 \qquad \text{and} \qquad v_2 = v$$

If we write

$$F = t^2 \qquad 9.10(4)$$

in 9.10(3) we obtain

$$f(t^2 | v) = \frac{\Gamma((1 + v)/2)(1 + t^2/v)^{-(v + 1)/2}t^{-1}}{\Gamma(v/2)} \qquad 9.10(5)$$

and differentiating 9.10(4) we get $dF = 2t\, dt$. Thus

$$\int_{F_0}^{\infty} f(F|1:v)\, dF = \int_{t_0}^{\infty} f(t^2|v) 2t\, dt$$

$$= 2 \int_{t_0}^{\infty} \frac{\Gamma((1+v)/2)(1+t^2/v)^{-(v+1)/2}\, dt}{\Gamma(v/2)} \qquad 9.10(6)$$

which is twice the integral of the Student t distribution between t_0 and ∞, where $t_0 = \sqrt{F_0}$ (from 9.10(4)). Thus

$$\int_{F_0}^{\infty} f(F|1:v)\, dF = \text{Prob. } F > F_0$$

$$= 2 \text{ Prob. } t > t_0$$

$$= 2 \int_{t_0}^{\infty} f(t|v)\, dt$$

Finally, if we have a symmetrical density distribution for the variable x as in Figure 9.10, then let the probability of

$$x > x_0 \text{ be given by } P_0 \qquad 9.10(7)$$

Since the function is symmetrical the probability of

$$x < -x_0 \text{ is also given by } P_0 \qquad 9.10(8)$$

If 9.10(7) is squared, then it is clear that the probability of $x^2 > x_0^2$ is also given by P_0. But if 9.10(8) is squared, then the equality is reversed, and we have $x^2 > x_0^2$, but the probability is the same as that for $x < -x_0$, that is P_0. Thus the probability of $x^2 > x_0^2$ is the sum of 9.10(7) and 9.10(8) and we have for a symmetrical distribution that

$$\text{Prob. } x^2 > x_0^2 = 2 \text{ Prob. } x > x_0$$

Since the Student distribution is symmetrical it follows that

$$\text{Prob. } t^2 > t_0^2 = 2 \text{ Prob. } t > t_0$$

thus completing the proof of the proposition 9.10(1).

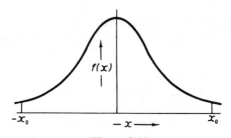

Figure 9.10

Relationship to χ^2 distribution

9.11   Let $F_p(v_1:v_2)$ indicate a value of F for v_1 and v_2, where p is the probability of this value being exceeded. Now

$$F_p(v_1:v_2) = \frac{s_{v_1}^2}{s_{v_2}^2} = \frac{\chi_{v_1}^2}{v_1} \bigg/ \frac{\chi_{v_2}^2}{v_2} \qquad\qquad 6.17(3)$$

Now $\chi^2/v = s^2/\sigma^2$ (see 6.16(4)) for a Gaussian distribution, and so as $v \to \infty$ and $s^2 \to \sigma^2$, $\chi^2/v \to 1$. Thus if $v_2 \to \infty$ in 6.17(3) we have

$$F_p(v_1:\infty) = \chi_{pv_1}^2/v_1 \qquad\qquad 9.11(1)$$

where $\chi_{pv_1}^2$ is the value of χ^2 for v_1 degrees of freedom, whose probability of being exceeded is p. Similarly

$$F_p(\infty:v_2) = \frac{v_2}{\chi_{1-p,\,v_2}^2} = \frac{1}{F_{1-p}(v_2:\infty)} \qquad\qquad 9.11(2)$$

from 6.20(1). It follows from 9.11(1) and 9.11(2) that $F(v_1:\infty) \to 1$ when $v_1 \to \infty$, and $F(\infty:v_2) \to 1$ when $v_2 \to \infty$.

F Test. Internal and External Consistency Tests

9.12   Besides being used to decide on the consistency of two sets of data as in paragraph 9.09, the F test can be used to decide whether a number of sets of observations are part of a larger normal population. If individual sets are affected by systematic errors then the complete set of observations will not be homogeneous, and the F test may be used to reveal this inhomogeneity.

9.13   Suppose that there are a total of N observations and that there are m separate sets, each containing n_q observations, where $\sum_{q=1}^{q=m} n_q = N$, and where the n_q are generally unequal, that is not equal to N/m. Let the mean of each set be $\bar{x}_q$ where

$$\bar{x}_q = \sum_{r=1}^{r=n_q} w_{rq} x_{rq} \bigg/ \sum_{r=1}^{r=n_q} w_{rq} \qquad\qquad 9.13(1)$$

where w_{rq} is the weight of each observation x_{rq}.

9.14   Let μ denote the mean value of the population and σ the standard deviation. Then

$$\sum_{q=1}^{q=m} \sum_{r=1}^{r=n_q} \frac{w_{rq}(x_{rq} - \mu)^2}{\sigma^2} = \sum_{q=1}^{q=m} \sum_{r=1}^{r=n_q} w_{rq}\{(x_{rq} - \bar{x}_q) + (\bar{x}_q - \bar{x}) + (\bar{x} - \mu)\}^2/\sigma^2$$

where $\bar{x}$ is the mean of all the observations and is given by 9.17(2) (see also paragraphs 2.62 and 2.66 for the derivation). Using the properties of the means it can be shown that

$$\underbrace{\sum_{q=1}^{q=m}\sum_{r=1}^{r=n_q}\frac{w_{rq}(x_{rq}-\mu)^2}{\sigma^2}}_{\chi_0^2} = \underbrace{\sum_{q=1}^{q=m}\sum_{r=1}^{r=n_q}\frac{w_{rq}(x_{rq}-\bar{x}_q)^2}{\sigma^2}}_{\chi_1^2}$$

$$+ \underbrace{\sum_{q=1}^{q=m}\sum_{r=1}^{r=n_q}\frac{w_{rq}(\bar{x}_q-\bar{x})^2}{\sigma^2}}_{\chi_2^2}$$

$$+ \underbrace{\sum_{q=1}^{q=m}\sum_{r=1}^{r=n_q}\frac{w_{rq}(\bar{x}-\mu)^{2}}{\sigma^2}}_{\chi_3^2}^{\dagger} \qquad 9.14(1)$$

Now each term of 9.14(1) is distributed as χ^2 independently of the others. The degrees of freedom of the four terms of this expression are respectively N, $\sum_{q=1}^{q=m}(n_q-1) = N - m \ (\equiv v_1)$, $m - 1 \ (\equiv v_2)$, and 1. If the whole set is homogeneous, then

$$F = \frac{\chi_1^2/v_1}{\chi_2^2/v_2} = \frac{\sum_{q=1}^{q=m}\sum_{r=1}^{r=n_q}w_{rq}(x_{rq}-\bar{x}_q)^2(m-1)}{\sum_{q=1}^{q=m}\sum_{r=1}^{r=n_q}w_{rq}(\bar{x}_q-\bar{x})^2(N-m)} = \frac{s_1^2}{s_2^2} \qquad 9.14(2)$$

will be distributed as F with $(N - m, m - 1)$ degrees of freedom.

9.15 The F test is used in this instance to compare two values of the standard deviation of the mean of the observations, one obtained by using the means of the m sets, and the other by using the mean of all the readings. The former is known as the standard error by internal consistency, whilst the latter is known as the standard error by external consistency.

9.16 The square of the standard error in the mean, by internal consistency, is given by

$$s_1^2 = \frac{\sum_{q=1}^{q=m}\sum_{r=1}^{r=n_q}w_{rq}(x_{rq}-\bar{x}_q)^2}{(N-m)\sum_{q=1}^{q=m}\sum_{r=1}^{r=n_q}w_{rq}} \qquad 9.16(1)$$

since there are $N - m$ degrees of freedom, that is there are a total of N observations and m values of the $\bar{x}_q$, giving $N - m$ independent variables. (See paragraphs 2.69 to 2.71 for the derivation of s_1^2.)

9.17 The square of the standard error in the mean by external consistency is given by

$$s_2^2 = \frac{\sum_{q=1}^{q=m}\sum_{r=1}^{r=n_q}w_{rq}(\bar{x}_q-\bar{x})^2}{(m-1)\sum_{q=1}^{q=m}\sum_{r=1}^{r=n_q}w_{rq}} \qquad 9.17(1)$$

† See Appendix III for proof.

(See paragraphs 2.67 to 2.68 for the derivation of s_2^2.) The weighted mean is given by

$$\bar{x} = \frac{\sum_{q=1}^{q=m} \sum_{r=1}^{r=n_q} w_{rq} x_{rq}}{\sum_{q=1}^{q=m} \sum_{r=1}^{r=n_q} w_{rq}} \qquad 9.17(2)$$

(See paragraphs 2.62 and 2.66 for the derivation of $\bar{x}$.)

Weights

9.18 Weights are usually chosen so that the largest weight is given to readings which have the smallest standard deviation. It is usually not possible to give an independent weight to each reading, but it is possible to give a weight to each set of n_q readings. We have seen from paragraph 2.58 that the appropriate weight for a set of readings is proportional to the reciprocal of their variance, and so we may put

$$w_{rq} = 1/s^2(x_q) \qquad 9.18(1)$$

where $s^2(x_q)$ is independent of r for given q, and is given by

$$s^2(x_q) = \sum_{r=1}^{r=n_q} (x_{rq} - \bar{x}_q)^2/(n_q - 1) \qquad 9.18(2)$$

(See paragraph 2.65 for the derivation of $s^2(x_q)$.) In general there will thus be m different weights, each corresponding to a set of readings n_q.

9.19 Alternatively, all the readings can be given equal weights, in which case w_{rq} can be put equal to unity. This case may be assumed if the separate $s^2(x_q)$ given by 9.18(2) do not differ significantly from one another. Consideration will now be given to the deriving of the sets of equations relevant to the two cases mentioned above.

Case 1. Readings given equal weights, that is $w_{rq} = 1$

9.20 The variance of each set of readings is given by 9.18(2). The mean of each set is given by

$$\bar{x}_q = \sum_{r=1}^{r=n_q} x_{rq}/n_q \qquad 9.20(1)$$

from 9.12(1). The variance of the mean by internal consistency s_1^2 is given by putting $w_{rq} = 1$ in 9.16(1) and leads to

$$s_1^2 = \sum_{q=1}^{q=m} \sum_{r=1}^{r=n_q} (x_{rq} - \bar{x}_q)^2/N(N - m) \qquad 9.20(2)$$

where s_1^2 has $N - m$ degrees of freedom. The variance of the mean by external consistency s_2^2 is given by substituting $w_{rq} = 1$ in 9.17(1). Thus

$$s_2^2 = \sum_{q=1}^{q=m} \frac{n_q(\bar{x}_q - \bar{x})^2}{N(m-1)} \qquad 9.20(3)$$

where s_2^2 has $m - 1$ degrees of freedom. The weighted mean is given by

$$\bar{x} = \sum_{q=1}^{q=m} \sum_{r=1}^{r=n_q} x_{rq}/N$$

$$= \sum_{q=1}^{q=m} n_q \bar{x}_q/N \qquad 9.20(4)$$

where in 9.17(2) w_{rq} is put equal to unity and 9.20(1) is used to replace $\sum_{r=1}^{r=n_q} x_{rq}$ by $n_q \bar{x}_q$. Thus

$$F = \frac{s_1^2}{s_2^2} = \frac{(m-1)\sum_{q=1}^{q=m}\sum_{r=1}^{r=n_q}(x_{rq} - \bar{x}_q)^2}{(N-m)\sum_{q=1}^{q=m} n_q(\bar{x}_q - \bar{x})^2} \qquad 9.20(5)$$

Case 2. *Each set of observations given a weight inversely proportional to its variance, that is $w_{rq} = 1/s^2(x_q)$*

9.21 The variance of each set of readings is again given by 9.18(2). The mean of each set is given by using 9.13(1) and putting

$$w_{rq} = 1/s^2(x_q) \qquad 9.21(1)$$

giving

$$\bar{x}_q = \sum_{r=1}^{r=n_q} \frac{x_{rq}/s^2(x_q)}{n_q/s^2(x_q)} = \sum_{r=1}^{r=n_q} \frac{x_{rq}}{n_q} \qquad 9.21(2)$$

9.22 The variance of the mean by internal consistency is given by substituting the value of w_{rq} given by 9.21(1) in 9.16(1). This gives

$$s_1^2 = \frac{\sum_{q=1}^{q=m}\sum_{r=1}^{r=n_q}(x_{rq} - \bar{x}_q)^2/s^2(x_q)}{(N-m)\sum_{q=1}^{q=m} n_q/s^2(x_q)}$$

Now by 9.18(2)

$$\sum_{r=1}^{r=n_q}(x_{rq} - \bar{x}_q)^2/s^2(x_q) = n_q - 1$$

Thus, substituting in s_1^2 for this expression we have

$$
s_1^2 = \frac{\sum_{q=1}^{q=m}(n_q - 1)}{(N - m)\sum_{q=1}^{q=m} n_q/s^2(x_q)}
$$

$$
= \frac{N - m}{(N - m)\sum_{q=1}^{q=m} n_q/s^2(x_q)}
$$

$$
= \frac{1}{\sum_{q=1}^{q=m} 1/s^2(\bar{x}_q)} \qquad 9.22(1)
$$

where

$$
s^2(\bar{x}_q) = s^2(x_q)/n_q \qquad 9.22(2)
$$

9.23   The variance of the mean by external consistency is given by substituting the value of w_{rq} given by 9.21(1) in 9.17(1). Thus

$$
s_2^2 = \frac{\sum_{q=1}^{q=m}\sum_{r=1}^{r=n_q}(\bar{x}_q - \bar{x})^2/s^2(x_q)}{(m - 1)\sum_{q=1}^{q=m}\sum_{r=1}^{r=n_q} 1/s^2(x_q)}
$$

$$
= \frac{\sum_{q=1}^{q=m} n_q(\bar{x}_q - \bar{x})^2/s^2(x_q)}{(m - 1)\sum_{q=1}^{q=m} n_q/s^2(x_q)}
$$

$$
= \frac{\sum_{q=1}^{q=m}(\bar{x}_q - \bar{x})^2/s^2(\bar{x}_q)}{(m - 1)\sum_{q=1}^{q=m} 1/s^2(\bar{x}_q)} \qquad 9.23(1)
$$

using 9.22(2).

9.24   The weighted mean is given by substituting for w_{rq} in 9.17(2). Thus

$$
\bar{x} = \frac{\sum_{q=1}^{q=m}\sum_{r=1}^{r=n_q} x_{rq}/s^2(x_q)}{\sum_{q=1}^{q=m} n_q/s^2(x_q)}
$$

$$
= \frac{\sum_{q=1}^{q=m} n_q\bar{x}_q/s^2(x_q)}{\sum_{q=1}^{q=m} n_q/s^2(x_q)}
$$

since

$$
\bar{x}_q = \sum_{r=1}^{r=n_q} x_{rq}/n_q \qquad 9.20(1)
$$

Thus finally

$$
\bar{x} = \frac{\sum_{q=1}^{q=m} \bar{x}_q/s^2(\bar{x}_q)}{\sum_{q=1}^{q=m} 1/s^2(\bar{x}_q)} \qquad 9.24(1)
$$

using 9.22(2). Hence

$$
F = \frac{s_1^2}{s_2^2} = \frac{(m - 1)}{\sum_{q=1}^{q=m}(\bar{x}_q - \bar{x})^2/s^2(\bar{x}_q)} \qquad 9.24(2)
$$

where s_1^2 has $N - m$ degrees of freedom and s_2^2 has $m - 1$.

9.25 If the results being compared cover a number of workers, the number of degrees of freedom of s_1 may not be known. If the number is thought to be large it is sometimes assumed to be infinite. Thus we have the case of

$$s_2^2/s_1^2 \leqslant 1$$

Now

$$F_p(\infty:v_2) = v_2/\chi_{(1-p), v_2}^2$$

$$= \frac{1}{F_{(1-p),(v_2; \, \infty)}} \qquad 9.11(2)$$

Thus

$$\frac{\chi_{(1-p)v_2}^2}{v_2} = F_{(1-p),(v_2; \, \infty)} = \frac{s_2^2}{s_1^2} \qquad 9.25(1)$$

where $s_2^2/s_1^2 \leqslant 1$, and so

$$\chi_{(1-p), v_2}^2 = v_2 \frac{s_2^2}{s_1^2} \qquad 9.25(2)$$

where $v_2 = m - 1$ is the number of degrees of freedom for s_2, p is the probability of the calculated value of χ^2 being exceeded, and so $1 - p$ is the probability of χ^2 being less than the calculated value. (See Table VI, Appendix I. The last four columns of this table should be consulted in order to fix the probability for the value χ^2 given by 9.25(2) being equal to or less than a tabulated value.)

$$s_2^2/s_1^2 \geqslant 1$$

This is the case when the value of F found from s_1^2/s_2^2 is less than unity, and has to be inverted, to provide a value of F larger than unity in order that Tables VIII and IX, Appendix I, may be used. Now

$$F_p(v_1:v_2) = \frac{v_2\chi_{v_1}^2}{v_1\chi_{v_2}^2} \qquad 6.17(3)$$

If F is inverted, then

$$F_p(v_2:v_1) = \frac{v_1\chi_{v_2}^2}{v_2\chi_{v_1}^2}$$

and so if $v_1 \to \infty$, we have

$$F_p(v_2:\infty) = \frac{\chi_{p, v_2}^2}{v_2} = \frac{s_2^2}{s_1^2} \qquad 9.25(3)$$

and thus

$$\chi_{p, v_2}^2 = v_2 s_2^2/s_1^2 \qquad 9.25(4)$$

where p is the probability of the calculated value of χ^2 being exceeded. In this case the first four columns of Table VI, Appendix I, should be consulted in order to fix the probability for the value of χ^2 given by 9.25(4) being equal to or greater than a tabulated value. If χ^2 is less than a tabulated value, then the readings are homogeneous.

9.26   When the number of degrees of freedom for both s_1 and s_2 are known, and F has been found either by 9.20(5) Case 1 and/or 9.24(2) Case 2 then Tables VIII and IX should be consulted to see if the values of F given for either case is reasonable. If the value of F obtained from the readings is greater than that given by Table VIII, then the probability of occurrence of the value of F obtained from the observations is less than 0.05. If the value from the observations is also greater than that obtained from Table IX, then the probability of occurrence of the value of F obtained is less than 0.01. If the probability of occurrence of F is found to be greater than 0.05 the readings can be assumed to be consistent, if the probability of F lies between 0.05 and 0.01 the consistency of the readings should be treated as suspect, whilst if the probability of F is less than 0.01 the observations should be regarded as inconsistent.

9.27   It should be noted that when Tables VIII and IX are used, the value of F should always be greater than unity, and in order to achieve this condition $s_1^2/s_2^2 = F$ may have to be inverted. A further point to note is that the v_1 of the tables always refers to the numerator of the fraction for F, whether F has been *inverted* or *not*.

Results Consistent

Weighted mean

9.28   If the results are found to be consistent, then the weight of each mean $\bar{x}_q$ is given by equation 9.18(1). If Case 1 is used the mean is given by

$$\bar{x} = \sum_{q=1}^{q=m} n_q \bar{x}_q / N \qquad\qquad 9.20(4)$$

whilst if Case 2 is used the weighted mean is given by

$$\bar{x} = \frac{\sum_{q=1}^{q=m} \bar{x}_q / s^2(\bar{x}_q)}{\sum_{q=1}^{q=m} 1/s^2(\bar{x}_q)} \qquad\qquad 9.24(4)$$

Case 1 of course gives equal weight to each reading, whilst Case 2 gives each reading of group q a weight of $1/s^2(\bar{x}_q)$, that is inversely proportional to the variance of each group.

Standard error of the mean

9.29 This is given generally by

$$s^2(\bar{x}) = \frac{\sum_{q=1}^{q=m}\sum_{r=1}^{r=n_q} w_{rq}(x_{rq} - \bar{x})^2}{(N-1)\sum_{q=1}^{q=m}\sum_{r=1}^{r=n_q} w_{rq}} \qquad 9.29(1)$$

(See paragraphs 2.61 to 2.64 for the derivation of $s^2(\bar{x})$ and $\bar{x}$.) Now

$$
\begin{aligned}
(x_{rq} - \bar{x})^2 &= x_{rq}^2 + \bar{x}^2 - 2x_{rq}\bar{x} \\
&= (x_{rq}^2 - 2x_{rq}\bar{x}_q + \bar{x}_q^2) + 2x_{rq}\bar{x}_q - \bar{x}_q^2 \\
&\quad + (\bar{x}_q^2 - 2\bar{x}_q\bar{x} + \bar{x}^2) - \bar{x}_q^2 + 2\bar{x}_q\bar{x} - 2x_{rq}\bar{x} \\
&= (x_{rq} - \bar{x}_q)^2 + (\bar{x}_q - \bar{x})^2 + 2\bar{x}_q(x_{rq} - \bar{x}_q) - 2\bar{x}(x_{rq} - \bar{x}_q) \ 9.29(2)
\end{aligned}
$$

Now

$$
\begin{aligned}
\sum_{r=1}^{r=n_q} w_{rq}\bar{x}_q(x_{rq} - \bar{x}_q) &= \bar{x}_q \sum_{r=1}^{r=n_q} w_{rq}x_{rq} - \bar{x}_q^2 \sum_{r=1}^{r=n_q} w_{rq} \\
&= \bar{x}_q\left(\sum_{r=1}^{r=n_q} w_{rq}x_{rq} - \bar{x}\sum_{r=1}^{r=n_q} w_{rq}\right) \\
&= 0 \qquad\qquad\qquad 9.29(3)
\end{aligned}
$$

since

$$\bar{x}_q = \frac{\sum_{r=1}^{r=n_q} w_{rq}x_{rq}}{\sum w_{rq}} \qquad 9.13(1)$$

Similarly

$$
\begin{aligned}
\sum_{r=1}^{r=n_q} w_{rq}\bar{x}(x_{rq} - \bar{x}_q) &= \bar{x}\left\{\sum_{r=1}^{r=n_q}(w_{rq}x_{rq} - w_{rq}\bar{x}_q)\right\} \\
&= \bar{x}\left(\sum_{r=1}^{r=n_q} w_{rq}x_{rq} - \bar{x}_q\sum_{r=1}^{r=n_q} w_{rq}\right) \\
&= 0 \qquad\qquad\qquad 9.29(4)
\end{aligned}
$$

again by 9.13(1). Thus

$$\sum_{q=1}^{q=m}\sum_{r=1}^{r=n_q} w_{rq}(x_{rq} - \bar{x})^2 = \sum_{q=1}^{q=m}\sum_{r=1}^{r=n_q} w_{rq}(x_{rq} - \bar{x}_q)^2 + \sum_{q=1}^{q=m}\sum_{r=1}^{r=n_q} w_{rq}(\bar{x}_q - \bar{x})^2 \ 9.29(5)$$

Thus

$$s^2(\bar{x}) = \frac{\sum_{q=1}^{q=m}\sum_{r=1}^{r=n_q} w_{rq}(x_{rq} - \bar{x})^2}{(N-1)\sum_{q=1}^{q=m}\sum_{r=1}^{r=n_q} w_{rq}} = \frac{\sum_{q=1}^{q=m}\sum_{r=1}^{r=n_q} w_{rq}(x_{rq} - \bar{x}_q)^2}{(N-1)\sum_{q=1}^{q=m}\sum_{r=1}^{r=n_q} w_{rq}}$$

$$+ \frac{\sum_{q=1}^{q=m}\sum_{r=1}^{r=n_q} w_{rq}(\bar{x}_q - \bar{x})^2}{(N-1)\sum_{q=1}^{q=m}\sum_{r=1}^{r=n_q} w_{rq}}$$

$$= \left(\frac{N-m}{N-1}\right)s_1^2 + \left(\frac{m-1}{N-1}\right)s_2^2 \qquad 9.29(6)$$

s_1^2 and s_2^2 are given in the general case by equations 9.16(1) and 9.17(1), respectively, and by equations 9.20(2) and 9.20(3), respectively, for Case 1 and by equations 9.22(1) and 9.23(1), respectively, for Case 2.

Results Inconsistent

9.30 If the results of the analysis of the grouped observations lead to the conclusion that the results are inconsistent, the best mean is given by taking the mean of the sum of the m means, in each case giving these means a weight of unity. Thus

$$\bar{x} = \sum_{q=1}^{q=m} \frac{\bar{x}_q}{m} \qquad 9.30(1)$$

The standard error of this mean is given by

$$s^2(\bar{x}) = \frac{1}{m(m-1)} \sum_{q=1}^{q=m} (\bar{x}_q - \bar{x})^2 \qquad 9.30(2)$$

It is to be noted that Case 1 leads to the same formulae if each set of measurements contain equal numbers of observations. (See equation 9.20(3).)

χ^2 Test to Check if a Hypothetical Distribution $F(x)$ Fits an Empirically Obtained Distribution

9.31 In order to carry out this test the abscissa (x axis) should be divided up into n intervals, each containing at least four values of the empirically observed frequency distribution. If, however, $n - 1$ is greater than about eight and the sample size greater than about forty, then it is permissible for the number of observations in an interval, in isolated cases, to be as low as one. Let each interval q contain f_q observations. If a sample observation should lie at a common boundary point of two intervals the value 0.5 should be added to each of the adjoining f_q. From the hypothetical distribution

$F(x)$ the probability p_q of an observation occurring in each interval should now be found. The total number of observations is $\sum_{q=1}^{q=n} f_q = N$ and the expected number of observations in each interval is Np_q.

9.32 The expression

$$\sum_{q=1}^{q=n} \frac{(f_q - Np_q)^2}{Np_q} \rightarrow \chi_v^2 \qquad 9.32(1)$$

when $v \rightarrow \infty$, and where v is the number of degrees of freedom, equal to $n - 1$ if no parameters have to be calculated to find the p_q of the formula given by 9.32(1). The relationship given by 9.32(1) was discovered by K Pearson.

9.33 The left-hand side of expression 9.32(1) is now evaluated using the empirical data, and the hypothetical distribution $F(x)$. If k parameters have to be estimated to find the fitted frequencies, then $v = n - 1 - k$. If m groups of samples, each containing n classes, are submitted to a χ^2 test based on 9.32(1), then the number of degrees of freedom is $v = (n - 1)(m - 1)$. If each sample requires k parameters to be found in order to calculate the p_q, then $v = (n - 1 - k)(m - 1)$.

Interpretation of test from Table VI, Appendix I

9.34 Having found the value of χ^2 using 9.32(1), Table VI is consulted and the appropriate row for v selected. If the value of χ_v^2 using 9.32(1) is less than the numbers in columns 1 to 4 of the tabulated values of χ^2, and greater than those in columns 5 to 8, then the function f can be assumed to fit the empirical data, since the probability of the value of χ^2 (9.32(1)) being exceeded is greater than 0.05, whilst the probability of a smaller value is also greater than 0.05.

9.35 If, however, χ_v^2 is greater than any or all of the values given in columns 1 to 4, then the probability of its being exceeded is less than the probability corresponding to the column which immediately exceeds it and greater than the probability of the column it immediately precedes. For example, if χ_v^2 lay between the values given by columns 1 and 2, then the probability of its being exceeded is less than 0.05 but greater than 0.025. In this particular case the fit would be regarded with some suspicion and further investigation made. If χ_v^2 lay between columns 2 and 3 or 3 and 4 then the fit given by F should be rejected.

Kolmogorov–Smirnov Statistic

9.36 The Kolmogorov–Smirnov statistic is used as a basis for two tests: (1) for the comparison of a known distribution with one defined by measured data, and (2) for the comparison of two distributions defined by measured data. For details, see Appendix I, Tables XX and XXI.

Examples on Chapter 9

Example 9.1   Two means for the value of an electrical resistance, obtained by using two different examples of the same type of equipment, are $10.103\,\Omega$ and $10.11\,\Omega$. The first value has an estimated standard deviation for the resistance of $0.03\,\Omega$, whilst the second has an estimated standard deviation of $0.06\,\Omega$. The first determination was the result of fifteen readings, whilst the second was the result of ten readings. Find between what probability limits the t test places the t_{v_d} calculated, and express an opinion on the consistency of the *two* means.

Answer: Using 9.03(1a)

$$s_v = \left(\frac{0.03^2 \times 14 + 0.06^2 \times 9}{23} \right)^{1/2} = 0.044\,23$$

Using 9.03(2) to find s_{v_d} we have

$$s_{v_d} = 0.044\,23 \left(\frac{1}{10} + \frac{1}{25} \right)^{1/2} = 0.016\,550$$

Therefore

$$t_{v_d} = \frac{\bar{x}_1 - \bar{x}_2}{s_{v_d}} = \frac{0.007}{0.016\,55} = 0.423$$

$$v_d = 10 + 15 - 2 = 23$$

From Table IV, we find that the t_v, for $v = 20$ or 25, are much greater than the value found, and thus the two means are consistent.

Example 9.2   Two laboratories have determined the mean value of a standard resistance. The first laboratory finds the mean value to be $1.000\,14\,\Omega$, with an estimated standard deviation of $0.000\,32\,\Omega$ from twelve readings; the second laboratory finds the mean value to be $0.999\,28\,\Omega$ with an estimated standard deviation of $0.000\,41\,\Omega$ from ten readings. Use the t test to determine whether the two means are consistent and state the value of t_{v_d}.

Answer: $t_{v_d} = 3.107$ and $v_d = 20$. The two means are inconsistent because the probability of $t_{v_d} \geqslant 3.107$ for $v_d = 20$ lies between 0.005 and 0.01 which is too small a probability for consistency.

Example 9.3 The mean value of the temperature found using an optical pyrometer is 1251.0°C with a standard deviation of 3.1°C from ten readings. A second optical pyrometer makes the mean temperature 1241.1°C with a standard deviation of 2.6°C from seven readings. Determine if the two means are consistent, and state t_{v_d} and the range of probabilities between which the probability of it being exceeded lies.

Answer: $t_{v_d} = 3.4$ and $v_d = 15$. The probability of t_{v_d} being exceeded lies between 0.005 and 0.001 and therefore the two means must be considered inconsistent.

Example 9.4 Two values of a resistance have been found using different types of equipment. If the value found, using the first type of equipment, gave a mean value of 1000.014 Ω with a standard deviation of 0.032 Ω from eleven readings, whilst the second type of equipment gave a mean value of 1000.054 Ω with a standard deviation of 0.074 Ω from sixteen readings, state if the two mean values are consistent. Assume that the two standard deviations are not equal.

Answer: The two means are consistent. $t_{v_d} = 1.917$ and $v_d = 21.85$ (round to 21).

Example 9.5 Two standard deviations have been found, one the result of sixteen readings has a value of 0.0002 V, whilst the other one the result of eleven readings has a standard deviation of 0.0005 V. Check to find if the two standard deviations are consistent.

Answer: The standard deviations are not consistent as they yield a value for F of 6.25, which has a probability of occurrence of less than 0.01.

Example 9.6 Check the two standard deviations given in Example 9.1 to see if they are consistent.

Answer: The probability of F being exceeded is approximately 0.01, and so one of the standard deviations should be regarded with considerable suspicion. As Example 9.1 suggests that the two means are consistent, the reason for the wide difference in the standard deviations could be the result of friction in one of the pieces of equipment.

Example 9.7 A standard resistance has been measured a number of times by different laboratories, and there is some difference in the mean values obtained and also in the standard errors in the means obtained. Use the F test to decide if the results are homogeneous.

Hint: Assume weights proportional to $1/s^2(x_q)$. Case 2 of test.

Mean value of resistance (Ω)	$v_r = n - 1$ (n = number of readings)	s_r Estimated standard deviation of mean
1.000 152	10	0.000 008
1.000 174	10	0.000 015
1.000 182	8	0.000 007
1.000 168	8	0.000 012
1.000 152	8	0.000 009
1.000 194	10	0.000 016
1.000 185	9	0.000 014
1.000 143	8	0.000 010
	$\sum v_r = 71$	

Answer:

$$F = \frac{s_2^2}{s_1^2} = 89.51 \qquad v_1 = 71 \qquad v_2 = 7$$

Note that the value of s_1^2/s_2^2 has had to be inverted to obtain a number larger than 1 for F. The results are not compatible as the value of F obtained has a very small probability of occurrence, that is much less than 0.01. The result implies either that some laboratories have inaccurate measuring equipment or that the resistance measured is unstable.

Example 9.8   The mean values of a number of measurements of the resistance of a particular transfer standard have been made, but the number of individual measurements for each of the means is not known. Use the χ^2 test (see paragraph 9.25) to ascertain if the group of measurements are homogeneous or compatible.

Mean value of resistance (Ω)	$\bar{s}$ Estimated standard deviation of mean
100.1256	0.0005
100.1281	0.0005
100.1272	0.0011
100.1295	0.0009
100.1262	0.0006
100.1274	0.0007
100.1251	0.0009
100.1252	0.0008
100.1262	0.0012
$v_2 = 8$	

Answer:

$$\frac{s_2^2}{s_1^2} = 6.874 \quad \chi^2 = 6.874 \times 8 = 54.992$$

The results are not homogeneous as the value of χ^2 is large and the probability of its occurrence is low, that is less than 0.005. The standard may be unstable or the measurements of some of the laboratories may be suspect, perhaps because of unaccounted-for uncertainties.

10

Method of Least Squares

10.01 In this chapter we shall apply the method of least squares[1] to a number of problems including curve fitting and the determination of a mean plane. The method is best illustrated by application to a particular problem.

Determination of a Mean Plane

10.02 This is a problem which occurs in metrology when a surface plate is calibrated, and it is required to find the plane that will best fit the empirically measured points. When this plane has been determined it is possible to determine the departures from flatness of individual points on the surface plate and to determine the overall error.

10.03 We will first derive the formula for the determination of the separate points which contribute to the mean plane. The surface to be measured is usually surveyed by measuring along certain lines. (See Figure 10.03.) The dots represent points at which the height of the test surface above some arbitrary line is calculated. The number of points chosen will depend on the size of the surface plate but the number along a diagonal may be between

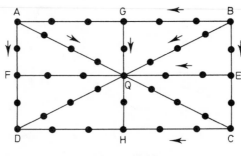

Figure 10.03

ten and twenty, the latter figure being suitable for a fairly large table of the order of 6 ft by 4 ft.

10.04 The readings are obtained by taking successive readings, with an auto-collimator, of the light reflected from a mirror mounted on a three-point suspension table. At each move the mirror mount is moved through a distance equal to the spacing of the third foot from the other two. The number of readings along any line should, for computational reasons, always be an even number. Alternatively the gradient can be read directly by using an instrument such as a Talyvel, which is also mounted on a three-point suspension table.

Readings along Diagonals

10.05 Let α_1 be the auto-collimator reading for the first position, α_2 for the second, etc., up to α_n where the αs are the measured angles converted to radians, and where n is the number of readings for a diagonal. If the spacing between the third foot and the other two is d_d, the height of the measuring points about the arbitrary line OO' (see Figure 10.05) is given by

$$z_p = \sum_{r=0}^{r=p} d_d \alpha_r \qquad 10.05(1)$$

(it should be noted that $\alpha_0 = 0$ by definition) where p is the pth position relative to zero ($x_p = pd_d$) and z_p is the height of the pth point above OO'.

10.06 If we now join the point O ($x = 0$) with the last point on the diagonal P (x_n, z_n) then z_n is given by $\sum_{r=0}^{r=n} d_d \alpha_r$. The height above OO' at the pth position (x_p) on the line OP is thus

$$(p/n) \sum_{r=0}^{r=n} d_d \alpha_r = z'_p$$

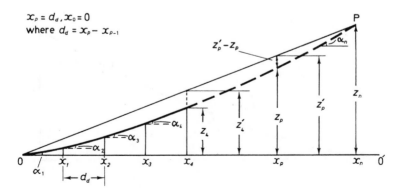

Figure 10.05

The difference between z'_p and z_p is given by

$$z'_p - z_p = (p/n) \sum_{r=0}^{r=n} d_d\alpha_r - \sum_{r=0}^{r=p} d_d\alpha_r \qquad\qquad 10.06(1)$$

Let us now draw a line RS parallel to OP through the point $Q(x_{n/2}, z_{n/2})$, and calculate the height of the points on the surface from this new line RQS (see Figure 10.06). The spacing between the straight lines OP and RS is seen to be $z'_{n/2} - z_{n/2}$ and thus the required height z''_p is seen to be given by

$$z''_p = z'_{n/2} - z_{n/2} - (z'_p - z_p)$$

$$= \tfrac{1}{2} \sum_{r=0}^{r=n} d_d\alpha_r - \sum_{r=0}^{r=n/2} d_d\alpha_r - (p/n) \sum_{r=0}^{r=n} d_d\alpha_r + \sum_{r=0}^{r=p} d_d\alpha_r$$

or

$$z''_p = d_d\left(\sum_{r=0}^{r=p} \alpha_r - (p/n) \sum_{r=0}^{r=n} \alpha_r + \tfrac{1}{2} \sum_{r=0}^{r=n} \alpha_r - \sum_{r=0}^{r=n/2} \alpha_r \right) \qquad 10.06(2)$$

This gives a series of points such that

$$z''_0 = z''_n = \tfrac{1}{2} \sum_{r=0}^{r=n} \alpha_r - \sum_{r=0}^{r=n/2} \alpha_r$$

and $z''_{n/2} = 0$.

10.07   If the two diagonals AC and BD of Figure 10.03 are calculated as above, then to complete the survey, the formula for the height of the parallel sides must be found. Let the value of z''_0 and z''_n for the diagonal AC be b and for the diagonal BD let the value be a, where a and b and the heights of the measured points on the two diagonals are all measured from the reference plane xy defined by the two reference lines RQS and R'QS', where R'QS' is the reference line for one diagonal and RQS is the reference line for the other, and Q is their point of intersection.

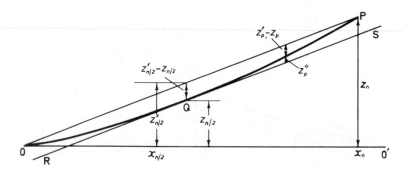

Figure 10.06

Parallel Sides

10.08 Let us consider the side BC of Figure 10.03. If we proceed as before the distances of points on the curve obtained below the line joining the starting and end points are given as before by

$$z'_p - z_p = (p/n) \sum_{r=0}^{r=n} d_d \alpha_r - \sum_{r=0}^{r=p} d_d \alpha_r \qquad 10.06(1)$$

(see Figure 10.08) where d_d is the spacing between the third foot and the other two feet of the mirror mount. The height of the curve RQS, given by the measured points above the $x'y'$ reference plane, is given by the difference between the height of the line RS above the reference plane $x'y'$ and $(z'_p - z_p)$. Thus

$$z'''_p = a + (b - a)(p/n) - (z'_p - z_p)$$

$$= \sum_{r=0}^{r=p} d_d \alpha_r - (p/n) \sum_{r=0}^{r=n} d_d \alpha_r + (p/n)(b - a) + a$$

$$= d_d \left(\sum_{r=0}^{r=p} \alpha_r - (p/n) \sum_{r=0}^{r=n} \alpha_r \right) + (p/n)(b - a) + a \qquad 10.08(1)$$

10.09 Equation 10.08(1) is used to calculate the height of the points, making up the parallel sides, above or below the reference plane xy. If desired, the magnitude of the negative point having the greatest magnitude can be added to all the readings, making all readings greater or equal to zero. This of course is only relevant if negative values exist. It is then a very simple matter to judge the departure of the plate from the flat, that is the xy reference

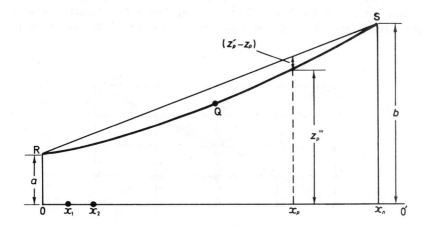

Figure 10.08

plane, since the z readings give this departure, that is z_p'' and z_p''' (see equations 10.06(2) and 10.08(1) respectively).

10.10 Referring to Figure 10.03, if the lines GH and FE are measured and are made to join up to the lines AB and DC, and AD and BC respectively, then in general the two lines GH and FE will not intersect where they cross, and neither will either of them intersect the two diagonals AC and BD at the centre Q. If an adjustment is required for this non-intersection of the lens GH and FE at the centre then this may be made as follows.

10.11 If the intersection height of the two diagonals at the centre Q above the chosen reference plane is z_1, and if the heights of the lines GH and FE at the centre are z_2 and z_3 respectively, then the mean intersection point can be taken as $(2z_1 + z_2 + z_3)/4 = \bar{z}$. The correction amount $(\bar{z} - z_1)(1 - 2s_1/w)$ should now be added to points on the two diagonals AQC and BQD, where s_1 is the distance from the centre Q along any diagonal to the corresponding corner, and w is the length of a diagonal. The corresponding correction for the paths QGA, QGB, QHD and QHC is $(\bar{z} - z_2)\{1 - 2s_2/(u + v)\}$ where s_2 is the distance from Q along either of the above paths, u is the length of $AB \equiv BC$ and v is the length of $AD \equiv BC$. Similarly the correction for the paths QFA, QFD, QEB and QEC is $(\bar{z} - z_3)\{1 - 2s_3/(u + v)\}$ where s_3 is the distance from Q along the paths just quoted. The s are of course always positive.

For the calculation of the uncertainty of the difference in height between any two points on a surface, see Appendix IV.

Improved Coverage of Plate

10.12 The surface plate in Figure 10.03 was divided into 49 datum points obtained from 46 measurements using a Union Jack configuration. A better coverage of the surface is obtained by using the layout scheme shown in Figure 10.12. This shows 25 datum points obtained from 48 measurements.

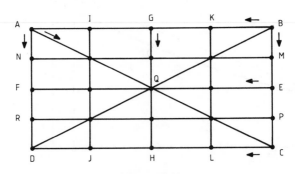

Figure 10.12

The arrows show the directions in which measurements should be taken starting with the position given by the arrow. The procedure for working up the results is initially the same as for Figure 10.03, i.e. the Union Jack is treated first, the diagonals made to cross in the middle with identical heights, and the heights of the ends of each diagonal made equal. The lines AB, CD, AD and BC are then hung from the ends of the diagonals at A, B, C and D. The lines GH and EF are hung from the centres of the outside lines of sight. This leads in general to three different values of the heights at the centre Q, say z_1, z_2 and z_3, where z_1 is the height of the crossing point of the diagonals and z_2 and z_3 are the heights of EF and GH at their centres at Q. The mean height at the centre is taken as before as $(2z_1 + z_2 + z_3)/4$ and the correction for each measurement point extended back along the sight lines to the corners using the expressions given in paragraph 10.11. The rest of the sight lines should be hung from the outside generators after the above corrections have been made, and any non-agreement at crossing points averaged.

10.13 For larger surface plates the number of measurement points may be increased by increasing the number of rectangles. The working up procedure remains the same. For a large 6 ft by 4 ft table, each sight line might have 13 datum points leading to 169 datum points in all with 336 measurements. The formula for working out the number of datum points and number of measurements for a layout like Figure 10.12 is

$$\text{datum points} = n^2$$

$$\text{measurements} = 2(n - 1)(n + 1)$$

where n = number of datum points on a side.

The number of measurements may be somewhat reduced for large tables by adopting the layout scheme shown in Figure 10.13. The accuracy obtained will be slightly reduced. The basic difference is that the number of datum points along the long sides only is doubled. The working up procedure is the same as for the previous layout. Corrections to the added points apart

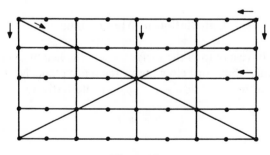

Figure 10.13

from the outside ones covered by the Union Jack procedure are optional. If required, the correction to the non-intersecting points excluding those on the outside sight lines is given by $((z'_r - z_r) + (z'_{r+1} - z_{r+1}))/2$, where z'_r and z'_{r+1} are the corrected values for two successive intersecting points on either side of a point to be corrected and z_r and z_{r+1} are the corresponding points on the appropriate long sight line. This value should be added, if positive, or subtracted, if negative, from the point to be corrected. The number of points for such a scheme is

$$\text{datum points} = \frac{n(n+1)}{2}$$

$$\text{measurements} = (n-1)(3n+7)/4$$

where n is the number of datum points along a long side, $(n+1)/2$ is the number of datum points along a short side and along a diagonal. If $n = 13$ say, for a 6 ft × 4 ft table then the number of datum points is 91 and the number of measurement points is 138, a considerable reduction on $n = 13$ for the full rectangle scheme of Figure 10.12 of 336.

10.14 A different approach to surface plate measurement using a least squares method has been developed at the National Physical Laboratory at Teddington†. For a rectangular table the layout of Figure 10.12 is advocated. This does lead to a large number of observations. For a large 6 ft × 4 ft table the number of datum points is 13 per side and for each diagonal the number of datum points is 169 with 364 observations. If n = number of datum points per side,

$$\text{number of datum points} = n^2$$

$$\text{number of observations} = 2n(n+1)$$

The method requires considerable computation being performed on the measurement data, using a specialized computer program. The method would seem to give reliable results.

Calculation of Mean Plane

10.15 We now have a map of the surface where the height of each measured point is given either above or below the xy reference plane. Now the equation of a plane is given by $a'x + b'y + c'z = f$ or alternatively

$$z = a''x + b''y + f' \qquad\qquad 10.15(1)$$

† See NPL reports MOM 5 1973 and MOM 9 1974.

The deviation of each point on the surface plate from this mean plane will be obtained by subtracting from each z'_j, (co-ordinates x_j, y_j) representing the height of each measured point above the reference plane xy, the value of z given by 10.15(1) corresponding to the height of the mean plane above the reference plane xy for the same co-ordinates x_j, y_j.

10.16 Thus the deviation of each point from the mean plane is given by

$$z'_j - z_j = z'_j - a''x_j - b''y_j - f' \qquad 10.16(1)$$

Now if each of these deviations from the mean plane is a random deviation, and all these random deviations belong to the same random population, then the probability of all the deviations found occurring together is given by

$$P = \frac{1}{\sigma^n(2\pi)^{n/2}} \prod_{j=1}^{j=n} \{e^{-(z'_j - z_j)^2/2\sigma^2}\} \qquad 10.16(2)$$

where Π implies the product of the term in the bracket with like terms, where j varies between 1 and n, n being the number of terms. This may be written alternatively as

$$P = \frac{1}{\sigma^n(2\pi)^{n/2}} \cdot \exp\left\{ -\sum_{j=1}^{j=n} \frac{(z'_j - z_j)^2}{2\sigma^2} \right\} \qquad 10.16(3)$$

Now the expression covered by the summation sign sigma is a function of a'', b'' and f', and in order that the set of deviations obtained should have occurred, P must be made a maximum by choosing the right values of a'', b'' and f'. It is easily seen that for P to be a maximum

$$W = \sum_{j=1}^{j=n} (z'_j - z_j)^2 \qquad 10.16(4)$$

must be a minimum, and thus we require the values of a'', b'', and f' corresponding to the three conditions

$$\frac{\partial W}{\partial a''} = 0, \qquad \frac{\partial W}{\partial b''} = 0 \qquad \text{and} \qquad \frac{\partial W}{\partial f'} = 0 \qquad 10.16(5)$$

Principle of least squares

10.17 It is seen that in applying the condition that P should be a maximum, known as the 'principle of maximum likelihood', that this in turn has led to the requirement that the sum of the squares of the deviations should be a minimum. This latter condition is known as the principle of least squares[2] and is often applied directly without using the principle of maximum likelihood.

10.18 Substituting for z_j in 10.16(4) we have, using 10.16(1), that

$$W = \sum_{j=1}^{j=n} (z'_j - a''x_j - b''y_j - f')^2 \qquad 10.18(1)$$

Using equation 10.18(1) we have

$$\frac{\partial W}{\partial a''} = -\sum_{j=1}^{j=n} x_j(z'_j - a''x_j - b''y_j - f') = 0 \qquad 10.18(2)$$

$$\frac{\partial W}{\partial b''} = -\sum_{j=1}^{j=n} y_j(z'_j - a''x_j - b''y_j - f') = 0 \qquad 10.18(3)$$

$$\frac{\partial W}{\partial f'} = -\sum_{j=1}^{j=n} (z'_j - a''x_j - b''y_j - f') = 0 \qquad 10.18(4)$$

Rewritten these become

$$a''\sum x_j^2 + b''\sum x_jy_j + f'\sum x_j = \sum x_jz'_j \qquad 10.18(5)$$

$$a''\sum x_jy_j + b''\sum y_j^2 + f'\sum y_j = \sum y_jz'_j \qquad 10.18(6)$$

and

$$a''\sum x_j + b''\sum y_j + nf' = \sum z'_j \qquad 10.18(7)$$

where $\sum$ stands for $\sum_{j=1}^{j=n}$. Eliminating f' between 10.18(5) and 10.18(7) by multiplying the former by n and the latter by $\sum x_j$ and subtracting 10.18(7) from 10.18(5) gives

$$a''\{n\sum x_j^2 - (\sum x_j)^2\} + b''\{n\sum x_jy_j - \sum x_j\sum y_j\} = n\sum x_jz'_j - \sum x_j\sum z'_j$$

$$10.18(8)$$

Similarly, eliminating f' between equations 10.18(6) and 10.18(7) by multiplying the former by n and the latter by $\sum y_j$ and subtracting 10.18(7) from 10.18(6) we have

$$a''\{n\sum x_jy_j - \sum x_j\sum y_j\} + b''\{n\sum y_j^2 - (\sum y_j)^2\} = n\sum y_jz'_j - \Sigma y_j\sum z'_j$$

$$10.18(9)$$

10.19   10.18(8) and 10.18(9) can be rewritten as

$$g_{xx}a'' + g_{xy}b'' = g_{xz'} \qquad 10.19(1)$$

and

$$g_{xy}a'' + g_{yy}b'' = g_{yz'} \qquad 10.19(2)$$

where

$$g_{rs} = (n\sum r_js_j - \sum r_j\sum s_j) \qquad 10.19(3)$$

Thus

$$a'' = \frac{g_{xz'}g_{yy} - g_{yz'}g_{xy}}{g_{xx}g_{yy} - g_{xy}^2} \qquad 10.19(4)$$

and

$$b'' = \frac{g_{yz'}g_{xx} - g_{xy}g_{xz'}}{g_{xx}g_{yy} - g_{xy}^2}$$

10.19(5)

Now from 10.18(7)

$$f' = \frac{\sum z_j'}{n} - \frac{a''\sum x_j}{n} - \frac{b''\sum y_j}{n}$$

10.19(6)

and thus substituting for a'' and b'' from 10.19(4) and 10.19(5) gives

$$f' = \frac{\sum z_j'}{n} - \left(\frac{g_{xz'}g_{yy} - g_{yz'}g_{xy}}{g_{xx}g_{yy} - g_{xy}^2}\right)\frac{\sum x_j}{n} - \left(\frac{g_{yz'}g_{xx} - g_{xy}g_{xz'}}{g_{xx}g_{yy} - g_{xy}^2}\right)\frac{\sum y_j}{n}$$

10.19(7)

The equation of the mean plane is thus

$$z = \left(\frac{g_{xz'}g_{yy} - g_{yz'}g_{xy}}{g_{xx}g_{yy} - g_{xy}^2}\right)x + \left(\frac{g_{yz'}g_{xx} - g_{xy}g_{xz'}}{g_{xx}g_{yy} - g_{xy}^2}\right)y + \frac{\sum z_j'}{n}$$
$$- \left(\frac{g_{xz'}g_{yy} - g_{yz'}g_{xy}}{g_{xx}g_{yy} - g_{xy}^2}\right)\frac{\sum x_j}{n} - \left(\frac{g_{yz'}g_{xx} - g_{xy}g_{xz'}}{g_{xx}g_{yy} - g_{xy}^2}\right)\frac{\sum y_j}{n}$$

10.19(8)

10.20 Now the co-ordinates of the centroid of all the points are

$$\frac{\sum x_j}{n}, \qquad \frac{\sum y_j}{n}, \qquad \frac{\sum z_j'}{n}$$

and if the co-ordinates of this point are substituted for x, y and z in 10.19(8) it is clearly seen that the expression is identically equal to zero, and thus the centroid lies in the mean plane.

Change of Axes

10.21 Let the centroid co-ordinates be given by

$$\bar{x} = \frac{\sum x_j}{n}, \qquad \bar{y} = \frac{\sum y_j}{n} \qquad \text{and} \qquad \bar{z} = \frac{\sum z_j'}{n}$$

10.21(1)

Now let us change the centre of our co-ordinate system to $\bar{x}$, $\bar{y}$ and $\bar{z}$. Thus if the new co-ordinates are represented by capital X_j, Y_j and Z_j', we have

$$X_j = x_j - \bar{x}_j, \qquad Y_j = y_j - \bar{y}_j \qquad \text{and} \qquad Z_j' = z_j' - \bar{z}_j'$$

10.21(2)

10.22 Expressed in these co-ordinates, the true mean plane becomes

$$Z = a'' X + b'' Y \qquad \text{10.22(1)}$$

from 10.19(8) and

$$W = \sum (Z'_j - Z_j)^2$$
$$= \sum_{j=1}^{j=n} (Z'_j - a'' X_j - b'' Y_j)^2 \qquad \text{10.22(2)}$$

The conditions for a minimum in this case reduce to two conditions, namely

$$\frac{\partial W}{\partial a''} = 0 \qquad \text{and} \qquad \frac{\partial W}{\partial b''} = 0 \qquad \text{10.22(3)}$$

from 10.16(4). Thus

$$\frac{\partial W}{\partial a''} = 0 = \sum X_j (Z'_j - a'' X_j - b'' Y_j) \qquad \text{10.22(4)}$$

and

$$\frac{\partial W}{\partial b''} = 0 = \sum Y_j (Z'_j - a'' X_j - b'' Y_j) \qquad \text{10.22(5)}$$

Thus

$$a'' \sum X_j^2 + b'' \sum X_j Y_j = \sum X_j Z'_j \qquad \text{10.22(6)}$$

and

$$a'' \sum X_j Y_j + b'' \sum Y_j^2 = \sum Y_j Z'_j \qquad \text{10.22(7)}$$

Eliminating b'' between 10.22(6) and 10.22(7) gives

$$a'' = \frac{\sum Y_j^2 \sum X_j Z'_j - \sum X_j Y_j \sum Y_j Z'_j}{\sum X_j^2 \sum Y_j^2 - (\sum X_j Y_j)^2} \qquad \text{10.22(8)}$$

Similarly

$$b'' = \frac{\sum X_j^2 \sum X_j Z'_j - \sum X_j Y_j \sum X_j Z'_j}{\sum X_j^2 \sum Y_j^2 - (\sum X_j Y_j)^2} \qquad \text{10.22(9)}$$

10.23 Either set of equations, that is 10.19(4), 10.19(5), 10.19(6) and 10.15(1) or 10.22(8), 10.22(9) and 10.22(1) can be used to find the mean plane, but the set using the centroid co-ordinates is decidedly the easier to calculate. Once $\bar{x}$, $\bar{y}$ and $\bar{z}'$ have been found and the new variables X, Y and Z' found using 10.21(2), a'' and b'' are easily found using 10.22(8) and 10.22(9).

10.24 Usually the overall departure from the mean plane is not much different from the overall difference of height obtained by using the analysis given in paragraphs 10.02 to 10.11. The mean plane is, however, useful if the standard deviation of the departure of the surface plate from the mean plane is required.

Standard Deviation from Mean Plane

10.25 The standard deviation of the points from the mean plane will give some indication of the statistical quality of the plate, and the value of this is found as follows. The equation of the mean plane may be written

$$\frac{a'' X + b'' Y - Z}{(a''^2 + b''^2 + 1)^{1/2}} = 0 \qquad 10.25(1)$$

The distance of any point on the plate X_j, Y_j, Z'_j from the mean plane is given by

$$\frac{a'' X_j + b'' Y_j - Z'_j}{(a''^2 + b''^2 + 1)^{1/2}} \qquad 10.25(2)$$

The standard deviation is thus given by

$$s = \left\{ \sum_{j=1}^{j=n} \frac{(a'' X_j + b'' Y_j - Z'_j)^2}{(n-2)(a''^2 + b''^2 + 1)} \right\}^{1/2} \qquad 10.25(3)$$

$n - 2$ appears in the denominator because two constants have had to be found, namely a'' and b''.

Curve Fitting: Fitting a Straight Line

10.26 There are many cases in which it is known that two variables, say x and y, are connected by a straight line relationship. Owing to measuring inaccuracies, if a graph is plotted of y against x, we get a series of points which do not lie on a straight line. The problem is to find the best straight line through the points obtained. In this analysis the values of the x will be assumed to be correct, whilst only the y values will be assumed to have uncertainties.

10.27 Let the straight line be

$$y = mx + c \qquad 10.27(1)$$

then if y'_i corresponds to the measured ys the sum of the squares of the deviations is given by

$$W = \sum_{i=1}^{i=n} (y'_i - y_i)^2 \qquad 10.27(2)$$

where there are n points. Substituting for y from 10.27(1) we have

$$\sum_{i=1}^{i=n} (y_i' - mx_i - c)^2 = W \qquad\qquad 10.27(3)$$

and thus, using the principle of least squares, we have, in order that W is a minimum

$$\frac{\partial W}{\partial m} = 0 \quad \text{and} \quad \frac{\partial W}{\partial c} = 0 \qquad\qquad 10.27(4)$$

Thus

$$\sum x_i(y_i' - mx_i - c) = 0 \qquad\qquad 10.27(5)$$

and

$$\sum (y_i' - mx_i - c) = 0 \qquad\qquad 10.27(6)$$

that is

$$m\sum x_i^2 + c\sum x_i = \sum x_i y_i' \qquad\qquad 10.27(7)$$

and

$$m\sum x_i + nc = \sum y_i' \qquad\qquad 10.27(8)$$

Eliminating c between 10.27(7) and 10.27(8) by taking 10.27(7) times n minus 10.27(8) times $\sum x_i$, we have

$$m\{n\sum x_i^2 - (\sum x_i)^2\} = n\sum x_i y_i' - \sum x_i \sum y_i'$$

or

$$m = \frac{n\sum x_i y_i' - \sum x_i \sum y_i'}{n\sum x_i^2 - (\sum x_i)^2} \qquad\qquad 10.27(9)$$

We obtain c by the operation 10.27(7) times $\sum x_i$ minus 10.27(8) times $\sum x_i^2$ giving

$$c\{(\sum x_i)^2 - n\sum x_i^2\} = \sum x_i \sum x_i y_i' - \sum x_i^2 \sum y_i'$$

or

$$c = \frac{\sum x_i^2 \sum y_i' - \sum x_i \sum x_i y_i'}{n\sum x_i^2 - (\sum x_i)^2} \qquad\qquad 10.27(10)$$

10.28   An alternative expression is obtained as follows. Let $\bar{x}$ be the mean of all the x_i and $\bar{y}'$ the mean of all the y_i'. Thus

$$\bar{x} = \sum_{i=1}^{i=n} \frac{x_i}{n} \qquad\qquad 10.28(1)$$

and

$$\bar{y}' = \sum \frac{y'_i}{n} \qquad 10.28(2)$$

Now let ε_i be the deviation of x_i from $\bar{x}$ and let τ_i be the deviation of y'_i from $\bar{y}'$, thus

$$x_i = \bar{x} + \varepsilon_i \qquad 10.28(3)$$

and

$$y'_i = \bar{y}' + \tau_i \qquad 10.28(4)$$

and

$$\sum x_i^2 = \sum (\bar{x} + \varepsilon_i)^2 = n\bar{x}^2 + \sum \varepsilon_i^2 + 2\bar{x} \sum \varepsilon_i = n\bar{x}^2 + \sum \varepsilon_i^2$$

$$10.28(5)$$

since $2\bar{x} \sum \varepsilon_i = 0$ because if we substitute for x_i from 10.28(3) in 10.28(1) we get $n\bar{x} = \sum (\bar{x} + \varepsilon_i) = n\bar{x} + \sum \varepsilon_i$ and therefore

$$\sum \varepsilon_i = 0 \qquad 10.28(6)$$

Also

$$\sum x_i y'_i = \sum (\bar{x} + \varepsilon_i)(\bar{y}' + \tau_i)$$
$$= \sum \bar{x}\bar{y}' + \sum \varepsilon_i \tau_i + \bar{y}' \sum \varepsilon_i + \bar{x} \sum \tau_i$$
$$= n\bar{x}\bar{y}' + \sum \varepsilon_i \tau_i \qquad 10.28(7)$$

since

$$\sum \varepsilon_i = 0 \qquad 10.28(6)$$

Further $\sum \tau_i = 0$, because on substituting for y'_i from 10.28(4) in 10.28(2) we have

$$\bar{y}'n = \sum (\bar{y}' + \tau_i) = n\bar{y}' + \sum \tau_i$$

and therefore

$$\sum \tau_i = 0 \qquad 10.28(1)$$

10.29 Thus, substituting in 10.27(9) for $\sum x_i y'_i$, $\sum x_i$, $\sum y'_i$, $\sum x_i^2$ and $(\sum x_i)^2$ we have

$$m = \frac{n(n\bar{x}\bar{y}' + \sum \varepsilon_i \tau_i) - n\bar{x}n\bar{y}'}{n(n\bar{x}^2 + \sum \varepsilon_i^2) - n^2 \bar{x}^2}$$

$$= \frac{\sum \varepsilon_i \tau_i}{\sum \varepsilon_i^2} \qquad 10.29(1)$$

whilst the same substitution in 10.27(10) gives

$$c = \frac{(n\bar{x}^2 + \sum \varepsilon_i^2)n\bar{y}' - n\bar{x}(n\bar{x}\bar{y}' + \sum \varepsilon_i \tau_i)}{n \sum \varepsilon_i^2}$$

$$= \frac{\bar{y}' \sum \varepsilon_i^2 - \bar{x} \sum \varepsilon_i \tau_i}{\sum \varepsilon_i^2} \qquad\qquad 10.29(2)$$

The required straight line is thus

$$y - \bar{y}' = \frac{\sum \varepsilon_i \tau_i}{\sum \varepsilon_i^2}(x - \bar{x}) \qquad\qquad 10.29(3)$$

or

$$\tau = m\varepsilon \qquad\qquad 10.29(4)$$

This is sometimes called the line of *regression* of y on x. A point to note is that the line of regression passes through $\bar{x}, \bar{y}'$. The two equations of condition 10.27(7) and (8) can be written as

$$m\left(\bar{x}^2 + \frac{\sum \varepsilon_i^2}{n}\right) + c\bar{x} = \bar{x}\bar{y}' + \frac{1}{n}\sum \varepsilon_i \tau_i \qquad\qquad 10.29(5)$$

and

$$m\bar{x} + c = \bar{y}' \qquad\qquad 10.29(6)$$

$(\sum \varepsilon_i^2)/n$ is sometimes written as σ_x^2 where σ_x is the standard deviation of x_i from its mean $\bar{x}$. Now 10.29(5) can be written as

$$\bar{x}(m\bar{x} + c - \bar{y}') + m\sum \frac{\varepsilon_i^2}{n} = \frac{1}{n}\sum \varepsilon_i \tau_i$$

which in view of 10.29(6) reduces to

$$m\sum \varepsilon_i^2 = \sum \varepsilon_i \tau_i$$

giving

$$m = \frac{\sum \varepsilon_i \tau_i}{\sum \varepsilon_i^2} \qquad\qquad 10.29(1)$$

$(\sum \tau_i^2)/n$ is sometimes written as σ_y^2 where σ_y is the standard deviation of y_i from its mean $\bar{y}'$.

Weighted Mean line

10.30 If each point x_i, y_i has a weight w_i, then the gradient m_w is easily shown to be

$$m_w = \frac{\sum w_i(x_i - \bar{x}_w)(y_i' - \bar{y}_w)}{\sum w_i(x_i - \bar{x}_w)^2} = \frac{\sum w_i \varepsilon_{iw} \tau_{iw}}{\sum w_i \varepsilon_{iw}^2} \qquad 10.30(1)$$

Similarly

$$c_w = \frac{\bar{y}_w' \sum w_i x_i^2 - \bar{x}_w \sum w_i x_i y_i'}{\sum w_i(x_i - \bar{x}_w)}$$

$$= \frac{\bar{y}_w' \sum w_i \varepsilon_{iw}^2 - \bar{x}_w \sum w_i \varepsilon_{iw} \tau_{iw}}{\sum w_i \varepsilon_{iw}^2} \qquad 10.30(2)$$

The correct value for the weight w_i is $1/\sigma_{yLi}^2$, where σ_{yLi} is the standard deviation of y_i' from the mean line. For single points σ_{yLi} is not generally known, but if each point is the mean of a number of observations, n_i, then each point is given by

$$x_i \text{ and } y_i' = \left(\sum_1^{n_1} y_i'' \right) \Big/ n_i$$

in which each $w_i = n_i/\sigma_{yLi}^2$, where σ_{yLi} is the standard deviation of the y_i'' from the mean line at the point x_i, y_i. Note that

$$\bar{x}_w = \sum w_i x_i / \sum w_i \qquad 10.30(3)$$

and

$$\bar{y}_w' = \sum w_i y_i' / \sum w_i \qquad 10.30(4)$$

where w_i is the value stated above. Also note that in the weighted equations $\varepsilon_{iw} = w_i(x_i - \bar{x}_w)$, $\tau_{iw} = w_i(y_i' - \bar{y}_w')$ and weighted $\sum \varepsilon_{iw}^2 = w_i(x_i - \bar{x}_w)^2$.

Standard Deviation and Standard Error in the Mean of y_i' From the Mean Line $\tau = m\varepsilon$

10.31 The sum of the squares of the residuals is equal to

$$\sum_{i=1}^{i=n} (y_i' - mx_i - c)^2 = \sum_{i=1}^{i=n} (\tau_i - m\varepsilon_i)^2 \qquad 10.31(1)$$

Since there are two equations of condition, and m and $\bar{x}$ are fixed, there are only $n - 2$ degrees of freedom. Thus the best approximation to the *square of the standard deviation* σ_{yL}^2 of the y_i' about the mean line is given by

$$s_{yL}^2 = \frac{\sum_{i=1}^{i=n} (\tau_i - m\varepsilon_i)^2}{n - 2} \qquad 10.31(2)$$

The weighted variance

$$s_{yLw}^2 = \frac{\sum_{i=1}^{i=n} w_i (\tau_{iw} - m_w \varepsilon_w)^2}{\sum_{i=1}^{i=n} w_i} \times \frac{n}{n-2} \qquad 10.31(3)$$

(See also paragraphs 10.71 to 10.76.) The square of the *standard error in the mean* of y_i' from the mean line is thus given by

$$s_{\bar{y}L}^2 = \frac{\sum_{i=1}^{i=n} (\tau_i - m \varepsilon_i)^2}{n(n-2)} \qquad 10.31(4)$$

or alternatively

$$= \frac{\sum (y_i' - mx_i - c)^2}{n(n-2)} \qquad 10.31(5)$$

The weighted variance is given by

$$s_{\bar{y}Lw}^2 = \frac{\sum_{i=1}^{i=n} w_i (\tau_{iw} - m_w \varepsilon_{iw})^2}{(n-2) \sum_{i=1}^{i=n} w_i} \qquad 10.31(6)$$

Standard Deviation σ_m of the Gradient m of the Mean Line

10.32 The value of m for *each* line through $\bar{x}, \bar{y}$ is given by

$$\frac{\tau_i}{\varepsilon_i} \qquad \text{that is} \qquad \frac{y_i' - \bar{y}}{x_i - \bar{x}} \qquad 10.32(1)$$

Now the mean value $\bar{x}_w$ of a set of values x_i, each of weight w_i, is given by

$$\bar{x}_w = \frac{\sum_1^n w_i x_i}{\sum_1^n w_i} \qquad 10.32(2)$$

Thus the mean value of the gradient for n lines through $\bar{x}, \bar{y}$ is given by

$$m = \sum_1^n w_i \left(\frac{\tau_i}{\varepsilon_i} \right) \bigg/ \sum_1^n w_i \qquad 10.32(3)$$

Now we know that

$$m = \frac{\sum_1^n \tau_i \varepsilon_i}{\sum_1^n \varepsilon_i^2} \qquad 10.29(1)$$

and in order that these two equations should be identical

$$w_i \equiv \varepsilon_i^2 \qquad 10.32(4)$$

Now the estimate of the standard deviation in the mean $\sigma_{\bar{z}}$ of a set of quantities z_i is given by

$$s_{\bar{z}}^2 = \frac{\sum w_i(z_i - \bar{z}_w)^2}{(n - q)\sum w_i} \qquad 10.32(5)$$

where $q =$ number of dependent variables, that is m and $\bar{x}$. Thus

$$s_m^2 = \frac{\sum_1^n \varepsilon_i^2(\tau_i/\varepsilon_i - m)^2}{(n - 2)\sum_1^n \varepsilon_i^2}$$

$$= \frac{\sum_1^n (\tau_i - \varepsilon_i m)^2}{(n - 2)\sum_1^n \varepsilon_i^2} = s_{yL}^2 \bigg/ \sum_1^n \varepsilon_i^2 \qquad 10.32(6)$$

since the *mean* gradient is equal to $m \equiv (z)_w$ and $z_i \equiv \tau_i/\varepsilon_i$. Alternatively $\sigma_{\bar{m}}^2$ may be written

$$s_m^2 = \frac{\sum_1^n (y_i' - mx_i - c)^2}{(n - 2)\sum_1^n (x_i - \bar{x})^2} \qquad 10.32(7)$$

The weighted estimate of the standard deviation of the weighted mean gradient m_w is given by

$$s_{m_w}^2 = \frac{\sum_1^n w_i(\tau_{iw} - m_w\varepsilon_{iw})^2}{(n - 2)\sum_1^n w_i\varepsilon_{iw}^2} \qquad 10.32(8)$$

Alternative Derivation for Standard Deviation σ_m of the Gradient m of the Mean Line

10.33

$$m = \frac{\sum_1^n \tau_i\varepsilon_i}{\sum_1^n \varepsilon_i^2} \qquad 10.29(1)$$

Since the ε_i are accurately known, and the errors are confined to the τ_i, we may write the standard deviation of m in terms of the standard deviation of the τ_i. Thus

$$s_m^2 = \sum_1^n s_{\tau_i}^2\left(\frac{\partial m}{\partial \tau_i}\right)^2 \qquad 10.33(1)$$

from

$$s_F^2 = \sum_1^n s_i^2\left(\frac{\partial F}{\partial x_i}\right)^2 \qquad 8.04(4)$$

Thus

$$s_m^2 = \sum_1^n s_{\tau_i}^2 \varepsilon_i^2 \Big/ \left(\sum_1^n \varepsilon_i^2 \right)^2 \qquad 10.33(2)$$

since

$$\frac{\partial m}{\partial \tau_i} = \varepsilon_i \Big/ \sum_1^n \varepsilon_i^2 \qquad 10.33(3)$$

Now

$$s_{\tau_i}^2 = s_\tau^2 = s_{yL}^2$$

since all y deviations are assumed to have the same standard deviation. Thus

$$s_m^2 = \frac{s_{yL}^2 \sum_1^n \varepsilon_i^2}{(\sum_1^n \varepsilon_i^2)^2} = \frac{s_{yL}^2}{\sum_1^n \varepsilon_i^2} = \frac{\sum_1^n (\tau_i - m\varepsilon_i)^2}{(n-2)\sum_1^n \varepsilon_i^2} \,\dagger \qquad 10.33(4)$$

as in 8.32(6). s_{yL}^2 is given by 10.31(2).

Standard Deviation of Intercept c

10.34   Now

$$\bar{y} = m\bar{x} + c \qquad 10.29(5)$$

since $\bar{x}$ and $\bar{y}$ are on the mean line. Now $\bar{y}$, m and c can be considered as variables, but since $\bar{x}$ is considered to be known accurately it can be considered to be a constant. From 8.04(4)

$$s_F^2 = \sum_{i=1}^{i=n} s_{x_i}^2 \left(\frac{\partial F}{\partial x_i} \right)^2 \qquad 8.04(4)$$

Now we write

$$c = \bar{y} - m\bar{x} \qquad 10.34(1)$$

and putting $F \equiv c$ we have

$$s_c^2 = \left(\frac{\partial c}{\partial \bar{y}} \right)^2 s_{\bar{y}l}^2 + \left(\frac{\partial c}{\partial m} \right)^2 s_m^2 \,\ddagger \qquad 10.34(2)$$

† For confidence and tolerance limits of y for given x, that is for $y|x$ for a mean line, see Appendix V.
‡ It should be noted that $s_{\bar{y}l}$ and s_m are uncorrelated. Formula 8.04(4) can only be used if the s_{x_i} are uncorrelated.

Now

$$\frac{\partial c}{\partial \bar{y}} = 1 \qquad \text{and} \qquad \frac{\partial c}{\partial m} = -\bar{x} \qquad\qquad 10.34(3)$$

therefore

$$s_c^2 = s_{\bar{y}l}^2 + \bar{x}^2 s_m^2 \qquad\qquad 10.34(4)$$

$$= \frac{\sum_{i=1}^{i=n}(\tau_i - m\varepsilon_i)^2}{n(n-2)} + \bar{x}^2 \frac{\sum_{i=1}^{i=n}(\tau_i - m\varepsilon_i)^2}{(n-2)\sum_{i=1}^{i=n}\varepsilon_i^2}$$

$$= \frac{\sum_{i=1}^{i=n}(\tau_i - m\varepsilon_i)^2}{(n-2)}\left(\frac{1}{n} + \frac{\bar{x}^2}{\sum_{i=1}^{i=n}\varepsilon_i^2}\right)$$

$$= \frac{\sum_{i=1}^{i=n}(\tau_i - m\varepsilon_i)^2}{n(n-2)\sum_{i=1}^{i=n}\varepsilon_i^2}\left(\sum_{i=1}^{i=n}\varepsilon_i^2 + n\bar{x}^2\right)$$

$$= \frac{\sum_{i=1}^{i=n}(\tau_i - m\varepsilon_i)^2 \sum_{i=1}^{i=n}x_i^2}{n(n-2)\sum_{i=1}^{i=n}\varepsilon_i^2} \qquad\qquad 10.34(5)$$

by 10.28(5). The weighted standard deviation of the mean of c is given by

$$s_{cw}^2 = \frac{\sum_{i=1}^{i=n}w_i(\tau_i - m_w\varepsilon_i)^2 \cdot \sum_{i=1}^{i=n}w_ix_i^2}{(n-2)\sum_{i=1}^{i=n}w_i \sum_{i=1}^{i=n}w_i\varepsilon_i^2} \qquad\qquad 10.34(6)$$

Alternative Derivation for Standard Deviation of Intercept c

10.35 Using 8.04(4) and 10.27(10) we have

$$s_c^2 = \sum_{i=1}^{i=n}\left(\frac{\partial c}{\partial y_i'}\right)^2 s_{y_i}^2 \qquad\qquad 10.35(1)$$

since c is a function of y_i' only, since the x_i are considered correct, that is constant. Thus

$$\frac{\partial c}{\partial y_i'} = \frac{\sum_{i=1}^{i=n}x_i^2 - x_i\sum_{i=1}^{i=n}x_i}{n\sum_{i=1}^{i=n}x_i^2 - (\sum_{i=1}^{i=n}x_i)^2} \qquad\qquad 10.35(2)$$

and therefore

$$\left(\frac{\partial c}{\partial y_i'}\right)^2 = \frac{\{(\sum_{i=1}^{i=n}x_i^2)^2 + x_i^2(\sum_{i=1}^{i=n}x_i)^2 - 2x_i\sum_{i=1}^{i=n}x_i\sum_{i=1}^{i=n}x_i^2\}}{\{n\sum_{i=1}^{i=n}x_i^2 - (\sum_{i=1}^{i=n}x_i)^2\}^2}$$

$$10.35(3)$$

Now we take all the s_{y_i} to be of equal weight, thus $s_{y_i} = s_{yL}$, and thus it follows that

$$s_c^2 = s_{yL}^2 \frac{\sum \{(\sum x_i^2)^2 + x_i^2(\sum x_i)^2 - 2x_i \sum x_i \sum x_i^2\}^\dagger}{\{n \sum x_i^2 - (\sum x_i)^2\}^2}$$

$$= s_{yL}^2 \frac{\{n(\sum x_i^2)^2 + \sum x_i^2(\sum x_i)^2 - 2(\sum x_i)^2 \sum x_i^2\}}{\{n \sum x_i^2 - (\sum x_i)^2\}^2}$$

$$= s_{yL}^2 \sum x_i^2 \frac{\{n \sum x_i^2 - (\sum x_i)^2\}}{\{n \sum x_i^2 - (\sum x_i)^2\}^2}$$

$$= \frac{s_{yL}^2 \sum x_i^2}{\{n \sum x_i^2 - (\sum x_i)^2\}} \equiv \frac{s_{yL}^2 \sum x_i^2}{n \sum \varepsilon_i^2} \qquad 10.35(4)$$

Now

$$\sum \varepsilon_i^2 = \sum (x_i - \bar{x})^2 = \sum x_i^2 - 2\bar{x} \sum x_i + n\bar{x}^2$$

$$= \sum x_i^2 - \frac{2(\sum x_i)^2}{n} + \frac{n(\sum x_i)^2}{n^2}$$

$$= \frac{n \sum x_i^2 - (\sum x_i)^2}{n} \qquad 10.35(5)$$

Therefore

$$s_c^2 = \frac{s_{yL}^2 \sum x_i^2}{n \sum x_i^2 - (\sum x_i)^2} \qquad 10.35(6)$$

Now

$$s_{yL}^2 = \frac{\sum (\tau_i - m\varepsilon_i)^2}{n - 2} \qquad 10.31(2)$$

Therefore

$$s_c^2 = \frac{\sum (\tau_i - m\varepsilon_i)^2 \sum x_i^2}{n(n - 2) \sum \varepsilon_i^2} \qquad \text{giving} \qquad 10.34(5)$$

Standard Deviation of the Intercept x_0

10.36   The mean line is given by

$$y = mx + c \qquad 10.36(1)$$

---

†where the limits of the sums have been omitted for the sake of clarity.

When $y = 0$

$$x = -\frac{c}{m} \equiv x_0 \qquad 10.36(2)$$

(See Figure 10.36.) Now

$$\bar{y} = m\bar{x} + c \qquad 10.29(5)$$

since $\bar{x}, \bar{y}$ is on the mean line, and thus

$$x_0 = -\frac{(\bar{y} - m\bar{x})}{m} = \bar{x} - \frac{\bar{y}}{m} \qquad 10.36(3)$$

Using 8.04(4) again, since $\bar{x}$ is known exactly and is thus a constant, we have

$$s_{x_0}^2 = s_{\bar{y}l}^2 \left(\frac{\partial x_0}{\partial \bar{y}}\right)^2 + s_m^2 \left(\frac{\partial x_0}{\partial m}\right)^2 \qquad 10.36(4)$$

Now

$$\frac{\partial x_0}{\partial \bar{y}} = -\frac{1}{m} \qquad \text{and} \qquad \frac{\partial x_0}{\partial m} = \frac{\bar{y}}{m^2} \qquad 10.36(5)$$

and therefore

$$s_{x_0}^2 = \frac{s_{\bar{y}l}^2}{m^2} + s_m^2 \frac{\bar{y}^2}{m^4}$$

$$= \frac{1}{m^2} \frac{\sum_{i=1}^{i=n} (\tau_i - m\varepsilon_i)^2}{n(n-2)} + \frac{\bar{y}^2}{m^4} \frac{\sum_{i=1}^{i=n} (\tau_i - m\varepsilon_i)^2}{(n-2)\sum_{i=1}^{i=n} \varepsilon_i^2}$$

$$= \frac{1}{m^2} \frac{\sum_{i=1}^{i=n} (\tau_i - m\varepsilon_i)^2}{(n-2)} \left\{ \frac{1}{n} + \frac{\bar{y}^2}{m^2 \sum_{i=1}^{i=n} \varepsilon_i^2} \right\} \qquad 10.36(6)$$

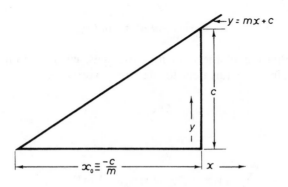

Figure 10.36

A Simple Treatment of the Straight Line $y = mx + c$

10.37 Before leaving the straight line it is worth noting a simplified treatment, which gives a good approximation to the least squares values for the mean line and also for the standard deviation in y_{yL} and in m (the gradient of the line).

10.38 Provided the number of readings does not exceed about twelve a reasonable approximation to the standard deviation in x is given by

$$\sigma_x \simeq (x_{max} - x_{min})/n \qquad \qquad 10.38(1)$$

where x is the measured quantity.

10.39 Let there be $2n$ values of x and y, and let these be divided into two sets, each containing n values of x, the values of x in the first set all being smaller than the smallest value of x in the second set. If the mean values of x and y for these two sets are found, that is $\bar{x}_1, \bar{y}_1$, and $\bar{x}_2, \bar{y}_2$, then a good approximation to the gradient of the mean line is given by

$$m = \frac{\bar{y}_2 - \bar{y}_1}{\bar{x}_2 - \bar{x}_1} \qquad \qquad 10.39(1)$$

The mean line is obtained by joining $\bar{x}_1, \bar{y}_1$ and $\bar{x}_2, \bar{y}_2$, and this is given by

$$y - \bar{y}_1 = \frac{\bar{y}_2 - \bar{y}_1}{\bar{x}_2 - \bar{x}_1}(x - \bar{x}_1) \qquad \qquad 10.39(2)$$

It is readily seen that this line passes through both $\bar{x}_1, \bar{y}_1$ and $\bar{x}_2, \bar{y}_2$ and also through the centre of gravity of all the points, given by

$$\bar{x} = \frac{\bar{x}_1 + \bar{x}_2}{2} \qquad \text{and} \qquad \bar{y} = \frac{\bar{y}_1 + \bar{y}_2}{2} \qquad \qquad 10.39(3)$$

Derivation of σ_y and σ_m

10.40 The division of the data into two sets, each containing the same number of points, is equivalent to the two expressions

$$\sum_{i=1}^{i=n} \{y_i - \bar{y}_1 - m(x_i - \bar{x}_1)\} = 0 \qquad \qquad 10.40(1)$$

and

$$\sum_{i=n+1}^{i=2n} \{y_i - \bar{y}_2 - m(x_i - \bar{x}_2)\} = 0 \qquad \qquad 10.40(2)$$

This formulation suggests that the *range* $\delta_{1i} = y_i - \bar{y}_1 - m(x_i - \bar{x}_i)$ may be used as a measure of the standard error in y_1, that is

$$\sigma_{y_1} \simeq (\delta_{i\,\text{max}} - \delta_{i\,\text{min}})/n \qquad 10.40(3)$$

and similarly for σ_{y_2}. Thus we may take the value of σ_y as

$$\sigma_y = \tfrac{1}{2}(\sigma_{y_1}^2 + \sigma_{y_2}^2)^{1/2} \qquad 10.40(4)$$

From 10.39(1) and using 8.04(4), with $s_F \equiv \sigma_{\bar{m}}$, $s_1 \equiv \sigma_{\bar{y}_1}$, $s_2 \equiv \sigma_{y_2}$, $F \equiv m$, $x_1 \equiv \bar{y}_1$ and $x_2 \equiv \bar{y}_2$, it immediately follows that

$$\sigma_m = \frac{\sqrt{(\sigma_{y_1}^2 + \sigma_{y_2}^2)}}{\bar{x}_2 - \bar{x}_1} \qquad 10.40(5)$$

since the xs are assumed to be without error and thus $\bar{x}_1$ and $\bar{x}_2$ are assumed constant.

10.41 If there is an odd number of readings, say $2n + 1$, the first set should contain n readings and the second set $n + 1$. $\sigma_{\bar{y}_1}$ is given by 10.40(3) as before, whilst $\sigma_{\bar{y}_2}$ is now given by

$$\sigma_{y_2} = (\delta_{i\,\text{max}} - \delta_{i\,\text{min}})/(n + 1) \qquad 10.41(1)$$

Also,

$$\bar{x}_1 = \sum_{i=1}^{i=n} \frac{x_i}{n} \qquad \text{and} \qquad \bar{y} = \sum_{i=1}^{i=n} \frac{y_i}{n} \qquad 10.41(2)$$

but

$$\bar{x}_2 = \sum_{i=n+1}^{i=2n+1} \frac{x_i}{n} \qquad \text{and} \qquad \bar{y} = \sum_{i=n+1}^{i=2n+1} \frac{y_i}{n} \qquad 10.41(3)$$

and 10.39(3) becomes

$$\bar{x} = \frac{n\bar{x}_1 + (n + 1)\bar{x}_2}{2n + 1} \qquad \text{and} \qquad \bar{y} = \frac{n\bar{y}_1 + (n + 1)\bar{y}_2}{2n + 1} \qquad 10.41(4)$$

10.42 The method† is sometimes attributed to J H Awbery, but according to H R Nettleton the method was discovered by Tobias Mayer in 1748 and rediscovered by N R Campbell in 1920, who called it the method of zero sum in view of the identities 10.40(1) and (2).

10.43 If desired, a slightly better approximation for σ_{y_1} and σ_{y_2} can be obtained from the extreme range data given in Table XVII, Appendix I, with $m = 1$ (see also paragraph 6.47).

† The method was communicated to the author by Dr J C E Jennings, who was also responsible for the method given for the estimation of σ_y and σ_m.

Correlation (Linear)

10.44 Referring to equation 10.29(3), the line of regression of y on x, this can be written

$$\frac{y - \bar{y}'}{\sqrt{(\sum \tau_i^2)}} = \frac{\sum \varepsilon_i \tau_i}{\sqrt{(\sum \varepsilon_i^2 \sum \tau_i^2)}} \cdot \frac{(x - \bar{x})}{\sqrt{(\sum \varepsilon_i^2)}} \qquad 10.44(1)$$

Writing

$$\frac{\sum \tau_i^2}{n} = \sigma_y^2 \qquad \text{and} \qquad \frac{\sum \varepsilon_i^2}{n} = \sigma_x^2 \qquad 10.44(2)$$

and

$$r = \frac{\sum \varepsilon_i \tau_i}{\sqrt{(\sum \varepsilon_i^2 \sum \tau_i^2)}} \qquad 10.44(3)$$

we have

$$\frac{y - \bar{y}'}{\sigma_y} = \frac{r(x - \bar{x})}{\sigma_x} \qquad 10.44(4)$$

where r is known as the *coefficient of correlation*. The weighted value of r is given by

$$r_w = \frac{\sum w_i \varepsilon_i \tau_i}{(\sum w_i \varepsilon_i^2 \sum w_i \tau_i^2)^{1/2}} \qquad 10.44(5)$$

whilst the weighted values of σ_x and σ_y are given by

$$\sigma_{xw}^2 = \frac{\sum w_i \varepsilon_i^2}{\sum w_i} \qquad 10.44(6)$$

$$\sigma_{yw}^2 = \frac{\sum w_i \tau_i^2}{\sum w_i} \qquad 10.44(7)$$

The best weighted estimates of these quantities are given by

$$s_{xw}^2 = \frac{\sum w_i \varepsilon_i^2}{\sum w_i} \cdot \frac{n}{(n - 2)} \qquad 10.44(8)$$

$$s_{yw}^2 = \frac{\sum w_i \tau_i^2}{\sum w_i} \cdot \frac{n}{(n - 2)} \qquad 10.44(9)$$

Note that

$$\sigma_y \not\equiv \sigma_{yL}$$

The weighted line of regression is given by

$$\frac{y - \bar{y}'_w}{\sigma_{yw}} = \frac{r_w(x - \bar{x}_w)}{\sigma_{xw}}$$

where $\bar{y}'_w$ and $\bar{x}_w$ are given respectively by 10.30(4) and 10.30(3). Equation 10.44(4) was obtained on the assumption that the xs were measured without uncertainty and that all the uncertainties occurred in the y values.

10.45 If we assume that all the uncertainties occur in the x and that the y are measured without uncertainty, then it is easily shown that the line obtained is given by

$$\frac{x - \bar{x}}{\sigma_x} = \frac{r(y - \bar{y})}{\sigma_y} \qquad\qquad 10.45(1)$$

This line represents the line of *regression* of x on y. When x changes by σ_x in 10.44(4), y/σ_y changes by r. Similarly in the line represented by 10.45(1), when y changes by σ_y, x/σ_x changes by r. Thus r is common to both the lines, each of which represent a hypothetical population, and is called the coefficient of correlation; it is taken as a measure of the extent to which the sets of numbers x and y are dependent on one another.

10.46 If the two lines of regression are coincident then $r = 1$, whilst if they are at right angles $r = 0$, that is $x = \bar{x}'$ and $y = \bar{y}$ and x and y are independent of one another.

10.47 Often it is correct to assume that only one of the variables involved is uncertain, in which case the best fit curve is given by equations 10.44(4) or 10.45(1), depending on which variable is assumed to be uncertain.

10.48 Let us now consider the general case in which two variables x and y are linearly related but the uncertainties which occur in each occur independently. Also let the standard deviation of the x uncertainties be σ_{xl} and that of the y errors be σ_{yl}. Thus the probability of an uncertainty $X_r - x_r$ is proportional to

$$\frac{e^{-(X_i - x_i)^2/2\sigma_{xl}^2}}{\sigma_{xl}\sqrt{(2\pi)}}$$

whilst the probability of an uncertainty $Y_i - y_i$ is proportional to

$$\frac{e^{-(Y_i - y_i)^2/2\sigma_{yl}^2}}{\sigma_{yl}\sqrt{(2\pi)}}$$

where (X_i, Y_i) is the point on the straight line to which (x_i, y_i) is an empirical approximation. If we have n such points, the probability of obtaining the total set of measurements is proportional to

$$\frac{\exp\left[-\sum_1^n \left\{(X_i - x_i)^2/2\sigma_{xl}^2 + (Y_i - y_i)^2/2\sigma_{yl}^2\right\}\right]}{(\sigma_{yl}\sigma_{xl}2\pi)^n} \qquad\qquad 10.48(1)$$

10.49 Using the method of maximum likelihood, the expression given by 10.48(1) will be a maximum when

$$\sum_1^n \{(X_i - x_i)^2/2\sigma_{xl}^2 + (Y_i - y_i)^2/2\sigma_{yl}^2\} \qquad 10.49(1)$$

is a minimum. Let us now transform the co-ordinates and write

and

$$\left. \begin{array}{cc} X_i = \sigma_{xl} X_i' & x_i = \sigma_{xl} x_i' \\ \\ Y_i = \sigma_{yl} Y_i' & y_i = \sigma_{yl} y_i' \end{array} \right\} \qquad 10.49(2)$$

Substituting in 10.49(1) we get

$$\tfrac{1}{2}\sum_1^n \{(X_i' - x_i')^2 + (Y_i' - y_i')^2\} \qquad 10.49(3)$$

which is to be a minimum. Now 10.49(3) represents the square of the distance of the point (x_i', y_i') from the corresponding point (X_i', Y_i') on the line

$$x' \cos \alpha + y' \sin \alpha = p \qquad 10.49(4)$$

Now unless (X_i', Y_i') is the foot of the perpendicular from (x_i', y_i') on to 10.49(4) the given expression 10.49(3) will not attain its least value. Now the length of the perpendicular from (x_i', y_i') on to 10.49(4) is equal to

$$x_i' \cos \alpha + y_i' \sin \alpha - p \qquad 10.49(5)$$

Thus we require that the sum of squares of all these perpendiculars shall be a minimum. Thus we require that

$$\sum_1^n (x_i' \cos \alpha + y_i' \sin \alpha - p)^2 \qquad 10.49(6)$$

shall be a minimum.

10.50 Differentiating with respect to p and α we have

$$\sum_1^n (x_i' \cos \alpha + y_i' \sin \alpha - p) = 0 \qquad 10.50(1)$$

and

$$\sum_1^n (x_i' \cos \alpha + y_i' \sin \alpha - p)(x_i' \sin \alpha - y_i' \cos \alpha) = 0 \qquad 10.50(2)$$

Equation 10.50(1) may be written as

$$\sum_1^n \frac{x_i'}{n} \cos \alpha + \sum_1^n \frac{y_i'}{n} \sin \alpha - p = 0 \qquad 10.50(3)$$

which shows that the mean position $(\bar{x}', \bar{y}')$ of the points (x_i', y_i') lies on 10.49(4) where

$$\bar{x}' = \sum_1^n \frac{x_i'}{n} \qquad \text{and} \qquad \bar{y}' = \sum_1^n \frac{y_i'}{n} \qquad \qquad 10.50(4)$$

If we write

$$\left. \begin{aligned} x_i' &= \bar{x}' + \varepsilon_i' \\[2ex] y_i' &= \bar{y} + \tau_i' \end{aligned} \right\} \qquad \qquad 10.50(5)$$

and

then

$$\sum_1^n x_i' = n\bar{x}_i' + \sum \varepsilon_i' \qquad \qquad 10.50(6)$$

and

$$\sum_1^n \varepsilon_i' = 0 \qquad \qquad 10.50(7)$$

since

$$\sum_1^n x_i' = n\bar{x}'$$

Similarly

$$\sum \tau_i' = 0 \qquad \qquad 10.50(8)$$

10.51   Substituting in equations 10.50(1) and (2) for x_i' and y_i' from 10.50(5) we have respectively

$$\bar{x}' \cos \alpha + \bar{y}' \sin \alpha - p = 0 \qquad \qquad 10.51(1)$$

and

$$\sum_1^n (\varepsilon_i' \cos \alpha + \tau_i' \sin \alpha + \bar{x}' \cos \alpha + \bar{y}' \sin \alpha - p)$$

$$\times (\varepsilon_i' \sin \alpha - \tau_i' \cos \alpha + \bar{x}' \sin \alpha - \bar{y}' \cos \alpha) = 0$$

or

$$\sum_1^n (\varepsilon_i' \cos \alpha + \tau_i' \sin \alpha)(\varepsilon_i' \sin \alpha - \tau_i' \cos \alpha + \bar{x}' \sin \alpha - \bar{y}' \cos \alpha) = 0$$

and finally

$$(\cos^2 \alpha - \sin^2 \alpha) \sum_1^n \varepsilon_i' \tau_i' = \sin \alpha \cos \alpha \sum_1^n (\varepsilon_i'^2 - \tau_i'^2)$$

whence

$$\tan 2\alpha = \frac{2\sum_1^n \varepsilon_i' \tau_i'}{\sum_1^n (\varepsilon_i'^2 - \tau_i'^2)} \qquad 10.51(2)$$

Thus the required line is given by

$$(x' - \bar{x}') \cos \alpha + (y' - \bar{y}') \sin \alpha = 0 \qquad 10.51(3)$$

obtained by substituting 10.51(1) in 10.49(5) for p; α is given by 10.51(2). Using 10.51(3) we have

$$\tan \alpha = -\frac{(x' - \bar{x}')}{(y' - \bar{y}')}$$

Now

$$\tan 2\alpha = \frac{2 \tan \alpha}{1 - \tan^2 \alpha} = \frac{2(x' - \bar{x}')(y' - \bar{y}')}{(x' - \bar{x}')^2 - (y' - \bar{y}')^2}$$

$$= \frac{2\sum_1^n \varepsilon_i' \tau_i'}{\sum_1^n (\varepsilon_i'^2 - \tau_i'^2)} \qquad 10.51(4)$$

Let us now revert to our original system of co-ordinates. Thus we write

$$\left.\begin{array}{l} x = x'\sigma_{xl} \\ y = y'\sigma_{yl} \\ \bar{x} = \bar{x}'\sigma_{xl} \\ \bar{y} = \bar{y}'\sigma_{yl} \\ \varepsilon_i = \varepsilon_i'\sigma_{xl} \\ \tau_i = \tau_i'\sigma_{yl} \end{array}\right\} \qquad 10.51(5)$$

Thus substituting in 10.51(2) for the dashed co-ordinates we have

$$\frac{(x - \bar{x})(y - \bar{y})}{\sigma_{yl}^2(x - \bar{x})^2 - \sigma_{xl}(y - \bar{y})^2} = \frac{\sum_1^n \varepsilon_i \tau_i}{\sum_1^n (\sigma_{yl}^2 \varepsilon_i^2 - \sigma_{xl}^2 \tau_i^2)} \qquad 10.51(6)$$

10.52 Substituting for $\sum_1^n \varepsilon_i \tau_i$ from 10.44(3) and using 10.44(2), 10.51(6) can be written as

$$\frac{(x - \bar{x})(y - \bar{y})}{\sigma_{yl}^2(x - \bar{x})^2 - \sigma_{xl}^2(y - \bar{y})^2} = \frac{r\sigma_x \sigma_y}{(\sigma_x \sigma_{yl} - \sigma_y \sigma_{xl})} \qquad 10.52(1)$$

where

$$r = \frac{\sum_1^n \varepsilon_i \tau_i}{\sqrt{(\sum \varepsilon_i^2 \tau_i^2)}} \qquad \qquad 10.44(3)$$

$$\left.\begin{aligned} \sigma_x^2 &= \sum_1^n \frac{(x_i - \bar{x})^2}{n} = \sum \frac{\varepsilon_i^2}{n} \\ \sigma_y^2 &= \sum_1^n \frac{(y_i - \bar{y})^2}{n} = \sum \frac{\tau_i^2}{n} \end{aligned}\right\} \qquad 10.44(2)$$

$$\bar{x} = \sum_1^n \frac{x_i}{n} \qquad \qquad 10.28(1)$$

$$\bar{y} = \sum_1^n \frac{y_i}{n} \qquad \qquad 10.28(2)$$

Now 10.52(1) represents a pair of lines, one of which gives a maximum value for the correlation between x and y and the other a minimum value.
10.53 This is easily shown by writing $x = x' + \bar{x}$ and $y = y' + \bar{y}$, and substituting in 10.52(1), giving

$$x'^2 k \sigma_{yl}^2 - x'y' - k\sigma_{xl}^2 y'^2 = 0 \qquad \qquad 10.53(1)$$

where

$$k = \frac{r\sigma_x \sigma_y}{(\sigma_x \sigma_{yl})^2 - (\sigma_y \sigma_{xl})^2} \qquad \qquad 10.53(2)$$

If 10.53(1) represents two straight lines, then it must have two real roots and thus

$$y'^2 + 4k^2 \sigma_{yl}^2 \sigma_{xl}^2 y'^2 \geqslant 0$$

which is manifestly true if all the quantities involved are real.
10.54 Solving 10.53(1) for x' we have

$$x' = y' \frac{\{1 \pm (1 + 4k^2 \sigma_{xl}^2 \sigma_{yl}^2)^{1/2}\}}{2k\sigma_{yl}^2} \qquad \qquad 10.54(1)$$

If $\sigma_{xl} = 0$ then $\sigma_{yl} = \sigma_{yL}$, that is no errors in the values of x, and 10.54(1) reduces to

$$x' = \frac{y'}{k\sigma_{yl}^2} \qquad \text{or} \qquad x' = 0$$

Thus

$$y' = x'k\sigma_{yl}^2 \qquad\qquad 10.54(2)$$

which on putting $\sigma_{xl} = 0$ in $k\,(10.53(2))$ yields $y - \bar{y} = (x - \bar{x})r(\sigma_y)/\sigma_x$ giving as one might expect equation 10.44(4).

10.55   The case of $\sigma_{yl} = 0$, that is $\sigma_{xl} = \sigma_{xL}$, is slightly more complicated. Put $\sigma_{yl} = \delta$, where δ is a small quantity; then 10.54(1) becomes

$$x' = y' \frac{\{1 \pm (1 + 4k^2\sigma_{xl}^2\delta^2)^{1/2}\}}{2k\delta^2}$$

$$= y' \frac{\{1 \pm (1 + 2k^2\sigma_{xl}^2\delta^2)\ldots\}}{2k\delta^2}$$

on expanding by the binomial theorem and neglecting terms higher than δ^2. Thus

$$x' = y' \frac{\{1 + k^2\sigma_{xl}^2\delta^2\}}{k\delta^2}$$

is one solution which gives $y' = 0$ or $y = \bar{y}$ when $\delta \to 0$. Alternatively

$$x' = y'(-k\sigma_{xl}^2) \qquad\qquad 10.55(1)$$

is the other solution. Substituting for k with $\sigma_{yl} = 0$ we get

$$y' = \frac{-x'}{k\sigma_{xl}^2} = \frac{-x'}{\sigma_{xl}^2(-r\sigma_x/\sigma_y\sigma_{xl}^2)}$$

$$= \frac{-x'\sigma_y}{r\sigma_x}$$

or

$$y - \bar{y} = \frac{(x - \bar{x})\sigma_y}{r\sigma_x}$$

which is equation 10.45(1).

10.56   Thus $\sigma_{xl} = 0$ yields the line of regression of y on x and $\sigma_{yl} = 0$ yields the line of regression of x on y. When σ_{xl} and σ_{yl} are both finite, the best fit line lies between the two lines of regression 10.44(4) and 10.45(1). If the standard deviations of x and y from the mean line are σ_{xl} and σ_{yl} then the best fit line is given by

$$x' = y' \frac{\{1 + (1 + 4k^2\sigma_{xl}^2\sigma_{yl}^2)^{1/2}\}}{2k\sigma_{yl}^2} \qquad\qquad 10.56(1)$$

or

$$x' = y' \frac{\{1 + (1 + 4k'^2 V^2)^{1/2}\}}{2k'} \qquad\qquad 10.56(2)$$

where

$$V = \frac{\sigma_{xl}}{\sigma_{yl}}$$

10.56(3)

and

$$k' = r\sigma_x\sigma_y/(\sigma_x^2 - V^2\sigma_y^2) = k\sigma_{yl}^2$$

10.56(4)

and k is given by 10.53(2). This cannot usually be evaluated because the ratio V is required. All that can be said is that the best line lies between 10.44(4) and 10.45(1).

10.57 If it is known that $\sigma_{yl} = \sigma_{xl}$ then 10.52(1) reduces to

$$\frac{(x - \bar{x})(y - \bar{y})}{(x - \bar{x})^2 - (y - \bar{y})^2} = \frac{r\sigma_x\sigma_y}{\sigma_x^2 - \sigma_y^2}$$

10.57(1)

The values of σ_{yl} and σ_{xl} can of course be found if repeated readings at one point x, y are taken, and σ_{xl} and σ_{yl} found from the corresponding spreads of x and y. Note that all the foregoing is based on the assumption that σ_{xl} and σ_{yl} are independent of one another.

Significance of r

10.58 Let us consider the following case which is an example of correlation. Suppose two riflemen, adjacent to one another, are firing at targets in a strong wind. The wind will be supposed to affect the shooting of the two men in much the same way, so that we may expect a certain amount of correlation between their records.

10.59 Let x denote that part of the deviation of the first man's bullet from the point aimed at, which is due to causes affecting him alone, that is all causes except the wind, and similarly let y denote that part of the second man's deviation which is due to causes affecting him only. Let $x + az$ denote the total deviation of the first man's bullet, and $y + bz$ the total deviation of the second man's bullet, where z is due to the wind. x, y and z are assumed to be independent of one another, and we assume that each occurs according to the normal law of frequency.

10.60 Thus the probability that x lies between x and $x + dx$ is

$$\frac{\exp(-x^2/2\sigma_x^2)\,dx}{\sigma_x\sqrt{(2\pi)}}$$

10.60(1)

and that y lies between y and $y + dy$ is

$$\frac{\exp(-y^2/2\sigma_y^2)\,dy}{\sigma_y\sqrt{(2\pi)}}$$

10.60(2)

and that z lies between z and $z + dz$ is

$$\frac{\exp(-z^2/2\sigma_z^2)\,dz}{\sigma_z\sqrt{(2\pi)}} \qquad 10.60(3)$$

10.61 The probability of these ranges for x, y and z occurring simultaneously is thus given by the product of 10.60(1), (2) and (3), namely

$$dP_{x,y,z} = \frac{1}{\sigma_x\sigma_y\sigma_z(2\pi)^{3/2}} \exp\left(-\frac{x^2}{2\sigma_x^2} - \frac{y^2}{2\sigma_y^2} - \frac{z^2}{2\sigma_z^2}\right) dx\,dy\,dz \qquad 10.61(1)$$

Now we require that $u = x + az$ should satisfy the inequality

$$u + du \geqslant x + az \geqslant u$$

whilst at the same time $v = x + bz$ satisfies the inequality

$$v + dv \geqslant y + bz > v$$

In order to satisfy these two inequalities, we must integrate 10.61(1) with respect to x between $u - az$ and $u + du - az$, with respect to y between $v - bz$ and $v + dv - bz$ and with respect to z from $-\infty$ to $+\infty$, since whatever values are chosen for x and y there will always exist a range of values for z which will satisfy the inequalities whilst u and v remain constant. The probability of u having a value between u and $u + du$ whilst v has a value between v and $v + dv$ is thus given by

$$dP_{u,v} = \frac{1}{\sigma_x\sigma_y\sigma_z(2\pi)^{3/2}} \int_{-\infty}^{\infty} e^{-z^2/2\sigma_z^2}\,dz \int_{u-az}^{u+du-az} e^{-x^2/2\sigma_x^2}\,dx \int_{v-bz}^{v+dv-bz} e^{-y^2/2\sigma_y^2}\,dy$$

$$= \frac{du\,dv}{\sigma_x\sigma_y\sigma_z(2\pi)^{3/2}} \int_{-\infty}^{\infty} \exp\left\{-\frac{z^2}{2\sigma_z^2} - \frac{(u-az)^2}{2\sigma_x^2} - \frac{(v-bz)^2}{2\sigma_y^2}\right\} dz \qquad 10.61(2)$$

10.62 This can be rewritten as

$$dP_{u,v} = \frac{du\,dv}{\sigma_x\sigma_y\sigma_z(2\pi)^{3/2}} \exp\left[-\frac{u^2}{2\sigma_x^2} - \frac{v^2}{2\sigma_y^2} + \frac{\{(au/2\sigma_x^2) + (by/2\sigma_y^2)\}^2}{(a^2/2\sigma_x^2) + (b^2/2\sigma_y^2) + (1/2\sigma_z^2)}\right]$$

$$\times \int_{-\infty}^{\infty} \exp\left[-\left(\frac{a^2}{2\sigma_x^2} + \frac{b^2}{2\sigma_y^2} + \frac{1}{2\sigma_z^2}\right)\left\{z - \frac{(au/2\sigma_x^2) + (by/2\sigma_y^2)}{(a^2/2\sigma_x^2) + (b^2/2\sigma_y^2) + (1/2\sigma_z^2)}\right\}^2\right] dz$$

Putting

$$\left(\frac{a^2}{2\sigma_x^2} + \frac{b^2}{2\sigma_y^2} + \frac{1}{2\sigma_z^2}\right)^{1/2}\left\{z - \frac{(au/2\sigma_x^2) + (bv/2\sigma_y^2)}{(a^2/2\sigma_x^2) + (b^2/2\sigma_y^2) + (1/2\sigma_z^2)}\right\} = \alpha$$

we have

$$dz = d\alpha \bigg/ \left(\frac{a^2}{2\sigma_x^2} + \frac{b^2}{2\sigma_y^2} + \frac{1}{2\sigma_z^2}\right)$$

The integral thus reduces to

$$\int_{-\infty}^{\infty} e^{-\alpha^2} \, d\alpha = \sqrt{\pi}$$

(see paragraphs 2.09 and 2.10) and the required probability is given by

$$dP_{u,v} = \exp\left\{ -u^2/2\sigma_x^2 - v^2/2\sigma_y^2 + \frac{\{(au/2\sigma_x^2 + bv/2\sigma_y^2)\}^2}{(a^2/2\sigma_x^2 + b^2/2\sigma_y^2 + 1/2\sigma_z^2)} \right\} du \, dv$$

$$\times \left[\pi\sigma_x\sigma_y\sigma_z 2^{3/2}\{(a^2/2\sigma_x^2) + (b^2/2\sigma_y^2) + (1/2\sigma_z^2)\} \right]^{-1} \qquad 10.62(1)$$

This is seen to be of the form

$$\varphi(uv) \, du \, dv = ce^{-p^2u^2 - q^2v^2 + 2suv} \, du \, dv \qquad 10.62(2)$$

The constant s involves the product of a and b, and thus represents the common influence acting on the two riflemen. If either a or b is zero, then s is zero, and there is no correlation and 10.62(2) then consists of the product of two independent factors $e^{-p^2u^2}$ and $e^{-q^2v^2}$.

10.63 The expression given by 10.62(2) is known as the normal law of frequency for two variables, and may be regarded as the extension to two variables u and v of the normal law for one variable u, that is

$$\frac{e^{-u^2/2\sigma^2}}{\sigma\sqrt{(2\pi)}} \, du$$

10.64 The constant c in the expression 10.62(2) may be determined from the condition that

$$\int_{-\infty}^{\infty}\int_{-\infty}^{\infty} \varphi(uv) \, du \, dv = 1 \qquad 10.64(1)$$

Writing $p^2u^2 - 2suv$ as

$$\left(pu - \frac{sv}{p} \right)^2 - s^2v^2/p^2$$

the integral 10.64(1) can be written as

$$c\int_{-\infty}^{\infty} e^{-q^2v^2 + s^2v^2/p^2} \, dv \int_{-\infty}^{\infty} e^{-\{pu - sv/p\}^2} \, du = 1$$

Putting

$$pu - \frac{sv}{p} = \alpha$$

we have $du = d\alpha/p$ and the integral becomes

$$c \int_{-\infty}^{\infty} e^{-q^2v^2 + s^2v^2/p^2} \frac{1}{p} \int_{-\infty}^{\infty} e^{-\alpha^2} d\alpha$$

$$= c \int_{-\infty}^{\infty} e^{-q^2v^2 + s^2v^2/p^2} dv \frac{\sqrt{\pi}}{p} = 1$$

(see paragraphs 2.09 and 2.10). Putting $v(q^2 - s^2/p^2)^{1/2} = \beta$ we have

$$dv = \frac{d\beta}{(q^2 - s^2/p^2)^{1/2}} = \frac{pd\beta}{(p^2q^2 - s^2)^{1/2}}$$

giving the final integral as

$$\frac{c\sqrt{\pi}}{(p^2q^2 - s^2)^{1/2}} \int_{-\infty}^{\infty} e^{-\beta^2} d\beta = 1$$

or

$$c = \frac{(p^2q^2 - s^2)^{1/2}}{\pi} \qquad \text{10.64(2)}$$

Thus the expression for the normal law of frequency for two variables becomes

$$\varphi(uv) \, du \, dv = \frac{(p^2q^2 - s^2)^{1/2}}{\pi} e^{-p^2u^2 - q^2v^2 + 2suv} \, du \, dv \qquad \text{10.64(3)}$$

Determination of the Constants *p*, *q* and *s* in a Normal Frequency Distribution with Two Variables (Determination from Measured Values of the Variables *u* and *v*)

10.65   Let u_p and v_p be values of u and v which are the most probable. Let the probability that the attribute A has a value between u and $u + du$, whilst the second attribute B has a value between v and $v + dv$, be $\varphi(uv) \, du \, dv$ where, by 10.64(3)

$$\varphi(uv) = \frac{(p^2q^2 - s^2)^{1/2}}{\pi} e^{-p^2(u - u_p)^2 - q^2(v - v_p)^2 + 2s(u - u_p)(v - v_p)} \qquad \text{10.65(1)}$$

10.66   Let it be required to determine the most probable values of p, q, s, u_p and v_p from a set of observations. Let the measures of the observed set of observations be $(u_1, v_1), (u_2, v_2), \ldots, (u_n, v_n)$. Then the *a priori* probability that the observations will yield these measures is

$$\delta u^n \delta v^n (p^2 q^2 - s^2)^{n/2} \exp\left\{ -p^2 \sum_1^n (u_i - u_p)^2 - q^2 \sum_1^n (v_i - v_p)^2 \right.$$
$$\left. + 2s \sum_1^n (u_i - u_p)(v_i - v_p) \right\} \qquad 10.66(1)$$

Using the principle of maximum likelihood, the most probable values of p, q, s, u_p and v_p are those which make the expression 10.65(1) a maximum when $(u_1, v_1), (u_2, v_2), \ldots, (u_n, v_n)$ are given.

10.67 Taking the logarithm of 10.66(1) we have

$$G = \tfrac{1}{2} n \log(p^2 q^2 - s^2) - p^2 \sum (u_i - u_p)^2 - q^2 \sum (v_i - v_p)^2$$
$$+ 2s \sum (u_i - u_p)(v_i - v_p) \qquad 10.67(1)$$

and this must be a minimum for 10.66(1) to be a maximum, and therefore

$$\frac{\partial G}{\partial u_p} = 0, \quad \frac{\partial G}{\partial v_p} = 0, \quad \frac{\partial G}{\partial p} = 0, \quad \frac{\partial G}{\partial q} = 0, \quad \frac{\partial G}{\partial s} = 0 \qquad 10.67(2)$$

The first two equations yield

$$2p^2 \sum (u_i - u_p) - 2s \sum (v_i - v_p) = 0 \qquad 10.67(3)$$

and

$$2q^2 \sum (v_i - v_p) - 2s \sum (u_i - u_p) = 0 \qquad 10.67(4)$$

These give $p^2 q^2 = s^2$ as one solution but this is not generally true. Thus in order to satisfy 10.67(3) and 10.67(4) in all cases

$$u_p = \sum_1^n \frac{u_i}{n} \qquad 10.67(5)$$

and

$$v_p = \sum_1^n \frac{v_i}{n} \qquad 10.67(6)$$

10.68 The other three equations are

$$\frac{npq^2}{p^2 q^2 - s^2} - 2p \sum (u_i - u_p)^2 = 0 \qquad 10.68(1)$$

$$\frac{np^2 q}{p^2 q^2 - s^2} - 2q \sum (v_i - v_p)^2 = 0 \qquad 10.68(2)$$

$$\frac{-ns^2}{p^2 q^2 - s^2} + 2 \sum (u_i - u_p)(v_i - v_p) = 0 \qquad 10.68(3)$$

Put

$$\frac{1}{n}\sum(u_i - u_p)^2 = \sigma_u^2 \qquad\qquad 10.68(4)$$

$$\frac{1}{n}\sum(v_i - v_p)^2 = \sigma_v^2 \qquad\qquad 10.68(5)$$

and

$$\frac{1}{n\sigma_u\sigma_v}\sum(u_i - u_p)(v_i - v_p) = r \qquad\qquad 10.68(6)$$

Thus σ_u, σ_v and r can be calculated from the observed values of u and v. The three equations 10.68(1), (2) and (3) can now be written as

$$\frac{q^2}{2\sigma_u^2} = \frac{p^2}{2\sigma_v^2} = \frac{s}{2\sigma_u\sigma_v r} = p^2 q^2 - s^2 \qquad\qquad 10.68(7)$$

Thus

$$\frac{p^2 q^2}{4\sigma_u^2 \sigma_v^2} = \frac{s^2}{4\sigma_u^2 \sigma_v^2 r^2} = (p^2 q^2 - s^2)^2 \qquad\qquad 10.68(8)$$

Therefore

$$\frac{p^2 q^2 - s^2}{4\sigma_u^2 \sigma_v^2 (1 - r^2)} = (p^2 q^2 - s^2)^2$$

yielding

$$p^2 q^2 - s^2 = \frac{1}{4\sigma_u^2 \sigma_v^2 (1 - r^2)} \qquad\qquad 10.68(9)$$

Thus

$$p^2 = \frac{1}{2(1 - r^2)\sigma_u^2} \qquad\qquad 10.68(10)$$

$$q^2 = \frac{1}{2(1 - r^2)\sigma_v^2} \qquad\qquad 10.68(11)$$

and

$$s = \frac{r}{2(1 - r^2)\sigma_u\sigma_v} \qquad\qquad 10.68(12)$$

10.69 Replacing p, q and s in 10.65(1), we see that the probability of attribute A having a value between u and $u + du$, whilst attribute B has a value between

v and $v + dv$ is given by

$\varphi(uv) \, du \, dv$

$$= \exp\left[-\frac{1}{2(1 - r^2)}\left\{\frac{(u - u_p)^2}{\sigma_u^2} + \frac{(v - v_p)^2}{\sigma_v^2} - \frac{2r(u - u_p)(v - v_p)}{\sigma_u \sigma_v}\right\}\right]$$
$$\times [2\pi\sigma_u\sigma_v(1 - r^2)^{1/2}]^{-1} \qquad\qquad 10.69(1)$$

and where the constants u_p, v_p, σ_u, σ_v and r are expressed in terms of the measured quantities by the equations

$$u_p = \sum_1^n \frac{u_i}{n} \qquad\qquad 10.67(5)$$

$$v_p = \sum_1^n \frac{v_i}{n} \qquad\qquad 10.67(6)$$

$$\sigma_u^2 = \sum_1^n \frac{(u_i - u_p)^2}{n} \qquad\qquad 10.68(4)$$

$$\sigma_v^2 = \sum_1^n \frac{(v_i - v_p)^2}{n} \qquad\qquad 10.68(5)$$

and

$$r = \frac{1}{n\sigma_u\sigma_v} \sum_1^n (u_i - u_p)(v_i - v_p) \qquad\qquad 10.68(6)$$

r is known as the correlation coefficient and is identical with the r of the lines of regression (see paragraphs 10.44 et seq.).

Derivation of Line of Regression from 10.69(1)

Frequency of variables taken singly

10.70 Let us first find the probability that attribute A lies between u and $u + du$, when the attribute B can take any value. This is given by integrating 10.69(1) with respect to v from $-\infty$ to $+\infty$. Thus the required probability is

$$w_u \, du = \frac{du}{2\pi\sigma_u\sigma_v(1 - r^2)^{1/2}} \int_{-\infty}^{\infty} \exp\left[-\frac{1}{2(1 - r^2)}\left\{\frac{(u - u_p)^2}{\sigma_u^2}\right.\right.$$
$$\left.\left. + \frac{(v - v_p)^2}{\sigma_v^2} - \frac{2(u - u_p)(v - v_p)}{\sigma_u\sigma_v}\right\}\right] dv \qquad 10.70(1)$$

Writing

$$\frac{(u - u_p)^2}{2(1 - r^2)\sigma_u^2} = a \qquad \qquad 10.70(2)$$

$$\frac{-(v - v_p)(u - u_p)r}{\sigma_u \sigma_v (1 - r^2)} = bx \qquad \qquad 10.70(3)$$

and

$$\frac{(v - v_p)^2}{2(1 - r^2)\sigma_v} = cx^2 \qquad \qquad 10.70(4)$$

where $v - v_p = x$, we see that the integral is of the form

$$\int_{-\infty}^{\infty} e^{-(a + bx + cx)^2} dx$$

Now

$$(a + bx + cx)^2 = c\left[\left(x + \frac{b}{2c}\right)^2 + \frac{a}{c} - \frac{b^2}{4c^2}\right]$$

Thus the required integral is

$$e^{-\{a - (b^2/4c)\}} \int_{-\infty}^{\infty} e^{-c\{x + (b/2c)\}^2} dx$$

Putting $\sqrt{c}\{x + (b/2c)\} = y$ we have $dx = dy/\sqrt{c}$ and the integral reduces to

$$e^{-\{a - (b^2/4c)\}} \int_{-\infty}^{\infty} e^{-y^2} dy$$

$$= \frac{\sqrt{\pi}}{\sqrt{c}} e^{-\{a - (b^2/4c)\}} \qquad \qquad 10.70(5)$$

(see paragraphs 2.09 and 2.10). Now

$$a - \frac{b^2}{4c} = \frac{(u - u_p)^2}{2(1 - r^2)\sigma_u^2} - \frac{(u - u_p)^2 r^2 2(1 - r^2)\sigma_v^2}{4\sigma_u^2 \sigma_v^2 (1 - r^2)^2}$$

$$= \frac{(u - u_p)^2 (1 - r^2)}{2(1 - r^2)\sigma_u^2} = \frac{(u - u_p)^2}{2\sigma_u^2} \qquad \qquad 10.70(6)$$

and

$$\frac{1}{\sqrt{c}} = \sqrt{2}(1 - r^2)^{1/2}\sigma_v \qquad \qquad 10.70(7)$$

Thus 10.70(1) reduces to

$$w_u \, du = \frac{e^{-(u-u_p)^2/2\sigma u^2} \, du}{\sigma_u \sqrt{(2\pi)}}$$

10.70(8)

This equation shows that σ_u is the standard deviation of attribute A when attribute B is ignored. Similarly σ_v is the standard deviation for the attribute B when the attribute A is ignored.

10.71 The probability that attribute A lies between u and $u + du$ whilst the attribute B lies between $v + dv$ is given by

$$\varphi(u, v) \, du \, dv$$

10.69(1)

Let us denote by $f(uv) \, dv$ the probability that B lies between v and $v + dv$ when A is known to lie between u and $u + du$, then

$$\varphi(uv) \, du \, dv = w_u f(uv) \, du \, dv$$

10.71(1)

Substituting for $\varphi(uv)$ from 10.69(1), and w_u from 10.70(8), we obtain

$$f(u, v) = \exp\left[-\frac{1}{2(1-r^2)\sigma_v^2}\left\{ (v - v_p) - \frac{\sigma_v}{\sigma_u} r(u - u_p) \right\}^2 \right]$$
$$\times \left[\sigma_v \sqrt{(2\pi)}(1 - r^2)^{1/2} \right]^{-1}$$

10.71(2)

Thus the probability that B lies between v and $v + dv$ when A is known to lie between u and $u + du$ is

$$f(u, v) \, dv$$

10.71(3)

This is a normal frequency distribution about the mean $v_p + \sigma_v r(u - u_p)/\sigma_u$ with a standard deviation of

$$\sigma_v (1 - r^2)^{1/2}$$

10.71(4)

Thus if we take a large number of observations and find the mean v_m of all the measured values of B for which the measured value of A lies between u and $u + du$ and if we then plot v_m against u, the plotted points will lie on the straight line

$$v_m - v_p = \frac{\sigma_v}{\sigma_u} r(u - u_p)$$

10.71(5)

10.72 If B lies between v and $v + dv$, when u can take any value, then the probability of this is given by

$$w_v \, dv = \frac{e^{-(v-v_p)^2/2\sigma_v^2} \, dv}{\sigma_v \sqrt{(2\pi)}}$$

10.72(1)

This result is obtained by using the same method as that used to derive 10.70(8).

10.73 Similarly the probability $g(uv)\,dv$ that A lies between u and $u + du$ when B is known to lie between v and $v + dv$ is given by

$$g(uv)\,du = \exp\left[-\frac{1}{2(1 - r^2)\sigma_u^2}\left\{ u - u_p - \frac{\sigma_u}{\sigma_v}r(v - v_p) \right\}^2 \right] du$$
$$\times \left[\sigma_u\sqrt{(2\pi)}(1 - r^2)^{1/2} \right]^{-1} \qquad\qquad 10.73(1)$$

by comparison with 10.71(2). This is a normal frequency distribution about the mean

$$u_p + \frac{\sigma_u}{\sigma_v}r(v - v_p) \qquad\qquad 10.73(2)$$

with a standard deviation of

$$\sigma_u(1 - r^2)^{1/2} \qquad\qquad 10.73(3)$$

Similarly if a large number of observations are taken of the attribute A and the mean u_m found for which the measured values of B lie between v and $v + dv$, then if u_m is plotted against v, the plotted points will lie on the straight line

$$v - v_p = \frac{\sigma_v}{r\sigma_u}(u_m - u_p) \qquad\qquad 10.73(4)$$

It is seen that 10.71(5) and 10.73(4) are the lines of regression 10.44(4) and 10.45(1) respectively where $v \equiv y$ and $u \equiv x$.

10.74 It should be noticed that 10.71(4) and 10.73(3) are respectively the standard deviations from the lines of regression given by 10.71(5) and 10.73(4) when the error in one of the variables is considered zero.

10.75 If s_x is the estimate of σ_x and s_y is the estimate of σ_y then

$$s_{yL}^2 = s_y^2(1 - r^2) \qquad\qquad 10.75(1)$$
$$s_{xL}^2 = s_x^2(1 - r^2) \qquad\qquad 10.75(2)$$

where

$$s_y^2 = \sum_1^n \tau_i^2/(n - 2) \qquad\qquad 10.75(3)$$

$$s_x^2 = \sum_1^n \varepsilon_i^2/(n - 2) \qquad\qquad 10.75(4)$$

and where τ_i and ε_i are given respectively by 10.28(4) and 10.28(3).

10.76 Expressions 10.75(1) and (2) can be derived alternatively as follows. Consider 10.31(2), thus

$$s_{yL}^2 = \sum_1^n \frac{(\tau_i - m\varepsilon_i)^2}{(n - 2)} = \frac{\sum \tau_i^2 + m^2\sum \varepsilon_i^2 - 2m\sum \varepsilon_i\tau_i}{(n - 2)} \qquad 10.76(1)$$

Now

$$r = \frac{\sum \varepsilon_i \tau_i}{(\sum \varepsilon_i^2 \sum \tau_i^2)^{1/2}} \qquad 10.44(4)$$

and

$$m = \frac{\sum \varepsilon_i \tau_i}{\sum \varepsilon_i^2} \qquad 10.29(1)$$

thus

$$\frac{r}{m} = \left(\frac{\sum \varepsilon_i^2}{\sum \tau_i^2}\right)^{1/2} = \frac{s_x}{s_y} = \frac{\sigma_x}{\sigma_y} \qquad 10.76(2)$$

Substituting for m in terms of r from 10.76(2) in 10.76(1) we have

$$s_{yL}^2 = \frac{\{\sum \tau_i^2 + r_i^2 \sum \tau_i^2 - 2(\sum \varepsilon_i \tau_i)^2 / \sum \varepsilon_i^2\}}{(n-2)}$$

$$= \frac{(\sum \tau_i^2 + r^2 \sum \tau_i^2 - 2r^2 \sum \tau_i^2)}{(n-2)}$$

using 10.44(3)

$$= \frac{\sum \tau_i^2(1 - r^2)}{n-2}$$

$$= s_y^2(1 - r^2) \qquad \text{from } 10.75(3) \qquad 10.76(3)$$

from 10.75(3). Similarly

$$s_{xL}^2 = s_x^2(1 - r^2) \qquad 10.76(4)$$

The weighted standard deviations s_{yLw}^2 and s_{xLw}^2 are given by

$$s_{yLw}^2 = \frac{\sum_1^n w_i \tau_i^2(1 - r_w^2)}{\sum w_i} \cdot \frac{n}{(n-2)} \qquad 10.76(5)$$

and

$$s_{xLw}^2 = \frac{\sum_1^n w_i \varepsilon_i^2(1 - r_w^2)}{\sum w_i} \cdot \frac{n}{(n-2)} \qquad 10.76(6)$$

Similarly

$$s_{\bar{y}L}^2 = s_{yL}^2/n, \qquad s_{\bar{x}L}^2 = s_{xL}^2/n, \text{ etc.} \qquad 10.76(7)$$

The weighted values of s_{xw}^2 and s_{yw}^2 are given by

$$s_{xw}^2 = \frac{\sum_1^n w_i \varepsilon_i^2}{\sum w_i} \cdot \frac{n}{(n-2)} \qquad 10.76(8)$$

and

$$s_{yw}^2 = \frac{\sum_1^n w_i \tau_i^2}{\sum w_i} \cdot \frac{n}{(n-2)}$$

10.76(9)

Analogously to 10.76(3) and 10.76(4)

$$s_{yLw}^2 = s_{yw}^2(1 - r_w^2)$$

10.76(10)

and

$$s_{xLw} = s_{xw}^2(1 - r_w^2)$$

10.76(11)

Note that

$$s_{\bar{y}Lw}^2 = \frac{s_{yw}^2(1 - r_w^2)}{n}$$

10.76(12)

and

$$s_{\bar{x}Lw} = \frac{s_{xw}^2(1 - r_w^2)}{n}$$

10.76(13)

10.77 It should be noted that if $\sigma_{xl} = 0$, then $\sigma_{yl} = \sigma_{yL}$ and the line of regression is

$$y - \bar{y} = r\frac{\sigma_y}{\sigma_x}(x - \bar{x})$$

10.44(4)

If $\sigma_{yl} = 0$ then $\sigma_{xl} = \sigma_{xL}$ and the line of regression is

$$y - \bar{y} = \frac{1}{r}\frac{\sigma_y}{\sigma_x}(x - \bar{x})$$

10.45(1)

If $\sigma_{yl} \neq 0$ and $\sigma_{xl} \neq 0$ then the mean line is given by 10.56(1), (2), (3) and (4) or by solving 10.52(1), which yields two solutions, a best line and worst one.

Note on the Value of *r*

10.78 Now

$$r = \sum \varepsilon_i \tau_i / (\sum \varepsilon_i^2 \sum \tau_i^2)^{1/2}$$

10.44(3)

but nothing has so far been said about the values that *r* can assume. From a perusal of 10.69(1) it would seem likely that *r* could reasonably go from 0 to ± 1, but the limits of ± 1 are not obvious. The proof that the limits of *r* are plus or minus unity is as follows.

10.79 Consider

$$(a_1 b_2 - a_2 b_1)^2 = a_1^2 b_2^2 + a_2^2 b_1^2 - 2a_1 a_2 b_1 b_2$$

whence

$$\sum_{s=1}^{s=n} \sum_{r=1}^{r=n} (a_r b_s - a_s b_r)^2 = 2\sum\sum a_r^2 b_s^2 - 2\sum\sum a_r b_r a_s b_s \qquad (r \neq s)$$

therefore

$$\sum_1^n \sum_1^n a_r^2 b_s^2 = \sum_1^n \sum_1^n a_r b_r a_s b_s + \sum_1^n \sum_1^n (a_r b_s - a_s b_r)^2/2 \qquad 10.79(1)$$

$r \neq s$ in any term, and thus

$$\sum\sum a_r^2 b_s^2 \geqslant \sum\sum a_r b_r a_s b_s \qquad 10.79(2)$$

Now

$$\sum_1^n a_r^2 \sum_1^n b_r^2 = \sum_1^n a_r^2 b_r^2 + \sum_1^n \sum_1^n a_r^2 b_s^2 \qquad (r \neq s) \qquad 10.79(3)$$

Also

$$\left(\sum a_r b_r\right)^2 = \sum_1^n a_r^2 b_r^2 + \sum\sum a_r b_r a_s b_s \qquad 10.79(4)$$

Thus using 10.79(2) we see that

$$\sum_1^n a_r^2 \sum_1^n b_r^2 \geqslant \left(\sum_1^n a_r b_r\right)^2 \qquad 10.79(5)$$

or

$$1 \geqslant \left(\sum_1^n a_r b_r\right)^2 \Big/ \sum_1^n a_r^2 \sum_1^n b_r^2 \qquad 10.79(6)$$

Thus, comparing with 10.44(3) we see that the right-hand side of 10.79(6) is equivalent to r^2. Therefore $1 \geqslant r^2$ or $1 \geqslant r \geqslant -1$.

Non-linear Correlation and Best Fit for a Parabola

10.80 The method used to find the coefficient of linear correlation given in paragraphs 10.27 to 10.29 and 10.44 can be immediately extended to the case of parabolic correlation.

If

$$x' = \frac{x - \bar{x}}{\sigma_x} \qquad \text{and} \qquad y' = \frac{y - \bar{y}}{\sigma_y} \qquad 10.80(1)$$

where

$$\bar{x} = \sum_{1}^{n} x_i/n, \qquad \bar{y} = \sum_{1}^{n} y_i/n \qquad\qquad 10.80(2)$$

and

$$\sigma_x^2 = \sum_{1}^{n} (x_i - \bar{x})^2/n, \qquad \sigma_y^2 = \sum_{1}^{n} (y_i - \bar{y})^2/n \qquad 10.80(3)$$

then the coefficient of parabolic correlation is given by λ in the equation

$$y' = \lambda x'^2 \qquad\qquad 10.80(4)$$

where this equation represents the best fit of a parabola to the points $(x'_1, y'_1), (x'_2, y'_2), \ldots, (x'_n, y'_n)$.

10.81 λ is thus found using the method of least squares, that is it is that value of λ which makes $\sum (y'_i - \lambda x'^2_i)^2$ a minimum, and therefore differentiating with respect to λ we have

$$\sum (y'_i - \lambda x'^2_i) x'^2_i = 0 \qquad\qquad 10.81(1)$$

Therefore

$$\lambda = \frac{\sum x'^2_i y'_i}{\sum x'^4_i}$$

$$= \sigma_x^2 \frac{\sum (x_i - \bar{x})^2 (y_i - \bar{y})}{\sigma_y \sum (x_i - \bar{x})^4} \qquad\qquad 10.81(2)$$

$$= \frac{\sum (x_i - \bar{x})^2 \sum (x_i - \bar{x})^2 (y_i - \bar{y})}{\{n \sum (y_i - \bar{y})^2\}^{1/2} \sum (x_i - \bar{x})^4} \qquad 10.81(3)$$

If the readings are weighted, then the sum of the squares of the deviations from the required parabola that are to be weighted is given by $\sum w_i (y'_i - \lambda x'^2_i)^2$. Differentiating as before leads to a value for λ given by

$$\lambda_w = \frac{\sum w_i x'^2_{iw} y'_{iw}}{\sum w_i x'^4_{iw}}$$

$$= \frac{\sigma^2_{xw} \sum w_i (x_i - \bar{x}_w)^2 (y_i - \bar{y}_w)}{\sigma_{yw} \sum w_i (x_i - \bar{x}_w)^4} \qquad\qquad 10.81(4)$$

where now

$$x'_{iw} = \frac{x_i - \bar{x}_w}{\sigma_{xw}} \qquad \text{and} \qquad y'_{iw} = \frac{y_i - \bar{y}_w}{\sigma_{yw}} \qquad 10.81(5)$$

and

$$\bar{x}_w = \frac{\sum w_i x_i}{\sum w_i} \qquad 10.30(3)$$

$$\bar{y}_w = \frac{\sum w_i y_i}{\sum w_i} \qquad 10.30(4)$$

$$\sigma_{xw}^2 = \frac{\sum w_i (x_i - \bar{x}_w)^2}{\sum w_i} \qquad 10.81(6)$$

$$\sigma_{yw}^2 = \frac{\sum w_i (y_i - \bar{y}_w)^2}{\sum w_i} \qquad 10.81(7)$$

Usually estimates of s_{xw} and s_{yw} will have to be used in place of σ_{xw} and σ_{yw} and these are given, respectively, by

$$s_{xw}^2 = \frac{\sum w_i (x_i - \bar{x}_w)^2}{\sum w_i} \cdot \frac{n}{(n-3)} \dagger \qquad 10.81(8)$$

$$s_{yw}^2 = \frac{\sum w_i (y_i - \bar{y}_w)^2}{\sum w_i} \cdot \frac{n}{n-3} \qquad 10.81(9)$$

The estimated standard deviation in the mean of y_i from the mean parabola is given by

$$s_{\bar{y}Lw}^2 = \frac{\sum w_i \{ y_i - \bar{y}_w - \lambda_w (\sigma_{yw}/\sigma_{xw}^2)(x_i - \bar{x}_w)^2 \}^2 \ddagger}{(n-3) \sum w_i}$$

$$= \frac{\sum w_i \sigma_{yw}^2 (y_i' - \lambda_w x_i'^2)^2}{(n-3) \sum w_i} \qquad 10.81(10)$$

Unweighted Values

If the readings are unweighted, that is belong to the same population, then

$$s_{\bar{y}L}^2 = \frac{\sum \sigma_y^2 (y_i' - \lambda x_i'^2)^2}{n(n-3)}$$

$$= \frac{\sum \{ y_i - \bar{y} - \lambda (\sigma_y/\sigma_x^2)(x_i - \bar{x})^2 \}^2 \S}{n(n-3)} \qquad 10.81(11)$$

---

† The factor $n - 3$ is used because three constants are involved for a parabola.
‡ See 10.81(13).
§ See 10.81(14).

and the standard deviation of a reading from the line is given by

$$s_{yL}^2 = \frac{\sum \sigma_y^2 (y_i' - \lambda x_i')^2}{(n-3)}$$

$$= \frac{\sum \{y_i - \bar{y} - \lambda(\sigma_y/\sigma_x^2)(x - \bar{x})^2\}^2 \dagger}{(n-3)} \qquad 10.81(12)$$

In this case y_i', x_i' and λ are given respectively by 10.80(1), 10.81(2), and σ_x and σ_y are replaced by their estimates

$$s_x^2 = \frac{n}{n-3} \sigma_x^2 \qquad \text{and} \qquad s_y^2 = \frac{n}{n-3} \sigma_y^2$$

in 10.81(12) and in λ (10.81(2)).

It is to be noted that

$$\lambda_w \frac{\sigma_{yw}}{\sigma_{xw}^2} = \lambda_w' \frac{s_{yw}}{s_{xw}^2} = \frac{\sum w_i (x_i - \bar{x}_w)^2 (y_i - \bar{y}_w)}{\sum w_i (x_i - \bar{x}_w)^4} \qquad 10.81(13)$$

where λ_w contains σ_{xw} and σ_{yw} and λ_w' contains s_{xw} and s_{yw}. Similarly

$$\lambda \frac{\sigma_y}{\sigma_x^2} = \lambda' \frac{s_y}{s_x^2} = \frac{\sum (x_i - \bar{x})^2 (y_i - \bar{y})}{\sum (x_i - \bar{x})^4} \qquad 10.81(14)$$

where λ contains σ_x and σ_y and λ' contains s_x and s_y.

The interpretation of λ is quite different from r in the case of linear correlation.

Alternative Derivation of Best Fit for a Parabola

10.82   The general equation for a parabola with vertical axis is

$$y = a + bx + cx^2 \qquad 10.82(1)$$

If n pairs of measurements are taken of x and y, where n is greater than three, then the method of least squares gives the best fit for a parabola. Let the n pairs of measurements be $(x_1, y_1); (x_2, y_2); \ldots ; (x_n, y_n)$ and let us assume that all the errors lie in the y values. Thus the deviation of the y value from the least squares parabola value of y for any point is given by

$$\delta y_i = y_i - a - bx_i - cx_i^2 \qquad 10.82(2)$$

The least squares parabola is thus obtained by solving the three equations

$$\frac{\partial(\sum \delta y_i^2)}{\partial a} = \frac{\partial(\sum \delta y_i^2)}{\partial b} = \frac{\partial(\sum \delta y_i^2)}{\partial c} = 0 \qquad 10.82(3)$$

† See 10.81(14).

Least Squares Fit for General Function

10.83 A general function of x can be written as

$$y = f(x, a_1, a_2, a_3, \ldots, a_n) \qquad 10.83(1)$$

where $a_1, a_2, \ldots, a_n$ are constants. If the values are assumed correct the error or deviation of any point x_r, y_r from the best least squares fit curve is given by $\delta y_i = y_i - f(x_i, a_1, a_2, \ldots, a_n)$. Thus the best least squares fit curve is obtained by solving the equations

$$\frac{\partial(\sum \delta y_i^2)}{\partial a_1} = \frac{\partial(\sum \delta y_i^2)}{\partial a_2} = \ldots = \frac{\partial(\sum \delta y_i^2)}{\partial a_n} = 0 \qquad 10.83(2)$$

for $a_1, a_2, \ldots, a_n$. The determination of the precision of the constants so obtained is found in a similar way to those obtained for the straight line (see paragraphs 10.31 to 10.36).

Criterion for Choice of Best Curve or Functional Relation to Fit Given Data Points

10.84 Let the two relations to be compared be $\varphi(x, a_1, a_2, \ldots, a_p)$ and $f(x, a_1, a_2, \ldots, a_q)$ where φ has p constants and f has q constants. As before, let all the errors be assumed to be in the y values. Thus the error

$$\delta_\varphi y_i = y_i - \varphi(x_i) \qquad 10.84(1)$$

and the error

$$\delta_f y_s = y_s - f(x_s) \qquad 10.84(2)$$

The variance of the y_i from the best fit line $\varphi(x_i)$ is thus given by

$$s_{y_\varphi}^2 = \sum_{r=1}^{r=n_\varphi} \frac{(\delta_\varphi y_i)^2}{n_\varphi - p} = \sum_{r=1}^{r=n_\varphi} \frac{\{y_i - \varphi(x_i)\}^2}{n_\varphi - p} \qquad 10.84(3)$$

whilst the variance of the y_i from the best fit line for $f(x_i)$ is given by

$$s_{y_f}^2 = \sum_{s=1}^{s=n_f} \frac{(\delta_f y_s)^2}{n_f - q} = \sum_{s=1}^{s=n_f} \frac{\{y_s - f(x_s)\}^2}{n_f - q} \qquad 10.84(4)$$

where n_φ measurements are made for the φ function and n_f measurements are made for the f function. The best fit function is usually that which has the smallest variance. The last statement however needs some qualification. If a set of data is fitted to a successive series of polynomials of increasing degree, then the variance of the dependent variable from the fitted curve will show a steady decrease and then settle down to a nominally constant value. It will later again decrease, but this decrease is usually caused by the polynomial beginning to fit the random variations of the dependent variable. The fit accepted should be the polynomial with the lowest degree of the nominally constant variance group.

The Use of Orthogonal Polynomials in Curve Fitting

10.85 We have seen from equation 10.16(1) that fitting a plane to the n measured points required 10.16(1) to be satisfied for each of the observed points. This cannot be done because the observed points do not all lie on a plane. Clearly the n equations of *condition* represented by 10.16(1) cannot be satisfied since we have only three unknowns, i.e. a'', b'' and f'. The best that can be done is to take the sum of the squares of the residuals, represented by 10.18(1), and find the conditions making this a minimum. Differentiating this three times, i.e. once with respect to a'', b'' and f', leads to three simultaneous equations in a'', b'' and f', the solution of which gives the best solution, i.e. in this case, to the least squares plane. If we wish to fit a polynomial of degree p to n experimental points, $n > p$, then we shall get n equations of *condition* and $p + 1$ *normal* equations after differentiating the sum of the squares of the residuals. For example if

$$y = c_0 + c_1 x + c_2 x^2 \dots c_p x^p \qquad\qquad 10.85(1)$$

then the n equations of condition are given by

$$\left.\begin{aligned}
y_1 &= c_0 + c_1 x_1 + c_2 x_1^2 + c_3 x_1^3 \dots c_p x_1^p \\
y_2 &= c_0 + c_1 x_2 + c_2 x_2^2 + c_3 x_2^3 \dots c_p x_2^p \\
&\vdots \\
y_n &= c_0 + c_1 x_n + c_2 x_n^2 + c_3 x_n^3 \dots c_p x_n^p
\end{aligned}\right\} \qquad 10.85(2)$$

The sum of the squares of the residuals is given by

$$R_p = \sum_{r=1}^{r=n} (y_r - c_0 - c_1 x_r + c_2 x_r^2 + c_3 x_r^3 \dots c_p x_r^p)^2 \qquad 10.85(3)$$

Differentiating with respect to $c_0, c_1, c_2, \dots, c_p$ gives $p + 1$ normal equations represented by the expressions

$$\sum_{r=1}^{r=n} x_r^q (y_r - c_0 + c_1 x_r - c_2 x_r^2 - c_3 x_r^3 \dots c_p x_r^p) = 0 \qquad 10.85(4)$$

where 'q' takes successively the values 0 to p.

10.86 If p is large, then a large number of simultaneous linear equations have to be solved, which will mean recourse to a computer and the writing of a fairly complex program. There is a further reason why the foregoing method is not entirely satisfactory and that is that usually the power series with the lowest degree that best represents the data will be required. The degree of this series is not easy to predict and in practice if one starts with $p \simeq n/2$, and then progressively reduces p, one will arrive at a value of p which gives a good representation of the data, using as a criterion the value obtained for the standard deviation of the measured points from the fitted line. As the degree of the fitted polynomial rises the standard deviation of the dependent variable will fall and then reach a more or less constant value, but will eventually fall again as p approaches n. The degree indicated by the first of the approximately constant standard deviations will be the degree to use for the fitted curve. The smaller standard deviation obtained with the much higher power means that the curve is beginning to fit the random noise of the data. The drawback to the method is that with each successive power it will be necessary to compute a new set of the coefficients c_r. This is because the coefficients c_r have some degree of correlation between them. A further drawback is that if the standard deviation of one of the coefficients is required it is found to be a function of the other coefficients. This makes computation of the variances of the coefficients somewhat complicated.

Orthogonal polynomials

10.87   The drawbacks of the foregoing method can be largely eliminated by the use of functions known as orthogonal functions. Before going into how they are used we shall first derive such a set of functions† and explain their properties. Let a set of orthogonal polynomials be denoted by $P_r(x)$ where r is the power of the polynomial, and

$$P_r(x) = x^r + a_{r,r-1}x^{r-1} + a_{r,r-2}x^{r-2} \ldots a_{r,1}x + a_{r,0} \qquad 10.87(1)$$

The first suffix of the coefficients denotes the power of the polynomial to which it belongs, whilst the second suffix denotes the number of the coefficient and the power to which x is raised. A set of polynomials $P_0, P_1, P_2, \ldots, P_q$ is said to be an orthogonal set with respect to a set of points $x_1, x_2, x_3, \ldots, x_n$, $n > q$ if

$$\sum_{k=1}^{k=n} P_i(x_k)P_j(x_k) = 0 \qquad \text{for } i \neq j \qquad 10.87(2)$$

where i takes the values 0 to $q - 1$, and for each value of i, j takes the values $i + 1$ to q, thus giving $q(q + 1)/2$ expressions like 10.87(2). Although it is

---

† The following treatment and derivation of orthogonal polynomials and their application to curve fitting is based on notes communicated to the author by Dr J C E Jennings. The method of making the polynomials orthogonal on a set of values of the independent variable rather than over an integral between set limits is taken from work by J H Cadwell and D E Williams. See Bibliography for reference.

legitimate to use y and x, where y is the dependent variable and x is the independent one, there are some advantages in calculating a fitted curve about the mean of x and y and also of normalizing the range of x. The advantages are that where there is symmetry some terms vanish and also the moments will tend to be smaller. Furthermore normalizing the range to -1 to $+1$ enables one to see the maximum effect of any one term on the value of the polynomial fitted. Probably the most important fact about using orthogonal polynomials is that the addition of the next higher power orthogonal polynomial to the fitted curve leaves all the previous polynomial coefficients unaffected, unlike the case when the fitted curve is written as a power series in x when every coefficient of each power of x will change.

10.88　To normalize the range we write

$$X = (2x - x_{max} - x_{min})/(x_{max} - x_{min}) \qquad 10.88(1)$$

and

$$\bar{X} = (2\bar{x} - x_{max} - x_{min})/(x_{max} - x_{min}) \qquad 10.88(2)$$

Thus we write

$$\xi = X - \bar{X} = 2(x - \bar{x})/(x_{max} - x_{min}) \qquad 10.88(3)$$

and also

$$\eta = y - \bar{y} \qquad 10.88(4)$$

where the bars above the letters denote the mean value. It is to be noted that if the values of x are equally spaced or symmetrically disposed about the mean $\bar{x}$ then $\bar{X} = 0$, $X_{max} = 1$, $X_{min} = -1$ and $\xi_{max} = 1$, $\xi_{min} = -1$ and $\bar{\xi} = 0$. Otherwise $\bar{X}$ will have a small negative or positive value and ξ_{max} and ξ_{min} will be slightly greater or less than 1 or -1 respectively. $\bar{\xi}$ will always be zero and the range X_{min} to X_{max} or ξ_{min} to ξ_{max} will always be equal to 2. Lastly, since $\eta = y - \bar{y}$

$$\sum_{1}^{n} \eta_k = \sum_{1}^{n} y_k - n\bar{y} = n\bar{\eta} = 0 \qquad 10.88(5)$$

since $\sum_{1}^{n} y_k = n\bar{y}$ by definition. Thus $\bar{\eta}$ is always zero. The expression for an orthogonal polynomial 10.87(1) can now be written

$$P_r(\xi) = \xi^r + a_{r,r-1}\xi^{r-1} + a_{r,r-2}\xi^{r-2}\ldots a_{r,1}\xi + a_{r,0} \qquad 10.88(6)$$

We will now calculate some of the coefficients of the orthogonal polynomials. From 10.88(6), if $r = 0$, then by definition

$$P_0(\xi) = 1 = a_{0,0} \qquad 10.88(7)$$

and if $r = 1$

$$P_1(\xi) = \xi + a_{1,0} \qquad 10.88(8)$$

Since $P_0(\xi)$ and $P_1(\xi)$ are orthogonal

$$\sum_{k=1}^{k=n} P_0(\xi_k)P_1(\xi_k) = 0$$

Therefore

$$\sum_{k=1}^{k=n} \xi_k + na_{1,0} = 0$$

or

$$a_{1,0} = -\sum_{k=1}^{k=n} \xi_k/n = -\bar{\xi} \qquad 10.88(9)$$

Now $\xi = X - \bar{X}$, therefore

$$\sum_{k=1}^{k=n} \xi_k = \sum_{k=1}^{k=n} (X_k - \bar{X}) = \sum_{k=1}^{k=n} X_k - n\bar{X} = 0 \qquad 10.88(10)$$

and so by 10.88(9), $a_{1,0} = 0 = \bar{\xi}$ and thus

$$P_1(\xi) = \xi \qquad 10.88(11)$$

If $r = 2$

$$P_2(\xi) = \xi^2 + a_{2,1}\xi + a_{2,0} \qquad 10.88(12)$$

Now since P_0 and P_2 are orthogonal

$$\sum_{k=1}^{k=n} P_0(\xi_k)P_2(\xi_k) = 0 = \sum_{k=1}^{k=n} (\xi_k^2 + a_{2,1}\xi_k + a_{2,0})$$

$$= \sum_{k=1}^{k=n} \xi_k^2 + a_{2,1}\sum_{k=1}^{k=n} \xi_k + na_{2,0}$$

Since $\sum_{k=1}^{k=n} \xi_k = 0$ by 10.88(10) we have

$$a_{2,0} = -\sum_{k=1}^{k=n} \xi_k^2/n \qquad 10.88(13)$$

Also since P_1 and P_2 are orthogonal

$$\sum_{k=1}^{k=n} P_1(\xi_k)P_2(\xi_k) = 0 = \sum_{k=1}^{k=n} (\xi_k^3 + a_{2,1}\xi_k^2 + a_{2,0}\xi_k)$$

$$= \sum_{k=1}^{k=n} \xi_k^3 + a_{2,1}\sum_{k=1}^{k=n} \xi_k^2 + a_{2,0}\sum_{k=1}^{k=n} \xi_k$$

As $\sum_{k=1}^{k=n} \xi_k = 0$ by 10.88(10)

$$a_{2,1} = -\sum_{k=1}^{k=n} \xi_k^3 \bigg/ \sum_{k=1}^{k=n} \xi_k^2 \qquad 10.88(14)$$

whence

$$P_2 = \xi^2 - \left(\sum_{k=1}^{k=n} \xi_k^3 \bigg/ \sum_{k=1}^{k=n} \xi_k^2 \right) \xi - \sum_{k=1}^{k=n} \xi_k^2 / n \qquad 10.88(15)$$

Determination of P_r values generally

The equations for the derivation of $a_{r,r-1}$ etc. for r higher than 2 are simultaneous ones and rapidly become complicated. Thus $P_3(\xi)$ contains $a_{3,2}$, $a_{3,1}$ and $a_{3,0}$ for which three simultaneous equations are obtained if one proceeds in the same way as for $P_2(\xi)$. However it has long been known that the P_r satisfy a recurrence relation

$$P_{r+1}(\xi) = (\xi - \beta_r)P_r(\xi) + \gamma_{r-1}P_{r-1}(\xi) \qquad 10.88(16)$$

This may be shown to be true by the method of induction. It is clear that equation 10.88(16) is true for $r = 1$ which gives $P_2(\xi) = \xi^2 - \beta_1\xi + \gamma_0$ which is identical to equation 10.88(12) if $\beta_1 = -a_{2,1}$ and $\gamma_0 = a_{2,0}$.

10.89   Consider now the general case in which P_r and P_{r-1} are orthogonal to one another and to all lower order polynomials of the set; we require to show that P_{r+1} as given by 10.88(16) is orthogonal to $P_r, P_{r-1}, \ldots, P_1, P_0$. Now the orthogonality of P_{r+1} to P_r and to P_{r-1} depends on β_r and γ_{r-1} taking appropriate values, whilst its orthogonality to P_{r-2} and lower orders is ensured by the conditions that have been imposed on P_r and P_{r-1}.

Thus if P_{r+1} is to be orthogonal to P_s, equation 10.88(16) gives

$$\sum P_{r+1}P_s = \sum \xi P_r P_s - \beta_r \sum P_r P_s + \gamma_{r-1} \sum P_{r-1}P_s = 0 \qquad 10.89(1)$$

If $s \leqslant r - 2$, this equation reduces to

$$\sum \xi P_r P_s = 0 \qquad 10.89(2)$$

because P_r and P_{r-1} are orthogonal to all other polynomials of lower order by definition. Now 10.88(16) is true because from equation 10.88(6) ξP_s can be expressed as a linear combination of P_{s+1} and lower order polynomials of the set. This can be done by putting

$$\xi P_s = A_{s+1}P_{s+1} + A_s P_s + A_{s-1}P_{s-1} \ldots A_1 P_1 + A_0 P_0 \qquad 10.89(3)$$

and equating coefficients of powers of ξ, providing $s + 2$ simultaneous equations for $s + 1$ constants. Thus $\sum \xi P_r P_s = 0$. Only if $s = r$ or $s = r - 1$ does 10.88(16) lead to a condition on β_r or γ_{r-1}, viz. if $s = r$.

$$0 = \sum \xi P_r^2 - \beta_r \sum P_r^2$$

Therefore

$$\beta_r = \sum_{k=1}^{k=n} \xi_k P_r^2(\xi_k) \Big/ \sum_{k=1}^{k=n} P_r^2(\xi_k) \qquad 10.89(4)$$

If $s = r - 1$, then

$$\sum \xi P_r P_{r-1} + \gamma_{r-1} \sum P_{r-1}^2 = 0$$

But ξP_{r-1} can be expressed as P_r plus a linear combination of lower orders. Hence

$$\gamma_{r-1} = - \sum_{k=1}^{k=n} P_r^2(\xi_k) \Big/ \sum_{k=1}^{k=n} P_{r-1}^2(\xi_k) \qquad 10.89(5)$$

since if we put $s = r - 1$ in 10.89(3) and the coefficients of ξ^r are equated, then $A_r = 1$. Also again putting $s = r - 1$ in 10.89(3) and multiplying by P_r and summing over all the P_s we have

$$\sum \xi P_r P_{r-1} = \sum P_r^2 + A_{r-1} \sum P_r P_{r-1} + A_{r-2} \sum P_r P_{r-2} + \ldots = 0$$

by the orthogonal property

Hence 10.89(5).

Let us write

$$P_r(\xi_k) = {}_k P_r \qquad 10.89(6)$$

Thus 10.88(16) can be rewritten as

$${}_k P_{r+1} = (\xi_k - \beta_r)_k P_r + \gamma_{r-1\,k} P_{r-1} \qquad 10.89(7)$$

Also

$$\beta_r = \sum_{k=1}^{k=n} \xi_{k\,k} P_r^2 \Big/ \sum_{k=1}^{k=n} {}_k P_r^2 \qquad 10.89(8)$$

and

$$\gamma_{r-1} = - \sum_{k=1}^{k=n} {}_k P_r^2 \Big/ \sum_{k=1}^{k=n} {}_k P_{r-1}^2 \qquad 10.89(9)$$

If only the numerical values of the ${}_k P_r$ are required, then these can be obtained by using the expressions 10.89(7), (8) and (9). We shall see that this will be sufficient to find the value of the dependent variable of any fitted curve for all the values of the independent variable at which measurements were made. If an explicit expression for the P_r in terms of their independent variable ξ is required then the constants of the P_r will need to be found.

Determination of a_{rq} constants of P_r given by 10.88(6) where q takes the values r − 1 to 0

10.90 Let us write equation 10.88(16) in terms of ξ and the constants of the P_r. Thus we have

$$\xi^{r+1} + a_{r+1,r}\xi^r + a_{r+1,r-1}\xi^{r-1}\ldots a_{r+1,1}\xi + a_{r+1,0}$$
$$= (\xi - \beta_r)(\xi^r + a_{r,r-1}\xi^{r-1} + a_{r,r-2}\xi^{r-2}\ldots a_{r,1}\xi + a_{r,0})$$
$$+ \gamma_{r-1}(\xi^{r-1} + a_{r-1,r-2}\xi^{r-2} + a_{r-1,r-3}\xi^{r-3}\ldots a_{r-1,1}\xi + a_{r-1,0})$$

$$10.90(1)$$

Equating coefficients of ξ, we have

$$a_{r+1,r} = a_{r,r-1} - \beta_r$$
$$a_{r+1,r-1} = a_{r,r-2} - \beta_r a_{r,r-1} + \gamma_{r-1}$$
$$a_{r+1,r-2} = a_{r,r-3} - \beta_r a_{r,r-2} + a_{r-1,r-2}\gamma_{r-1}$$
$$a_{r+1,r-3} = a_{r,r-4} - \beta_r a_{r,r-3} + a_{r-1,r-3}\gamma_{r-1}$$
$$\vdots \qquad\qquad\qquad 10.90(2)$$
$$a_{r+1,2} = a_{r,1} - \beta_r a_{r,2} + a_{r-1,2}\gamma_{r-1}$$
$$a_{r+1,1} = a_{r,0} - \beta_r a_{r,1} + a_{r-1,1}\gamma_{r-1}$$
$$a_{r+1,0} = \qquad - \beta_r a_{r,0} + a_{r-1,0}\gamma_{r-1}$$

It is easily seen that these expressions may be written as

$$a_{r+1,r-p} = a_{r,r-p-1} - \beta_r a_{r,r-p} + \gamma_{r-1}a_{r-1,r-p} \qquad 10.90(3)$$

where p takes the values $p = 0$ to $p = r$.

It is to be noted that $a_{0,0} = P_0 = 1$ (10.88(7)), and that

$$a_{1,0} = -\bar{\xi} = 0 \qquad\qquad 10.88(9)$$

and that

$$\beta_0 = \sum \xi P_0^2 / \sum P_0^2 = \sum \xi/n = \bar{\xi} = 0 \qquad \text{(from 10.89(4)–10.90(4))}$$

and

$$\gamma_0 = -\sum P_1^2 / \sum P_0^2 = -\sum \xi^2/n \qquad \text{(from 10.89(5)–10.90(5))}$$

Fitting of observed points to a polynomial

10.91 Let the fitted polynomial be expressed in terms of the orthogonal polynomials just discussed. Thus we may write

$$\eta = \sum_{r=0}^{r=p} b_r P_r(\xi) \qquad\qquad 10.91(1)$$

where p is the power of the leading polynomial.

Expressed in a conventional way as a power series in ξ

$$\eta = \sum_{r=0}^{r=p} c_r \xi^r \qquad\qquad 10.91(2)$$

We shall discuss later how the c_r may be expressed in terms of the b_r.

According to the principle of least squares, the b_r equation 10.91(1) may be determined by minimizing the expression

$$R_p = \sum_{k=1}^{k=n} \left[\eta_k - \sum_{r=0}^{r=p} b_r P_r(\xi_k) \right]^2 \qquad\qquad 10.91(3)$$

with respect to the coefficients $b_0, b_1, b_2, \ldots, b_p$. The expression for R is known as the residuals. Thus differentiating with respect to b_s ($s = 0$ to p) we have

$$\sum_{k=1}^{k=n} \left[\eta_k - \sum_{r=0}^{r=p} b_r P_r(\xi_k) \right] P_s(\xi_k) = 0$$

or

$$\sum_{k=1}^{k=n} \eta_k P_s(\xi_k) = \sum_{k=1}^{k=n} \left(P_s(\xi_k) \sum_{r=0}^{r=p} b_r P_r(\xi_k) \right) \qquad\qquad 10.91(4)$$

Since the P_rs are orthogonal polynomials, 10.91(4) reduces to

$$\sum_{k=1}^{k=n} \eta_k P_s(\xi_k) = \sum_{k=1}^{k=n} b_s P_s^2(\xi_k)$$

giving

$$b_s = \sum_{k=1}^{k=n} \eta_k P_s(\xi_k) \bigg/ \sum_{k=1}^{k=n} P_s^2(\xi_k) \qquad\qquad 10.91(5)$$

This may also be written as

$$b_s = \sum_{k=1}^{k=n} \delta_{s-1,k} P_s(\xi_k) \bigg/ \sum_{k=1}^{k=n} P_s^2(\xi_k) \qquad\qquad 10.91(5a)$$

where

$$\delta_{s-1,k} = \eta_k - \sum_{r=0}^{r=s-1} b_r P_r(\xi_k)$$

whence

$$\sum_{k=1}^{k=n} \delta_{s-1,k} P_s(\xi_k) = \sum_{k=1}^{k=n} \eta_k P_s(\xi_k) - \sum_{k=1}^{k=n} \underbrace{\left[P_s(\xi_k) \sum_{r=0}^{r=s-1} b_r P_r(\xi_k) \right]}_{0}$$

because $\sum_{k=1}^{k=n} P_s(\xi_k) P_r(\xi_k) = 0$ for $r = 0$ to $r = s - 1$, since P_s and P_r are orthogonal. Thus all the b_s coefficients may be obtained by putting s successively equal to 0 up to p.

Relation between constants a_r, b_r and c_r

10.92 If the fitted curve is to be expressed as a power series in descending powers of ξ, i.e.

$$\eta = c_p \xi^p + c_{p-1} \xi^{p-1} + c_{p-2} \xi^{p-2} \dots c_1 \xi + c_0 \qquad \text{10.92(1)}$$

or

$$\eta = \sum_{r=0}^{r=p} c_r \xi^r \qquad \text{10.92(2)}$$

then we must find the c coefficients in terms of the a and b coefficients. Taking 10.90(1) and writing it out in expanded form, and writing the orthogonal polynomials in terms of ξ and the a coefficients, we have

$$\eta = b_p(\xi^p + a_{p,p-1}\xi^{p-1} + a_{p,p-2}\xi^{p-2} \dots a_{p,1}\xi + a_{p,0})$$
$$+ b_{p-1}(\xi^{p-1} + a_{p-1,p-2}\xi^{p-2} + a_{p-1,p-3}\xi^{p-3} \dots a_{p-1,1}\xi + a_{p-1,0})$$
$$+ b_{p-2}(\xi^{p-2} + a_{p-2,p-3}\xi^{p-3} + a_{p-2,p-4}\xi^{p-4} \dots a_{p-2,1}\xi + a_{p-2,0})$$
$$\vdots$$
$$+ b_1(\xi + a_{1,0})$$
$$\downarrow$$
$$0$$
$$+ b_0 \qquad \text{10.92(3)}$$

Equating the coefficients of equal powers of ξ in 10.92(1) and 10.92(3) we have

$$c_p = b_p$$

$$c_{p-1} = b_p a_{p,p-1} + b_{p-1}$$

$$c_{p-2} = b_p a_{p,p-2} + b_{p-1} a_{p-1,p-2} + b_{p-2}$$

$$c_{p-3} = b_p a_{p,p-3} + b_{p-1} a_{p-1,p-3} + b_{p-2} a_{p-2,p-3} + b_{p-3}$$

$$c_{p-4} = b_p a_{p,p-4} + b_{p-1} a_{p-1,p-4}$$

$$+ b_{p-2} a_{p-2,p-4} + b_{p-3} a_{p-3,p-4} + b_{p-4}$$

$$\vdots$$

$$\text{10.92(4)}$$

An inspection of the above expressions shows that a general expression for the c coefficients is possible, namely

$$c_{p-s} = \sum_{r=0}^{r=s-1} (b_{p-r} a_{p-r,p-s}) + b_{p-s} \qquad \text{10.92(4a)}$$

where s takes the values 1 to p for the summation term and 0 to p for the single term, in order to cover all the coefficients. Alternatively 10.92(4a) may be written

$$c_{p-s} = \sum_{r=0}^{r=s} (b_{p-r} a_{p-r,p-s}) \qquad \text{10.92(5)}$$

where s takes the values 0 to p and $a_{p-s,p-s} = 1$. If $s = p - 1$ and p we have for the last two terms

$$c_1 = b_p, a_{p,1} + b_{p-1} a_{p-1,1} + b_{p-2} a_{p-2,1} \ldots b_2 a_{2,1} + b_1$$

$$c_0 = b_p, a_{p,0} + b_{p-1} a_{p-1,0} + b_{p-2} a_{p-2,0} \ldots b_2 a_{2,0} + b_1 a_{1,0} + b_0$$

Standard deviation and standard error of coefficients of b_r

10.93 From 10.91(5)

$$b_{s,p} = \sum_{k=1}^{k=n} \eta_k P_s(\xi_k) \bigg/ \sum_{k=1}^{k=n} P_s^2(\xi_k) \qquad \text{10.91(5)}$$

where the added suffix p to b denotes the power of the fitted polynomial. Differentiating 10.91(5) with respect to $b_{s,p}$ and η_k we have

$$\frac{\partial b_{s,p}}{\partial \eta_k} = P_s(\xi_k) \bigg/ \sum_{k=1}^{k=n} P_s^2(\xi_k) \qquad \text{10.93(1)}$$

Now comparing with

$$s_F^2 = \sum_{i=1}^{i=n} s_{x_i}^2 \left(\frac{\partial F}{\partial x_i}\right)^2$$

8.04(4)

derived from $F(x_1, x_2, x_3, \ldots, x_n) = F$, then $\partial b_{s,p}/\partial \eta_k$ is equivalent to $(\partial F/\partial x_i)$ and η_k is equivalent to x_i, thus

$$s_{b_{s,p}}^2 = \sum_{k=1}^{k=n} s_{\eta_{p,k}}^2 P_s^2(\xi_k) \bigg/ \left(\sum_{k=1}^{k=n} P_s^2(\xi_k)\right)^2$$

Now if all the $s_{\eta_{p,k}}$ are considered to be equal, we put

$$s_{\eta_p} = s_{\eta_{p,k}}, \qquad k = 1, 2, 3, \ldots, n$$

and the expression for $s_{b_{s,p}}^2$ becomes

$$s_{b_{s,p}}^2 = s_{\eta_p}^2 \sum_{k=1}^{k=n} P_s^2(\xi_k) \bigg/ \left(\sum_{k=1}^{k=n} P_s^2(\xi_k)\right)^2$$

$$= s_{\eta_p}^2 \bigg/ \sum_{k=1}^{k=n} P_s^2(\xi_k)$$

10.93(2)

The standard error in the mean of $b_{s,p}$ is thus given by

$$s_{\bar{b}_{s,p}}^2 = s_{b_{s,p}}^2/n$$

10.93(3)

Now

$$s_{\eta,p}^2 = \sum_{k=1}^{k=n} \left[\eta_k - \sum_{r=0}^{r=p} b_r P_r(\xi_k)\right]^2 \bigg/ (n - p - 1)$$

10.93(4)

from 10.91(3), where the suffix p has been added to show the degree of the fitted polynomial.

Writing

$$\delta_{p,k} = \left[\eta_k - \sum_{r=0}^{r=p} b_r P_r(\xi_k)\right]$$

10.93(5)

$$s_{\eta,p}^2 = \sum_{k=1}^{k=n} \delta_{pk}^2/(n - p - 1)$$

10.93(6)

Thus substituting for $s_{\eta,p}^2$ in 10.93(2) we have

$$s_{b_{s,p}}^2 = \sum_{k=1}^{k=n} \left[\eta_k - \sum_{r=0}^{r=p} b_r P_r(\xi_k)\right]^2 \bigg/ (n - p - 1) \sum_{k=1}^{k=n} P_s^2(\xi_k)$$

10.93(7)

$$= \sum_{k=1}^{k=n} \delta_{p,k}^2 \bigg/ (n - p - 1) \sum_{k=1}^{k=n} P_s^2(\xi_k)$$

10.93(8)

giving the variance of $b_{s,p}$. The variance of the standard error of $b_{s,p}$ is given by

$$\bar{s}_{b_{s,p}}^2 = s_{b_{s,p}}^2/n$$

10.93(9)

Standard deviation and standard error of coefficients of c_r

10.94 From the expression for the c coefficients, namely

$$c_{p-s} = \sum_{r=0}^{r=s-1} (b_{p-r}a_{p-r,p-s}) + b_{p-s} \qquad 10.92(3)$$

where s takes the values 1 to p for the summation and 0 to p for the single term, we can find the standard deviation of c_{p-s} as follows. Firstly it is to be noted that as are to be considered as constants, since they are derived solely from the xs or ξs which are without error. Thus again comparing with 8.04(4), i.e.

$$s_F^2 = \sum_{i=1}^{i=n} s_{x_i}^2 \left(\frac{\partial F}{\partial x_i} \right)^2 \qquad 8.04(4)$$

we have

$$\frac{\partial c_{p-s}}{\partial b_{p-r}} = a_{p-r,p-s} \qquad 10.94(1)$$

which we compare with $\partial F/\partial x_i$, whilst b_{p-r} we compare with x_i. Hence the standard deviation of c_{p-s} is given by

$$s_{c_{p-s}}^2 = \sum_{r=0}^{r=s-1} s_{b_{p-r}}^2 a_{p-r,p-s}^2 + s_{b_{p-s}}^2 \qquad 10.94(2)$$

where as before s goes from 1 to p for the summation term and from 0 to p for the single term. Alternatively this may be written as

$$s_{c_{p-s}}^2 = \sum_{r=0}^{r=s} s_{b_{p-r}}^2 a_{p-r,p-s}^2 \qquad 10.94(3)$$

where s goes from 0 to p and $a_{p-s,p-s} = 1$.

The standard error in the mean of c_{p-s} is given by

$$s_{\bar{c}_{p-s}}^2 = s_{c_{p-s}}^2/n \qquad 10.94(4)$$

Derivatives

Consider the first derivative of η, i.e. which we will write as $^1\eta = d\eta/d\xi$. From equation 10.91(1) we have

$$^1\eta = \sum_{r=0}^{r=p} b_r \, ^1P_r(\xi) \qquad 10.94(5)$$

where

$$^1P_r(\xi) = dP_r/d\xi \qquad 10.94(5a)$$

Also

$$^q\eta = \sum_{r=0}^{r=p} b_r{}^q P(\xi) = \frac{d^q\eta}{d\xi^q} \qquad\qquad 10.94(5b)$$

From the recurrence relation 10.88(16) the first derivative is given by

$$^1P_{r+1}(\xi) = P_r(\xi) + (\xi - \beta_r)^1 P_r(\xi) + \gamma_{r-1}{}^1 P_{r-1}(\xi) \qquad 10.94(6)$$

Taking successive derivatives it is soon apparent that

$$^q P_{r+1}(\xi) = q^{q-1} P(\xi) + (\xi - \beta_r)^q P_r(\xi) + \gamma_{r-1}{}^q P_{r-1}(\xi) \qquad 10.94(7)$$

Now

$$\eta = \sum_{r=0}^{r=p} b_r P_r(\xi) \qquad\qquad 10.91(1)$$

Comparing with

$$s_F^2 = \sum_{i=0}^{i=n} s_{x_i}^2 \left(\frac{\partial F}{\partial x_i}\right)^2 \qquad\qquad 8.04(4)$$

where $F = F(x_0, x_1, x_2, x_3, \ldots, x_n)$ we have

$$s_{\eta_p}^2 = \sum_{r=0}^{r=p} s_{b_r}^2 P_r^2(\xi) \qquad\qquad 10.94(8)$$

Similarly

$$s_{{}^q\eta_p}^2 = \sum_{r=0}^{r=p} s_{b_r}^2 ({}^q P(\xi))^2 \qquad\qquad 10.94(9)$$

The derivatives of η can also be obtained from equation 10.91(2) giving

$$\cdot\ {}^q\eta = \sum_{r=0}^{r=p-q} \frac{(q + r)!}{r!} c_{q+r} \xi^r \qquad\qquad 10.94(10)$$

where q is the number of differentiations and p is the power of η. The variance of η and its derivatives cannot be found from the variances of the c_r of 10.91(2) without taking into account the correlation between the c_r. The variance of η and its derivatives are best found using 10.94(9), since the b_r are uncorrelated.

Orthogonal function in general

10.95 A set of functions $f_r(x)$, $r = 0, 1, 2, \ldots, p$, may be orthogonalized with respect to the set of points x_k, $k = 1, 2, 3, \ldots, n$. The rth function

$$F_r(x) = f_r(x) + a_{r,r-1} F_{r-1}(x) + \ldots + a_{r,1} F_1(x) + a_{r,0}$$

$$10.95(1)$$

is orthogonal to all $F_s(x), s < r$, including $F_0(x) = 1$, if $\sum_{k=1}^{k=n} F_r(x_k)F_s(x_k) = 0$ for all $r \neq s$. This relationship is satisfied if

$$a_{r,s} = -\sum_{k=1}^{k=n} f_r(x_k)F_s(x_k) \bigg/ \sum_{k=1}^{k=n} F_s^2(x_k)$$

This is easily proven by multiplying 10.95(1) by $F_s(x_k)$ and summing for k from 0 to n. Although formally simple this relationship requires rather more computation than the recurrence relation used for the polynomials previously. A set of observed $y(x)$ may be fitted to

$$y(x) = \sum_{r=0}^{r=p} a_r F_r(x) \qquad\qquad 10.95(2)$$

using the orthogonal property, giving

$$a_s = \sum_{k=1}^{k=n} y(x_k)F_s(x_k) \bigg/ \sum_{k=1}^{k=n} F_s^2(x_k) \qquad\qquad 10.95(3)$$

where the F_r are found from 10.95(1) with $y_r(x) \equiv f_r(x)$ and 10.95(3) is obtained by multiplying 10.95(2) by $F_s(x)$ and summing over $k = 1$ to n, i.e.

$$\sum_{k=1}^{k=n} y(x_k)F_s(x_k) = \sum_{k=1}^{k=n} \sum_{r=0}^{r=p} a_r F_r(x_k)F_s(x_k)$$

$$= \sum_{k=1}^{k=n} a_s F_s^2(x_k)$$

whence

$$a_s = \sum_{k=1}^{k=n} y(x_k)F_s(x_k) \bigg/ \sum_{k=1}^{k=n} F_s^2(x_k) \qquad\qquad 10.95(3)$$

Worked Examples

10.96 We will first recapitulate the values of the first three orthogonal polynomials, namely

$$P_0 = 1 \qquad\qquad 10.88(7)$$

$$P_1 = \xi \qquad\qquad 10.88(11)$$

and

$$P_2 = \xi^2 - \left(\sum_{k=1}^{k=n} \xi_k^3 \bigg/ \sum_{k=1}^{k=n} \xi_k^2\right)\xi - \sum_{k=1}^{k=n} \xi_k^2/n \qquad\qquad 10.88(15)$$

Let us first consider the equation for a straight line, namely $y = mx + c$ or $\eta = m\xi$ when written in the normalized notation we have been using. Now the equation for a straight line when written in terms of orthogonal polynomials becomes

$$\eta = b_1 P_1 + b_0 P_0$$
$$= b_1 \xi + b_0 \qquad\qquad 10.96(1)$$

Now from 10.91(5)

$$\left. \begin{array}{l} b_0 = \sum_1^n \eta_k/n - \bar{\eta} = 0 \\[3mm] b_1 = \sum_1^n \eta_k P_1(\xi_k) \Big/ \sum_1^n P_1^2(\xi_k) \end{array} \right\} \text{(from 10.88(5))} \qquad 10.96(2)$$

Hence equating $m = b_1$ we have the expression for the gradient of a straight line as

$$\eta = \xi \sum_1^n \eta_k \xi_k \Big/ \sum_1^n P_k^2 \qquad\qquad 10.96(3)$$

Standard deviation and standard error of η and of m the gradient of the mean line

The standard deviation of η from the mean line is given by 10.93(4), which with $p = 1$ yields

$$s_{\eta_1}^2 = \sum_{k=1}^{k=n} [\eta_k - b_1 P_1(\xi_k)]^2/(n-2)$$
$$= \sum_{k=1}^{k=n} [\eta_k - b_1 \xi_k]^2/n - 2 \qquad\qquad 10.96(4)$$

and

$$s_{\bar{\eta}}^2 = s_\eta^2/n \qquad\qquad 10.96(5)$$

where $b_1 = m$, given by 10.96(2).

Now from 10.93(2)

$$s_{b_{1,1}}^2 = s_m^2 = s_\eta^2 \Big/ \sum_{k=1}^{k=n} P_1^2(\xi_k)$$
$$= s_\eta^2 \Big/ \sum_{k=1}^{k=n} \xi_k^2 \qquad\qquad 10.96(6)$$

Therefore

$$s_{\bar{m}}^2 = s_m^2/n = \sum_{k=1}^{k=n} [\eta_k - m\xi_k]^2/n(n-2) \sum_{k=1}^{k=n} \xi_k^2 \qquad 10.96(7)$$

Note that $s_{\bar{m}}$ is the standard error of the gradient.

10.97 If the equation to a straight line is written in the form $y = mx + c$, then we must replace ξ by x in the expressions for P_r, thus

$$P_r(x) = x^r + a_{r,r-1}x^{r-1} + a_{r,r-2}x^{r-2} \ldots a_{r,1}x + a_{r,0}$$

or

$$= \sum_{j=0}^{j=r} a_{r,j}x^j \qquad\qquad 10.97(1)$$

where $a_{r,r} = 1$.

As before

$$P_0(x) = 1$$
$$P_1(x) = x + a_{1,0}$$

Since the P_rs are orthogonal

$$\sum_{k=0}^{k=n} P_0(x)P_1(x) = 0 = \sum_{k=1}^{k=n} (x_k + a_{1,0}) = n\bar{x}_n + na_{1,0}$$

Therefore

$$a_{1,0} = -\bar{x}_n \qquad \text{and} \qquad P_1(x) = x - \bar{x}_n \qquad\qquad 10.97(2)$$

If $r = 2$

$$P_2(x) = x^2 + a_{2,1}x + a_{2,0}$$

Multiplying by $P_0(x)$ and summing we have

$$\sum_{k=1}^{k=n} P_0(x)P_2(x) = 0 = \sum_{k=1}^{k=n} x_k^2 + a_{2,1}\sum_{k=1}^{k=n} x_k + a_{2,0}n \qquad\qquad 10.97(3)$$

Multiplying $P_2(x)$ by $P_1(x)$ and summing we have

$$\sum_{k=1}^{k=n} P_1(x)P_2(x) = 0 = \sum_{k=1}^{k=n} (x_k - \bar{x}_n)(x_k^2 + a_{2,1}x_k + a_{2,0})$$

$$= \sum_{k=1}^{k=n} x_k^2(x_k - \bar{x}_n) + a_{2,1}\sum_{k=1}^{k=n} x_k(x_k - \bar{x}_n)$$

$$+ a_{2,0}\sum_{k=1}^{k=n} (x_k - \bar{x}_n)$$

The term involving $a_{2,0}$ is zero, since $\sum_{k=1}^{k=n} x_k = n\bar{x}_n$. Therefore

$$a_{2,1} = - \sum_{k=1}^{k=n} x_k^2(x_k - \bar{x}_n) \bigg/ \sum_{k=1}^{k=n} x_k(x_k - \bar{x}_n) \qquad\qquad 10.97(4)$$

From 10.97(3)

$$a_{2,0} = -\left(\sum_{k=1}^{k=n} x_k^2 + a_{2,1} \sum_{k=1}^{k=n} x_k\right)\bigg/ n = -\left(\sum_{k=1}^{k=n} x_k^2 + a_{2,1} n\bar{x}_n\right)\bigg/ n$$

Substituting for $a_{2,1}$ and simplifying we have

$$a_{2,0} = \bar{x}_n \sum_{k=1}^{k=n} x_k^2(x_k - \bar{x}_n)\bigg/ \sum_{k=1}^{k=n} x_k(x_k - \bar{x}_n) - \sum_{k=1}^{k=n} x_k^2/n \qquad 10.97(5)$$

The equation for a straight line thus becomes

$$y = b_1 P_1 + b_0 P_0 \qquad 10.97(6)$$

or

$$y = b_1(x - \bar{x}_n) + b_0 \qquad 10.97(7)$$

Thus

$$\sum_{k=1}^{k=n} y_k P_0 = b_1 \underbrace{\sum_{k=1}^{k=n} P_1 P_0} + b_0 \sum P_0^2$$

giving

$$\sum_{k=1}^{k=n} y_k = \qquad 0 \qquad + b_0 n$$

or

$$b_0 = \sum_{k=1}^{k=n} y_k/n \qquad 10.97(8)$$

Multiplying 10.97(6) by P_1 and summing we have

$$\sum_{k=1}^{k=n} y_k P_1 = b_1 \sum_{k=1}^{k=n} P_1^2 + b_0 \underbrace{\sum_{k=1}^{k=n} P_1 P_0}_{0}$$

Therefore

$$\sum_{k=1}^{k=n} y_k(x_k - \bar{x}_n) = b_1 \sum_{k=1}^{k=n} (x_k - \bar{x}_n)^2$$

or

$$b_1 = \sum_{k=1}^{k=n} y_k(x_k - \bar{x}_n)\bigg/ \sum_{k=1}^{k=n} (x_k - \bar{x}_n)^2$$

Thus the equation to a straight line becomes

$$y = x \sum_{k=1}^{k=n} y_k(x_k - \bar{x}_n)\bigg/ \sum_{k=1}^{k=n} (x_k - \bar{x}_n)^2 + \sum_{k=1}^{k=n} y_k/n$$

$$- \bar{x} \sum_{k=1}^{k=n} y_k(x_k - \bar{x}_n)\bigg/ \sum_{k=1}^{k=n} (x_k - \bar{x}_n)^2$$

which on simplifying

$$= x \sum_{k=1}^{k=n} y_k(x_k - \bar{x}_n) \bigg/ \sum_{k=1}^{k=n} (x_k - \bar{x}_n)^2$$

$$+ \left(\sum_{k=1}^{k=n} y_k \sum_{k=1}^{k=n} x_k^2 - \sum_{k=1}^{k=n} x_k \sum_{k=1}^{k=n} x_k y_k \right) \bigg/ \left(n \sum_{k=1}^{k=n} x_k^2 - \left(\sum_{k=1}^{k=n} x_k \right)^2 \right)$$

10.97(9)

since

$$\sum_{k=1}^{k=n} (x_k - \bar{x}_n)^2 = \sum x_k^2 - \left(\sum x_k \right)^2 / n$$

whence

$$m = \sum_{k=1}^{k=n} y_k(x_k - \bar{x}_n) \bigg/ \sum_{k=1}^{k=n} (x_k - \bar{x}_n)^2 \qquad 10.97(10a)$$

$$= \sum_{k=1}^{k=n} y_k \left(nx_k - \sum_{k=1}^{k=n} x_k \right) \bigg/ \left(\sum_{k=1}^{k=n} nx_k^2 - \left(\sum_{k=1}^{k=n} x_k \right)^2 \right) \qquad 10.97(10b)$$

and

$$c = \left(\sum_{k=1}^{k=n} y_k \sum_{k=1}^{k=n} x_k^2 - \sum_{k=1}^{k=n} x_k \sum_{k=1}^{k=n} x_k y_k \right) \bigg/ \left(n \sum_{k=1}^{k=n} x_k^2 - \left(\sum_{k=1}^{k=n} x_k \right)^2 \right)$$

10.97(11)

Invoking 8.04(4) again, i.e.

$$s_F^2 = \sum_{i=0}^{i=n} s_{x_i}^2 \left(\frac{\partial F}{\partial x_i} \right)^2$$

we have on differentiating m with respect to y_k that

$$\frac{\partial m}{\partial y_k} = (x_k - \bar{x}_n) \bigg/ \sum_{k=1}^{k=n} (x_k - \bar{x}_n)^2 \qquad 10.97(12)$$

whence

$$s_m^2 = \sum_{k=1}^{k=n} s_{y_k}^2 (x_k - \bar{x}_n)^2 \bigg/ \left[\sum_{k=1}^{k=n} (x_k - \bar{x}_n)^2 \right]^2$$

$$= s_y^2 \sum_{k=1}^{k=n} (x_k - \bar{x}_n)^2 \bigg/ \left[\sum_{k=1}^{k=n} (x_k - \bar{x}_n)^2 \right]^2$$

$$= s_y^2 \bigg/ \sum_{k=1}^{k=n} (x_k - \bar{x}_n)^2 \qquad 10.97(13)$$

since $s_{y_k}^2 = s_y^2$, i.e. all y_k are assumed to have the same standard deviation. Also

$$s_y^2 = \sum_{k=1}^{k=n} (y_k - mx_k - c)^2/(n-2)$$

10.97(14)

Similarly

$$s_c^2 = s_y^2 \left(\sum_{k=1}^{k=n} x_k^2 \right)^2 \bigg/ \left(n \sum_{k=1}^{k=n} x_k^2 - \left(\sum_{k=1}^{k=n} x_k \right)^2 \right)^2$$

10.97(15)

The standard errors of s_m and s_c are of course given by

$$s_{\bar{m}}^2 = s_m^2/n$$

10.97(16)

and

$$s_{\bar{c}}^2 = s_c^2/n$$

10.97(17)

Second order curves

10.98   We write a second order curve as

$$\eta = c_2 \xi^2 + c_1 \xi + c_0$$

10.98(1)

The orthogonal version is thus

$$\eta = b_2 P_2(\xi) + b_1 P_1(\xi) + b_0$$

10.98(2)

From 10.91(5)

$$b_2 = \sum_{k=1}^{k=n} \eta_k P_2(\xi_k) \bigg/ \sum_{k=1}^{k=n} P_2^2(\xi_k)$$

10.98(3)

$$b_1 = \sum_{k=1}^{k=n} \eta_k P_1(\xi_k) \bigg/ \sum_{k=1}^{k=n} P_1^2(\xi_k) = \sum_{k=1}^{k=n} \eta_k \xi_k \bigg/ \sum_{k=1}^{k=n} \xi_k^2$$

10.98(4)

$$b_0 = \sum_{k=1}^{k=n} \eta_k P_0(\xi) \bigg/ \sum_{k=1}^{k=n} P_0^2(\xi_k) = \sum_{k=1}^{k=n} \eta_k/n = \bar{\eta}_n = 0$$

10.98(5)

and where

$$P_2 = \xi^2 - \left(\sum_{k=1}^{k=n} \xi_k^3 \bigg/ \sum_{k=1}^{k=n} \xi_k^2 \right) \xi - \sum_{k=1}^{k=n} \xi_k^2/n$$

10.88(15)

Thus

$$\eta = b_2 \xi^2 + (b_2 a_{2,1} + b_1)\xi + (b_2 a_{2,0} + b_1 a_{1,0} + b_0)$$

10.98(6)

from 10.92(3). Now from the expressions for P_0, P_1 and P_2, 10.88(7), 10.88(11) and 10.88(15)

$$a_{2,1} = - \sum_{k=1}^{k=n} \xi_k^3 \Big/ \sum_{k=1}^{k=n} \xi_k^2, \qquad a_{2,0} = - \sum_{k=1}^{k=n} \xi_k^2 / n \qquad a_{1,0} = 0$$

$$10.98(5)$$

Thus finally, substituting for the suffixed a and b terms we have

$$\eta = \xi^2 \sum_{k=1}^{k=n} \eta_k P_2(\xi_k) \Big/ \sum_{k=1}^{k=n} P_2^2(\xi_k)$$

$$- \xi \left[\left(\sum_{k=1}^{k=n} \eta_k P_2(\xi_k) \Big/ \sum_{k=1}^{k=n} P_2^2(\xi_k) \right) \left(\sum_{k=1}^{k=n} \xi_k^3 \Big/ \sum_{k=1}^{k=n} \xi_k^2 \right) \right.$$

$$\left. - \sum_{k=1}^{k=n} \eta_k \xi_k \Big/ \sum_{k=1}^{k=n} \xi_k^2 \right] - \left[\left(\sum_{k=1}^{k=n} \eta_k P_2(\xi_k) \Big/ \sum_{k=1}^{k=n} P_2^2(\xi_k) \right) \sum_{k=1}^{k=n} \xi_k^2 / n \right]$$

$$10.98(8)$$

The gradient at ξ is given by

$$\frac{d\eta}{d\xi} = 2b_2 \xi + (b_2 a_{2,1} + b_1) \qquad 10.98(9)$$

Standard deviation of η

10.99 The standard deviation of η is given by

$$s_{\eta_2}^2 = \sum_{k=1}^{k=n} (\eta - b_2 \xi^2 - (b_2 a_{2,1} + b_1)\xi - b_2 a_{2,0})^2 / n - 3 \qquad 10.99(1)$$

Note that the divisor is now $n - 3$ because p, the degree, is now 2 (see 10.93(4)).

As before

$$s_{\bar{\eta}}^2 = s_{\eta}^2 / n \qquad 10.99(2)$$

Standard deviation of c_2, c_1 and c_0

Now $c_2 = b_2$, $c_1 = (b_2 a_{2,1} + b_1)$, $c_0 = b_2 a_{2,0}$ from 10.98(6), and since $a_{1,0} = 0$ and $b_0 = 0$ (see 10.98(7) and 10.98(5) respectively). Thus

$$s_{c_2}^2 = s_{b_2}^2 \qquad 10.99(3)$$

$$s_{c_1}^2 = s_{b_2}^2 a_{2,1}^2 + s_{b_1}^2 \qquad 10.99(4)$$

$$s_{c_0}^2 = s_{b_2}^2 a_{2,0}^2 \qquad 10.99(5)$$

where the expression for the a is given in 10.98(7) and s_{b_2} and s_{b_1} are derived from 10.93(7) giving

$$s_{b_2}^2 = \sum_{k=1}^{k=n} [\eta_k - b_2 P_2(\xi_k) - b_1 P_1(\xi_k)]^2 \Big/ (n-3) \sum_{k=1}^{k=n} P_2^2(\xi_k)$$

10.99(6)

and

$$s_{b_1}^2 = \sum_{k=1}^{k=n} [\eta_k - b_2 P_2(\xi_k) - b_1 P_1(\xi_k)]^2 \Big/ (n-3) \sum_{k=1}^{k=n} \xi_k^2$$

10.99(7)

The standard errors are given by

$$s_{\bar{c}_2}^2 = s_{b_2}^2/n \qquad\qquad 10.99(8)$$

$$s_{\bar{c}_1}^2 = s_{b_2}^2 a_{2,1}^2/n + s_{b_1}^2/n \qquad\qquad 10.99(9)$$

$$s_{\bar{c}_0}^2 = s_{b_2}^2 a_{2,0}^2/n \qquad\qquad 10.99(10)$$

Worked Example to Illustrate the Method of Curve Fitting Described and to Assess its Efficiency

10.100 The following example† shows how the method is used and illustrates the efficiency of the method, based on number tests. In order to show with some clarity the efficiency of the method, a cubic equation, namely $y = 1 - \xi + \xi^3$ was chosen, where $\xi = x - \bar{x}$. 21 values of y were calculated using the expression $y = 1 - \xi + \xi^3 + R$, $x = -1(0.1)1$, giving the 21 points where R is a Gaussian random number for each value of x or ξ. R was chosen so that $\bar{R} = 0$, and $\sigma_R = 0.01$.

The value of R was generated from two random numbers generated from a rectangular distribution with range 0 to 1. It could also have been generated from a single random number obtained from a rectangular distribution with range -0.5 to 0.5. The formulae for generating random numbers are given in Appendix VII, numbers 3, 4 and 5.

Having generated 21 values of y from $y = 1 - \xi + \xi^3 + R$, the results were fitted to $\eta = b_1\xi + b_2 P_2 + b_3 P_3$, where $\eta = y - \bar{y}$ and $P_2 = \xi^2 - \sum_{k=1}^{k=21} \xi_k^2/21$ and

$$P_3 = \xi \left[P_2 - \sum_{k=1}^{k=21} P_2^2(\xi_k) \Big/ \sum_{k=1}^{k=n} \xi_k^2 \right]$$

(using 10.88(6), 10.89(4) and 10.89(5)) also $P_1 = \xi$, $P_0 = 0$, $b_0 = 0$.

---

† I am indebted to Dr J C E Jennings for the use of this example and the subsequent analysis.

To satisfy the least squares fit

$$
\left.\begin{aligned}
b_1 &= \sum_{k=1}^{k=21} \eta_k \xi_k \bigg/ \sum_{k=1}^{k=n} \xi_k^2 \\
b_2 &= \sum_{k=1}^{k=21} \eta_k \left(\xi_k^2 - \sum_{k=1}^{k=21} \xi_k^2/21 \right) \bigg/ \sum_{k=1}^{k=n} \left[\xi_k^2 - \sum_{k=1}^{k=21} \xi_k^2/21 \right]^2 \text{ and} \\
b_3 &= \sum \eta_k P_3(\xi_k) / \sum P_3^2(\xi_k)
\end{aligned}\right\}
$$

$$10.100(1)$$
$$(\text{from } 10.91(5))$$

It is probably better arithmetically (although formally equivalent) to write

$$
b_2 = \sum_{k=1}^{k=21} \delta_{1,k} P_2(\xi_k) \bigg/ \sum_{k=1}^{k=21} P_2^2(\xi_k) \qquad 10.100(2)
$$

and

$$
b_3 = \sum_{k=1}^{k=21} \delta_{2,k} P_3(\xi_k) \bigg/ \sum_{k=1}^{k=21} P_3^2(\xi_k) \qquad 10.100(3)
$$

where

$$
\delta_{1,k} = \eta_k - \sum_{r=0}^{r=1} b_r P_r(\xi_k)
$$

$$
= \eta_k - b_1 \xi_k \qquad \text{since } b_0 = 0
$$

and

$$
\delta_{2,k} = \eta_k - \sum_{r=0}^{r=2} b_r P_r(\xi_k) = \eta_k - b_1 \xi_k - b_2 P_2(\xi_k)
$$

$$
= \delta_{1,k} - b_2 P_2(\xi_k)
$$

If we revert to the non-orthogonal form of the equation

$$
\eta = c_0 + c_1 \xi + c_2 \xi^2 + c_3 \xi^3 \qquad 10.100(4)
$$

From equations 10.92(4a)

$$
\left.\begin{aligned}
c_3 &= b_3 \\
c_2 &= b_3 a_{3,2} + b_2 \\
c_1 &= b_3 a_{3,1} + b_2 a_{2,1} + b_1 \\
c_0 &= b_3 a_{3,0} + b_2 a_{2,0} + b_1 a_{1,0} + b_0
\end{aligned}\right\}
\qquad 10.100(5)
$$

Now by inspection of the Ps

$$a_{3,2} = 0 \qquad a_{3,1} = - \sum_{k=1}^{k=21} P_2^2(\xi_k) \left/ \sum_{k=1}^{k=21} \xi_k^2 - \sum_{k=1}^{k=21} \xi_k^2/21 \right\} \text{from } P_3$$

$$a_{3,0} = 0$$

$$a_{2,1} = 0 \qquad a_{2,0} = - \sum_{k=1}^{k=21} \xi_k^2/21 \qquad\qquad\qquad \text{from } P_2$$

$$a_{1,0} = 0 \qquad \text{from } P_1$$

$$\hspace{11cm} 10.100(6)$$

Thus

$$c_3 = b_3 = \sum_{k=1}^{k=21} \delta_{2,k} P_3(\xi_k) \left/ \sum_{k=1}^{k=21} P_3^2(\xi_k) \right. \hspace{2cm} 10.100(7)$$

$$c_2 = b_2 = \sum_{k=1}^{k=21} \delta_{1,k} P_2(\xi_k) \left/ \sum_{k=1}^{k=21} P_2^2(\xi_k) \right. \hspace{2cm} 10.100(8)$$

$$c_1 = b_1 - \sum_{k=1}^{k=21} P_2^2(\xi_k) \left/ \sum_{k=1}^{k=21} \xi_k^2 - \sum_{k=1}^{k=21} \xi_k^2/21 \right.$$

$$= \sum_{k=1}^{k=21} \delta_{0,k} P_1(\xi_k) \left/ \sum_{k=1}^{k=21} P_1^2(\xi_k) - \sum_{k=1}^{k=21} P_2^2(\xi_k) \right/ \sum_{k=1}^{k=21} \xi_k^2$$

$$- \sum_{k=1}^{k=21} \xi_k^2/21 \hspace{6cm} 10.100(9)$$

$$c_0 = -b_2 \sum_{k=1}^{k=21} \xi_k^2/21 \hspace{5cm} 10.100(10)$$

Using 10.100(5), 10.95(2) and 10.100(6) we have

$$S_{c_3} = S_{b_3}$$

$$S_{c_2} = S_{b_2}$$

$$S_{c_1} = \left[s_{b_1}^2 + s_{b_3}^2 \left(\sum_{k=1}^{k=21} P_2^2(\xi_k) \left/ \sum_{k=1}^{k=21} \xi_k^2 + \sum_{k=1}^{k=21} \xi_k^2/21 \right)^2 \right]^{1/2} \right\}$$

$$S_{c_0} = S_{b_2} \sum_{k=1}^{k=21} \xi_k^2/21$$

$$\hspace{11cm} 10.100(11)$$

where

$$s_{b_1}^2 = s_{\eta_3}^2 \bigg/ \sum_{k=1}^{k=21} P_1^2(\xi_k) = s_{\eta_3}^2 \bigg/ \sum_{k=1}^{k=21} \xi_k^2 \qquad 10.100(12)$$

$$s_{b_2}^2 = s_{\eta_3}^2 \bigg/ \sum_{k=1}^{k=21} P_2^2(\xi_k) \qquad 10.100(13)$$

$$s_{b_3}^2 = s_{\eta_3}^2 \bigg/ \sum_{k=1}^{k=21} P_3^2(\xi_k) \qquad 10.100(14)$$

All the s_b^2s are derived from 10.93(2).

$$s_{\eta_3}^2 = \sum_{k=1}^{k=21} \delta_{3,k}/(21 - 3 - 1) \equiv \sum_{k=1}^{k=n} \left\{ \eta_k - \sum_{r=0}^{r=p} b_r P_r(\xi_k)^2 \right\} \bigg/ 17$$

$$10.100(15)$$

obtained from 10.93(4) and 10.93(6).

In a typical trial $s_{\eta_1} = 0.1826$, $s_{\eta_2} = 0.1876$, $s_{\eta_3} = 0.0110$. The values of η used for the curve fitting were obtained from $y = 1 - x + x^3$. y was replaced by η, using $\eta = y - \bar{y}$, and x was replaced by $\xi = x - \bar{x}$. Since the values of x were given by $x = -1(0.1)1$, the value of $\bar{x} = 0$, and $\bar{y} = 1$. Thus

$$\eta = 1 - \xi + \xi^3 - 1 = -\xi + \xi^3 = b_1 P_1 + b_2 P_2 + b_3 P_3$$

The randomized equation from which the empirical values of η were obtained was $y = 1 - \xi + \xi^3 + R$, with the standard deviation of R being put equal to 0.01. Using the F test to compare the two standard deviations, for the case when the assumed standard deviation has infinite degrees of freedom (see paragraph 9.25) we find

$$\left(\frac{s_2}{s_1} \right)^2 = \frac{\chi_{p,v_2}^2}{v_2} = \left(\frac{0.011}{0.01} \right)^2 = 1.21$$

giving

$$\chi_{p,v_2}^2 = 17 \times 1.21 = 20.57$$

where

$$v_2 = 17 \text{ and } v_1 = \infty$$

The upper and lower limits for χ^2 for a probability of 0.05 are 27.6 and 8.67. The value obtained, namely 0.011, is thus quite consistent with the assumed value of 0.01. The expected values of the coefficients of ξ are $b_{0E} = 0$, $b_{1E} = -1$, $b_{2E} = 0$, $b_{3E} = 1$, where E denotes expected values.

Fourteen trials were now made, providing $14 \times 21 = 294$ separate points and fourteen values of s_{η_3} whose average value was 0.009 70. The standard deviation of the fourteen standard deviations so obtained was found to be

0.001 70. Using the formula for the standard deviation of the standard deviation, namely $\sigma_{std} = \bar{\sigma}/\sqrt{(2n)}$ (2.34(8)) with $n = 21$ and $\sigma_{\eta,3E} = 0.010$ gave $\sigma_{std} = 0.001\,54$, a very satisfactory result. Further it is to be noticed that $\bar{s}_{\eta_3}$ differs from the estimated value $\sigma_{\eta,3E} = 0.010$ by only 0.000 30, which is less than the standard error of $\bar{\sigma}_{\eta,3} = 0.001\,70/\sqrt{14} = 0.000\,45$. Each of the fourteen trials gave a value for c_i ($i = 0, 1, 2, 3$).

The following table gives the average value for each c_i, the expected value c_{iE} for each c_i, the standard deviation s_{c_i} of the c_i calculated from the fourteen values, plus an expected value for each s_{c_i} ($s_{c_{iE}}$) calculated using $\sigma_{\eta,3} = 0.010$ and the mean estimated $_{est}\bar{s}_{c_i}$ calculated from taking the mean of the s_{c_i} found from the fourteen values of the s_{c_i} obtained using 10.100(7) to (10). Finally $(\bar{c}_i - c_{iE})/(s_{c_{iE}}/\sqrt{14})$ is given.

$i =$	0	1	2	3
$\bar{c}_i$	0.000 84	−0.995 15	−0.002 29	0.995 53
c_{iE}	0	−1	0	1
s_{c_i}	0.002 78	0.006 84	0.007 59	0.010 04
$s_{c_{iE}}$	0.002 45	0.009 08	0.006 68	0.012 67
$_{est}\bar{s}_{c_i}$	0.002 37	0.008 81	0.006 47	0.012 28
$\dfrac{\bar{c}_i - c_{iE}}{s_{c_{iE}}/\sqrt{14}}$	1.28	2.00	−1.28	−1.32

Note that the mean estimated $\bar{s}_{c_i}$ are averages over the fourteen samples of the s_{c_i} calculated from internal consistency, i.e. from the goodness of fit to the orthogonal cubic in each trial. We should expect $\bar{c}_i - c_{iE}$ to be distributed about zero with uncertainties $s_{c_{iE}}/\sqrt{14}$. Thus using the values found we have

i	0	1	2	3	
$\bar{c}_i - c_{iE}$	84	485	−229	−447	$\left.\right\} \times 10^{-5}$
$s_{c_{iE}}/\sqrt{14}$	65	243	179	339	

If we assign weights $w = (1/(s_{c_{iE}}/\sqrt{14}))^2$ to each of the $\bar{c}_i - c_{iE}$ we find the average value of the $(\bar{c}_i - c_{iE})$s to be $58.1 \equiv \bar{x}$ and $\sum_1^4 w(x_i - \bar{x})^2 = 8.0$, where $x_i = \bar{c}_i - c_{iE}$, and where the $\sum$ term may be compared with $\chi^2(3)$, which exceeds 8.0 in about 5% of cases, which is an acceptable result.

To complete the analysis, data obtained for the third degree polynomial were fitted to a fourth degree polynomial. Five trials of 21 points were run and the following values obtained for $\sigma_{\eta i}$ ($i = 0, 1, 2, 3, 4$) which are shown in the table below.

Trial	Values of σ_{η_i}				
i	1	2	3	4	5
1	0.1879	0.1834	0.1777	0.1762	0.1772
2	0.1930	0.1885	0.1826	0.1810	0.1821
3	0.0092	0.0113	0.0105	0.0114	0.0094
4	0.0084	0.0113	0.0108	0.0116	0.0090

It will be seen that the third degree and fourth degree curve both fit the data equally well, witness that the standard deviations of η_3 and η_4 are all of the order of 0.010, the standard deviation of the random Gaussian variable R used. The fall of the standard deviation to an approximately constant value of 0.010 for η_3 and η_4 is an instance of what generally happens (see paragraph 10.86).

There is a small correction that could be made, but it is probably of not much significance. σ_{n_3E} was equated with σ_R.

Actually

$$\sigma_{yE} = \sigma_R \qquad \text{and} \qquad \eta = y - \bar{y}$$

and so

$$\sigma_\eta = (\sigma_y^2 + \sigma_{\bar{y}}^2)$$

if y and $\bar{y}$ are treated as being negligibly correlated.

Therefore

$$\sigma_{n_3E} = \sigma_R(1 + \tfrac{1}{21})^{1/2}$$

$$\simeq \sigma_R[1 + \tfrac{1}{42}] \simeq 1.15$$

Further

$$\frac{\sigma_\sigma}{\sigma} = \frac{1}{\sqrt{(2n)}} = \frac{1}{\sqrt{(42)}} \simeq 15\%$$

When reduced by $\sqrt{14}$ from fourteen groups this value is reduced to about 4%.

It is to be noted that if a group of g readings is taken of a variable v_i, which has a standard deviation of σ_r, and a mean of μ, where the latter two quantities are based on an infinite number of readings, then the standard deviation s_r

found from g readings will be

$$s_r = \sum_{i=1}^{i=g} ((v_i - \bar{v})^2/(g-1))$$

where $\bar{v} = \sum_{i=1}^{i=g} v_i/g$, where each reading of the variable is v_i. Now s_r will be subject to fluctuation such that if we take h groups of g readings we shall find that

$$\sigma_{s_r} = \underset{h \to \infty}{\text{Lim}} \left(\sum_{j=1}^{j=h} (s_{rj} - \bar{s}_r)/(h-1) \right)$$

where

$$\bar{s}_r = \sum_{j=1}^{j=h} s_{rj}/h$$

σ_{s_r} is also given by $\sigma_r/\sqrt{(2g)}$.

However, if we do not let $h \to \infty$, this estimate of σ_{s_r} will itself be spread, viz. its one standard deviation range will be

$$s_{s_r} = \sigma_{s_r} \left(1 \pm \frac{1}{\sqrt{(2h)}} \right)$$

In the example given, $g = 21$, $h = 14$ and $\sigma_r = 0.010$, $\bar{s}_r = 0.009\,70$, $\sigma_{s_r} = 0.01/\sqrt{42} = 0.001\,54$ and $s_{s_r} = 0.001\,54\,(1 \pm (1/\sqrt{28})) = 0.001\,54 \pm 0.000\,29$. The value of s_{s_r} estimated by external consistency among the h groups was $0.001\,70$, in comfortable agreement with the foregoing.

Weights

10.101 If the various ordinates are of differing accuracy then it will be necessary to weight the experimental values by $\omega_k = 1/\sigma_k^2$ where σ_k is the standard deviation of the kth ordinate. The least squares function will now become

$$R_p = \sum_{k=1}^{k=n} \omega_k \left[\eta_k - \sum_{r=0}^{r=p} b_r P_r(\xi_k) \right]^2 \qquad 10.101(1)$$

It is to be noted that the ω_k may be multiplied by any common factor without altering the effect on the curve fitting. σ_k is a measure of the accuracy of the kth y, but any factor which is proportional to the accuracy of the y_k may be used. For instance if it is known that the absolute percentage accuracy of the y_k is proportional to y_k then $\omega_k = 1/y_k^2$.

The need to apply weights often arises through having transformed the dependent variable prior to curve fitting. Thus if η is the dependent variable being worked with, which is really a function of an original dependent variable y, then the appropriate weights to use are

$$\omega_k = 1 \left/ \left(\frac{d\eta}{dy} \right)^2 \right. \qquad 10.101(2)$$

evaluated at the kth data point, assuming the original data values to be of equal weight. For example, if $\eta = \ln y$ then

$$\frac{d\eta}{dy} = \frac{1}{y}$$

and so

$$\omega_k = y_k^2 \qquad 10.101(3)$$

If the original data of y had differing weights $_0\omega_k$ say then 10.101(3) would become

$$\omega_k = {_0\omega_k} y_k^2 \qquad \text{or} \qquad {_0\omega_k} \left/ \left(\frac{d\eta}{dy} \right)^2 \right. \qquad 10.101(4)$$

Selection of Data Points

10.102   As mentioned in paragraph 10.36, the maximum degree of polynomials selected for curve fitting should not exceed half the number of points. Since the orthogonal method enables the addition of an extra degree by evaluating only the corresponding constant, it is very easy to calculate $s_{\eta,p}$, the standard deviation of each polynomial of degree p. If taking equally spaced points does not provide an acceptable fit, then it may be worth trying unequally spaced values of ξ, the independent variable, given by $\cos(\pi k/n)$ where $k = 0$ to n and where n is the number of data points. ξ is of course related to the original data by equations 10.88(1) to 10.88(3) and $\eta = y - \bar{y}$ (10.88(4)) where the y are the original dependent variable.

There is a relationship between the fitted coefficients of the orthogonal polynomials and the standard deviation of the η for the fitted curve. A sudden drop in the values of the bs will be found to be related to the approximate constancy of the $\sigma_{\eta,p}$. The degree of the fitted polynomial, which begins the approximate constancy of succeeding $\sigma_{\eta,p}$, will be found to correspond to the degree of the orthogonal polynomial immediately preceding the one which produces the large drop in the value of its b. The degree of the polynomial selected should be that corresponding to one preceding that for which the drop in b occurs. The drop in the value of b can thus be used as a guide for choosing the degree of the polynomial required for fitting.

References

1[**10.01**] ⎫ For more information on the method of least squares, see E T
Whittaker and G Robinson *Calculus of Observations* Chapter IX
and
2[**10.17**] ⎭ B W Lindgren *Statistical Theory* p 446 et seq.

Examples on Chapter 10

Example 10.1   The method described in paragraphs 10.03 to 10.09 has been used to provide the readings for the calculation of the topography of a surface plate. The area covered by the survey is about 4 ft 7 in by 2 ft 6 in. The Talyvel readings which are shown below give the gradient in units of 0.000 01 in/in, equivalent to being expressed in radians, and the text following these readings gives the steps in the calculation of the shape of the surface.

Gradient in units of 0.000 01 in/ 1 in	Diagonal A–C	Diagonal B–D	Side A–D	Side B–C	Side B–A	Side C–D	Middle side E–F	Middle side G–H
α_0	0.0	0.0	0.0	0.0	0.0	0.0	0.0	0.0
α_1	0.0	0.0	0.0	0.0	0.0	0.0	0.0	0.0
α_2	0.0	−3.0	1.0	−5.0	−2.0	3.0	−1.0	2.0
α_3	1.5	−5.0	3.0	2.0	−1.0	0.0	−3.0	1.0
α_4	3.0	−5.0	5.0	−2.0	1.0	6.0	−1.0	3.0
α_5	2.0	−4.0	4.0	1.0	2.0	8.0	−2.0	3.0
α_6	2.0	−4.0	1.0	5.0	0.0	7.0	0.0	2.0
α_7	2.0	−2.0			0.0	4.0	−1.0	
α_8	0.5	−3.0			1.0	2.0	−2.0	
α_9	1.5	−2.0			2.0	3.0	−2.0	
α_{10}	3.0	−2.0			2.0	2.0	−1.0	
α_{11}	2.0	−1.0						
α_{12}	3.0	−5.0						

$d_d = 5.5$ in for AC, BD, BA and CD. $d_d = 5.0$ in for AD, GH and BC.

Formula 10.06(2) is now used to calculate z_p'' for the diagonal sides (columns 2 and 3) and formula 10.08(1) to calculate the points for the remaining sides. By way of illustration, when the points for the side A → D are found, the a of the formula 10.08(1) corresponds to the z_0'' of the diagonal A → C and the b to the z_n'' of the diagonal B → D. Similarly, when calculating the points for GH a is the calculated height for the mid-point of the side B → A and b is the calculated height of the mid-point of the side C → D. From what has been said it will be seen that the order in which the heights are calculated is predetermined: first the diagonals AC and BD, then the sides BA, CD, AD and BC in any order, and then the two middle paths GH and FE. Having found the heights, the magnitude of the greatest negative point has been added to all the heights, making all the readings of the heights positive or zero above the reference plane xy. The first table below gives the values of z_p'' and z_p''' and the second table gives all the positive values of the heights of the points above the reference plane xy.

Table of heights z relative to xy plane

Heights z in units of 0.0001 in	Diagonal A–C	Diagonal B–D	Side A–D	Side B–C	Side B–A	Side C–D	Middle side E–F	Middle side G–H
z_0	0.96	1.65	0.96	1.65	1.65	0.96	−0.44	−0.07
z_1	0.02	3.30	−0.09	1.45	1.31	−0.89	0.30	−0.80
z_2	−0.92	3.30	−0.64	−1.25	−0.14	−1.10	0.49	−0.53
z_3	−1.03	2.20	−0.19	−0.44	−1.03	−2.96	−0.42	$\boxed{-0.77}$
z_4	−0.32	1.10	1.25	−1.64	−0.83	−1.51	−0.23	−0.00
z_5	−0.16	0.55	2.20	−1.34	−0.07	1.03	$\boxed{-0.59}$	0.76
z_6	$\boxed{0.00}$	$\boxed{0.00}$	1.65	0.96	−0.41	3.02	0.15	1.03
z_7	0.16	0.55			−0.76	3.37	0.34	
z_8	−0.50	0.55			−0.55	2.61	−0.02	
z_9	−0.62	1.10			0.21	2.41	−0.38	
z_{10}	0.09	1.65			0.96	1.65	−0.19	
z_{11}	0.25	2.75						
z_{12}	0.96	1.65						

The values z_6 for AC and BD are arbitrarily made to be zero, and the values z_3 of GH and z_5 of FE should also be zero, because they also correspond to the centre point of the measurement paths, that is the point Q on Figure Ex. 10.1 of the measurement path. The discrepancy is a measure of the

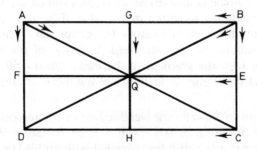

Figure Ex. 10.1 Shows the relation between the paths of the measured points. The arrows give the order in which the measurements were made

cumulative error inherent in the measurement process. It is left to the reader to apply the averaging process described in paragraphs 10.10 and 10.11.

Table of heights above *xy* plane

Heights z above xy plane in units of 0.0001 in	Diagonal A–C	Diagonal B–D	Side A–D	Side B–C	Side B–A	Side C–D	Middle side E–F	Middle side G–H
z_0	3.92	4.61	3.92	4.61	4.61	3.92	2.51	2.89
z_1	2.97	6.26	2.87	4.41	4.26	2.06	3.25	2.15
z_2	2.04	6.26	2.31	1.71	2.82	1.86	3.44	2.42
z_3	1.93	5.16	2.76	2.51	1.92	0.00	2.53	2.19
z_4	2.63	4.06	4.21	1.31	2.13	1.44	2.72	2.95
z_5	2.80	3.51	5.16	1.62	2.89	3.99	2.36	3.72
z_6	2.96	2.96	4.61	3.92	2.54	5.98	3.10	3.99
z_7	3.12	3.51					6.32	3.29
z_8	2.45	3.51					5.57	2.93
z_9	2.34	4.06					5.36	2.57
z_{10}	3.05	4.61					4.61	2.76
z_{11}	3.21	5.71						
z_{12}	3.92	4.61						

The outlined values, which correspond to the mid-point Q of the diagram of the measurement paths, should all be the same value, but as just mentioned, this has not occurred because of the build-up of measurement uncertainties.

Example 10.2 A standard cell has been measured by several different laboratories over a period of time covering approximately ten months. The readings were made at approximately 37.5°C, but the readings given below have been corrected to an exact temperature of 37.5°C. The data on which the readings were made is also given, as is the claimed uncertainty of each laboratory. Calculate the weighted mean value of the cell voltage, using as a weighting factor for each laboratory the reciprocal of the square of its uncertainty. Calculate also the standard deviation of the mean weighted voltage, together with the gradient of the mean linear drift line of voltage against time, and the standard deviation of the mean gradient of this line.

Hint: Use equation 10.30(4) for the weighted mean voltage, equation 10.31(6) for the standard deviation of this mean voltage, equation 10.30(1) for the mean gradient of the mean drift line together with 10.30(3) and 10.30(4) for the means involved, and equation 10.32(8) for the standard deviation of the mean gradient. τ_r and ε_r which will also be needed are given by equations 10.28(3) and 10.28(4).

Table of standard cell voltage readings

Laboratory	Corrected voltage	Claimed uncertainty in μV	Number of days after first measurement
1	1.017 6859	±3	0
2	1.017 6878	±5	40
3	1.017 6866	±5	60
4	1.017 6843	±5	82
5	1.017 6840	±8	109
6	1.017 6853	±3	145
7	1.017 6820	±15	173
8	1.017 6822	±5	190
9	1.017 6817	±4	217
10	1.017 6850	±35	237
11	1.017 6904	±15	252
12	1.017 6842	±3	274
13	1.017 6840	±7	295

Answer:
Weighted mean voltage = 1.017 6851 V.
Standard deviation of mean voltage = 0.34 V.
Gradient of mean drift line (voltage against time) = -0.0067 μV/day.
Standard deviation of gradient of mean drift line = $+0.0028$ μV/day.

Example 10.3 Given below are the deviations in ohms from a nominal value of 2×10^5 Ω for a standard resistor when it was measured by a number of laboratories. Also given are the claimed uncertainties of each laboratory and the data of each set of measurements. Find the weighted mean value of the resistor, together with the standard deviation of this mean. Also calculate the gradient of the mean drift line with time and its standard deviation. State also the uncertainty of the weighted mean and of the gradient of the mean line for a confidence probability of 0.997. Take the weight of each laboratory's reading as proportional to the reciprocal of the square of its uncertainty.

Laboratory	Deviation of resistor from 2×10^5 Ω in Ω	Claimed uncertainty of readings in Ω	Numbers of days after first measurement
1	+0.63	±3	0
2	+2.5	±2.6	74
3	+4.0	±100	77
4	+1.2	±2	79
5	+5.0	±6	109
6	+2.0	±6	126
7	+3.0	±20	147
8	−9.0	±30	165
9	0.0	±6	194
10	−0.4	±6	226
11	0.0	±10	246
12	9.99	±40	290

Answer:

Weighted mean = 200 00.685 Ω.

Standard deviation of mean = 0.486 Ω.

Gradient of mean line = -0.0171 Ω/day.

Standard deviation of mean line gradient = 0.007 77 Ω/day.

Uncertainty of the weighted mean value for a confidence probability of 0.997 = 1.458 Ω.

Uncertainty of gradient in mean line for a confidence probability of 0.997 = 0.0233 Ω/day.

Example 10.4 The values given in the following table have been found for the capacitance of a condenser when it has been measured at 1 kHz by a number of laboratories. Find the weighted mean value of the capacitance, the weighted mean time, and determine the weighted standard error of deviations from the mean line and the confidence limits for deviations from the mean line for a confidence probability of 0.997. Find also the gradient of the line of regression, that is the mean line (capacitance against time), the standard deviation of the gradient of this line, and the coefficient of correlation. Find the equation of the mean line and the mean value of the capacitance 185 days from the time the first measurement was made. Take the weight of each capacitor as proportional to the reciprocal of the square of each laboratory's claimed uncertainty.

Laboratory	Capacitance in pF at 1 kHz	Uncertainty claimed in pF	Number of days after first measurement
1	699.70	± 0.1	0
2	699.82	± 0.07	34
3	699.669	± 0.035	75
4	699.8	± 0.2	96
5	700.72	± 1.4	111
6	699.689	± 0.15	160
7	699.97	± 0.1	182
8	699.90	± 0.35	195
9	700.03	± 0.10	234
10	699.7	± 0.4	245

Answer:

Weighted mean value of the capacitance = 699.745 pF.

Weighted mean time = 87.17 days.

Weighted standard deviation of the mean from the mean line = 0.0316 pF.

Confidence limits (0.997 confidence probability) = 0.0948 pF.

Gradient of mean line (regression line) = 0.0013 pF/days.

Standard deviation of gradient of mean line = 0.00053 pF/days.
Coefficient of correlation = 0.6559.
Equation of mean line is $y = 0.0013x + 699.625$ where $y \equiv$ capacitance in
  pF and $x \equiv$ time in days from first measurement of the capacitor.
Mean value of capacitance after 185 days = 699.865 pF.

Example 10.5 In the previous example calculate the value of the parabolic
coefficient of correlation, and state the equation of the parabola. Calculate
also the weighted standard error of the deviations from the mean parabola.

Answer:
Coefficient of correlation = 0.3949.
The equation of the mean parabola is given by

$$y = 1.189 \times 10^{-5} x^2 - 2.073 \times 10^{-3} x + 699.835$$

where x is measured in days from the time of the first measurement and y
is measured in pF. The weighted standard error of deviations from the mean
parabola = 0.0327 pF.

Standard deviation of combination of two non-independent variables

Example 10.6 Two variables u and v, each of which has a normal
distribution, are not independent. If the coefficient of correlation is 'p' and
the standard deviation of each variable is respectively σ_1 and σ_2, find
the standard deviation and density function of the variable z, where
$z = a(u - u_p) + b(v - v_p)$, where v_p is the mean value of v and u_p the mean
value of u.

From equation 10.69(1) the probability that u lies between $u + du$ and u,
whilst v lies between $v + dv$ and v, is given by

$$\exp - \left[\frac{1}{2(1 - r^2)} \left\{ \frac{(u - u_p)^2}{\sigma_1^2} - \frac{2r(u - u_p)(v - v_p)}{\sigma_1 \sigma_2} + \frac{(v - v_p)^2}{\sigma_2^2} \right\} \right] du \, dv$$

$$\times \left[2\pi \sigma_1 \sigma_2 \sqrt{(1 - r^2)} \right]^{-1} \qquad \text{Ex. 10.6(1)}$$

Put $u - u_p = x$ and $v - v_p = y$ and therefore $du = dx$ and $dv = dy$. Thus
$z = ax + by$.
  Now we require to know the probability that

$$z + dz \geqslant x + y \geqslant z \qquad \text{Ex. 10.6(2)}$$

because this will give the density function of z and hence the standard
deviation. In order to satisfy the inequality Ex. 10.6(2) we must integrate
expression Ex. 10.6(1) (with x and y substituted for u and v), with respect to

y between the limits y and $y + dy$ and with respect to x between the limits $-\infty$ to $+\infty$. Now

$$y = \frac{z - ax}{b} \quad \text{and} \quad y + \delta y = \frac{z + dz - ax}{b}$$

whilst

$$\delta y = \frac{dz}{b}$$

The required probability that z should lie between z and $z + dz$ is thus given by

$$P_z \, dz = \frac{1}{2\pi\sigma_1\sigma_2\sqrt{(1 - r^2)}} \int_{-\infty}^{\infty} dx \int_{y}^{y+\delta y} \exp -\left[\frac{1}{2(1 - r^2)} \left\{ \frac{x^2}{\sigma_1^2} - \frac{2rxy}{\sigma_1\sigma_2} + \frac{y^2}{\sigma_2^2} \right\} \right] dy$$

$$= \frac{dz}{b2\pi\sigma_1\sigma_2\sqrt{(1 - r^2)}} \int_{-\infty}^{\infty} \exp -\left[\frac{1}{2(1 - r^2)} \left\{ \frac{x^2}{\sigma_1^2} - \frac{2rx(z - ax)}{\sigma_1\sigma_2 b} + \frac{(z - ax)^2}{b^2\sigma_2^2} \right\} \right] dx$$

Ex. 10.6(3)

The part of the index of the exponential in curly brackets, namely

$$\frac{x^2}{\sigma_1^2} - \frac{2rx(z - ax)}{\sigma_1\sigma_2 b} + \frac{(z - ax)^2}{b^2\sigma_2^2}$$

can be written as

$$x^2 \left(\frac{1}{\sigma_1^2} + \frac{2ra}{b\sigma_1\sigma_2} + \frac{a^2}{b^2\sigma_2^2} \right) - \frac{2zx}{b\sigma_2} \left(\frac{r}{\sigma_1} + \frac{a}{b\sigma_2} \right) + \frac{z^2}{b^2\sigma_2^2}$$

If we put

$$\left.\begin{array}{l} f = \dfrac{b^2\sigma_2^2 + a^2\sigma_1^2 + 2rab\sigma_1\sigma_2}{b^2\sigma_1^2\sigma_2^2}, \\[4mm] g = -\dfrac{2z}{b\sigma_2} \left(\dfrac{r}{\sigma_1} + \dfrac{a}{b\sigma_2} \right) \quad \text{and} \quad c = \dfrac{z^2}{b^2\sigma_2^2} \end{array}\right\}$$

Ex. 10.6(4)

then it can be written as $fx^2 + gx + c$ which can be put equal to

$$f(x + h)^2 + k = fx^2 + 2fhx + fh^2 + k$$

Equating coefficients of x^2, x and of the constant terms in the two above expressions, we have $g = 2fh$, and $c = fh^2 + k$, giving

and

$$\left.\begin{array}{l} h = g/2f \\[2mm] k = c - g^2/4f \end{array}\right\}$$

Ex. 10.6(5)

Substituting in equations Ex. 10.6(5) for c, f and g from Ex. 10.6(4) gives

$$h = \frac{-z\sigma_1(br\sigma_2 + a\sigma_1)}{(a^2\sigma_1^2 + b^2\sigma_2^2 + 2rab\sigma_1\sigma_2)}$$

and

$$k = \frac{z^2}{b^2\sigma_2^2} - \frac{z^2(br\sigma_2 + a\sigma_1)^2}{b^2\sigma_2^2(a^2\sigma_1^2 + b^2\sigma_2^2 + 2rab\sigma_1\sigma_2)}$$

$$= \frac{z^2(1 - r^2)}{(a^2\sigma_1^2 + b^2\sigma_2^2 + 2rab\sigma_1\sigma_2)}$$

The required probability $P_z\, dz$ is now given by

$$P_z\, dz = \frac{dz}{b2\pi\sigma_1\sigma_2\sqrt{(1 - r^2)}} \exp - \left[\frac{k}{2(1 - r^2)}\right] \int_{-\infty}^{\infty} \exp - \left[\frac{f(x + h)^2}{2(1 - r^2)}\right] dx$$

$$= \frac{dz}{b2\pi\sigma_1\sigma_2\sqrt{(1 - r^2)}} \exp - \left[\frac{z^2}{2(a^2\sigma_1^2 + b^2\sigma_2^2 + 2rab\sigma_1\sigma_2)}\right]$$

$$\times \int_{-\infty}^{\infty} \exp - \left[\frac{(a^2\sigma_1^2 + b^2\sigma_2^2 + 2rab\sigma_1\sigma_2)}{2(1 - r^2)b^2\sigma_1^2\sigma_2^2}\left\{x - \frac{z\sigma_1(br\sigma_2 + a\sigma_1)}{a^2\sigma_1^2 + b^2\sigma_2^2 + 2rab\sigma_1\sigma_2}\right\}^2\right] dx$$

If we write

$$\frac{\sqrt{f}(x + h)}{\sqrt{\{2(1 - r^2)\}}} = W$$

then

$$dx = dW\frac{\sqrt{\{2(1 - r^2)\}}}{\sqrt{f}}$$

and the integral becomes

$$\int_{-\infty}^{\infty} \frac{\sqrt{\{2(1 - r^2)\}}}{\sqrt{f}} e^{-W^2}\, dW = \frac{\sqrt{\{2\pi(1 - r^2)\}}}{\sqrt{f}}$$

$$= \frac{b\sigma_1\sigma_2\sqrt{\{2\pi(1 - r^2)\}}}{(a^2\sigma_1^2 + b^2\sigma_2^2 + 2rab\sigma_1\sigma_2)^{1/2}}$$

Thus finally

$$P_z\, dz = \frac{\exp - [z^2/2(a^2\sigma_1^2 + b^2\sigma_2^2 + 2rab\sigma_1\sigma_2)]\, dz}{\sqrt{\{2\pi(a^2\sigma_1^2 + b^2\sigma_2^2 + 2rab\sigma_1\sigma_2)\}}}$$

The required density function is the coefficient of dz on the right-hand side. Comparison with the standard form of the normal distribution shows that the distribution for z is normal and that it has a standard deviation of $(a^2\sigma_1^2 + b^2\sigma_2^2 + 2rab\sigma_1\sigma_2)^{1/2}$. It can be shown that this expression is also true when the distributions of u and v are non-normal. It also applies to the case of discrete or stochastic distributions.

Example 10.7 If u and v are the same variables as in the previous example, find the density distribution and standard deviation of $z = a(u - u_p) - b(v - v_p)$.

Answer:

$$P_z = \frac{\exp - [z^2/2(a^2\sigma_1^2 + b^2\sigma_2^2 - 2rab\sigma_1\sigma_2)]}{\sqrt{\{2\pi(a^2\sigma_1^2 + b^2\sigma_2^2 - 2rab\sigma_1\sigma_2)\}}}$$

Standard deviation of $z = (a^2\sigma_1^2 + b^2\sigma_2^2 - 2rab\sigma_1\sigma_2)^{1/2}$.

11

Theorems of Bernoulli and Stirling and the Binomial, Poisson and Hypergeometric Distributions

Bernoulli's Theorem

11.01 Suppose we have a population of twelve balls, eight black ones and four white ones. Then the probability of selecting at random a white ball from this population is equal to $4/12 = 1/3$. Similarly the probability of selecting a black ball is equal to $8/12 = 2/3$. If p is the probability of choosing a white ball, and q the probability of choosing a black ball, then $p + q = 1$, and thus $q = 1 - p$. The eight black balls and the four white balls are called sub-classes of the population of twelve balls. Let the sub-class of eight black balls be designated by a, and the sub-class of the four white balls by b. Suppose we now select, at random and in succession, four balls from the population of twelve, replacing each ball after its selection before removing the next one. We now ask the question, what is the probability that of the four balls selected, three belong to the sub-class a. Let us consider an actual case. The probability of the first ball belonging to the sub-class a is $2/3$, and the probability of the second and third balls belonging to the sub-class a is also $2/3$. Thus the probability of the first three balls belonging to the sub-class a is $2/3 \times 2/3 \times 2/3$. Now the fourth ball selected must not belong to the sub-class a, and the probability of this is $1 - 2/3 = 1/3$. Thus the probability of the first three balls selected belonging to the sub-class a and the next one to the sub-class b is thus $2/3 \times 2/3 \times 2/3 \times 1/3 = 8/81$.

11.02 But this is not the probability of any four balls selected at random, with replacement after each choice, consisting of three balls of class a and one ball of class b, because the order in which the balls are selected can differ. In fact it is easily seen that the same selection can be made in four ways so that the total probability is

$$\frac{4 \times 8}{81} = 0.395$$

11.03 Let us now generalize. Suppose we have a population which contains two sub-classes a and b, such that the probability of an article being in class a is p, whilst the probability of it being in sub-class b is $q = 1 - p$. Thus the probability that in any n members of the population r belong to sub-class a and $n - r$ to sub-class b is obtained as follows. If we make a selection with replacement as before, then the probability that the first r articles belong to sub-class a is p^r, whilst the probability that the next $n - r$ belong to sub-class b is $(1 - p)^{n-r} = q^{n-r}$.

11.04 Thus the probability of this double event is $p^r(1 - p)^{n-r}$. Now this selection can be performed in several ways. The number of ways we can arrange n different things among themselves is $n! = n(n - 1)(n - 2)\ldots3.\,2.\,1$. The number of ways r different things can be arranged amongst themselves is thus $r!$ Now if r of the n things are identical the number of arrangements is reduced in the ratio of $1/r!$ Similarly, if the remaining $n - r$ things are identical, the number of arrangements is reduced further by $1/(n - r)!$ Thus the number of different ways of arranging n things amongst themselves which consist of one set of r like things and a second set of $n - r$ like things different from the first set of r like things is equal to

$$\frac{n!}{r!(n - r)!} = {}^nC_r \qquad\qquad 11.04(1)$$

11.05 Put another way, this is the number of different ways r like things and $n - r$ different like things can be selected from a population of n things containing both sets of things. Thus the required probability is equal to $p^r(1 - p)^{n-r}$, which is the probability of any one selection containing just r things of sub-class a out of the n chosen, multiplied by the number of ways this selection can be made, that is

$$ {}^nC_r p^r(1 - p)^{n-r} \qquad\qquad 11.05(1)$$

This probability is known as Bernoulli's theorem.

11.06 Two theorems follow directly from Bernoulli's theorem, the first is

Theorem 1

If p is the probability that a member of a population should belong to a specified sub-class, then the probability that out of n members of the population not more than r belong to this sub-class is

$$^nC_0(1-p)^n + {}^nC_1 p(1-p)^{n-1} \ldots {}^nC_r p^r (1-p)^{n-r}\dagger$$

The second theorem is as follows.

Theorem 2

The probability that not less than r members out of n members belong to the sub-class is

$$^nC_r p^r (1-p)^{n-r} + {}^nC_{r+1} p^{r+1}(1-p)^{n-r-1} \ldots {}^nC_n p^n \ddagger$$

11.07 It is to be noted that each term of the two series is contained in the expansion of

$$(q+p)^n \qquad\qquad 11.07(1)$$

where $q = 1 - p$. Since the expansion of 11.07(1) is known as the binomial theorem, and as each term of 11.07(1) gives the probability of finding r things of a sub-class in a group of n ($r = 0, 1, 2, \ldots, n$) the distribution given by 11.07(1) is known as the binomial distribution, or Bernoulli distribution.
11.08 Bernoulli's theorem can be generalized as follows.

First generalization of Bernoulli's theorem

Let a population be divisible into sub-classes $a_1, a_2, a_3, \ldots, a_i$, the probabilities associated with each sub-class being $p_1, p_2, \ldots, p_t$. Then the probability that a group of n members of the population, otherwise unspecified, should contain r_1 members of sub-class a_1, r_2 members of sub-class $a_2, \ldots, r_t$ members of sub-class a_t is

$$\frac{n!}{r_1! r_2! \ldots r_t!} \cdot p_1^{r_1} \cdot p_2^{r_2} \cdot p_3^{r_3} \ldots p_t^{r_t} \qquad\qquad 11.08(1)$$

11.09 This is shown as follows. The probability of r_1 members of a group of n members belonging to the sub-class a_1 is $p_1^{r_1}$, the probability of r_2 members of the same group belonging to the sub-class a_2 is $p_2^{r_2}$ etc. Thus the probability of the combined event is $p_1^{r_1} p_2^{r_2} p_3^{r_3} \ldots p_t^{r_t}$. The required probability is the product of this probability and the number of ways in which this probability can be

---

† For an approximation to this sum see paragraphs 11.22 and 11.23.
‡ For an approximation to this sum see equation 11.24(1) in which z_2 is put equal

to ∞, giving the required probability as $\frac{1}{2}\left\{ 1 - \left(\dfrac{P}{-k_1 \text{ to } k_1} \right)_{k_1 = z_1} \right\}$ where z is given by

11.23(1) and $r \equiv x$.

formed, subject to the condition that the total number of members is n. Thus we have to multiply $p_1^{r_1} p_2^{r_2} p_3^{r_3} \dots p_t^{r_t}$ by the number of ways in which a term of this type can arise from n combinations of all such ps. This number is equal to the coefficient of $p_1^{r_1} p_2^{r_2} p_3^{r_3} \dots p_t^{r_t}$ in the expansion of $(p_1 + p_2 + p_3 \dots p_t)^n$, that is

$$\frac{n!}{r_1! r_2! r_3! \dots r_t!}$$

Average and Most Probable Value of r for Bernoulli's Expansion

11.10 From 11.07(1) we know that the probability of obtaining r events, whose initial probability is p, is given by the rth term in the expansion of $(q + p)^n$ in ascending powers of p, where $q = 1 - p$. The sum of all the probabilities is of course equal to unity. The average value of r is given by

$$\bar{r} = \frac{\sum_{r=0}^{r=n} {}^nC_r p^r q^{n-r} \cdot r}{\sum_{r=0}^{r=n} {}^nC_r p^r q^{n-r}}$$

$$= \sum_{r=0}^{r=n} {}^nC_r p^r q^{n-r} \cdot r \qquad\qquad 11.10(1)$$

since

$$\sum_{r=0}^{r=n} {}^nC_r p^r q^{n-r} = (q + p)^n = 1 \qquad\qquad 11.10(2)$$

Thus differentiating 11.10(2) with respect to p and putting $p + q = 1$, we have

$$n = \sum_{r=0}^{r=n} {}^nC_r p^{r-1} q^{n-r} \cdot r$$

Multiplying by p on both sides we have

$$np = \sum_{r=0}^{r=n} {}^nC_r p^r q^{n-r} \cdot r = \bar{r} \qquad\qquad 11.10(3)$$

thus the average value of r is equal to np.

11.11 Now, to determine the value of r for which the Bernoulli probability

$$B(n, r) = {}^nC_r p^r (1 - p)^{n-r} \qquad\qquad 11.11(1)$$

has its greatest value, we require to find the greatest value of this function for integral r (if it exists) in the range $0 \leqslant r \leqslant n$. Thus we require to find the value of r such that

$$B(n, r - 1) \leqslant B(n, r) \leqslant B(n, r + 1)$$

Substituting 11.11(1) into the above, and cancelling common factors, we have

$$np + p \geqslant r \geqslant np - (1 - p) \qquad 11.11(2)$$

If np is integral, then since p is less than unity, then in this case $r = np$. If np is not integral, but equal to $s + u$, where s is integral and u is fractional, then $s + u + p \geqslant r \geqslant s + u - 1 + p$. If $u + p < 1$, then $r = s$, the integral part of np. If $u + p \geqslant 1$, then $r = s + 1$. The required value of r is designated as $[np]$, that is the most probable integral value of r. Thus by 11.10(3) the average value of r is approximately equal to the most probable value of r, which gives the maximum value of $B(n, r)$.

Standard Deviation of Bernoulli Distribution

11.12 The deviation of any value of r from its mean value np is $np - r$, and the standard deviation of r is thus given by

$$\sigma_r^2 = \sum_{r=0}^{r=n} {}^nC_r p^r q^{n-r}(np - r)^2$$

$$= n^2 p^2 \sum_{r=0}^{r=n} {}^nC_r p^r q^{n-r} - 2np \sum_{r=0}^{r=n} {}^nC_r p^r q^{n-r} \cdot r + \sum {}^nC_r p^r q^{n-r} \cdot r^2 \qquad 11.12(1)$$

Now

$$(p + q)^n = \sum_{r=0}^{r=n} {}^nC_r p^r q^{n-r} \qquad 11.10(2)$$

Thus, differentiating with respect to p and multiplying both sides by p we get

$$np(p + q)^{n-1} = \sum_{r=0}^{r=n} {}^nC_r p^r q^{n-r} \cdot r \qquad 11.12(2)$$

yielding

$$np = \sum_{r=0}^{r=n} {}^nC_r p^r q^{n-r} \cdot r \qquad 11.10(3)$$

since $(p + q) = 1$. Differentiating 11.12(2) with respect to p we have

$$(n - 1)np(p + q)^{n-2} + n(p + q)^{n-1} = \sum_{r=0}^{r=n} {}^nC_r p^{r-1} q^{n-r} \cdot r^2$$

and on multiplying all through by p we obtain

$$(n - 1)np^2(p + q)^{n-2} + np(p + q)^{n-1} = \sum_{r=0}^{r=n} {}^nC_r p^r q^{n-r} \cdot r^2 \qquad 11.12(3)$$

Putting $p + q = 1$ we have finally

$$n(n-1)p^2 + np = \sum_{r=0}^{r=n} {}^nC_r p^r q^{n-r} \cdot r^2 \qquad 11.12(4)$$

Using 11.10(2) with $p + q = 1$, 11.10(3) and 11.12(4), we can replace, respectively, the three sums involved in 11.12(1). This gives

$$\sigma_r^2 = n^2p^2 - 2n^2p^2 + n(n-1)p^2 + np$$

$$= np(1-p) = npq \qquad 11.12(5)$$

and therefore

$$\sigma_r = \sqrt{(npq)} \qquad 11.12(6)$$

Bernoulli's Limit Theorem

11.13 We have just seen that the average value of $|r - np|$ is equal to $\sqrt{(npq)}$. As n tends to infinity, $\sqrt{(npq)} \rightarrow \infty$ and so the probability of obtaining a deviation which is less than any assigned number tends to zero.

11.14 At the same time, the mean value of $|(r/n) - p|$ is equal to $\sqrt{(pq/n)}$. Thus we have the following result.

Theorem

When the number n of trials is increased indefinitely, the probability that $|(r/n) - p|$ will remain less than any assigned number tends to unity.

Stirling's Theorem

11.15 This theorem enables an approximation to $n!$ to be made, when n is very large. Now

$$\ln n! = \ln n + \ln(n-1) + \cdots + \ln 2$$

Let us now consider the curve represented by the function $y = \ln x$. If ordinates are erected at $x = 1, 2, \ldots, n$, then the area represented by the sum of the areas of the trapezia determined by successive pairs of ordinates will be less than the total area under the curve from $x = 1$ to $x = n$ (see Figure 11.15). The area under the curve is given by

$$\int_1^n \ln x \, dx = [x \ln x - x]_1^n = n \ln n - n + 1$$

The area of the trapezia is equal to

$$\tfrac{1}{2}(\ln 1 + \ln 2) + \tfrac{1}{2}(\ln 2 + \ln 3) + \cdots + \tfrac{1}{2}(\ln(n-1) + \ln n)$$

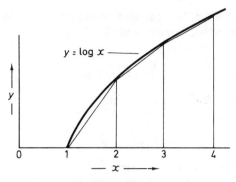

Figure 11.15

Thus

$$n \ln n - n + 1 > \ln 2 + \ln 3 + \cdots + \ln n - \tfrac{1}{2} \ln n$$

or

$$\ln(n^n e^{-n+1}) > \ln(n! \, n^{-1/2})$$

and so

$$n! < n^{n+1/2} e^{-n+1} \qquad\qquad 11.15(1)$$

11.16 In order to obtain a closer approximation to $n!$ a more exact estimate of $n!/n^{n+1/2} e^{-n+1}$ is required. Let us write

$$u_n = \ln(n!/n^{n+1/2} e^{-n+1})$$

Thus

$$u_{n+1} - u_n = \ln\left\{\frac{(n+1)! n^{n+1/2} e^{-n}}{n! (n+1)^{n+3/2} e^{-n-1}}\right\}$$

$$= 1 + \ln\left(\frac{n}{n+1}\right)^{n+1/2}$$

$$= 1 - (n + \tfrac{1}{2})\ln\left(1 + \frac{1}{n}\right)$$

$$= 1 - (n + \tfrac{1}{2})\left(\frac{1}{n} - \frac{1}{2n^2} + \frac{1}{3n^2} - \frac{1}{4n^4} + \cdots\right)$$

$$= -\frac{1}{12n^2} + \frac{1}{12n^3} - \frac{3}{40n^4} + \frac{1}{15n^5} - \cdots$$

$$\simeq -\frac{1}{12}\left(\frac{1}{n^2} - \frac{1}{n^3} + \frac{1}{n^4} - \frac{1}{n^5} + \cdots\right)$$

$$\simeq -\frac{1}{12n(n+1)} = \frac{1}{12}\left(\frac{1}{n+1} - \frac{1}{n}\right)$$

since the bracketed series in the penultimate line is a geometric progression, and so we may write that approximately

$$u_n = A + \frac{1}{12n} \qquad\qquad 11.16(1)$$

where A is an unspecified constant. Thus

$$\frac{n!}{n^{n+1/2}\,e^{-n}} = B\,e^{1/12n} \qquad\qquad 11.16(2)$$

where B is another unspecified constant.

11.17 B can be found by putting in a specified value for n, that is putting $n = 8$ yields

$$B \simeq \frac{8!}{8^8 2\sqrt{2}\,e^{-8}}$$

Alternatively B can be found as follows:

$$\frac{\sin \pi x}{\pi x} = \left(1 - \frac{x^2}{1}\right)\left(1 - \frac{x^2}{2}\right)\left(1 - \frac{x^2}{3}\right)^{\dagger} \cdots$$

When $x = \frac{1}{2}$ we have

$$\frac{2}{\pi} = \left(1 - \frac{1}{2^2}\right)\left(1 - \frac{1}{4^2}\right)\left(1 - \frac{1}{6^2}\right)$$

$$= \frac{1 \cdot 3}{2^2} \cdot \frac{3 \cdot 5}{4^2} \cdot \frac{5 \cdot 7}{6^2} \cdots = \frac{1^2 \cdot 3^2 \cdot 5^2 \cdot 7^2 \cdots}{2^2 \cdot 3^2 \cdot 6^2 \cdots} = \frac{1^2 \cdot 2^2 \cdot 3^2 \cdot 4^2 \cdot 5^2 \cdot 6^2 \cdot 7^2 \cdots}{2^4 \cdot 4^4 \cdot 6^4 \cdot 7}$$

thus

$$\frac{2}{\pi} = \lim_{n \to \infty} \frac{\{(2n+1)!\}^2}{2^{4n}(n!)^4(2n+1)}$$

Thus substituting for the two factorial terms from 11.16(2) we have

$$\frac{2}{\pi} = \lim_{n \to \infty} \frac{\{B(2n+1)^{2n+3/2}\,e^{(-2n-1)}\,e^{1/12(2n+1)}\}^2}{2^{4n}\{Bn^{n+1/2}\,e^{-n}\,e^{1/12n}\}^4(2n+1)}$$

$$= \lim_{n \to \infty} \exp\left\{-4n - 2 + 4n + \frac{1}{6(2n+1)} - \frac{1}{3n}\right\} \times \frac{(2n+1)^{4n+2}}{2^{4n}n^{4n+2}}\bigg/ B^2$$

$$= \frac{1}{B^2 e^2} \lim_{n \to \infty} 2^2\left(\frac{2n+1}{2n}\right)^{4n+2} = \frac{4}{B^2 e^2} \lim_{n \to \infty} \left(1 + \frac{1}{2n}\right)^{4n}\left(1 + \frac{1}{2n}\right)^2$$

$$= \frac{4}{B^2}$$

† See Whittaker and Watson, pp 136–7, paragraph 7.5.

Therefore $B = \sqrt{(2\pi)}$. The approximate formula for $n!$ for large values of n is thus given by

$$n! = \sqrt{(2\pi)}n^{n+1/2}e^{-n+(1/12n)}$$

$$= \sqrt{(2\pi)}n^{n+1/2}e^{-n}\left(1 + \frac{1}{12n}\right) \quad \text{approximately} \quad 11.17(1)$$

The $1/12n$ term is very often neglected, giving

$$n! = \sqrt{(2\pi)}n^{n+1/2}e^{-n} \quad\quad\quad 11.17(2)$$

Bernoulli's Theorem: Approximation using Stirling's Theorem

11.18 We shall now consider the case of a population where the probability of a certain sub-class of that population is p. From paragraph 11.11 it was shown that out of a sample size n, the most probable number of members of the sub-class is np if np is an integer, or is equal to the integral part of np or the least integer greater than np. We will denote the most probable number of members of the sub-class by $[np]$†. We shall now find the probability, for a sample of large size n, that r members of the sub-class occur, where r differs from its most probable value of $[np]$† by the integer x.

11.19 The probability of just r members occurring is by 11.05(1) given by

$$P = {}^nC_r p^r(1-p)^{n-r} = \frac{n!}{r!(n-r)!}p^r(1-p)^{n-r} \quad\quad 11.05(1)$$

Let us now write $r = pn + x$, thus

$$n - r = (1-p)n - x \quad\quad\quad 11.19(1)$$

Since n is to be considered large, then r will also be large, provided x is small compared with np. We now substitute for the factorial members of 11.05(1) above by means of Stirling's Theorem (see 11.17(2)), giving

$$P = \frac{n^{n+1/2}e^{-n}p^r(1-p)^{n-r}}{(2\pi)^{1/2}r^{r+1/2}e^{-r}(n-r)^{n-r+1/2}e^{-n+r}}$$

$$= \frac{n^{n+1/2}p^r(1-p)^{n-r}}{(2\pi)^{1/2}r^{r+1/2}(n-r)^{n-r+1/2}}$$

We now substitute for r and $n - r$ by means of the expressions in 11.19(1). Thus

$$P = \frac{n^{n+1/2}p^{pn+x}(1-p)^{(1-p)n-x}}{(2\pi)^{1/2}(pn+x)^{(pn+x+1/2)}\{(1-p)n-x\}^{\{(1-p)n-x+1/2\}}}$$

† See paragraph 11.11 for derivation of $[np]$.

Taking logarithms of both sides, we get

$$\log p = -\tfrac{1}{2}\log(2\pi) + (np + x)\log p + \{(1-p)n - x\}\log(1-p)$$
$$+ (n + \tfrac{1}{2})\log n - (pn + x + \tfrac{1}{2})\log(pn + x)$$
$$- \{(1-p)n - x + \tfrac{1}{2}\}\log\{(1-p)n - x\}$$

Now

$$\log(pn + x) = \log pn + \log\left(1 + \frac{x}{pn}\right)$$

$$= \log pn + \frac{x}{pn} - \frac{x^2}{2p^2n^2}$$

and so

$$-(pn + x + \tfrac{1}{2})\log(pn + x) = -(pn + x + \tfrac{1}{2})\log pn - x - \frac{x}{2pn} - \frac{x^2}{2pn}$$

Similarly

$$-\{(1-p)n - x + \tfrac{1}{2}\}\log\{(1-p)n - x\} = -\{(1-p)n - x + \tfrac{1}{2}\}\log(1-p)n$$

$$+ x + \frac{x}{2(1-p)n} - \frac{x^2}{2(1-p)n}$$

where terms above x^2 are neglected as are terms involving $(1-p)^2n^2$ and p^2n^2 in the denominator. Thus

$$\log P = -\tfrac{1}{2}\log(2\pi) + \log p\{np + x - np - x - \tfrac{1}{2}\}$$
$$+ \log(1-p)\{(1-p)n - x - (1-p)n + x - \tfrac{1}{2}\}$$
$$- \log n\{pn + x + \tfrac{1}{2} + (1-p)n - x + \tfrac{1}{2} - n - \tfrac{1}{2}\}$$
$$- \frac{x^2}{2np(1-p)} - \frac{x(1-2p)}{2n(1-p)p}$$

$$= -\tfrac{1}{2}\log\{2\pi p(1-p)n\} - \frac{1}{2n}\left\{\frac{x^2}{p(1-p)} + \frac{x(1-2p)}{p(1-p)}\right\}$$

Thus finally we have

$$P = \exp\left[-\frac{1}{2n}\left\{\frac{x^2}{p(1-p)} + \frac{x(1-2p)}{p(1-p)}\right\}\right]$$
$$\times \{2\pi p(1-p)n\}^{-1/2} \qquad\qquad 11.19(2)$$

If $|x|$ is much greater than $|1-2p|$, then we may neglect the term $x(1-2p)$ in comparison with x^2 in the exponent. We then obtain the approximation for P given by

$$P = \frac{\exp\{-x^2/2np(1-p)\}}{\{2\pi p(1-p)n\}^{1/2}}$$ 11.19(3)

the probability that a sample of large size n will contain $[pn] + x$ members of the sub-class whose probability is p where

$$pn \gg |x| \gg |1 - 2p|$$ 11.19(4)

If $pn \gg |x|$ but $|x|$ is of the same order as $(1 - 2p)$, then P is given by 11.19(2). If $x = 0$, then

$$P = \frac{1}{\{2\pi p(1-p)n\}^{1/2}}$$

the probability that the number of articles of the sub-class should be $[pn]$, the most probable number.

11.20 The probability that a sample of size n will contain a number of members of the sub-class lying in the range $pn - s$ to $pn + s$ is the sum of the probabilities that the sample will have precisely

$$[pn] + s, [pn] + s - 1, \ldots, [pn] - s$$

members of the sub-class. Thus the required probability is given by

$$P = \sum_{x=-s}^{x=s} \exp\left[-\left\{\frac{x^2 + (1 - 2p)x}{2np(1-p)}\right\}\right] \Big/ \{2\pi p(1-p)n\}^{1/2}$$

Now

$$\sum_{x=-s}^{x=s} \exp\left[-\left\{\frac{x^2 + (1 - 2p)x}{2np(1-p)}\right\}\right]$$

$$= \sum_{x=-s}^{x=s} \exp\left\{-\frac{x^2}{2np(1-p)}\right\} \cdot \left\{1 - \frac{(1 - 2p)x}{2p(1-p)n} + \cdots\right\}$$

and since to every positive value of x there is an equal negative value, the summation made up of the second term of the bracket vanishes. Terms in the bracket having higher powers than x are neglected in view of 11.19(4). Thus

$$P = \sum_{x=-s}^{x=s} \exp\left\{-\frac{x^2}{2np(1-p)}\right\} \Big/ \{2\pi p(1-p)n\}^{1/2}$$ 11.20(1)

If we write

$$y = x/\sqrt{\{np(1-p)\}^{1/2}}$$ 11.20(2)

then we may write

$$y + \delta y = (x + 1)/\sqrt{\{np(1-p)\}^{1/2}}$$

since x increases by unit increments. Therefore

$$\frac{\delta y}{\sqrt{\pi}} = \frac{1}{\sqrt{\{\pi np(1-p)\}}} \qquad 11.20(3)$$

Thus, replacing x by y in 11.20(1) we have

$$P = \sum_{-s/\sqrt{\{np(1-p)\}}}^{s/\sqrt{\{np(1-p)\}}} \frac{\mathrm{e}^{-y^2/2} \, \delta y}{\sqrt{(2\pi)}} = \frac{1}{\sqrt{(2\pi)}} \int_{-s/\sqrt{\{np(1-p)\}}}^{s/\sqrt{\{np(1-p)\}}} \mathrm{e}^{-y^2} \, dy \quad \text{approximately}$$

$$= \underset{-k_1 \text{ to } k_1}{P} \qquad 11.20(4)$$

where $k_1 = s/\{np(1-p)\}^{1/2}$ † and $\underset{-k_1 \text{ to } k_1}{P}$ is the integral of the Gaussian function given in Table II. The *standard deviation* is equal to $\{np(1-p)\}^{1/2}$ (see 11.12(6)), and P is the probability that in a selection of n members of a population, the number of members of the sub-class, whose probability is p, will lie between $pn - s$ to $pn + s$.

Alternative Derivation of an Approximation to Binomial Probabilities for Moderate Values of p, the Probability of a Single Event (i.e. $p \simeq 1/2$)

Central limit theorem

11.21   The central limit theorem states that if $X_1, X_2, \ldots, X_n$ is a sequence of identically distributed random variables, each of which has a mean μ and standard deviation σ, where

$$S_n = \sum_1^n X_i$$

then for each z

$$\lim_{n \to \infty} P\left(\frac{S_n - n\mu}{\sqrt{n}\sigma} \leqslant z\right) = \frac{1}{\sqrt{(2\pi)}} \int_{-\infty}^z \mathrm{e}^{-u^2/2} \, du \qquad 11.21(1)$$

where P is the probability of $(S_n - n\mu)/\sqrt{n}\sigma$ being less than or equal to z, that is from 0 to z. In particular if we put $z = (y - n\mu)/\sqrt{n}\sigma$ and substitute

---

† When calculating k_1 use $[np]$ in place of np, where $[np]$ has been found as in paragraph 11.11.

in 11.21(1) we have

$$P(S_n \leqslant y) = \frac{1}{\sqrt{(2\pi)}} \int_{-\infty}^{\frac{y - n\mu}{\sqrt{n}\sigma}} e^{-u^2/2} \, du \qquad 11.21(2)$$

Since μ is the mean of the X_i, $n\mu$ is the mean of the sum of the n variables, that is of S_n. Further, since σ is the standard deviation of the X_i, $\sqrt{n}\sigma$ is the standard deviation of the sum.

Application to binomial distribution

11.22 If $X_i = 0$ with a probability of $1 - p$ and $X_i = 1$ with a probability of p, then the sum of the X_i has a binomial distribution, with mean np and standard deviation $\sqrt{\{np(1 - p)\}}$ (see 11.10(3) and 11.12(6), respectively). According to the central limit theorem, the distribution of the sum S_n and hence of the binomial distribution is asymptotically normal for large n. Putting $\sqrt{n}\sigma = \sqrt{\{np(1 - p)\}}$, $n\mu = np$ and $S_n = x$, where x is the number of successes, we have from 11.21(1)

$$\lim_{n \to \infty} P\left(\frac{x - np}{\sqrt{\{np(1 - p)\}}} \leqslant z\right) = \frac{1}{\sqrt{(2\pi)}} \int_{-\infty}^{\frac{x - np}{\sqrt{\{np(1 - p)\}}}} e^{-u^2/2} \, du \quad 11.22(1)$$

where P is the probability that the number of successes will lie between 0 and x. The integrand is the density function of the normal distribution, with mean equal to zero and standard deviation equal to unity.

Better approximation to binomial probabilities

11.23 It can be shown that a better approximation† to the probability P is given by putting

$$z = (x + \tfrac{1}{2} - np)/\sqrt{\{np(1 - p)\}} \qquad 11.23(1)$$

that is

$$P\left(\frac{x + \tfrac{1}{2} - np}{\sqrt{\{np(1 - p)\}}} \leqslant z\right) = \frac{1}{\sqrt{(2\pi)}} \int_{-\infty}^{z} e^{-u^2/2} \, du \qquad 11.23(2)$$

The value of the integral

$$\frac{1}{\sqrt{(2\pi)}} \int_{-\infty}^{z} e^{-u^2/2} \, du \equiv I$$

† See *Statistical Theory* B W Lindgren, p 159.

can be found from Table II, Appendix I, as follows. If

$$z \equiv \frac{x + \frac{1}{2} - np}{\sqrt{\{np(1-p)\}}}$$

is *positive* then

$$I = \left[1 + \underset{-k_1 \text{ to } k_1}{P} \right] \Big/ 2 \qquad k_1 \equiv z \qquad\qquad 11.23(3)$$

If z is *negative*, then

$$I = \left[1 - \underset{-k_1 \text{ to } k_1}{P} \right] \Big/ 2 \qquad k_1 \equiv |z| \qquad\qquad 11.23(4)$$

Note that the probability of 0 successes is given by $z = (\frac{1}{2} - np)/\sqrt{\{np(1-p)\}}$.

Probability of number of successes x lying between x_1 and x_2

11.24   The probability of x lying between x_1 and x_2, $x_2 > x_1$, is given by

$$\frac{1}{\sqrt{(2\pi)}} \left[\int_{-\infty}^{z_2} e^{-u^2/2} \, du - \int_{-\infty}^{z_1} e^{-u^2/2} \, du \right]$$

$$= \int_{z_1}^{z_2} e^{-u^2/2} \, du$$

$$= \left[\underset{-k_1 \text{ to } k_1}{P} \right]_{k_1 = z_2} \Big/ 2 - \left[\underset{-k_1 \text{ to } k_1}{P} \right]_{k_1 = z_1} \Big/ 2 \qquad 11.24(1)$$

If we put $x = np \pm r$, then

$$z_2 = \frac{r + \frac{1}{2}}{\sqrt{\{np(1-p)\}}} \qquad\qquad 11.24(2)$$

and

$$z_1 = \frac{(-r + \frac{1}{2})}{\sqrt{\{np(1-p)\}}} \qquad\qquad 11.24(3)$$

Note that in this case the maximum value of r is given by $x = 0$, or $r = np$, giving

$$z_1 = \frac{-np + \frac{1}{2}}{\sqrt{\{np(1-p)\}}}$$

and

$$z_2 = \frac{np + \frac{1}{2}}{\sqrt{\{np(1-p)\}}} \qquad\qquad 11.24(4)$$

Uncertainty in sampling

11.25 If n articles are selected at random from a large population and x of these are found to lie in a certain sub-class, then the approximate probability p of that sub-class is given by x/n. It is often required to know the uncertainty in p. Let us first assume that the value of p found is the correct value. The probability P of the number x of articles in the sub-class lying between x_1 and x_2 in a sample of size n, where $x_1 = np + r_1$ and $x_2 = np + r_2$, is thus given by

$$P = \frac{1}{\sqrt{(2\pi)}} \int_{\frac{r_1}{\sqrt{\{np(1-p)\}}}}^{\frac{r_2}{\sqrt{\{np(1-p)\}}}} e^{-u^2/2} \, du$$

$$= \left[P \atop -k_1 \text{to} k_1 \right]_{k_1 \equiv z_2} \Big/ 2 - \left[P \atop -k_1 \text{to} k_1 \right]_{k_1 \equiv z_1} \Big/ 2 \qquad 11.25(1)$$

where

$$z_2 = \frac{r_2}{\sqrt{\{np(1-p)\}}} \qquad \text{and} \qquad z_1 = \frac{r_1}{\sqrt{\{np(1-p)\}}} \qquad 11.25(1a)$$

Dividing x_1 and x_2 by n we have

$$p_1 = \frac{x_1}{n} = p + \frac{r_1}{n}$$

and

$$p_2 = \frac{x_2}{n} = p + \frac{r_2}{n} \qquad 11.25(2)$$

Putting

$$r_1 = np\alpha_1$$

and

$$r_2 = np\alpha_2 \qquad 11.25(3)$$

and substituting for r_1 and r_2 in 11.25(2) we have

$$p_1 = p(1 + \alpha_1) \qquad \text{and} \qquad p_2 = p(1 + \alpha_2) \qquad 11.25(4)$$

P of 11.25(1) is thus the probability of the fractional uncertainty in p being $\alpha_2 - \alpha_1$. If r_1 is negative, then $r_1 = -np\alpha_1$ and the second term of the integral is added to the first one, where $k_1 \equiv |r_1| = np\alpha_1$. If, as is often the case, $r_1 = -r_2$, where r_2 is positive, the required probability P of the fraction p of articles in the sub-class lying between p_1 and p_2, or $p \pm r_2/n \equiv p(1 \pm \alpha_2)$, is given by

$$P = \frac{1}{\sqrt{(2\pi)}} \int_{-\frac{r_2}{\sqrt{\{np(1-p)\}}}}^{\frac{r_2}{\sqrt{\{np(1-p)\}}}} e^{-u^2/2}\, du = \left[P \atop {-k_1 \text{ to } k_1} \right]_{k_1 \equiv z_2} \qquad 11.25(5)$$

If the more accurate form of the integral is used, we have

$$P = \frac{1}{\sqrt{(2\pi)}} \int_{-\frac{r_2+\frac{1}{2}}{\sqrt{\{np(1-p)\}}}}^{\frac{r_2+\frac{1}{2}}{\sqrt{\{np(1-p)\}}}} e^{-u^2/2}\, du = \left[P \atop {-k_1 \text{ to } k_1} \right]_{k_1 \equiv \beta_2} \Big/ 2 + \left[P \atop {-k_1 \text{ to } k_1} \right]_{k_1 \equiv \beta_1} \Big/ 2$$

$$11.25(6)$$

where

$$\beta_2 = \frac{r_2 + \frac{1}{2}}{\sqrt{\{np(1-p)\}}}$$

and

$$\beta_1 = \frac{r_2 - \frac{1}{2}}{\sqrt{\{np(1-p)\}}}$$

Poisson Distribution

11.26   Bernoulli's Theorem 11.05(1) states that the probability of r members of a sub-class appearing as a result of n choices is given by

$$P = {}^nC_r p^2 (1-p)^{n-r} \qquad 11.05(1)$$

where p is the probability of a single event. Let us write $\varepsilon = pn$, which is approximately the most probable value of r. We shall now assume that p is very small compared with unity and that n is large so that ε can have an appreciable value. Substituting for p in 11.05(1) we have

$$P = {}^nC_r \left(\frac{\varepsilon}{n}\right)^r \left(1 - \frac{\varepsilon}{n}\right)^{n-r}$$

$$= \frac{n(n-1)(n-2)\cdots(n-r+1)}{r!} \left(\frac{\varepsilon}{n}\right)^r \left(1 - \frac{\varepsilon}{n}\right)^n \Big/ \left(1 - \frac{\varepsilon}{n}\right)^r$$

$$= \frac{\varepsilon^r}{r!}\left(1 - \frac{\varepsilon}{n}\right)^n \frac{n(n-1)(n-2)\cdots(n-r+1)}{n^r} \Big/ \left(1 - \frac{\varepsilon}{n}\right)^r$$

$$= \frac{\varepsilon^r}{r!}\left(1 - \frac{\varepsilon}{n}\right)^n \left(1 - \frac{1}{n}\right)\left(1 - \frac{2}{n}\right)\cdots\left(1 - \frac{r-1}{n}\right) \Big/ \left(1 - \frac{\varepsilon}{n}\right)^r \qquad 11.26(1)$$

Now if n is large $(1 - \varepsilon/n)^n$ is approximately equal to $e^{-\varepsilon}$.

11.27 Consider the product

$$\left(1 - \frac{1}{n}\right)\left(1 - \frac{2}{n}\right)\left(1 - \frac{3}{n}\right) \cdots \left(1 - \frac{r-1}{n}\right)$$

$$= 1 - \sum_{t=1}^{t=r-1} \frac{t}{n} + \text{terms in } \frac{1}{n^2} - \text{terms in } \frac{1}{n^3}$$

etc. Thus since n is large, the product lies between 1 and

$$1 - \sum_{t=1}^{t=r-1} \frac{t}{n} = 1 - \frac{r(r-1)}{2n}$$

since

$$\sum_{t=1}^{t=r-1} \frac{t}{n}$$

is an arithmetic series. We now assume that $r^2/2n$ is small; thus $r(r-1)/2n$ is small and so the product tends to unity.

11.28 Now

$$\left(1 - \frac{\varepsilon}{n}\right)^r = \left(1 - \frac{\varepsilon}{n}\right)^{nr/n}$$

which tends to $e^{-\varepsilon r/n} \to 1$ if r is small compared with n. Since $r^2/2n$ is small it follows that r/n is even smaller for $r \geqslant 2$.

11.29 Thus, substituting the approximations we have obtained for the various components of 11.26(1) in it we obtain

$$P = \frac{\varepsilon^r}{r!} e^{-\varepsilon} = e^{-np}(np)^r/r! \qquad \text{11.29(1)}$$

provided $r^2/2n$ is small. The probability distribution given by 11.29(1) is known as Poisson's law of distribution, and is applicable to cases where p is very small†, that is, to so-called rare events.

Note on Poisson's Law of Distribution

11.30 Poisson's distribution does not represent a true distribution, because the sum

$$\sum P = \sum_{r=0}^{r=n} e^{-\varepsilon}\varepsilon^r/r!$$

which should be unity is not. That is

$$\sum P = e^{-\varepsilon}\left(1 + \frac{\varepsilon}{1!} + \frac{\varepsilon^2}{2!} + \frac{\varepsilon^3}{3!} + \cdots + \frac{\varepsilon^n}{n!}\right)$$

$$= e^{-\varepsilon}\left(e^{\varepsilon} - \sum_{r=n+1}^{r=\infty} \frac{\varepsilon^r}{r!}\right) = 1 - e^{-\varepsilon}\sum_{r=n+1}^{r=\infty} \frac{\varepsilon^r}{r!}$$

† Poisson's distribution is a useful approximation to the binomial distribution even when p is as large as 0.2.

However, if n is large

$$\sum_{r=n+1}^{r=\infty} \varepsilon^r/r!$$

is very small and thus $\sum P$ is approximately equal to unity.

Average Value of r

11.31

$$\bar{r} = \frac{e^{-\varepsilon} \sum_{r=0}^{r=n} (\varepsilon^r/r!)r}{e^{-\varepsilon} \sum_{r=0}^{r=n} (\varepsilon^r/r!) \to 1 \text{ from above}} \simeq e^{-\varepsilon} \sum_{r=0}^{r=\infty} \frac{\varepsilon^r}{(r-1)!}$$

$$= \varepsilon e^{-\varepsilon} \sum_{r=0}^{r=\infty} \frac{\varepsilon^{r-1}}{(r-1)!}$$

$$= \varepsilon \, e^{-\varepsilon} \, e^{\varepsilon}$$

$$= \varepsilon \qquad\qquad\qquad\qquad 11.31(1)$$

where $\varepsilon = pn$.

Standard Deviation of Poisson's Distribution

11.32

$$\sigma^2 = e^{-\varepsilon} \sum_{r=0}^{r=n} \frac{\varepsilon^r}{r!} (r - \varepsilon)^2$$

$$\simeq e^{-\varepsilon} \sum_{r=0}^{r=\infty} \frac{\varepsilon^r}{r!} (r - \varepsilon)^2$$

$$= e^{-\varepsilon} \left\{ \sum_{r=0}^{r=\infty} \frac{\varepsilon^r}{r!} (r^2 - 2\varepsilon r + \varepsilon^2) \right\}$$

$$= e^{-\varepsilon} \left\{ \sum_{r=0}^{r=\infty} \frac{r\varepsilon^r}{(r-1)!} - 2\varepsilon \sum_{r=0}^{r=\infty} \frac{\varepsilon^r}{(r-1)!} + \varepsilon^2 \sum_{r=0}^{r=\infty} \frac{\varepsilon^r}{r!} \right\}$$

$$= e^{-\varepsilon} \left\{ \sum_{r=0}^{r=\infty} \frac{\{(r-1)+1\}}{(r-1)!} \varepsilon^r - 2\varepsilon^2 \sum_{r=0}^{r=\infty} \frac{\varepsilon^{r-1}}{(r-1)!} + \varepsilon^2 \sum_{r=0}^{r=\infty} \frac{\varepsilon^r}{r!} \right\}$$

$$= e^{-\varepsilon} \left\{ \varepsilon^2 \sum_{r=0}^{r=\infty} \frac{\varepsilon^{r-2}}{(r-2)!} + \varepsilon \sum_{r=0}^{r=\infty} \frac{\varepsilon^{r-1}}{(r-1)!} - 2\varepsilon^2 \sum_{r=0}^{r=\infty} \frac{\varepsilon^{r-1}}{(r-1)!} + \varepsilon^2 \sum_{r=0}^{r=\infty} \frac{\varepsilon^r}{r!} \right\}$$

$$= e^{-\varepsilon} (\varepsilon^2 \, e^{\varepsilon} + \varepsilon \, e^{\varepsilon} - 2\varepsilon^2 \, e^{\varepsilon} + \varepsilon^2 \, e^{\varepsilon})$$

and so

$$\sigma^2 = \varepsilon \, e^{-\varepsilon} \, e^{\varepsilon}$$

$$= \varepsilon \qquad\qquad 11.32(1)$$

The standard deviation of a Poisson distribution is thus given by

$$\sigma = \sqrt{\varepsilon} = \sqrt{(pn)} \qquad\qquad 11.32(2)$$

The Hypergeometric Function

11.33   This function is closely related to the binomial function

$$^nC_r p^r (1 - p)^{n-r} \qquad\qquad 11.05(1)$$

which gives the probability of selecting r members of a sub-class in n choices from a population, for which the probability of selecting a member of the sub-class in question is p. It is to be noted that p is constant. If the population is infinite then it is easy to see that p is constant, but if the population is finite, equal say to N, and if it contains m members of the sub-class, then $p = m/N$. Thus the only way under these circumstances that p can be a constant is for each member of the population to be replaced before another choice is made. This is called selection by replacement.

11.34   If, however, the n members are selected without replacement, the p will be a variable and under these circumstances the probability of selecting r members of a sub-class in n choices from a population N in number will not be equal to expression 11.05(1). This type of selection is known as selection without replacement and leads to the hypergeometric function for the required probability.

11.35   The function is derived as follows. Let the population consist of N members, which contains M members of a sub-class. Let n members of the population be selected at random, and let it be required to find the probability that 'r' of these n members selected belong to the required sub-class.

11.36   The number of different ways of selecting n things from N is given by

(a) $\qquad\qquad ^NC_n$ or as it is often written $\dbinom{N}{n}$

where the absence of the solidus distinguishes the binomial coefficient from N divided by n.

The number of different ways of selecting r things from M is given by

(b) $\qquad\qquad ^MC_r \equiv \dbinom{M}{r}$

Lastly the number of different ways of selecting $n - r$ things from $N - M$ is equal to

(c) $$^{N-M}C_{n-r} \equiv \begin{pmatrix} N - M \\ n - r \end{pmatrix}$$

Let the M members of (b) be members of the required sub-class, whilst the $N - M$ members of (c) are made up of the remaining members of the population of N. Now each of the ways of selecting a member of (b) can be combined with each way of selecting a member of (c) to obtain the total number of mutually exclusive ways of selecting r members of the sub-class in a selection of n members, that is

$$\begin{pmatrix} M \\ r \end{pmatrix} \begin{pmatrix} N - M \\ n - r \end{pmatrix}$$

Since (a) is the total number of different ways of selecting n things from N, then dividing the product of (b) and (c) by (a) gives the probability of obtaining r members of the sub-class of M members in a selection of n members drawn from a total population of N members. The probability function is thus given by

$$f(r; N, M, n) = \begin{pmatrix} M \\ r \end{pmatrix} \begin{pmatrix} N - M \\ n - r \end{pmatrix} \bigg/ \begin{pmatrix} N \\ n \end{pmatrix} \qquad 11.36(1)$$

A random variable with the probability function just given is said to have a hypergeometric distribution.

Standard Deviation and Mean Value of the Hypergeometric Distribution

11.37 In order to derive these quantities use will be made of the expectation operator E. This operator is defined such that when it operates on a variable it yields the expected value of that variable, that is $E(X) = c$, where c is the expected value of the variable X. For discrete distributions

$$E(X) = \sum_{i=1}^{i=n} x_i f(x_i) \qquad 11.37(1)$$

and for continuous distributions

$$E(X) = \int_a^b x f(x) \, dx \qquad 11.37(2)$$

where $f(x)$ denotes the probability function or density function as appropriate. It is clear that $E(X) = \mu$, the mean value (from 11.37(1)). The variance is given as

$$E[(X - \mu)^2] = E[(X - E(X))^2]$$
$$= E[X^2 + (E(X))^2 - 2XE(X)]$$

Thus

$$\text{Var } X = E(X^2) - (EX)^2 \qquad\qquad 11.37(3)$$

since

$$E[2X \, E(X)] = 2(EX)^2$$

where Var denotes variance, i.e. σ^2. Note that E can only operate once on a variable, that is

$$E[E(X)] = E(X)$$

or

$$E^r[E(X)] = E(X) \qquad\qquad 11.37(4)$$

11.38 It is now convenient to calculate the factorial moments of the hypergeometric distribution from which the standard deviation and mean value can be found. The kth factorial moment is defined as

$$E[(x)_k] = \sum_{r=k}^{r=n} (r)_k \binom{M}{r}\binom{N - M}{n - r} \bigg/ \binom{N}{n} \qquad 11.38(1)$$

where

$$(r)_k \equiv r(r - 1)(r - 2) \cdots (r - k + 1)$$

Consider $(r)_k \binom{M}{r}$; when written out fully this becomes

$$\frac{[r(r-1)(r-2)\cdots(r-k+1)][M(M-1)\cdots(M-k+1)(M-k)\cdots(M-r+1)]}{r(r-1)(r-2)\cdots(r-k+1)(r-k)\cdots 3\cdot 2\cdot 1}$$

where $r \geqslant k$. If common terms are cancelled and the first k terms in M are written as $(M)_k$, we have

$$(r)_k \binom{M}{r} = (M)_k \frac{(M - k)(M - k - 1)\cdots(M - r + 1)}{(r - k)(r - k - 1)\cdots 3\cdot 2\cdot 1}$$

$$= (M)_k \frac{(M - k)!}{(r - k)!(M - r)!}$$

$$= (M)_k \frac{(M - k)!}{(r - k)![(M - k) - (r - k)]!}$$

$$= (M)_k \binom{M - k}{r - k} \qquad\qquad 11.38(2)$$

We can now replace $(r)_k \binom{M}{r}$ in 11.38(1) by $(M)_k \binom{M-k}{r-k}$ and so

$$E\{(X)_k\} = \sum_{r=k}^{r=n} (r)_k \binom{M}{r} \binom{N-M}{n-r} \bigg/ \binom{N}{n}$$

$$= \sum_{r=k}^{r=n} (M)_k \binom{M-k}{r-k} \binom{N-M}{n-r} \bigg/ \binom{N}{n}$$

$$= (M)_k \sum_{j=0}^{j=n-k} \binom{M-k}{j} \binom{(N-k)-(M-k)}{(n-k)-j} \bigg/ \binom{N}{n}$$

where $r - k$ has been put equal to j

$$= \left[(M)_k \binom{N-k}{n-k} \bigg/ \binom{N}{n} \right] \left[\sum_{j=0}^{j=n-k} \binom{M-k}{j} \binom{(N-k)-(M-k)}{(n-k)-j} \right) \bigg/$$

$$\binom{N-k}{n-k} \right]$$

$$= \left[(M)_k \binom{N-k}{n-k} \bigg/ \binom{N}{n} \right] \sum_{j=0}^{j=n-k} f(j; N-k, M-k, n-k)$$

$$= (M)_k \binom{N-k}{n-k} \bigg/ \binom{N}{n}$$

since $\sum_{j=0}^{j=n-k} f(j; N-k, M-k, n-k) = 1$

$$= \frac{(M)_k (N-k)! \, n! \, (N-n)!}{(n-k)! \, (N-n)! \, N!}$$

$$= \frac{(M)_k (n)_k}{(N)_k} = E\{(X)_k\} \qquad\qquad 11.38(3)$$

The meaning of k in $E\{(X)_k\}$

If we put $k = 1$ we have

$$E(X) = \frac{Mn}{N} = \mu = np \qquad\qquad 11.38(4)$$

where

$$p = \frac{M}{N} \qquad\qquad 11.38(4a)$$

Putting $k = 2$ we have

$$E[X(X-1)] = \frac{M(M-1)n(n-1)}{N(N-1)}$$

Now

$$\text{Var } X = E(X^2) - (E(X))^2$$

from 11.37(3)

$$= E[X(X-1)] + E(X) - (E(X))^2$$

$$= \frac{M(M-1)n(n-1)}{N(N-1)} + \frac{Mn}{N} - \frac{M^2 n^2}{N^2}$$

$$= \frac{nM}{N} \cdot \frac{N-M}{N} \cdot \frac{N-n}{N-1}$$

$$= npq \frac{N-n}{N-1} \qquad \qquad 11.38(5)$$

where

$$p = \frac{M}{N} \qquad \qquad 11.38(4a)$$

the probability at each choice that the object selected belongs to the class of which there are initially M in number $q = 1 - p$.

The standard deviation of the hypergeometric distribution is thus equal to

$$\{np(1-p)(N-n)/(N-1)\}^{1/2} \qquad \qquad 11.38(6)$$

11.39 Comparing with the Bernoullian distribution, it is seen that the mean μ is given by the same value of r, whilst the standard deviation of the hypergeometric function is equal to that given for the Bernoulli distribution multiplied by the factor

$$\left(\frac{N-n}{N-1}\right)^{1/2}$$

Approximation to the Hypergeometric Probability Function

11.40 The hypergeometric probability function

$$f(r; N, M, n) \equiv \binom{M}{r}\binom{N-M}{n-r} \bigg/ \binom{N}{n} \equiv \frac{(M)_r (N-M)_{n-r} n!}{r!(n-r)!(N)_n} \qquad 11.40(1)$$

using 11.36(1), which when expanded becomes

$$\frac{M(M-1)(M-2)\cdots(M-r+1)}{N(N-1)(N-2)\cdots(N-r+1)}$$

$$\times \frac{(N-M)(N-M-1)\cdots(N-M-n+r+1)}{(N-r)(N-r-1)\cdots(N-n+1)} \times \binom{n}{r}$$

which in turn becomes

$$\frac{M}{N}\left(\frac{M}{N} - \frac{1}{N}\right)\left(\frac{M}{N} - \frac{2}{N}\right)\cdots\left(\frac{M}{N} - \frac{(r-1)}{N}\right)$$

$$\times\left[\left(1 - \frac{1}{N}\right)\left(1 - \frac{2}{N}\right)\cdots\left(1 - \frac{(r-1)}{N}\right)\right]^{-1}$$

$$\times\left(1 - \frac{M}{N}\right)\left(1 - \frac{(M+1)}{N}\right)\cdots\left(1 - \frac{(M+n-r-1)}{N}\right)$$

$$\times\left[\left(1 - \frac{r}{N}\right)\left(1 - \frac{(r+1)}{N}\right)\cdots\left(1 - \frac{(n-1)}{N}\right)\right]^{-1} \times \binom{n}{r}$$

where each term in the numerator and denominator is divided by N. If M and N are large compared with r and n, and we put

$$\frac{M}{N} = p, \qquad \text{and} \qquad 1 - p = q,$$

we have

$$f(r; N, M, n) \rightarrow p^r(1 - p)^{n-r}\binom{n}{r} = p^r q^{n-r}\binom{n}{r} \qquad 11.40(2)$$

which is equal to the binomial probability (see 11.05(1)).

Approximation to the Sum of the Hypergeometric Probabilities

11.41   In exactly the same way as for the binomial probability distribution, it can be shown that the probability of the sum of the hypergeometric function variables X_i lying between 0 and x, where x is the number of successes, is given by

$$P(S_n \leqslant x) = P\left(\frac{S_n - n\mu}{\sigma_{S_n}} \leqslant \frac{x - np}{\sqrt{\{np(1 - p)(N - n)/(N - 1)\}}}\right)$$

$$\simeq \frac{1}{\sqrt{(2\pi)}}\int_{-\infty}^{z} e^{-u^2/2}\,du \qquad 11.41(1)$$

where σ_{S_n} is the standard deviation of S_n

$$z = \frac{x - np}{\sqrt{\{np(1 - p)(N - n)/(N - 1)\}}}$$

and

$$\sigma_{S_n} = \sqrt{\left\{\frac{np(1 - p)(N - n)}{N - 1}\right\}}$$

the standard deviation of S_n†, and $n\mu = np$, the mean value of S_n. (See 11.38(5) and 11.38(4), respectively.)

Better approximation

11.42 Just as in the binomial case, a better approximation to P is given by putting

$$z = \frac{x + \frac{1}{2} - np}{\sqrt{\{np(1 - p)(N - n)/(N - 1)\}}} \qquad 11.42(1)$$

Examples on Chapter 11

Example 11.1 A batch of 200 parts have been produced on a lathe, and it is known that the standard deviation of the parts produced on this lathe is 0.001 in. What is the probable number of parts which exceed the mean by more than 2σ, that is 0.002 in. Find the probability of just this number occurring, and find also the probability of 2, 3, 4, 5, 6 and 8 parts exceeding the mean by 2σ or more.

The probability of parts exceeding the mean by more than 2σ is

$$\frac{(1 - 0.954)}{2} = 0.023$$

that is half the probability that parts will lie outside the range $\bar{x} - 2\sigma$ to $\bar{x} + 2\sigma$, where $\bar{x}$ is the mean value. The probable number of parts exceeding $\bar{x} + 2\sigma$ is thus $0.023 \times 200 = 4.6$. Using the method of paragraph 11.11 the most probable number of parts exceeding $\bar{x} + 2\sigma$ is seen to be 4. The probability of just this number of parts exceeding $\bar{x} + 2\sigma$ is given by Bernoulli's theorem, that is

$$P = \frac{n!}{r!(n - r)!} \cdot p^r(1 - p)^{n-r}$$

with $r = np = 4.6$, $n = 200$, $p = 0.023$. Using the approximation given by using Stirling's theorem and putting $x = 0$ (see paragraph 11.14) the required probability is

$$\frac{1}{\{2\pi p(1 - p)n\}^{1/2}} = 0.188$$

† $S_n = \sum_{i=1}^{i=n} X_i$, the sum of the hypergeometric variables, where $X_i = 1$ with a probability of p, and $X_i = 0$ with a probability of $1 - p$. The variance of S_n is not equal to the sum of the variances of the X_i since the latter are *not* independent.

where $p = 0.023$ and $n = 200$. To find the probability of 2, 3, 4 etc. parts exceeding the mean by 2σ or more, it is appropriate to use Poisson's distribution and equation 11.29(1), since $np = 4.6$, $p = 0.023$ and $r^2/2n$ is small. The required probabilities of occurrence of 2, 3, 4, 5, 6 and 8 parts exceeding $\bar{x} + 2\sigma$ are thus given by using

$$P = e^{-np}(np)^r/r! \,\dagger \qquad\qquad 11.29(1)$$

with $r = 2, 3, 4, 5,$ and 6.

Answer:

$$P_2 = 0.106, \qquad P_3 = 0.163, \qquad P_4 = 0.187\ddagger$$

$$P_5 = 0.173, \qquad P_6 = 0.132, \qquad P_8 = 0.050$$

Example 11.2   The standard deviation of a measuring instrument is 0.0003 in. 500 parts are checked for length on this machine, each length measurement being the result of two measurements, one at each end of each part. What is the most probable number of parts for which the uncertainty of length measurement exceeds 0.0009 in? What is the probability that just ten parts will have an uncertainty in their length measurement in excess of 0.0009 in? What is the probability that the number of parts, with an uncertainty in their length measurement in excess of 0.0009 in, will lie between 12 and 20?

Answer: Since each part is measured twice, the standard deviation of a length measurement is $\sqrt{2} \times 0.0003$ in. Assuming a Gaussian distribution, the tolerance factor k corresponding to a tolerance interval of 0.0009 in is given by

$$k\sigma = 0.0009 \text{ in} = k\sqrt{2} \times 0.0003 \text{ in}$$

or

$$k = 2.1213$$

Consulting Table II of Appendix I, we find that the probability associated with k is 0.966 35. The probability of an uncertainty in excess of 0.0009 in is thus equal to $(1 - 0.066\,35) = 0.035\,65$, and the most probable number of parts, for which the length measurement uncertainty is in excess of 0.0009 in, is given as follows:

$$np = 500 \times 0.033\,65 = 16.825$$

Then using 11.11(2) we have

$$16.825 + 0.033\,65 > r > 15.825 + 0.033\,65$$

and thus the most probable number of parts is 16.

---

† See Appendix I, Table XIX, for values of $\sum_{r=0}^{r=k} e^{-np}(np)^r/r!$.
‡ Notice that the most probable number of parts expected to exceed $\bar{x} + 2\sigma$ is 4, which one might have expected since $np = 4.6$.

To find the probability of just ten parts having length uncertainties in excess of 0.0009 in we use Poisson's equation 11.24(1) for which $P = 0.033\,65$, $n = 500$ and $r = 10$. The required probability is thus

$$P = e^{-500 \times 0.033\,65}(500 \times 0.033\,65)^{10}/10!$$

$$= e^{-16.825}\,(16.825)^{10}/10!$$

$$= 0.0247$$

To find the probability that the number of parts with uncertainties in their length measurements in excess of 0.0009 in will lie between 12 and 20, we use the limiting form of Bernoulli's Theorem given by equation 11.20(4).

The most probable number is as we have seen equal to 16. Now we require the probability that the number of parts will lie between 12 and 20, that is 16 ± 4, and so taking $pn = 16$ and $s = 4$, the quantity

$$s/\sqrt{\{np(1 - p)\}} = 4/\sqrt{\{16(0.966\,35)\}}$$

$$= 1.017\,26$$

Using Table II, Appendix I, we find the required probability corresponding to $k = 1.017\,26$ to be 0.690\,96.

Example 11.3   Parts are manufactured on a machine tool and the standard deviation of these parts, from past experience, is known to be 0.0003 in. The parts are now measured for size on a checking machine, with known standard deviation of 0.0001 in. Each part is measured once. What is the measured standard deviation of the parts likely to be? If 1000 parts are manufactured, what is the probability of parts differing from the mean value by more than 0.0010 in? What is the probable number of parts that will differ from the mean by more than 0.0010 in? What is the probability that up to six parts per thousand will differ from the mean by more than 0.001 in?

Answer: The measured standard deviation will be

$$(0.0003^2 + 0.0001^2)^{1/2} = 0.000\,3162 \text{ in}$$

If k is the tolerance factor, then $k \times 0.000\,3162 = 0.001$ giving $K = 3.162$. The associated probability (by Table II, Appendix I) is equal to 0.9985 and so the probability of parts exceeding the mean by more than 0.001 in is $(1 - 0.9985) = 0.0015$. The most probable number of parts differing by more than 0.001 from the mean is given as follows:

$$np = 1.5$$

Then using 11.11(2) we have

$$1.5 + 0.0015 > r > 0.5 + 0.0015$$

giving the most probable number of parts differing by more than 0.001 from the mean as 1. This is confirmed by the Poisson probabilities.

Using Poisson's equation 11.24(1), we calculate the probability of 0, 1, 2, 3, 4, and 6 parts differing by more than 0.001 in from the mean. With $np = 1.5$ and $r = 1, 2, 3, 4, 5$ and 6 we obtain

$$P_0 = 0.2231, \qquad P_1 = 0.3346, \qquad P_2 = 0.2510, \qquad P_3 = 0.1255$$

$$P_4 = 0.0471, \qquad P_5 = 0.0141, \qquad P_6 = 0.0035$$

The probability of up to six parts per thousand exceeding the mean by more than 0.001 in is thus

$$\sum_0^6 P_r = 0.9907$$

or the probability of seven or more parts per thousand differing from the mean by more than 0.001 in is less than 1%.

Example 11.4 A factory makes and tests five hundred voltmeters per week. Each test certificate states that the uncertainty in the full-scale deflection is not greater than 1% to a tolerance probability of 0.9973. What is the probable number of instruments which will have an uncertainty in their full-scale deflection of more than 1%? What is the probability that more than five instruments per five hundred will have an uncertainty in their full-scale deflection of more than 1%?

Answer: One instrument; required probability is 0.002 45.

Example 11.5 The standard deviation of a measuring instrument is 0.0005 in. It is used to measure the size of various articles, and as these are required to a high degree of accuracy, the readings for each measuring position are repeated ten times, in order that the mean value can be found. On average, in each set of ten readings, how many times will the total range of readings exceed 2σ, 4σ, and 6σ? What is the probability of a range of 4σ or more in any ten readings?

Answer: The number of times the total range will on average exceed 2σ, 4σ, and 6σ for a set of ten readings is 1.58, 0.227, 0.0135. The required probability is 0.077.

Example 11.6 A machine turns out parts at the rate of one thousand per day. The standard deviation of the machine is 0.0002 in, and parts outside two standard deviations are scrapped. Assuming an approximate Gaussian distribution, find the expected number of parts produced as scrap each day. One day the machine produces sixty pieces of scrap. Find the probability of producing sixty or more pieces of scrap.

Answer: Expected number of scrap pieces is 45. Probability of producing sixty or more pieces of scrap is 0.011. As this is rather a small value, the machine should be checked.

Appendix I

Tables

Tables I to XXI are adapted from the tables given respectively on pages 31, 29, 32 to 35, 43, 38 to 39, 47, 40 to 41, 44, 45, 42, 47 and 48 of *Documenta Geigy—Scientific Tables* (Geigy Pharmaceutical Company Ltd, Manchester). Publication of the tables is by kind permission of J R Geigy S A, Basle. For more complete or comprehensive tables than those given below see *Documenta Geigy*.

The following list headed 'Choice of Table' enables the appropriate table to be selected, depending on the known parameters, and the particular type of uncertainty, distribution constant or probability required.

Choice of Table

For a discussion of the particular distributions represented by the tables, see Chapter 6 and Index to Uncertainties, paragraph 6.43.

1. *Given the Probability of a Single Uncertainty to find the Associated Uncertainty Range*

   (a) Without confidence probabilities.†

Known parameters	Number of readings n	Constant	Table	Range
μ, σ or $\bar{x}, s$	$n > 200$	k_1	III	$\mu - k_1\sigma < x < \mu + k_1\sigma$ or $\bar{x} - k_1 s < x < \bar{x} + k_1 s$ see paragraph 6.25
$\bar{x}, \sigma$	$n < 100$	k_4	X	$\bar{x} - k_4\sigma < x < \bar{x} + k_4\sigma$ see paragraphs 6.26 to 6.28
$\bar{x}, s$	$n < 100$	k_5	XI	$\bar{x} - k_5 s < x < \bar{x} + k_5 s$ see paragraphs 6.29 and 6.30
μ, s	$n < 100$	k_2	IV	$\mu - k_2 s < x < \mu + k_2 s$ see paragraph 6.31

† Without confidence probabilities corresponds to a confidence probability of 0.5, that is the ranges are the most probable ones. Put another way, if the set of observations were repeated a large number of times, on average half the ranges would exceed the mean value calculated and half would be less.

401

(b)   With confidence probabilities.

Known parameters	*Number of readings n*	*Constant*	*Table*	*Range*
$\bar{x}, \sigma$	$n < 100$	k_6	XII	$\bar{x} - k_6\sigma < x < \bar{x} + k_6\sigma$ see paragraphs 6.32 and 6.33
$\bar{x}, s$	$n < 1000$	k_7	XIII	$\bar{x} - k_7\sigma < x < \bar{x} + k_7\sigma$ see paragraphs 6.34 and 6.35
μ, s	$n < 100$	k_8	XIV	$\mu - k_8 s < x < \mu + k_8 s$ see paragraphs 6.36 to 6.38

2.   *Given the Uncertainty Range in the Form*

$$\left.\begin{cases} (\mu - k\sigma < x < \mu + k\sigma) \\ (\bar{x} - ks < x < \bar{x} + ks) \end{cases}\right\}$$

to find the Associated Probability of the Range

Known parameters	*Number of readings n*	*Constant*	*Table*	*Range*
μ, σ or $\bar{x}, s$	$n > 200$	k_1	II	$\mu - k_1\sigma < x < \mu + k_1\sigma$ see paragraphs 6.23 and 6.24 $\bar{x} - k_1 s < x < \bar{x} + ks$

3.   *Given the Probability of an Uncertainty in the Mean (Confidence Probability) to find the Associated Uncertainty Range*

Known parameters	*Number of readings n*	*Constant*	*Table*	*Range*
$\sigma, \bar{x}$ or $s, \bar{x}$	$n > 200$	k_1	III	$\bar{x} - \dfrac{k_1}{\sqrt{n}}\sigma < \mu < \bar{x} + \dfrac{k_1}{\sqrt{n}}\sigma$ $\bar{x} - \dfrac{k_1}{\sqrt{n}}s < \mu < \bar{x} + \dfrac{k_1}{\sqrt{n}}s$ see paragraph 6.39
$\bar{x}, s$	$n < 200$	k_2	IV	$\bar{x} - k_2\dfrac{s_v}{\sqrt{n}} < \mu < \bar{x} + k_2\dfrac{s_v}{\sqrt{n}}$ see paragraph 6.40
		k_3	V	$\bar{x} - k_3 s_v < \mu < \bar{x} + k_3 s_v$

Note: If σ is known, then although $\bar{x}$ may be derived from a small number of readings, the range for a chosen probability is given by

$$\bar{x} - \frac{k_1}{\sqrt{n}}\sigma < \mu < \bar{x} + \frac{k_1}{\sqrt{n}}\sigma$$

4.   *Given the Probability of an Uncertainty in the Mean (Confidence Probability) to find the Associated Uncertainty Range from a Small Number of Readings using an Instrument with a Precalibrated Standard Deviation*

   (a)   Expected or most probable uncertainty range for mean value.
Let s_n be the standard deviation of a piece of measuring equipment found

from n readings, $n \geqslant 20$. Further let p be the number of observations made on a component by the measuring equipment, $n \gg p$. p will normally range from 3 to 7.

The required uncertainty range for the mean $\bar{x}_p$ of the p readings is then given by

$$\bar{x}_p - k_1 s_n/\sqrt{p} < \mu < \bar{x}_p + k_1 s_n/\sqrt{p} \qquad \text{I(1)}$$

where μ is the true mean and k_1 is the tolerance factor corresponding to the chosen probability P. (See Table III.)

(b) Uncertainty range for mean with stated frequency of it being exceeded. If more security is required for the uncertainty range of the mean given by I(1) then Table VII may be used to obtain a maximum value of the standard deviation s_n for a given probability. The value of $_{\max}k_q$ should be selected for the appropriate value of n and the chosen bracketed confidence probability β_{cp} used. Note that $(1 - \beta_{cp})$ gives the probability that σ will exceed $_{\max}k_q s_n$. The uncertainty in the mean is now given by

$$\bar{x}_p - k_{1.\max}k_p s_n/\sqrt{p} < \mu < \bar{x}_p + k_{1.\max}k_q s_n/\sqrt{p} \qquad \text{I(2)}$$

where k_1 is the normal tolerance coefficient corresponding to a chosen probability P. k_1 is obtained from Table III.

The meaning to be attached to I(2) is that the mean value of p readings of x will lie within the uncertainty range given, with a probability P corresponding to the value of k_1, derived from Table III. Further if 100 sets of p readings of x were taken, then in $100 \cdot \beta_{cp}$ cases the uncertainty range found would be expected to be withhin that given by I(2). In the case of I(1) out of 100 sets of p readings only 50 would be expected to lie within the range given by I(1), i.e. I(1) gives the most probable result.

Use of Table III.

5. *Given the Probability of an Uncertainty in σ (Confidence Probability) to find the Associated Uncertainty Range*

Known parameter s_v. For confidence range see Table VII. For maximum or minimum values of σ for a given probability see Table VI.

Table I

This table gives values of

$$\frac{e^{-c^2/2}}{\sqrt{(2\pi)}} = f(c)$$

for values of c, where $c = (x - \mu)/\sigma$, $x - \mu =$ uncertainty, $\sigma =$ standard deviation and $\mu =$ mean of an infinite number of observations $c = 1$ corresponds to one standard deviation.

$|c|$	$f(c)$	$|c|$	$f(c)$	$|c|$	$f(c)$	$|c|$	$f(c)$
0.00	0.398 94	0.35	0.375 24	0.70	0.312 25	1.05	0.229 88
0.01	0.398 92	0.36	0.373 91	0.71	0.310 06	1.06	0.227 47
0.02	0.398 86	0.37	0.372 55	0.72	0.307 85	1.07	0.225 06
0.03	0.398 76	0.38	0.371 15	0.73	0.305 63	1.08	0.222 65
0.04	0.398 62	0.39	0.369 73	0.74	0.303 39	1.09	0.220 25
0.05	0.398 44	0.40	0.368 27	0.75	0.301 14	1.1	0.217 85
0.06	0.398 22	0.41	0.366 78	0.76	0.298 87	1.2	0.194 19
0.07	0.397 97	0.42	0.365 26	0.77	0.296 59	1.3	0.171 37
0.08	0.397 67	0.43	0.363 71	0.78	0.294 31	1.4	0.149 73
0.09	0.397 33	0.44	0.362 13	0.79	0.292 00	1.5	0.129 52
0.10	0.396 95	0.45	0.360 53	0.80	0.289 69	1.6	0.110 92
0.11	0.396 54	0.46	0.358 89	0.81	0.287 37	1.7	0.094 05
0.12	0.396 08	0.47	0.357 23	0.82	0.285 04	1.8	0.078 95
0.13	0.395 59	0.48	0.355 53	0.83	0.282 69	1.9	0.065 62
0.14	0.395 05	0.49	0.353 81	0.84	0.280 34	2.0	0.053 99
0.15	0.394 48	0.50	0.352 07	0.85	0.277 98	2.1	0.043 98
0.16	0.393 87	0.51	0.350 29	0.86	0.275 62	2.2	0.035 47
0.17	0.393 22	0.52	0.348 49	0.87	0.273 24	2.3	0.028 33
0.18	0.392 53	0.53	0.346 67	0.88	0.270 86	2.4	0.022 39
0.19	0.391 81	0.54	0.344 82	0.89	0.268 48	2.5	0.017 53
0.20	0.391 04	0.55	0.342 94	0.90	0.266 09	2.6	0.013 58
0.21	0.390 24	0.56	0.341 05	0.91	0.263 69	2.7	0.010 42
0.22	0.389 40	0.57	0.339 12	0.92	0.261 29	2.8	0.007 92
0.23	0.388 53	0.58	0.337 18	0.93	0.258 88	2.9	0.005 95
0.24	0.387 62	0.59	0.335 21	0.94	0.256 47	3.0	0.004 43
0.25	0.386 67	0.60	0.333 22	0.95	0.254 06	3.1	0.003 27
0.26	0.385 68	0.61	0.331 21	0.96	0.251 64	3.2	0.002 38
0.27	0.384 66	0.62	0.329 18	0.97	0.249 23	3.3	0.001 72
0.28	0.383 61	0.63	0.327 13	0.98	0.246 81	3.4	0.001 23
0.29	0.382 51	0.64	0.325 06	0.99	0.244 39	3.5	0.000 87

$\lvert c\rvert$	$f(c)$	$\lvert c\rvert$	$f(c)$	$\lvert c\rvert$	$f(c)$	$\lvert c\rvert$	$f(c)$
0.30	0.381 39	0.65	0.322 97	1.00	0.241 97	3.6	0.000 61
0.31	0.380 23	0.66	0.320 86	1.01	0.239 55	3.7	0.000 42
0.32	0.379 03	0.67	0.318 74	1.02	0.237 13	3.8	0.000 29
0.33	0.377 80	0.68	0.316 59	1.03	0.234 71	3.9	0.000 20
0.34	0.376 54	0.69	0.314 43	1.04	0.232 30	4.0	0.000 13

Table II

This table gives values of the integral

$$\frac{1}{\sqrt{(2\pi)}} \int_{-k_1}^{k_1} e^{-c^2/2} \, dc = P_{-k_1 \text{ to } k_1}$$

the probability of an uncertainty occurring over the range $-k_1$ to $+k_1$, for incremental values of k, where

$$c = \frac{\lvert x - \mu \rvert}{\sigma} = k_1$$

where $\lvert x - \mu \rvert$ is the deviation from the mean, σ = standard deviation of the distribution and μ = mean of the distribution. $c = 1$ corresponds to one standard deviation σ. $(P_r - P_{r-1}) \equiv$ difference between successive values of $P_{-k_1 \text{ to } k_1}$.

k_1	$P_{-k_1 \text{ to } k_1}$	$(P_r - P_{r-1})$	k_1	$P_{-k_1 \text{ to } k_1}$	$(P_r - P_{r-1})$	k_1	$P_{-k_1 \text{ to } k_1}$	$(P_r - P_{r-1})$
0.00	0.000 00		0.35	0.273 66	0.007 52	0.7	0.516 07	0.006 26
0.01	0.007 98	0.007 98	0.36	0.281 15	0.007 49	0.8	0.576 29	0.060 22
0.02	0.015 96	0.007 98	0.37	0.288 62	0.007 47	0.9	0.631 88	0.055 59
0.03	0.023 93	0.007 97	0.38	0.296 05	0.007 43			
0.04	0.031 91	0.007 98	0.39	0.303 46	0.007 41			
0.05	0.039 88	0.007 97	0.40	0.310 84	0.007 38	1.0	0.682 69	0.050 81
0.06	0.047 84	0.007 96	0.41	0.318 19	0.007 35	1.1	0.728 67	0.045 98
0.07	0.055 81	0.007 97	0.42	0.325 51	0.007 32	1.2	0.769 86	0.041 ''
0.08	0.063 76	0.007 95	0.43	0.332 80	0.007 29	1.3	0.806 40	0.0.6 54
0.09	0.071 71	0.007 95	0.44	0.340 06	0.007 26	1.4	0.838 49	0.032 09
0.10	0.079 66	0.007 95	0.45	0.347 29	0.007 23	1.5	0.866 39	0.027 90
0.11	0.087 59	0.007 93	0.46	0.354 48	0.007 19	1.6	0.890 40	0.024 01
0.12	0.095 52	0.007 93	0.47	0.361 64	0.007 16	1.7	0.910 87	0.020 47
0.13	0.103 43	0.007 91	0.48	0.368 77	0.007 13	1.8	0.928 14	0.017 27
0.14	0.111 34	0.007 91	0.49	0.375 87	0.007 10	1.9	0.942 57	0.014 43

k_1	$P_{-k_1 \text{to} k_1}$	$(P_r - P_{r-1})$	k_1	$P_{-k_1 \text{to} k_1}$	$(P_r - P_{r-1})$	k_1	$P_{-k_1 \text{to} k_1}$	$(P_r - P_{r-1})$
0.15	0.119 24	0.007 90	0.50	0.382 92	0.007 05	2.0	0.954 50	0.011 93
0.16	0.127 12	0.007 88	0.51	0.389 95	0.007 03	2.1	0.964 27	0.009 77
0.17	0.134 99	0.007 87	0.52	0.396 94	0.006 99	2.2	0.972 19	0.007 92
0.18	0.142 85	0.007 86	0.53	0.403 89	0.006 95	2.3	0.978 55	0.006 36
0.19	0.150 69	0.007 84	0.54	0.410 80	0.006 91	2.4	0.983 60	0.005 05
0.20	0.158 52	0.007 83	0.55	0.417 68	0.006 88	2.5	0.987 58	0.003 98
0.21	0.166 33	0.007 81	0.56	0.424 52	0.006 84	2.6	0.990 68	0.003 10
0.22	0.174 13	0.007 80	0.57	0.431 32	0.006 80	2.7	0.993 07	0.002 39
0.23	0.181 91	0.007 78	0.58	0.438 09	0.006 77	2.8	0.994 89	0.001 82
0.24	0.189 67	0.007 76	0.59	0.444 81	0.006 72	2.9	0.996 27	0.001 38
0.25	0.197 41	0.007 74	0.60	0.451 49	0.006 68	3.0	0.997 30	0.001 03
0.26	0.205 14	0.007 73	0.61	0.458 14	0.006 65	3.1	0.998 06	0.000 76
0.27	0.212 84	0.007 70	0.62	0.464 74	0.006 60	3.2	0.998 63	0.000 57
0.28	0.220 52	0.007 68	0.63	0.471 31	0.006 57	3.3	0.999 03	0.000 40
0.29	0.228 18	0.007 66	0.64	0.477 83	0.006 52	3.4	0.999 33	0.000 30
0.30	0.235 82	0.007 64	0.65	0.484 31	0.006 48	3.5	0.999 53	0.000 20
0.31	0.243 44	0.007 62	0.66	0.490 75	0.006 44	3.6	0.999 68	0.000 15
0.32	0.251 03	0.007 59	0.67	0.497 14	0.006 39	3.7	0.999 78	0.000 10
0.33	0.258 60	0.007 57	0.68	0.503 50	0.006 36	3.8	0.999 86	0.000 08
0.34	0.266 14	0.007 54	0.69	0.509 81	0.006 31	3.891	0.999 90	

Table III

Values of the variable k_1 for incremental values of the tolerance probability

$$P_{-k_1 \text{to} k_1} = \frac{1}{\sqrt{(2\pi)}} \int_{-k_1}^{k_1} e^{-c^2/2} \, dc$$

The uncertainty or tolerance range associated with $P_{-k_1 \text{to} k_1}$ is then given by $k_1\sigma$, where σ is the standard deviation of the distribution.

$P_{-k_1 \text{to} k_1}$	k_1	$\delta = (_1k_r - {}_1k_{r-1})$	$P_{-k_1 \text{to} k_1}$	k_1	$\delta = (_1k_r - {}_1k_{r-1})$	$P_{-k_1 \text{to} k_1}$	k_1	$\delta = (_1k_r - {}_1k_{r-1})$
0.00	0.0000		0.35	0.4538	0.0139	0.70	1.036	0.021
0.01	0.0125	0.0125	0.36	0.4677	0.0140	0.71	1.058	0.022
0.02	0.0251	0.0126	0.37	0.4817	0.0141	0.72	1.080	0.022
0.03	0.0376	0.0125	0.38	0.4959	0.0142	0.73	1.103	0.023
0.04	0.0502	0.0126	0.39	0.5101		0.74	1.126	0.023
0.05	0.0627	0.0125	0.40	0.5244	0.0143	0.75	1.150	0.024
0.06	0.0753	0.0126	0.41	0.5388	0.0144	0.76	1.175	0.025
0.07	0.0878	0.0125	0.42	0.5534	0.0146	0.77	1.200	0.025
0.08	0.1004	0.0126	0.43	0.5681	0.0147	0.78	1.227	0.027
0.09	0.1130	0.0126	0.44	0.5828	0.0147	0.79	1.254	0.027

P $-k_1$ to k_1	k_1	$\delta =$ $({}_1k_r - {}_1k_{r-1})$	P $-k_1$ to k_1	k_1	$\delta =$ $({}_1k_r - {}_1k_{r-1})$	P $-k_1$ to k_1	k_1	$\delta =$ $({}_1k_r - {}_1k_{r-1})$
0.10	0.1257	0.0127	0.45	0.5978	0.0150	0.80	1.282	0.028
0.11	0.1383	0.0126	0.46	0.6128	0.0150	0.81	1.311	0.029
0.12	0.1510	0.0127	0.47	0.6280	0.0152	0.82	1.341	0.030
0.13	0.1637	0.0127	0.48	0.6433	0.0153	0.83	1.372	0.031
0.14	0.1764	0.0127	0.49	0.6588	0.0155	0.84	1.405	0.033
0.15	0.1891	0.0127	0.50	0.6745	0.0157	0.85	1.440	0.035
0.16	0.2019	0.0128	0.51	0.6903	0.0158	0.86	1.476	0.036
0.17	0.2147	0.0128	0.52	0.7063	0.0160	0.87	1.514	0.038
0.18	0.2275	0.0128	0.53	0.7225	0.0162	0.88	1.555	0.041
0.19	0.2404	0.0129	0.54	0.7388	0.0163	0.89	1.598	0.043
0.20	0.2533	0.0129	0.55	0.7554	0.0166	0.90	1.645	0.047
0.21	0.2663	0.0130	0.56	0.7722	0.0168	0.91	1.695	0.050
0.22	0.2793	0.0130	0.57	0.7892	0.0170	0.92	1.751	0.056
0.23	0.2924	0.0131	0.58	0.8064	0.0172	0.93	1.812	0.061
0.24	0.3055	0.0131	0.59	0.8239	0.0175	0.94	1.881	0.069
0.25	0.3186	0.0133	0.60	0.8416	0.0177	0.95	1.960	0.079
0.26	0.3319	0.0134	0.61	0.8596	0.0180	0.96	2.054	0.094
0.27	0.3451	0.0134	0.62	0.8779	0.0183	0.97	2.170	0.116
0.28	0.3585	0.0134	0.63	0.8965	0.0186	0.98	2.326	0.156
0.29	0.3719	0.0134	0.64	0.9154	0.0189	0.99	2.576	0.250
0.30	0.3853	0.0136	0.65	0.9346	0.0192	0.995	2.807	0.231
0.31	0.3989	0.0136	0.66	0.9542	0.0196	0.999	3.291	0.484
0.32	0.4125	0.0136	0.67	0.9741	0.0199			
0.33	0.4261	0.0138	0.68	0.9945	0.0204			
0.34	0.4399	0.0139	0.69	1.015	0.0205			

Table IV. Student t Distribution†: μ and s known

The table gives values of the range

$$t_v = \frac{x - \mu}{s_v} = k_2$$

for various values of the tolerance probability

$$\beta_{tp} = \int_{-t_v}^{t_v} f(t)\, dt$$

for integral values of v the number of degrees of freedom, where t_v is the standardized student variable and s_v is the estimated value of the standard deviation. s_v is given by the expression

$$s_v^2 = \sum_1^n (x - \bar{x})^2/(n - 1)$$

† See paragraphs 6.02 to 6.12.

and μ is the mean for an infinite number of observations. The integral β_{tp} is normalized, that is it is equal to unity for infinite limits and represents the probability of an uncertainty between the limits $-k_2$ and $+k_2$ for n readings. The tolerance range is thus $\mu \pm k_2 s_v$. v is equal to $n - 1$, where $n =$ number of observations of a single variable.

$\beta_{tp} = \int_{-t_v}^{t_v} f(t_v)\,dt_v$	0.90	0.95	0.98	0.99	0.995	0.999
$v = n - 1$	k_2	k_2	k_2	k_2	k_2	k_2
1	6.3138	12.706	31.821	63.657	127.32	636.619
2	2.9200	4.3027	6.965	9.9248	14.089	31.598
3	2.3534	3.1825	4.541	5.8409	7.4533	12.924
4	2.1318	2.7764	3.747	4.6041	5.5976	8.610
5	2.0150	2.5706	3.365	4.0321	4.7733	6.869
6	1.9432	2.4469	3.143	3.7074	4.3168	5.959
7	1.8946	2.3646	2.998	3.4995	4.0293	5.408
8	1.8595	2.3060	2.896	3.3554	3.8325	5.041
9	1.8331	2.2622	2.821	3.2498	3.6897	4.781
10	1.8125	2.2281	2.764	3.1693	3.5814	4.587
11	1.7959	2.2010	2.718	3.1058	3.4966	4.437
12	1.7823	2.1788	2.681	3.0545	3.4284	4.318
13	1.7709	2.1604	2.650	3.0123	3.3725	4.221
14	1.7613	2.1448	2.624	2.9768	3.3257	4.140
15	1.7530	2.1315	2.602	2.9467	3.2860	4.073
16	1.7459	2.1199	2.583	2.9208	3.2520	4.015
17	1.7396	2.1098	2.567	2.8982	3.2225	3.965
18	1.7341	2.1009	2.552	2.8784	3.1966	3.922
19	1.7291	2.0930	2.539	2.8609	3.1737	3.883
20	1.7247	2.0860	2.528	2.8453	3.1534	3.850
25	1.7081	2.0595	2.485	2.7874	3.0782	3.725
30	1.6973	2.0423	2.457	2.7500	3.0298	3.646
35	1.6996	2.0301	2.438	2.7239	2.9962	3.5915
40	1.6839	2.0211	2.423	2.7045	2.9713	3.5511
45	1.6794	2.0141	2.412	2.6896	2.9522	3.5207
50	1.6759	2.0086	2.403	2.6778	2.9370	3.4965
60	1.6707	2.0003	2.390	2.6603	2.9146	3.4606
70	1.6669	1.9945	2.381	2.6480	2.8988	3.4355
80	1.6641	1.9901	2.374	2.6388	2.8871	3.4169
90	1.6620	1.9867	2.368	2.6316	2.8779	3.4022
100	1.6602	1.9840	2.364	2.6260	2.8707	3.3909
150	1.6551	1.9759	2.351	2.6090	2.8492	3.3567
200	1.6525	1.9719	2.345	2.6006	2.8386	3.3400

Table V. Confidence Limits for the Mean μ†

The table gives values of the confidence factor k_3 defining the confidence limits $\bar{x} \pm k_3 s$ for the mean μ. Values of k_3 are given for two confidence probabilities β_{cp}, that is 0.95 and 0.99, and for integral values of n, the number of observations for the variable x. $k_3 = k_2/\sqrt{n}$ for corresponding values of n.

Note: that s is the estimated standard deviation, *not* the estimated standard error. $\bar{x}$ is the mean of n readings.

$$s = \sum \frac{(x - x)^2}{n - 1}$$

	Confidence probability $\beta_{cp} = 0.95$						Confidence probability $\beta_{cp} = 0.99$				
n	k_3	n	k_3	n	k_3	n	k_3	n	k_3	n	k_3
1		30	0.3734	95	0.2037	1		30	0.5033	95	0.2698
2	8.9845	31	0.3668	100	0.1964	2	4.5012	31	0.4939	100	0.2627
3	2.4842	32	0.3605	110	0.1890	3	5.7301	32	0.4851	110	0.2500
4	1.5913	33	0.3546	120	0.1808	4	2.9205	33	0.4767	120	0.2390
5	1.2461	34	0.3489	130	0.1735	5	2.0590	34	0.4688	130	0.2293
6	1.0494	35	0.3435	140	0.1671	6	1.6461	35	0.4612	140	0.2207
7	0.9248	36	0.3384	150	0.1674	7	1.4013	36	0.4540	150	0.2131
8	0.8360	37	0.3334	160	0.1561	8	1.2373	37	0.4471	160	0.2061
9	0.7687	38	0.3287	170	0.1514	9	1.1185	38	0.4405	170	0.1998
10	0.7154	39	0.3242	180	0.1471	10	1.0277	39	0.4342	180	0.1941
11	0.6718	40	0.3198	190	0.1431	11	0.9556	40	0.4282	190	0.1888
12	0.6354	41	0.3156	200	0.1394	12	0.8966	41	0.4224	200	0.1839
13	0.6043	42	0.3116	250	0.1240	13	0.8472	42	0.4168	250	0.1629
14	0.5774	43	0.3078	300	0.1132	14	0.8051	43	0.4115	300	0.1487
15	0.5538	44	0.3040	350	0.1048	15	0.7686	44	0.4063	350	0.1377
16	0.5329	45	0.3004	400	0.0980	16	0.7367	45	0.4013	400	0.1288
17	0.5142	46	0.2970	450	0.0924	17	0.7084	46	0.3966	450	0.1214
18	0.4973	47	0.2936	500	0.0877	18	0.6831	47	0.3919	500	0.1152
19	0.4820	48	0.2904	550	0.0836	19	0.6604	48	0.3875	550	0.1098
20	0.4680	49	0.2872	600	0.0800	20	0.6397	49	0.3832	600	0.1052
21	0.4552	50	0.2842	650	0.0769	21	0.6209	50	0.3790	650	0.1010
22	0.4434	55	0.2703	700	0.0741	22	0.6037	55	0.3600	700	0.0974
23	0.4324	60	0.2583	750	0.0716	23	0.5878	60	0.3436	750	0.0941
24	0.4223	65	0.2478	800	0.0693	24	0.5730	65	0.3293	800	0.0911
25	0.4128	70	0.2385	850	0.0672	25	0.5594	70	0.3166	850	0.0884
26	0.4039	75	0.2301	900	0.0653	26	0.5467	75	0.3053	900	0.0859
27	0.3956	80	0.2225	950	0.0636	27	0.5348	80	0.2951	950	0.0836
28	0.3878	85	0.2157			28	0.5236	85	0.2859		
29	0.3804	90	0.2095	1000	0.620	29	0.5131	90	0.2775	1000	0.0815

† See paragraph 6.40.

Table VI. Chi-square χ^2 Distribution†

The table gives the probability of χ^2 exceeding the values of χ^2 plotted against v and also the probability of χ^2 being less than the values of χ^2 plotted against v. In each case six probabilities are used, that is 0.20, 0.10, 0.05, 0.025, 0.01 and 0.005.

$$\chi_v^2 = \sum_{i=1}^{i=n} \left(\frac{x_i - \mu}{\sigma} \right)^2$$

No of degrees of freedom v	Probability of tabulated values of χ_v^2 being exceeded					
	Prob. 0.20	Prob. 0.10	Prob. 0.05	Prob. 0.025	Prob. 0.01	Prob. 0.005
1	1.642	2.706	3.841	5.024	6.635	7.879
2	3.219	4.605	5.991	7.378	9.210	10.597
3	4.642	6.251	7.815	9.348	11.345	12.838
4	5.989	7.779	9.488	11.143	13.277	14.860
5	7.289	9.236	11.070	12.832	15.086	16.750
6	8.558	10.645	12.592	14.449	16.812	18.548
7	9.803	12.017	14.067	16.013	18.475	20.278
8	11.030	13.362	15.587	17.535	20.090	21.955
9	12.242	14.684	16.919	19.023	21.666	23.589
10	13.442	15.987	18.307	20.483	23.209	25.188
11	14.631	17.275	19.675	21.920	24.725	26.757
12	15.812	18.549	21.026	23.336	26.217	28.300
13	16.985	19.812	22.362	24.736	27.688	29.819
14	18.151	21.064	23.685	26.119	29.141	31.319
15	19.311	22.307	24.996	27.488	30.578	32.801
16	20.465	23.542	26.296	28.745	32.000	34.267
17	21.615	24.769	27.587	30.191	33.409	35.718
18	22.760	25.989	28.869	31.526	34.805	37.156
19	23.900	27.204	30.144	32.852	36.191	38.582
20	25.038	28.412	31.410	34.170	37.566	39.997
25	30.675	34.382	37.652	40.646	44.314	46.928
30	36.250	40.256	43.773	46.979	50.892	53.672
35	41.778	46.059	49.802	53.203	57.342	60.275
40	47.269	51.805	55.758	59.342	63.691	66.766
45	52.729	57.505	61.656	65.410	69.957	73.166
50	58.164	63.167	67.505	71.420	76.154	79.490
60	68.972	74.397	79.082	83.298	88.379	91.952
70	79.715	85.527	90.531	95.023	100.425	104.215
80	90.405	96.578	101.879	106.629	112.329	116.321
90	101.054	107.565	113.145	118.126	124.116	128.299
100	111.667	118.498	124.342	129.561	135.806	140.169
150	164.349	172.581	179.581	185.800	193.207	198.360
200	216.609	226.021	233.994	241.058	249.445	255.264

† See paragraphs 6.13 to 6.15.

No of degrees of freedom *v*	Probability of χ^2_v being less than the values tabulated					
	Prob. 0.20	Prob. 0.10	Prob. 0.05	Prob. 0.025	Prob. 0.01	Prob. 0.005
1	0.0642	0.0158	0.003 93	0.000 982	0.000 157	0.0000393
2	0.446	0.211	0.103	0.0506	0.0201	0.0100
3	1.005	0.584	0.352	0.216	0.115	0.0717
4	1.649	1.064	0.711	0.484	0.297	0.207
5	2.343	1.610	1.145	0.831	0.554	0.412
6	3.070	2.204	1.635	1.237	0.872	0.676
7	3.822	2.833	2.167	1.690	1.239	0.989
8	4.594	3.490	2.733	2.180	1.646	1.344
9	5.380	4.168	3.325	2.700	2.088	1.735
10	6.179	4.865	3.940	3.247	2.558	2.156
11	6.989	5.578	4.575	3.816	3.053	2.603
12	7.807	6.304	5.226	4.404	3.571	3.074
13	8.634	7.042	5.892	5.009	4.107	3.565
14	9.467	7.790	6.571	5.629	4.660	4.075
15	10.307	8.547	7.261	6.262	5.229	4.601
16	11.152	9.312	7.962	6.908	5.812	5.142
17	12.002	10.085	8.672	7.564	6.408	5.697
18	12.857	10.865	9.390	8.231	7.015	6.265
19	13.716	11.651	10.117	8.907	7.633	6.844
20	14.578	12.443	10.851	9.951	8.260	7.434
25	18.940	16.473	14.611	13.120	11.524	10.520
30	23.364	20.599	18.493	16.791	14.953	13.787
35	27.836	24.797	22.465	20.569	18.509	17.192
40	32.345	29.051	26.509	24.433	22.164	20.707
45	36.884	33.350	30.612	28.361	25.901	24.311
50	41.449	37.689	34.764	32.357	29.707	27.991
60	50.641	46.459	43.188	40.482	37.485	35.535
70	59.898	55.329	51.739	48.758	45.442	43.275
80	69.207	64.278	60.391	57.153	53.540	51.172
90	78.558	73.291	69.126	65.647	61.754	59.196
100	87.945	82.358	77.930	74.222	70.065	67.328
150	135.263	128.275	122.692	117.985	112.668	109.143
200	183.003	174.835	168.279	162.728	156.436	152.241

Table VII

Confidence intervals for σ the standard deviation. The table gives values of k_9, the confidence factor, for integral values of n, the number of observations, for four values of the confidence probability β_{cp}. The probabilities in brackets are the values for the upper limit ($_{max}k_9$) not being exceeded and of k_9 not being less than $_{min}k_9$. That is β_{cp} unbracketed gives the probability that $_{min}k_9 s_v \leqslant \sigma \leqslant _{max}k_9 s_v$, whilst β_{cp} bracketed gives the probability that

$$\sigma \leqslant _{max}k_9 s_v \qquad \text{or} \qquad \sigma \geqslant _{min}k_9 s_v$$

where s_v is the estimated standard deviation for v degrees of freedom.

$k_9 \dagger$

n	$\beta_{cp} 0.90$ (0.95) $_{min}k_9 \quad _{max}k_9$	$\beta_{cp} = 0.95$ (0.975) $_{min}k_9 \quad _{max}k_9$	$\beta_{cp} = 0.98$ (0.99) $_{min}k_9 \quad _{max}k_9$	$\beta_{cp} = 0.99$ (0.995) $_{min}k_9 \quad _{max}k_9$
2	0.5102–15.947	0.4463–31.910	0.3882–79.789	0.3562–159.58
3	0.5777–4.416	0.5207–6.285	0.4660–9.974	0.4344–14.124
4	0.6196–2.920	0.5665–3.729	0.5142–5.111	0.4834–6.468
5	0.6493–2.372	0.5991–2.874	0.5489–3.669	0.5188–4.396
6	0.6720–2.089	0.6242–2.453	0.5757–3.003	0.5464–3.485
7	0.6903–1.915	0.6444–2.202	0.5974–2.623	0.5688–2.980
8	0.7054–1.797	0.6612–2.035	0.6155–2.377	0.5875–2.660
9	0.7183–1.711	0.6755–1.916	0.6310–2.204	0.6036–2.439
10	0.7293–1.645	0.6878–0.286	0.6445–2.076	0.6177–2.278
11	0.7391–1.593	0.6987–1.755	0.6564–1.977	0.6301–2.154
12	0.7477–1.551	0.7084–1.698	0.6670–1.898	0.6412–2.056
13	0.7555–1.515	0.7171–1.651	0.6765–1.833	0.6512–1.976
14	0.7625–1.485	0.7250–1.611	0.6852–1.799	0.6603–1.910
15	0.7688–1.460	0.7321–1.577	0.6931–1.733	0.6686–1.854
16	0.7747–1.437	0.7387–1.548	0.7004–1.694	0.6762–1.806
17	0.7800–1.418	0.7448–1.522	0.7071–1.659	0.6833–1.764
18	0.7850–1.400	0.7504–1.499	0.7133–1.629	0.6899–1.727
19	0.7896–1.384	0.7556–1.479	0.7191–1.602	0.6960–1.695
20	0.7939–1.370	0.7604–1.461	0.7246–1.578	0.7018–1.666
25	0.8118–1.316	0.7808–1.391	0.7473–1.487	0.7258–1.558
30	0.8255–1.280	0.7964–1.344	0.7647–1.426	0.7444–1.487
35	0.8364–1.253	0.8089–1.310	0.7788–1.382	0.7594–1.435
40	0.8454–1.232	0.8192–1.284	0.7904–1.349	0.7718–1.397
45	0.8529–1.215	0.8279–1.263	0.8002–1.323	0.7823–1.366

† See paragraph 6.16 for derivation of k_9.

k_9 †

	$\beta_{cp}0.90$ (0.95)		$\beta_{cp} = 0.95$ (0.975)		$\beta_{cp} = 0.98$ (0.99)		$\beta_{cp} = 0.99$ (0.995)	
n	$_{min}k_9$	$_{max}k_9$	$_{min}k_9$	$_{max}k_9$	$_{min}k_9$	$_{max}k_9$	$_{min}k_9$	$_{max}k_9$
50	0.8594–1.202		0.8353–1.246		0.8087–1.301		0.7914–1.341	
55	0.8651–1.190		0.8419–1.232		0.8161–1.283		0.7994–1.320	
60	0.8701–1.180		0.8476–1.220		0.8227–1.268		0.8065–1.303	
65	0.8746–1.172		0.8528–1.209		0.8286–1.255		0.8128–1.287	
70	0.8786–1.165		0.8574–1.200		0.8339–1.243		0.8185–1.274	
75	0.8822–1.158		0.8616–1.192		0.8387–1.233		0.8237–1.263	
80	0.8855–1.152		0.8655–1.184		0.8431–1.224		0.8284–1.252	
85	0.8885–1.147		0.8690–1.178		0.8471–1.216		0.8328–1.243	
90	0.8913–1.142		0.8722–1.172		0.8508–1.209		0.8368–1.235	
95	0.8939–1.138		0.8752–1.167		0.8543–1.202		0.8405–1.227	
100	0.8963–1.134		0.8780–1.162		0.8575–1.196		0.8440–1.220	
∞	1.0000		1.0000		1.0000		1.0000	

† See paragraph 6.16 for derivation of k_9.

Table VIII. *F* Test. Upper Limits for *F*. Probability of *F* exceeding these Values = 0.05

v_2	v_1 1	2	3	4	5	6	7	8	9	10	15	20	24	30	40	50	60	80	100
1	161.44	200	216	225	230	234	237	239	241	242	246	248	249	250	251	252	252	252	253
2	18.51	19.0	19.2	19.2	19.3	19.3	19.4	19.4	19.4	19.4	19.4	19.4	19.5	19.5	19.5	19.5	19.5	19.5	19.5
3	10.13	9.55	9.28	9.12	9.01	8.94	8.89	8.85	8.81	8.79	8.70	8.66	8.64	8.62	8.59	8.58	8.57	8.56	8.55
4	7.71	6.94	6.59	6.39	6.26	6.16	6.09	6.04	6.00	5.96	5.86	5.80	5.77	5.75	5.72	5.70	5.69	5.67	5.66
5	6.61	5.79	5.41	5.19	5.05	4.95	4.88	4.82	4.77	4.74	4.62	4.56	4.53	4.50	4.46	4.44	4.43	4.41	4.41
6	5.99	5.14	4.76	4.53	4.39	4.28	4.21	4.15	4.10	4.06	3.94	3.87	3.84	3.81	3.77	3.75	3.74	3.72	3.71
7	5.59	4.74	4.35	4.12	3.97	3.87	3.79	3.73	3.68	3.64	3.51	3.44	3.41	3.38	3.34	3.32	3.30	3.29	3.27
8	5.32	4.46	4.07	3.84	3.69	3.58	3.50	3.44	3.39	3.35	3.22	3.15	3.12	3.08	3.04	3.02	3.01	2.99	2.97
9	5.12	4.26	3.86	3.63	3.48	3.37	3.29	3.23	3.18	3.14	3.01	2.94	2.90	2.86	2.83	2.80	2.79	2.77	2.76
10	4.96	4.10	3.71	3.48	3.33	3.22	3.14	3.07	3.02	2.98	2.85	2.77	2.74	2.70	2.66	2.64	2.62	2.60	2.59
15	4.54	3.68	3.29	3.06	2.90	2.79	2.71	2.64	2.59	2.54	2.42	2.33	2.31	2.25	2.20	2.18	2.16	2.14	2.12
20	4.35	3.49	3.10	2.87	2.71	2.60	2.51	2.45	2.39	2.35	2.20	2.12	2.08	2.04	1.99	1.97	1.95	1.92	1.91
25	4.24	3.39	2.99	2.76	2.60	2.49	2.40	2.34	2.28	2.24	2.11	2.02	1.96	1.92	1.87	1.84	1.82	1.80	1.78
30	4.17	3.32	2.92	2.69	2.53	2.42	2.33	2.27	2.21	2.16	2.01	1.93	1.89	1.85	1.79	1.76	1.74	1.71	1.70
40	4.08	3.23	2.84	2.61	2.45	2.34	2.25	2.18	2.12	2.08	1.92	1.84	1.81	1.74	1.69	1.66	1.64	1.62	1.59
50	4.03	3.18	2.79	2.56	2.40	2.29	2.20	2.13	2.07	2.03	1.87	1.78	1.74	1.69	1.63	1.60	1.58	1.54	1.52
60	4.00	3.15	2.76	2.53	2.37	2.25	2.17	2.10	2.04	1.99	1.84	1.75	1.70	1.65	1.59	1.56	1.53	1.50	1.48
80	3.96	3.11	2.72	2.49	2.33	2.21	2.13	2.06	2.00	1.95	1.79	1.70	1.65	1.60	1.54	1.51	1.48	1.45	1.43
100	3.94	3.09	2.70	2.46	2.31	2.19	2.10	2.03	1.97	1.93	1.77	1.68	1.63	1.57	1.52	1.48	1.45	1.41	1.39

Probability P of table $= \int_{F_0}^{\infty} f(F)\, dF =$ probability of $F > F_0$

$$= 0.05$$

Tabulated values are of F_0 for a range of values of v_1 and v_2 where $F = s_{v_1}^2 / s_{v_2}^2$ and v_1 and v_2 are the number of degrees of freedom for each sample. For a single variable $v_1 = n_1 - 1$ and $v_2 = n_2 - 1$, where n_1 and n_2 are the number of observations for each sample.

Table IX. F Test. Upper Limits for F. Probability of F exceeding these Values = 0.01

v_1

v_2	1	2	3	4	5	6	7	8	9	10	15	20	24	30	40	50	60	80	100
2	98.50	99.0	99.2	99.2	99.3	99.4	99.4	99.4	99.4	99.4	99.4	99.4	99.5	99.5	99.5	99.5	99.5	99.5	99.5
3	34.12	30.8	29.5	28.7	28.2	27.9	27.7	27.5	27.3	27.2	26.9	26.7	26.6	26.5	26.4	26.4	26.3	26.3	26.2
4	21.20	18.0	16.7	16.0	15.5	15.2	15.0	14.8	14.7	14.5	14.2	14.0	13.9	13.9	13.7	13.7	13.7	13.6	13.6
5	16.26	13.3	12.1	11.4	11.0	10.7	10.5	10.3	10.2	10.1	9.72	9.55	9.47	9.38	9.29	9.24	9.20	9.16	9.13
6	13.75	10.9	9.78	9.15	8.75	8.47	8.26	8.10	7.98	7.87	7.56	7.40	7.31	7.23	7.14	7.09	7.06	7.01	6.99
7	12.25	9.55	8.45	7.85	7.46	7.19	6.99	6.84	6.72	6.62	6.31	6.16	6.07	5.99	5.91	5.86	5.82	5.78	5.75
8	11.26	8.65	7.59	7.01	6.63	6.37	6.18	6.03	5.91	5.81	5.52	5.36	5.28	5.20	5.12	5.07	5.03	4.99	4.96
9	10.56	8.02	6.99	6.42	6.06	5.80	5.61	5.47	5.35	5.26	4.96	4.81	4.73	4.65	4.57	4.52	4.48	4.44	4.42
10	10.04	7.56	6.55	5.99	5.64	5.39	5.20	5.06	4.94	4.85	4.56	4.41	4.33	4.25	4.17	4.12	4.08	4.04	4.01
15	8.68	6.36	5.42	4.89	4.56	4.32	4.14	4.00	3.89	3.80	3.52	3.37	3.29	3.21	3.13	3.08	3.05	3.00	2.98
20	8.10	5.85	4.94	4.43	4.10	3.87	3.70	3.56	3.46	3.37	3.09	2.94	2.86	2.78	2.69	2.64	2.61	2.56	2.54
25	7.77	5.57	4.68	4.18	3.86	3.63	3.46	3.32	3.22	3.13	2.85	2.70	2.62	2.54	2.45	2.40	2.36	2.32	2.29
30	7.56	5.39	4.51	4.02	3.70	3.47	3.30	3.17	3.07	2.98	2.70	2.55	2.47	2.39	2.30	2.25	2.21	2.16	2.13
40	7.31	5.18	4.31	3.83	3.51	3.29	3.12	2.99	2.89	2.80	2.52	2.37	2.29	2.20	2.11	2.06	2.02	1.97	1.94
50	7.17	5.06	4.20	3.72	3.41	3.19	3.02	2.89	2.79	2.70	2.42	2.27	2.18	2.10	2.01	1.95	1.91	1.86	1.82
60	7.08	4.98	4.13	3.65	3.34	3.12	2.95	2.82	2.72	2.63	2.35	2.20	2.12	2.03	1.94	1.88	1.84	1.78	1.75
80	6.96	4.88	4.04	3.56	3.26	3.04	2.87	2.74	2.64	2.55	2.27	2.12	2.03	1.94	1.85	1.79	1.75	1.69	1.66
100	6.90	4.82	3.98	3.51	3.21	2.99	2.82	2.69	2.59	2.50	2.22	2.07	1.98	1.89	1.80	1.73	1.69	1.63	1.60

Probability P of table $= \int_{F_0}^{\infty} f(F)\, dF =$ probability of $F > F_0$

$$= 0.01$$

Tabulated values are of F_0 for a range of values of v_1 and v_2 where $F = s_{v_1}^2 / s_{v_2}^2$ and v_1 and v_2 are the number of degrees of freedom for each sample. For a single variable $v_1 = n_1 - 1$ and $v_2 = n_2 - 1$, where n_1 and n_2 are the number of observations for each sample.

Table X. $\bar{x}$ and σ known

The table gives values of k_4 defining the tolerance limits $\bar{x} \pm k_4\sigma$ for two values of the tolerance probability β_{tp}, for integral values of n, the number of observations. The values of k_4† are without confidence probabilities.

$$k_4 = k_1\left(1 + \frac{1}{n}\right)^{1/2}$$

where k_1 is the tolerance factor of Tables II or III. The tolerance probability for k_4 is the same as that of the corresponding k_1.

	Observations k_4	
n	$\beta_{tp} = 0.95$	$\beta_{tp} = 0.99$
2	2.4005	3.1547
3	2.2632	2.9743
4	2.1913	2.8799
5	2.1470	2.8217
6	2.1170	2.7822
7	2.0953	2.7537
8	2.0789	2.7321
9	2.0660	2.7152
10	2.0556	2.7016
11	2.0471	2.6904
12	2.0400	2.6810
13	2.0340	2.6731
14	2.0288	2.6662
15	2.0242	2.6603
16	2.0203	2.6551
17	2.0168	2.6505
18	2.0137	2.6464
19	2.0109	2.6427
20	2.0084	2.6394
25	1.9988	2.6268
30	1.9924	2.6184
35	1.9878	2.6124
40	1.9843	2.6078
45	1.9816	2.6043
50	1.9795	2.6015

† See paragraphs 6.26 to 6.27.

Observations

	k_4	
n	$\beta_{tp} = 0.95$	$\beta_{tp} = 0.99$
55	1.9777	2.5991
60	1.9762	2.5972
70	1.9739	2.5942
80	1.9722	2.5919
90	1.9708	2.5901
100	1.9697	2.5887
∞	1.9600	2.5758

Table XI. $\bar{x}$ and s known

The table gives values of k_5† defining the tolerance limits $\bar{x} \pm k_5 s$ for two values of the tolerance probability β_{tp}, for integral values of n, the number of observations. These are 'on average' values and do not carry confidence probabilities.

$$k_5 = k_2 \left(1 + \frac{1}{n} \right)^{1/2}$$

for corresponding values of n, and tolerance probabilities β_{tp}, where k_2 is the tolerance factor of Table IV and n = number of observations.

	k_5	
n	$\beta_{tp} = 0.95$	$\beta_{tp} = 0.99$
2	15.562	77.964
3	4.9683	11.460
4	3.5581	6.5303
5	3.0414	5.0435
6	2.7766	4.3552
7	2.6158	3.9634
8	2.5080	3.7118
9	2.4307	3.5369
10	2.3726	3.4084
11	2.3272	3.3102
12	2.2909	3.2326
13	2.2610	3.1698
14	2.2362	3.1180

† See paragraphs 6.29 to 6.31.

	k_5	
n	$\beta_{tp} = 0.95$	$\beta_{tp} = 0.99$
15	2.2151	3.0744
16	2.1971	3.0374
17	2.1814	3.0055
18	2.1676	2.9776
19	2.1555	2.9532
20	2.1447	2.9315
25	2.1048	2.8523
30	2.0790	2.8020
35	2.0611	2.7671
40	2.0478	2.7415
45	2.0377	2.7220
50	2.0296	2.7066
55	2.0230	2.6942
60	2.0176	2.6839
65	2.0130	2.6753
70	2.0092	2.6679
75	2.0058	2.6616
80	2.0029	2.6560
85	2.0003	2.6512
90	1.9980	2.6468
95	1.9960	2.6430
100	1.9942	2.6396
∞	1.9600	2.5758

Table XII. $\bar{x}$ and σ known

This table gives values of k_6 † defining the tolerance limits $\bar{x} \pm k_6\sigma$ for three values of the tolerance probability β_{tp} and two values of the confidence probability β_{cp}, for integral values of n, the number of observations.

	k_6			
	$\beta_{cp} = 0.95$		$\beta_{cp} = 0.99$	
n	$\beta_{tp} = 0.90$	$\beta_{tp} = 0.95$	$\beta_{tp} = 0.95$	$\beta_{tp} = 0.99$
2	2.667	3.031	3.466	4.147
3	2.415	2.776	3.132	3.813
4	2.265	2.626	2.933	3.614

† See paragraphs 6.32 and 6.33.

$$k_6$$

n	$\beta_{cp} = 0.95$		$\beta_{cp} = 0.99$	
	$\beta_{tp} = 0.90$	$\beta_{tp} = 0.95$	$\beta_{tp} = 0.95$	$\beta_{tp} = 0.99$
5	2.165	2.525	2.797	3.478
6	2.093	2.450	2.698	3.370
7	2.038	2.394	2.620	3.301
8	1.995	2.349	2.558	3.238
9	1.961	2.313	2.507	3.186
10	1.932	2.283	2.465	3.143
11	1.909	2.258	2.428	3.105
12	1.889	2.236	2.397	3.073
13	1.872	2.218	2.369	3.044
14	1.856	2.201	2.345	3.018
15	1.843	2.186	2.324	2.996
16	1.832	2.174	2.309	2.976
17	1.821	2.162	2.287	2.958
18	1.812	2.153	2.272	2.941
19	1.804	2.143	2.258	2.926
20	1.796	2.134	2.245	2.912
25	1.767	2.101	2.193	2.856
30	1.747	2.079	2.158	2.816
35	1.733	2.063	2.132	2.786
40	1.722	2.051	2.112	2.763
45	1.714	2.040	2.096	2.745
50	1.707	2.033	2.083	2.729
55	1.702	2.026	2.073	2.717
60	1.697	2.021	2.064	2.706
65	1.693	2.016	2.056	2.696
70	1.689	2.012	2.049	2.689
75	1.686	2.009	2.044	2.681
80	1.684	2.006	2.039	2.676
85	1.682	2.004	2.034	2.670
90	1.680	2.001	2.030	2.665
95	1.678	1.999	2.026	2.661
100	1.676	1.997	2.023	2.657
∞	1.645	1.960	1.960	2.576

Table XIII. $\bar{x}$ and s known

The table gives values of k_7 † defining the tolerance limits $\bar{x} \pm k_7 s$ for four values of the tolerance probability β_{tp} and three values of the confidence probability β_{cp} for integral values of n, the number of observations.

n	$\beta_{cp} = 0.90$				$\beta_{cp} = 0.95$				$\beta_{cp} = 0.99$			
	$\beta_{tp} = 0.90$	$\beta_{tp} = 0.95$	$\beta_{tp} = 0.99$	$\beta_{tp} = 0.999$	$\beta_{tp} = 0.90$	$\beta_{tp} = 0.95$	$\beta_{tp} = 0.99$	$\beta_{tp} = 0.999$	$\beta_{tp} = 0.90$	$\beta_{tp} = 0.95$	$\beta_{tp} = 0.99$	$\beta_{tp} = 0.999$
2	15.978	18.800	24.167	30.227	32.019	37.674	48.430	60.573	160.193	188.491	242.300	303.054
3	5.847	6.919	8.974	11.309	8.380	9.916	12.861	16.208	18.930	22.401	29.055	36.616
4	4.166	4.943	6.440	8.149	5.369	6.370	8.259	10.502	9.398	11.150	14.527	18.383
5	3.404	4.152	5.423	6.879	4.275	5.079	6.634	8.415	6.612	7.855	10.260	13.015
6	3.131	3.723	4.870	6.188	3.712	4.414	5.775	7.337	5.337	6.345	8.301	10.545
7	2.902	3.452	4.521	5.750	3.369	4.007	5.248	6.676	4.613	5.488	7.187	9.142
8	2.743	3.264	4.278	5.446	3.136	3.732	4.891	6.226	4.147	4.936	6.468	8.234
9	2.626	3.125	4.098	5.220	2.967	3.532	4.631	5.899	3.822	4.550	5.966	7.600
10	2.535	3.018	3.959	5.046	2.839	3.379	4.433	5.649	3.582	4.265	5.594	7.129
11	2.463	2.933	3.849	4.906	2.737	3.259	4.277	5.452	3.397	4.045	5.308	6.766
12	2.404	2.863	3.758	4.792	2.655	3.162	4.150	5.291	3.250	3.870	5.079	6.477
13	2.355	2.805	3.682	4.697	2.587	3.081	4.044	5.158	3.130	3.727	4.893	6.240
14	2.314	2.756	3.618	4.615	2.529	3.012	3.955	5.045	3.029	3.608	4.737	6.043
15	2.278	2.713	3.562	4.545	2.480	2.954	3.878	4.949	2.945	3.507	4.605	5.876
16	2.246	2.676	3.514	4.484	2.437	2.903	3.812	4.865	2.872	3.421	4.492	5.732
17	2.219	2.643	3.471	4.430	2.400	2.858	3.754	4.791	2.808	3.345	4.393	5.607
18	2.194	2.614	3.433	4.382	2.366	2.819	3.702	4.725	2.753	3.279	4.307	5.497
19	2.172	2.588	3.399	4.339	2.337	2.784	3.656	4.667	2.703	3.221	4.230	5.399

20	2.152	2.564	3.368	4.300	2.310	2.752	3.615	4.614	2.659	3.168	4.161	5.312
25	2.077	2.474	3.251	4.151	2.208	2.631	3.457	4.413	2.494	2.972	3.904	4.955
30	2.025	2.413	3.170	4.049	2.140	2.549	3.350	4.278	2.385	2.841	3.733	4.768
35	1.988	2.368	3.112	3.974	2.090	2.490	3.272	4.179	2.306	2.748	3.611	4.611
40	1.959	2.334	3.066	3.917	2.052	2.445	3.213	4.104	2.247	2.677	3.518	4.493
45	1.935	2.306	3.030	3.871	2.021	2.408	3.165	4.042	2.200	2.621	3.444	4.399
50	1.916	2.284	3.001	3.833	1.996	2.379	3.126	3.993	2.162	2.576	3.385	4.323
55	1.901	2.265	2.976	3.801	1.976	2.354	3.094	3.951	2.130	2.538	3.335	4.260
60	1.887	2.248	2.955	3.774	1.958	2.333	3.066	3.916	2.103	2.506	3.293	4.206
65	1.875	2.235	2.937	3.751	1.943	2.315	3.042	3.886	2.080	2.478	3.257	4.160
70	1.865	2.222	2.920	3.730	1.929	2.299	3.021	3.859	2.060	2.454	3.225	4.125
75	1.856	2.211	2.906	3.712	1.917	2.285	3.002	3.835	2.042	2.433	3.197	4.084
80	1.848	2.202	2.894	3.696	1.907	2.272	2.986	3.814	2.026	2.414	3.173	4.053
85	1.841	2.193	2.882	3.682	1.897	2.261	2.971	3.795	2.012	2.397	3.150	4.024
90	1.834	2.185	2.872	3.669	1.889	2.251	2.958	3.778	1.999	2.382	3.130	3.999
95	1.828	2.178	2.863	3.657	1.881	2.241	2.945	3.763	1.987	2.368	3.112	3.976
100	1.822	2.172	2.854	3.646	1.874	2.233	2.934	3.748	1.977	2.355	3.096	3.954
120	1.804	2.150	2.826	3.610	1.850	2.205	2.898	3.702	1.942	2.314	3.041	3.665
140	1.791	2.134	2.804	3.582	1.833	2.184	2.870	3.666	1.916	2.283	3.000	3.885
160	1.780	2.121	2.787	3.561	1.819	2.167	2.848	3.638	1.896	2.259	2.968	3.792
180	1.771	2.111	2.774	3.543	1.808	2.154	2.831	3.616	1.879	2.239	2.942	3.759
200	1.764	2.102	2.762	3.529	1.798	2.143	2.816	3.597	1.865	2.222	2.921	3.731

† See paragraphs 6.34 and 6.35.

Table XIV. μ and s known

The table gives values of k_8 † defining the tolerance limits $\mu \pm k_8 s$ for three values of the tolerance probability β_{tp} and two values of the confidence probability β_{cp} for integral values of n, the number of observations.

	k_8			
	$\beta_{cp} = 0.95$		$\beta_{cp} = 0.99$	
n	$\beta_{tp} = 0.90$	$\beta_{tp} = 0.95$	$\beta_{tp} = 0.95$	$\beta_{tp} = 0.99$
2	26.231	31.256	156.38	205.52
3	7.263	8.654	19.550	25.694
4	4.803	5.723	10.018	13.166
5	3.902	4.650	7.191	9.451
6	3.437	4.095	5.887	7.736
7	3.151	3.754	5.141	6.756
8	2.956	3.522	4.659	6.122
9	2.814	3.354	4.320	5.678
10	2.706	3.225	4.069	5.348
11	2.620	3.122	3.875	5.093
12	2.551	3.039	3.720	4.889
13	2.492	2.970	3.593	4.722
14	2.443	2.911	3.487	4.583
15	2.401	2.861	3.397	4.464
16	2.364	2.817	3.319	4.361
17	2.332	2.778	3.252	4.274
18	2.303	2.744	3.192	4.196
19	2.277	2.714	3.140	4.126
20	2.524	2.686	3.092	4.064
25	2.165	2.580	2.914	3.830
30	2.105	2.508	2.795	3.674
35	2.061	2.455	2.710	3.561
40	2.026	2.415	2.644	3.475
45	1.999	2.382	2.593	3.407
50	1.977	2.355	2.550	3.352
55	1.958	2.333	2.515	3.305
60	1.942	2.314	2.485	3.266
65	1.928	2.297	2.459	3.232
70	1.916	2.282	2.437	3.203
75	1.905	2.270	2.417	3.176

† See paragraphs 6.36 to 6.38.

	k_8			
	$\beta_{cp} = 0.95$		$\beta_{cp} = 0.99$	
n	$\beta_{tp} = 0.90$	$\beta_{tp} = 0.95$	$\beta_{tp} = 0.95$	$\beta_{tp} = 0.99$
80	1.895	2.258	2.399	3.153
85	1.886	2.248	2.383	3.132
90	1.878	2.238	2.369	3.113
95	1.871	2.230	2.356	3.096
100	1.865	2.222	2.344	3.080
∞	1.645	1.960	1.960	2.576

Table XV. Probabilities Associated with Student t^2 being Exceeded

The table gives the probability of a given value of t^2 being exceeded for various probabilities and values of v. The probabilities are given by

$$P_r = \int_{t_0^2}^{\infty} f(t^2)\, dt^2 = \text{Probability } (t^2 > t_0^2) = 2 \text{ Prob}(t > t_0)\dagger$$

$v \equiv$ number of degrees of freedom, usually equal to $n - 1$ where n is the number of observations.

v

P_r	0.20	0.10	0.05	0.02	0.01	0.005	0.001
1	9.474	39.86	161.44	1012.6	4052.2	16 210.4	405 283.8
2	3.557	8.526	18.513	48.511	98.502	198.500	998.434
3	2.683	5.538	10.128	20.621	34.116	55.55	167.030
4	2.350	4.545	7.708	14.040	21.198	31.333	74.132
5	2.179	4.060	6.608	11.323	16.258	22.784	47.183
6	2.074	3.776	5.987	9.878	13.745	18.635	35.510
7	2.002	3.590	5.591	8.988	12.247	16.235	29.246
8	1.952	3.458	5.318	8.387	11.259	14.688	25.412
9	1.913	3.360	5.118	7.958	10.561	13.614	22.858
10	1.882	3.285	4.964	7.640	10.044	12.826	21.041
11	1.858	3.225	4.844	7.388	9.646	12.226	19.687
12	1.839	3.177	4.747	7.188	9.330	11.754	18.645
13	1.823	3.136	4.667	7.023	9.074	11.374	17.817
14	1.809	3.102	4.600	6.885	8.861	11.060	17.140

† See paragraph 9.10.

v

P_r	0.20	0.10	0.05	0.02	0.01	0.005	0.001
15	1.798	3.073	4.543	6.770	8.683	10.798	16.589
16	1.788	3.048	4.494	6.672	8.531	10.576	16.120
17	1.777	3.026	4.451	6.589	8.400	10.385	15.721
18	1.769	3.007	4.414	6.513	8.285	10.218	15.382
19	1.764	2.990	4.381	6.447	8.185	10.072	15.078
20	1.756	2.975	4.351	6.391	8.096	9.944	14.823
21	1.750	2.961	4.325	6.340	8.017	9.829	14.585
22	1.745	2.948	4.301	6.290	7.946	9.727	14.379
23	1.740	2.937	4.280	6.250	7.881	9.635	14.190
24	1.737	2.927	4.260	6.210	7.823	9.551	14.025
25	1.732	2.918	4.242	6.175	7.770	9.475	13.876
26	1.729	2.908	4.225	6.145	7.721	9.406	13.742
27	1.727	2.901	4.210	6.116	7.677	9.342	13.616
28	1.724	2.894	4.196	6.086	7.636	9.284	13.498
29	1.719	2.887	4.183	6.061	7.598	9.229	13.388
30	1.716	2.881	4.171	6.037	7.563	9.180	13.293
35	1.706	2.855	4.121	5.941	7.420	8.977	12.899
40	1.698	2.836	4.085	5.871	7.314	8.829	12.610
45	1.692	2.820	4.057	5.817	7.234	8.715	12.395
50	1.687	2.809	4.034	5.775	7.171	8.626	12.226
55	1.683	2.799	4.016	5.740	7.120	8.555	12.090
60	1.679	2.791	4.001	5.712	7.077	8.495	11.976
65	1.677	2.785	3.989	5.689	7.042	8.445	11.882
70	1.674	2.779	3.978	5.668	7.012	8.403	11.803
75	1.672	2.774	3.969	5.651	6.986	8.367	11.733
80	1.670	2.769	3.960	5.636	6.963	8.335	11.675
85	1.668	2.766	3.953	5.622	6.943	8.308	11.622
90	1.667	2.762	3.947	5.611	6.925	8.282	11.575
95	1.665	2.759	3.941	5.600	6.910	8.261	11.536
100	1.664	2.756	3.936	5.590	6.895	8.241	11.498
110	1.662	2.752	3.928	5.574	6.871	8.207	11.435
120	1.661	2.748	3.920	5.560	6.851	8.179	11.381
130	1.659	2.745	3.914	5.549	6.834	8.156	11.339
140	1.658	2.742	3.909	5.539	6.819	8.136	11.300
150	1.657	2.739	3.904	5.530	6.807	8.118	11.267
160	1.656	2.737	3.900	5.522	6.796	8.103	11.241
170	1.655	2.735	3.897	5.516	6.787	8.089	11.215
180	1.655	2.734	3.894	5.510	6.778	8.078	16.193
190	1.654	2.732	3.891	5.505	6.770	8.067	11.173
200	1.653	2.731	3.888	5.500	6.763	8.058	11.156

Table XVI. σ from Extreme Range

σ is obtained as a fraction of the mean extreme range. The table gives values of $1/\bar{W}_n$, for different values of n, the number of observations in each set which, when multiplied by the mean extreme range, gives an estimate of the standard deviation. The mean extreme range is given by

$$\bar{w}_{mn} = \sum_{q=1}^{q=m} \frac{(x_{qn} - x_{q1})}{m}, \ \sigma = \frac{\bar{w}_{mn}}{\bar{W}_n}$$

where

$x_{qn} \equiv$ largest value of x for qth set of observations

$x_{q1} \equiv$ smallest value of x for qth set of observations

$n =$ number of observations in each set

$m =$ number of sets of observations

The best estimate of σ is given when n lies between five and ten, and when m is *large*. However, even with $m = 1$, provided n lies between five and ten a reasonable estimate of σ is obtained. The estimated value of σ obtained using this method is greater than that obtained from the root mean square of the residuals. The accuracy of the method *falls as n increases* above ten. The value of σ obtained by this method is unbiased.

n	$1/\bar{W}_n$
2	0.886 23
3	0.590 82
4	0.485 73
5	0.429 94
6	0.394 57
7	0.369 77
8	0.351 22
9	0.336 70
10	0.324 94
15	0.288 03
20	0.267 74
25	0.254 41
30	0.244 77
35	0.237 35
40	0.231 40
45	0.226 48
50	0.222 31

Table XVII. $\sigma_{\bar{x}}$ from Extreme Range

$\sigma_{\bar{x}}$ is obtained as a fraction of the mean extreme range. The table gives values of $1/\bar{W}_n\sqrt{(mn)}$ for different values of n, the number of observations, and of m, the number of groups of observations. When multiplied by the mean extreme range $\bar{w}_{mn}$, $1/\bar{W}_n\sqrt{(mn)}$ gives an estimate of $\sigma_{\bar{x}}$ the standard error.

$$\bar{w}_{mn} = \sum_{q=1}^{q=m} \frac{(x_{qn} - x_{q1})}{m}, \quad \sigma_{\bar{x}} = \frac{w_{mn}}{\bar{W}_n\sqrt{(mn)}}$$

where

x_{qn} = largest value of x for qth set of observations

x_{q1} = smallest value of x for qth set of observations

n = number of observations in each set

m = number of sets of observations

The best estimate of $\sigma_{\bar{x}}$ is given when n lies between five and ten, and when m is large. However, even with $m = 1$, provided that n lies between five and ten, a reasonable estimate of $\sigma_{\bar{x}}$ is obtained. The estimated value of $\sigma_{\bar{x}}$ obtained using this method is greater than that obtained using the root mean square of the sum of the residuals. The accuracy of the method falls as n increases above ten.

n

m	2	3	4	5	6	7	8	9	10	11
1	0.626 66	0.341 11	0.242 87	0.192 27	0.161 08	0.139 76	0.124 18	0.112 23	0.102 75	0.095 03
2	0.443 11	0.241 20	0.171 73	0.135 96	0.113 90	0.098 83	0.087 81	0.079 36	0.072 66	0.067 20
3	0.361 80	0.196 94	0.140 22	0.111 01	0.093 00	0.080 69	0.071 69	0.064 80	0.059 32	0.054 86
4	0.313 33	0.170 55	0.121 43	0.096 14	0.080 54	0.069 55	0.062 09	0.056 12	0.051 38	0.047 51
5	0.280 25	0.152 55	0.108 61	0.085 99	0.072 04	0.062 50	0.055 53	0.050 19	0.045 95	0.042 50
6	0.255 83	0.139 26	0.099 15	0.078 50	0.065 76	0.057 06	0.050 70	0.045 82	0.041 95	0.038 79
7	0.236 85	0.128 93	0.091 79	0.072 67	0.060 88	0.052 83	0.046 93	0.042 42	0.038 84	0.035 92
8	0.221 56	0.120 60	0.085 87	0.067 98	0.056 95	0.049 41	0.043 90	0.039 68	0.036 33	0.033 60
9	0.208 89	0.133 70	0.080 96	0.064 09	0.053 69	0.046 59	0.041 39	0.037 41	0.034 25	0.031 68
10	0.198 17	0.107 87	0.076 80	0.060 80	0.050 94	0.044 20	0.039 27	0.035 49	0.032 49	0.030 05
15	0.161 80	0.088 07	0.062 71	0.049 65	0.041 59	0.036 09	0.032 06	0.028 98	0.026 53	0.024 54
20	0.140 12	0.076 27	0.054 31	0.042 99	0.036 02	0.031 25	0.027 77	0.025 10	0.022 98	0.021 25

n

m	12	13	14	15	16	17	18	19	20
1	0.088 59	0.083 14	0.078 45	0.074 37	0.070 78	0.067 60	0.064 75	0.062 19	0.059 87
2	0.062 64	0.058 79	0.055 47	0.052 59	0.050 05	0.047 80	0.045 79	0.043 98	0.042 33
3	0.051 15	0.048 00	0.045 29	0.042 94	0.040 87	0.039 03	0.037 38	0.035 91	0.034 57
4	0.044 30	0.041 57	0.039 23	0.037 18	0.035 39	0.033 80	0.032 38	0.031 10	0.029 93
5	0.039 62	0.037 18	0.035 08	0.033 26	0.031 66	0.030 23	0.028 96	0.027 81	0.026 77
6	0.036 17	0.033 94	0.032 03	0.030 36	0.028 90	0.027 60	0.026 44	0.025 39	0.024 44
7	0.033 48	0.031 42	0.029 65	0.028 11	0.026 75	0.025 50	0.024 47	0.023 51	0.022 63
8	0.031 32	0.029 39	0.027 74	0.026 29	0.025 03	0.023 90	0.022 89	0.021 99	0.021 17
9	0.029 53	0.027 71	0.026 15	0.024 79	0.023 59	0.022 53	0.021 59	0.020 73	0.019 96
10	0.028 02	0.026 29	0.024 81	0.023 52	0.022 38	0.021 38	0.020 48	0.019 67	0.018 93
15	0.022 87	0.021 47	0.020 26	0.019 20	0.018 28	0.017 45	0.016 72	0.016 06	0.015 46
20	0.019 81	0.018 59	0.017 54	0.016 63	0.015 83	0.015 12	0.014 48	0.013 91	0.013 39

Table XVIII

The table gives values for k_9, the bias corrector for s. An unbiased estimate of σ is obtained by multiplying s by k_9. This is necessary because although s^2 is an unbiased estimate of σ^2, $\sqrt{(s^2)}$ is not an unbiased estimated of σ. k_9 is given for integral values of n, the number of observations. $\sigma = k_9 s$.

n	k_9
2	1.2533
3	1.1284
4	1.0854
5	1.0638
6	1.0509
7	1.0424
8	1.0362
9	1.0317
10	1.0281
15	1.0180
20	1.0132
25	1.0105
30	1.0087
40	1.0064
50	1.0051
60	1.0043
70	1.0036
80	1.0032
90	1.0028
100	1.0025

Table XIX. Poisson Distribution Function or Accumulated Poisson Probabilities.

$$F(k) = \sum_{r=0}^{r=k} e^{-np}(np)^r/r!$$

$np \equiv \varepsilon$, the expected value.

Note that the value of $e^{-np}(np)^k/k! = F(k) - F(k-1)$.

k	0.02	0.04	0.06	0.08	0.10	0.15	0.20	0.25	0.30	0.35	0.40	0.45	0.50	0.55
								ε						
0	0.980	0.961	0.942	0.923	0.905	0.861	0.819	0.779	0.741	0.705	0.670	0.638	0.607	0.577
1	1.000	0.999	0.998	0.997	0.995	0.990	0.982	0.974	0.963	0.951	0.938	0.925	0.910	0.894
2		1.000	1.000	1.000	1.000	0.999	0.999	0.998	0.996	0.994	0.992	0.989	0.986	0.982
3					1.000	1.000	1.000	1.000	1.000	1.000	0.999	0.999	0.998	0.998
4											1.000	1.000	1.000	1.000

k	0.60	0.65	0.70	0.75	0.80	0.85	0.90	0.95	1.00	1.10	1.20	1.30	1.40	1.50
								ε						
0	0.549	0.522	0.497	0.472	0.449	0.427	0.407	0.387	0.368	0.333	0.301	0.273	0.247	0.223
1	0.878	0.861	0.844	0.827	0.809	0.791	0.772	0.754	0.736	0.699	0.663	0.627	0.592	0.558
2	0.977	0.972	0.966	0.959	0.953	0.945	0.937	0.929	0.920	0.900	0.879	0.857	0.833	0.809
3	0.997	0.996	0.994	0.993	0.991	0.989	0.987	0.984	0.981	0.974	0.966	0.957	0.946	0.934
4	1.000	0.999	0.999	0.999	0.999	0.998	0.998	0.997	0.996	0.995	0.992	0.989	0.986	0.981
5		1.000	1.000	1.000	1.000	1.000	1.000	1.000	0.999	0.999	0.998	0.998	0.997	0.996
6									1.000	1.000	1.000	1.000	0.999	0.999
7													1.000	1.000

	ε													
k	3.80	3.60	3.40	3.20	3.00	2.80	2.60	2.40	2.20	2.00	1.90	1.80	1.70	1.60
0	0.022	0.027	0.033	0.041	0.050	0.061	0.074	0.091	0.111	0.135	0.150	0.165	0.183	0.202
1	0.107	0.126	0.147	0.171	0.199	0.231	0.267	0.308	0.355	0.406	0.434	0.463	0.493	0.525
2	0.269	0.303	0.340	0.380	0.423	0.469	0.518	0.570	0.623	0.677	0.704	0.731	0.757	0.783
3	0.473	0.515	0.558	0.603	0.647	0.692	0.736	0.779	0.819	0.857	0.875	0.891	0.907	0.921
4	0.668	0.706	0.744	0.781	0.815	0.848	0.877	0.904	0.928	0.947	0.956	0.964	0.970	0.976
5	0.816	0.844	0.871	0.895	0.916	0.935	0.951	0.964	0.975	0.983	0.987	0.990	0.992	0.994
6	0.909	0.927	0.942	0.955	0.966	0.976	0.983	0.988	0.993	0.995	0.997	0.997	0.998	0.999
7	0.960	0.969	0.977	0.983	0.988	0.992	0.995	0.997	0.998	0.999	0.999	0.999	1.000	1.000
8	0.984	0.988	0.992	0.994	0.996	0.998	0.999	0.999	1.000	1.000	1.000	1.000		
9	0.994	0.996	0.997	0.998	0.999	0.999	1.000	1.000						
10	0.998	0.999	0.999	1.000	1.000	1.000								
11	0.999	0.999	1.000											
12	1.000													

Table XIX. (continued)

ε

k	4.0	4.2	4.4	4.6	4.8	5.0	5.2	5.4	5.6	5.8	6.0	6.2	6.4	6.6
0	0.018	0.015	0.012	0.010	0.008	0.007	0.006	0.005	0.004	0.003	0.002	0.002	0.002	0.001
1	0.092	0.078	0.066	0.056	0.048	0.040	0.034	0.029	0.024	0.021	0.017	0.015	0.012	0.010
2	0.238	0.210	0.185	0.163	0.143	0.125	0.109	0.095	0.082	0.072	0.062	0.054	0.046	0.040
3	0.433	0.395	0.359	0.326	0.294	0.265	0.238	0.213	0.191	0.170	0.151	0.134	0.119	0.105
4	0.629	0.590	0.551	0.513	0.476	0.440	0.406	0.373	0.342	0.313	0.285	0.259	0.235	0.213
5	0.785	0.753	0.720	0.686	0.651	0.616	0.581	0.546	0.512	0.478	0.446	0.414	0.384	0.355
6	0.889	0.867	0.844	0.818	0.791	0.762	0.732	0.702	0.670	0.638	0.606	0.574	0.542	0.511
7	0.949	0.936	0.921	0.905	0.887	0.867	0.845	0.822	0.797	0.771	0.744	0.716	0.687	0.658
8	0.979	0.972	0.964	0.955	0.944	0.932	0.918	0.903	0.886	0.867	0.847	0.826	0.803	0.780
9	0.992	0.989	0.985	0.980	0.975	0.968	0.960	0.951	0.941	0.929	0.916	0.902	0.886	0.869
10	0.997	0.996	0.994	0.992	0.990	0.986	0.982	0.977	0.972	0.965	0.957	0.949	0.939	0.927
11	0.999	0.999	0.998	0.997	0.996	0.995	0.993	0.990	0.988	0.984	0.980	0.975	0.969	0.963
12	1.000	1.000	0.999	0.999	0.999	0.998	0.997	0.996	0.995	0.993	0.991	0.989	0.986	0.982
13			1.000	1.000	1.000	0.999	0.999	0.999	0.998	0.997	0.996	0.995	0.994	0.992
14						1.000	1.000	1.000	0.999	0.999	0.997	0.998	0.997	0.997
15									1.000	1.000	0.999	0.999	0.999	0.999
16											1.000	1.000	1.000	0.999
17														1.000

ε

k	6.8	7.0	7.2	7.4	7.6	7.8	8.0	8.5	9.0	9.5	10	10.5	11	11.5
0	0.001	0.001	0.001	0.001	0.001	0.000	0.000	0.000	0.000	0.000	0.000	0.000	0.000	0.000
1	0.009	0.007	0.006	0.005	0.004	0.004	0.003	0.002	0.001	0.001	0.000	0.000	0.000	0.000
2	0.034	0.030	0.025	0.022	0.019	0.016	0.014	0.009	0.006	0.004	0.003	0.002	0.001	0.001
3	0.093	0.082	0.072	0.063	0.055	0.048	0.042	0.030	0.021	0.015	0.010	0.007	0.005	0.003
4	0.192	0.173	0.156	0.140	0.125	0.112	0.100	0.074	0.055	0.040	0.029	0.021	0.015	0.011
5	0.327	0.301	0.276	0.253	0.231	0.210	0.191	0.150	0.116	0.089	0.067	0.050	0.038	0.028
6	0.480	0.450	0.420	0.392	0.365	0.338	0.313	0.256	0.207	0.165	0.130	0.102	0.079	0.060
7	0.628	0.599	0.569	0.539	0.510	0.481	0.453	0.386	0.324	0.269	0.220	0.179	0.143	0.114
8	0.755	0.729	0.703	0.676	0.648	0.620	0.543	0.523	0.456	0.392	0.333	0.279	0.232	0.191
9	0.850	0.830	0.810	0.788	0.765	0.741	0.717	0.653	0.587	0.522	0.458	0.397	0.341	0.289
10	0.915	0.901	0.887	0.871	0.854	0.835	0.816	0.763	0.706	0.645	0.583	0.521	0.460	0.402
11	0.955	0.947	0.937	0.926	0.915	0.902	0.888	0.849	0.803	0.752	0.697	0.639	0.579	0.520
12	0.978	0.973	0.967	0.961	0.954	0.945	0.936	0.909	0.876	0.836	0.792	0.742	0.689	0.633
13	0.990	0.987	0.984	0.980	0.976	0.971	0.966	0.949	0.926	0.898	0.864	0.825	0.781	0.733
14	0.996	0.994	0.993	0.991	0.989	0.986	0.983	0.973	0.959	0.940	0.917	0.888	0.854	0.815
15	0.998	0.998	0.997	0.996	0.995	0.993	0.992	0.986	0.978	0.967	0.951	0.932	0.907	0.878
16	0.999	0.999	0.999	0.998	0.998	0.997	0.996	0.993	0.989	0.982	0.973	0.960	0.944	0.924
17	1.000	1.000	0.999	0.999	0.999	0.999	0.998	0.997	0.995	0.991	0.986	0.978	0.968	0.954
18			1.000	1.000	1.000	1.000	0.999	0.999	0.998	0.996	0.993	0.988	0.982	0.974
19							1.000	0.999	0.999	0.998	0.997	0.994	0.991	0.986
20								1.000	1.000	0.999	0.998	0.997	0.995	0.992
21										1.000	0.999	0.999	0.998	0.996
22											1.000	0.999	0.999	0.998
23												1.000	1.000	0.999
24														1.000

This table is adapted from Table XII of *Statistical Theory* B W Lindgren.

Table XX. Acceptance Limits for the Kolmogorov–Smirnov Test of Goodness of Fit

Acceptance limits for the Kolmogorov–Smirnov test of the goodness of fit between a known distribution and one defined by measured data are given in Table XX. The Kolmogorov–Statistic is defined as $D_n = |F_n(x) - F(x)|$ where $F(x)$ is the distribution function of a known distribution and $F_n(x)$ is the distribution function obtained from measured data (see Figure I.XX). The maximum value of this statistic is found by comparison of the measured data with the known distribution. If D_n found from the measured data exceeds the value of D_n from the table for a given significance level, and given number of observations, then it is adduced that the data do not fit the known distribution at the level of significance considered. If, however, the D_n from the data is less than that from the table, then the measured data are assumed to fit the known distribution at the significance level considered. The significance levels give the probability of the tabulated values of D_n being exceeded. When the sample size or number of observations exceeds 100, the asymptotic formulae given for D_n at the foot of the table should be used. When comparing measured data, the alignment of the known distribution with the distribution defined by the data is accomplished by calculating the estimated mean value of the data, and also the estimate of the standard deviation. $F_n(x)$ is then found and compared with corresponding values of $F(x)$ where both distributions have a common standard deviation and mean. The test, when used under these circumstances to test for normality, is more powerful than the chi-square test.

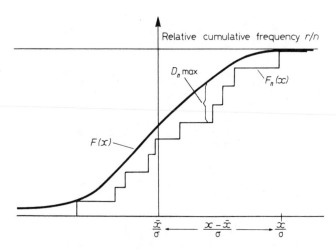

Figure I.XX

Number of observations	Values of D_n for five significance levels P_s, where P_s is the probability of the tabulated values of D_n being exceeded				
n	$P_s = 0.20$	$P_s = 0.15$	$P_s = 0.10$	$P_s = 0.05$	$P_s = 0.01$
1	0.900	0.925	0.950	0.975	0.995
2	0.684	0.726	0.776	0.842	0.929
3	0.565	0.597	0.642	0.708	0.829
4	0.494	0.525	0.564	0.624	0.734
5	0.446	0.474	0.510	0.563	0.669
6	0.410	0.436	0.470	0.521	0.618
7	0.381	0.405	0.438	0.486	0.577
8	0.358	0.381	0.411	0.457	0.543
9	0.339	0.360	0.388	0.432	0.514
10	0.322	0.342	0.368	0.409	0.486
11	0.307	0.326	0.352	0.391	0.468
12	0.295	0.313	0.338	0.375	0.450
13	0.284	0.302	0.325	0.361	0.433
14	0.274	0.292	0.314	0.349	0.418
15	0.266	0.281	0.304	0.338	0.404
16	0.258	0.274	0.295	0.328	0.391
17	0.250	0.266	0.286	0.318	0.380
18	0.244	0.259	0.278	0.309	0.370
19	0.237	0.252	0.272	0.301	0.361
20	0.231	0.246	0.264	0.294	0.352
25	0.21	0.22	0.24	0.264	0.32
30	0.19	0.20	0.22	0.242	0.29
35	0.18	0.19	0.21	0.23	0.27
40				0.21	0.25
50				0.19	0.23
60				0.17	0.21
70				0.16	0.19
80				0.15	0.18
90				0.14	
100				0.14	
Asymptotic formulae	$\dfrac{1.07}{\sqrt{n}}$	$\dfrac{1.14}{\sqrt{n}}$	$\dfrac{1.22}{\sqrt{n}}$	$\dfrac{1.36}{\sqrt{n}}$	$\dfrac{1.63}{\sqrt{n}}$

For further details, see B W Lindgren *Statistical Analysis* page 329 ff.

Note: The value of $F_n(x)$ for a particular value of x is the cumulative frequency given by r/n where n is the total number of readings and r the number of readings from $-\infty$ to the position x.

Table XXI. Acceptance Limits for the Kolmogorov–Smirnov Test of the Identity of Two Distributions, that is that $F_1(x) \equiv F_2(x)$ where $F(x)$ is a Continuous Function

The table enables two distributions to be compared for identity when each is represented by a finite number of readings. If the observations are of the value of a variable x, then the mean and the standard deviation should be found for each set, and the normalized cumulative frequency r/n tabulated for each set of observations against the normalized variable $(x - \bar{x})/\sigma$, where n is the total number of readings for each function and r is the number of readings from $-\infty$ to the position x. The maximum value of $D_n = |F_{n_2}(x) - F_{n_1}(x)|$† is found from the tabulation just described and compared with the appropriate value tabulated for n_1 and n_2 observations. For acceptance rules, see the note immediately above the table.

A diagram of the tabulation of $F_{n_2}(x)$ and $F_{n_1}(x)$ described above would look typically as shown in Figure I.XXI.

Find $D_n = |F_{n_2}(x) - F_{n_1}(x)|max$ and reject the identity of two distributions if the tabulated value is exceeded at both levels. If the value found is between the two tabulated values, then the identity is doubtful. If the value found is

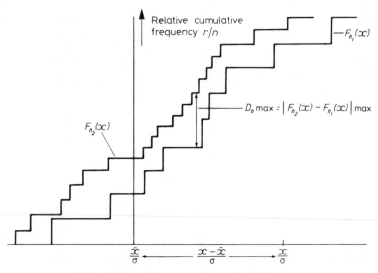

Note: Both means are superposed at $\bar{x}$

Figure I.XXI

† Note: The cumulative frequency is to be taken as constant between successive values of the variable for each $F(x)$. The maximum value of $|F_{n_2}(x) - F_{n_1}(x)|$ is given when the distance between the horizontal part of a step on $F_{n_2}(x)$ and the corresponding part of a step on $F_{n_1}(x)$ vertically above or below the first step is a maximum. See figure above.

less than the tabulated value at the higher significance value then the two distributions are assumed identical.

Values of D_n for two significance levels P_s where P_s is the probability of the tabulated values of D_n being exceeded. The upper value of D_n in each window is for $P_s = 0.05$, whilst the lower value is for $P_s = 0.01$.

Number of observations n_1 for $F_1(x)$

n_2	1	2	3	4	5	6	7	8	9	10	12	15
1	*	*	*	*	*	*	*	*	*	*	*	*
	*	*	*	*	*	*	*	*	*	*	*	*
2		*	*	*	*	*	*	0.875	0.889	0.900		
		*	*	*	*	*	*	*	*	*		
3		*	*	0.800	0.833	0.857	0.750	0.778			0.750	
		*	*	*	*	*	*	0.889			0.917	
4				0.750	0.800	0.750	0.750	0.750	0.750	0.700	0.667	
				*	*	0.833	0.857	0.875	0.889	0.800	0.833	
5					0.800	0.667	0.714	0.675	0.689	0.700		0.667
					0.800	0.833	0.857	0.800	0.800	0.800		0.733
6						0.667	0.690	0.667	0.667	0.633	0.583	
						0.833	0.833	0.750	0.778	0.733	0.75	
7							0.714	0.625	0.635	0.614		
							0.714	0.750	0.746	0.757		
8								0.625	0.625	0.548	0.583	
								0.750	0.750	0.700	0.667	
9									0.555	0.578	0.555	
									0.666	0.689	0.666	
10										0.600		0.500
										0.700		0.633
12											0.500	0.500
											0.583	0.583
15												0.467
												0.533

Number of observations n_2 for $F_2(x)$

Note: 1. Where a star appears in the table, do not reject at this level.

     2. For large values of n_1 and n_2, the following approximate formulae may be used

$P_s = 0.05$ $D_n = 1.36\sqrt{[(n_1 + n_2)/n_1 n_2]}$

$P_s = 0.01$ $D_n = 1.63\sqrt{[(n_1 + n_2)/n_1 n_2]}$

Table XXII

A table of measuring machine uncertainties and their effect on the achievement of required tolerances for manufactured parts. The distribution function of the manufactured parts and that of the measuring machine are assumed to be Gaussian.

Nomenclature

s_p = standard deviation of manufactured parts

s_m = standard deviation of measurement equipment

T_p = required tolerance range for parts $\equiv 2k_p s_p$

t_m = tolerance range adopted in order to minimize the acceptance of oversize or undersize parts $T_p > t_m$

i = inset of t_m on each side with respect to T_p $(i = (T_p - t_m)/2 = k_m s_m)$

a = offset of mean of manufactured parts with respect to the middle of the tolerance range T_p

k_p = tolerance factor for Gaussian distribution of manufactured parts, i.e. $k_p = T_p/2s_p$

k_m = Gaussian tolerance factor for measuring machine, i.e. $k_m = i/s_m \equiv (T_p - t_m)/2s_m$

Note that in Figure I.XXII

$$c_1 = -k_p s_p, \qquad c_2 = k_p s_p$$
$$b_1 = -k_p s_p + k_m s_m, \qquad b_2 = k_p s_p - k_m s_m$$

In the following table the values in columns 1, 2, 4 and 6 are given as percentages of the tolerance range T_p, whilst the values in columns 3 and 5 give the Gaussian tolerance factors for the tolerance range of the manufactured parts T_p and the inset i for the measuring machine, to minimize the acceptance of oversize and undersize parts. The tolerance range T_p in the tables is thus taken as 100 and thus $T_p/2 \equiv k_p s_p = 50\%$.

To use the table all relevant measured data should be converted to a percentage of T_p.

Thus since $i = k_m s_m$, the value of 'i' in the table is given by $i/T_p.100$, 's_m' in the table is given by $s_m/T_p.100$ and 's_p' in the table is given by $s_p/T_p.100$ where all values in this paragraph for variables are actual values.

The figures quoted in columns 7 to 18 give the percentages of parts expected to lie in the categories defined. Where parts are accepted, they are accepted as lying within the T_p tolerance range. Where parts are rejected, they are

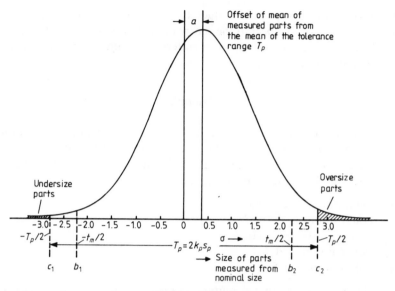

Figure I.XXII

rejected as lying outside the tolerance range T_p. All the values in columns 7 to 18 are given as percentages of the total number of parts manufactured.

C = total correct parts made
C_A = correct parts accepted
C_R = correct parts rejected
U = total undersized parts made
U_A = undersize parts accepted
U_R = undersize parts rejected
O = total oversize parts made
O_A = oversize parts accepted
O_R = oversize parts rejected
W = total wrong parts made
W_A = wrong parts accepted
W_R = wrong parts rejected

It is to be noted that

$$C = C_A + C_R$$
$$O = O_A + O_R$$
$$U = U_A + U_R$$
$$W = U + O$$
$$W_A = U_A + O_A$$
$$W_R = U_R + O_R$$

The table gives only positive values for the offset 'a' of the mean of the measured parts with respect to the centre of the tolerance range T_p. To obtain the values for negative values of 'a' interchange undersize part values for oversize part values.

Example

As an example in the use of the tables, if it is required to obtain a value of less than 5% for reject parts, with a value not exceeding 0.1% for wrong parts accepted, then the following conditions need to apply, where the percentages given are in terms of T_p the tolerance range

a should not exceed 15%
s_p should not exceed 16.67%
k_p should not be less than 3
s_m should not exceed 5%
i should not exceed 5%

It should also be noted that the number of correct parts rejected rises rapidly with increasing s_m.

It is to be *noted* that the initial tabulations for i, for a up to 20, are negative. Thus i under these circumstances is an offset rather than an inset, i.e. $b_2 > c_2$ and $b_1 < c_1$. There are some conditions where it may be advantageous to use a negative value for i in order to reduce the number of correct parts rejected. See Example 5.6 at the end of Chapter 5.

a	$-i=$ $k_m s_m$	k_m	s_m	k_p	s_p	C	C_A	C_R	U	U_A	U_R	O	O_A	O_R	W	W_A	W_R
0	1	1	1	3	16.67	99.73	99.73	0	0.13	0.02	0.11	0.13	0.02	0.11	0.27	0.05	0.22
0	1.5	1	1.5	3	16.67	99.73	99.72	0.01	0.13	0.03	0.1	0.13	0.03	0.1	0.27	0.07	0.2
0	2	1	2	3	16.67	99.73	99.72	0.01	0.13	0.04	0.09	0.13	0.04	0.09	0.27	0.09	0.18
0	2.5	1	2.5	3	16.67	99.73	99.72	0.01	0.13	0.05	0.08	0.13	0.05	0.08	0.27	0.1	0.17
0	5	1	5	3	16.67	99.73	99.69	0.04	0.13	0.07	0.06	0.13	0.07	0.06	0.27	0.15	0.12
0	7.5	1	7.5	3	16.67	99.73	99.66	0.07	0.13	0.09	0.05	0.13	0.09	0.05	0.27	0.17	0.1
0	10	1	10	3	16.67	99.73	99.6	0.13	0.13	0.09	0.04	0.13	0.09	0.04	0.27	0.19	0.08
0	12.5	1	12.5	3	16.67	99.73	99.52	0.21	0.13	0.1	0.04	0.13	0.1	0.04	0.27	0.2	0.07
0	15	1	15	3	16.67	99.73	99.4	0.33	0.13	0.1	0.03	0.13	0.1	0.03	0.27	0.2	0.07
0	17.5	1	17.5	3	16.67	99.73	99.23	0.5	0.13	0.1	0.03	0.13	0.1	0.03	0.27	0.21	0.06
0	20	1	20	3	16.67	99.73	99.01	0.72	0.13	0.1	0.03	0.13	0.1	0.03	0.27	0.21	0.06
0	22.5	1	22.5	3	16.67	99.73	98.74	0.99	0.13	0.11	0.03	0.13	0.11	0.03	0.27	0.21	0.06
0	25	1	25	3	16.67	99.73	98.42	1.31	0.13	0.11	0.03	0.13	0.11	0.03	0.27	0.21	0.06
0	1	1	1	2.5	20	98.76	98.74	0.02	0.62	0.09	0.54	0.62	0.09	0.54	1.24	0.17	1.07
0	1.5	1	1.5	2.5	20	98.76	98.73	0.02	0.62	0.12	0.5	0.62	0.12	0.5	1.24	0.24	1
0	2	1	2	2.5	20	98.76	98.72	0.03	0.62	0.15	0.47	0.62	0.15	0.47	1.24	0.31	0.93
0	2.5	1	2.5	2.5	20	98.76	98.72	0.04	0.62	0.18	0.44	0.62	0.18	0.44	1.24	0.37	0.88
0	5	1	5	2.5	20	98.76	98.66	0.1	0.62	0.29	0.33	0.62	0.29	0.33	1.24	0.58	0.66
0	7.5	1	7.5	2.5	20	98.76	98.58	0.18	0.62	0.35	0.27	0.62	0.35	0.27	1.24	0.71	0.53
0	10	1	10	2.5	20	98.76	98.47	0.28	0.62	0.4	0.23	0.62	0.4	0.23	1.24	0.79	0.45
0	12.5	1	12.5	2.5	20	98.76	98.34	0.42	0.62	0.42	0.2	0.62	0.42	0.2	1.24	0.84	0.4
0	15	1	15	2.5	20	98.76	98.16	0.6	0.62	0.44	0.18	0.62	0.44	0.18	1.24	0.88	0.36
0	17.5	1	17.5	2.5	20	98.76	97.94	0.82	0.62	0.45	0.17	0.62	0.45	0.17	1.24	0.91	0.33
0	20	1	20	2.5	20	98.76	97.68	1.08	0.62	0.46	0.16	0.62	0.46	0.16	1.24	0.93	0.31
0	22.5	1	22.5	2.5	20	98.76	97.38	1.38	0.62	0.47	0.15	0.62	0.47	0.15	1.24	0.94	0.3
0	25	1	25	2.5	20	98.76	97.04	1.72	0.62	0.48	0.14	0.62	0.48	0.14	1.24	0.96	0.29
0	1	1	1	2	25	95.45	95.41	0.04	2.28	0.22	2.06	2.28	0.22	2.06	4.55	0.44	4.11
0	1.5	1	1.5	2	25	95.45	95.39	0.06	2.28	0.32	1.96	2.28	0.32	1.96	4.55	0.63	3.92
0	2	1	2	2	25	95.45	95.37	0.08	2.28	0.41	1.87	2.28	0.41	1.87	4.55	0.81	3.74
0	2.5	1	2.5	2	25	95.45	95.35	0.1	2.28	0.49	1.78	2.28	0.49	1.78	4.55	0.98	3.57
0	5	1	5	2	25	95.45	95.23	0.22	2.28	0.84	1.44	2.28	0.84	1.44	4.55	1.67	2.88
0	7.5	1	7.5	2	25	95.45	95.09	0.36	2.28	1.08	1.2	2.28	1.08	1.2	4.55	2.15	2.4
0	10	1	10	2	25	95.45	94.91	0.54	2.28	1.25	1.03	2.28	1.25	1.03	4.55	2.5	2.05
0	12.5	1	12.5	2	25	95.45	94.71	0.74	2.28	1.37	0.9	2.28	1.37	0.9	4.55	2.74	1.81
0	15	1	15	2	25	95.45	94.48	0.97	2.28	1.46	0.81	2.28	1.46	0.81	4.55	2.93	1.62
0	17.5	1	17.5	2	25	95.45	94.21	1.24	2.28	1.53	0.74	2.28	1.53	0.74	4.55	3.07	1.48
0	20	1	20	2	25	95.45	93.91	1.54	2.28	1.59	0.69	2.28	1.59	0.69	4.55	3.17	1.38
0	22.5	1	22.5	2	25	95.45	93.58	1.87	2.28	1.63	0.65	2.28	1.63	0.65	4.55	3.26	1.29
0	25	1	25	2	25	95.45	93.23	2.22	2.28	1.66	0.61	2.28	1.66	0.61	4.55	3.32	1.23
0	1	1	1	1.5	33.33	86.64	86.57	0.07	6.68	0.4	6.28	6.68	0.4	6.28	13.36	0.81	12.55
0	1.5	1	1.5	1.5	33.33	86.64	86.54	0.1	6.68	0.59	6.09	6.68	0.59	6.09	13.36	1.19	12.17
0	2	1	2	1.5	33.33	86.64	86.5	0.14	6.68	0.78	5.9	6.68	0.78	5.9	13.36	1.55	11.81
0	2.5	1	2.5	1.5	33.33	86.64	86.47	0.17	6.68	0.95	5.73	6.68	0.95	5.73	13.36	1.91	11.46
0	5	1	5	1.5	33.33	86.64	86.28	0.36	6.68	1.73	4.96	6.68	1.73	4.96	13.36	3.45	9.91
0	7.5	1	7.5	1.5	33.33	86.64	86.07	0.57	6.68	2.35	4.33	6.68	2.35	4.33	13.36	4.69	8.67
0	10	1	10	1.5	33.33	86.64	85.84	0.8	6.68	2.85	3.83	6.68	2.85	3.83	13.36	5.69	7.67
0	12.5	1	12.5	1.5	33.33	86.64	85.59	1.05	6.68	3.25	3.43	6.68	3.25	3.43	13.36	6.5	6.86
0	15	1	15	1.5	33.33	86.64	85.31	1.32	6.68	3.57	3.11	6.68	3.57	3.11	13.36	7.15	6.21
0	17.5	1	17.5	1.5	33.33	86.64	85.02	1.62	6.68	3.84	2.84	6.68	3.84	2.84	13.36	7.68	5.69
0	20	1	20	1.5	33.33	86.64	84.71	1.93	6.68	4.05	2.63	6.68	4.05	2.63	13.36	8.11	5.25
0	22.5	1	22.5	1.5	33.33	86.64	84.39	2.25	6.68	4.23	2.45	6.68	4.23	2.45	13.36	8.46	4.9
0	25	1	25	1.5	33.33	86.64	84.04	2.59	6.68	4.38	2.3	6.68	4.38	2.3	13.36	8.76	4.6
5	1	1	1	3	16.67	99.6	99.6	0.01	0.05	0.01	0.04	0.35	0.06	0.29	0.4	0.07	0.33
5	1.5	1	1.5	3	16.67	99.6	99.59	0.01	0.05	0.01	0.04	0.35	0.08	0.26	0.4	0.1	0.3
5	2	1	2	3	16.67	99.6	99.59	0.01	0.05	0.02	0.03	0.35	0.1	0.24	0.4	0.12	0.28
5	2.5	1	2.5	3	16.67	99.6	99.59	0.02	0.05	0.02	0.03	0.35	0.12	0.23	0.4	0.14	0.25
5	5	1	5	3	16.67	99.6	99.56	0.05	0.05	0.03	0.02	0.35	0.18	0.16	0.4	0.21	0.18
5	7.5	1	7.5	3	16.67	99.6	99.51	0.09	0.05	0.03	0.02	0.35	0.22	0.13	0.4	0.25	0.15
5	10	1	10	3	16.67	99.6	99.45	0.16	0.05	0.03	0.01	0.35	0.24	0.11	0.4	0.27	0.12
5	12.5	1	12.5	3	16.67	99.6	99.35	0.25	0.05	0.04	0.01	0.35	0.25	0.1	0.4	0.28	0.11
5	15	1	15	3	16.67	99.6	99.22	0.39	0.05	0.04	0.01	0.35	0.26	0.09	0.4	0.29	0.1
5	17.5	1	17.5	3	16.67	99.6	99.03	0.57	0.05	0.04	0.01	0.35	0.26	0.08	0.4	0.3	0.09
5	20	1	20	3	16.67	99.6	98.81	0.8	0.05	0.04	0.01	0.35	0.27	0.08	0.4	0.31	0.09
5	22.5	1	22.5	3	16.67	99.6	98.53	1.08	0.05	0.04	0.01	0.35	0.27	0.08	0.4	0.31	0.09
5	25	1	25	3	16.67	99.6	98.2	1.4	0.05	0.04	0.01	0.35	0.27	0.07	0.4	0.31	0.08

a	$-i=$ $k_m s_m$	k_m	s_m	k_p	s_p	C	C_A	C_R	U	U_A	U_R	O	O_A	O_R	W	W_A	W_R
5	1	1	1	2.5	20	98.48	98.46	0.02	0.3	0.04	0.25	1.22	0.16	1.07	1.52	0.2	1.32
5	1.5	1	1.5	2.5	20	98.48	98.45	0.03	0.3	0.06	0.24	1.22	0.22	1	1.52	0.28	1.24
5	2	1	2	2.5	20	98.48	98.44	0.04	0.3	0.08	0.22	1.22	0.28	0.94	1.52	0.36	1.16
5	2.5	1	2.5	2.5	20	98.48	98.43	0.05	0.3	0.09	0.21	1.22	0.34	0.88	1.52	0.43	1.09
5	5	1	5	2.5	20	98.48	98.36	0.12	0.3	0.14	0.15	1.22	0.55	0.68	1.52	0.69	0.83
5	7.5	1	7.5	2.5	20	98.48	98.28	0.2	0.3	0.17	0.12	1.22	0.67	0.55	1.52	0.85	0.67
5	10	1	10	2.5	20	98.48	98.16	0.32	0.3	0.19	0.1	1.22	0.76	0.46	1.52	0.95	0.57
5	12.5	1	12.5	2.5	20	98.48	98.02	0.46	0.3	0.21	0.09	1.22	0.81	0.41	1.52	1.02	0.5
5	15	1	15	2.5	20	98.48	97.83	0.65	0.3	0.21	0.08	1.22	0.85	0.37	1.52	1.07	0.45
5	17.5	1	17.5	2.5	20	98.48	97.6	0.88	0.3	0.22	0.08	1.22	0.88	0.34	1.52	1.1	0.42
5	20	1	20	2.5	20	98.48	97.34	1.14	0.3	0.23	0.07	1.22	0.9	0.32	1.52	1.13	0.39
5	22.5	1	22.5	2.5	20	98.48	97.03	1.45	0.3	0.23	0.07	1.22	0.92	0.3	1.52	1.15	0.37
5	25	1	25	2.5	20	98.48	96.68	1.8	0.3	0.23	0.07	1.22	0.93	0.29	1.52	1.17	0.36
5	1	1	1	2	25	95.02	94.98	0.04	1.39	0.14	1.25	3.59	0.32	3.27	4.98	0.46	4.52
5	1.5	1	1.5	2	25	95.02	94.96	0.06	1.39	0.21	1.18	3.59	0.47	3.13	4.98	0.67	4.31
5	2	1	2	2	25	95.02	94.93	0.08	1.39	0.26	1.13	3.59	0.6	2.99	4.98	0.87	4.12
5	2.5	1	2.5	2	25	95.02	94.91	0.11	1.39	0.32	1.07	3.59	0.73	2.86	4.98	1.05	3.93
5	5	1	5	2	25	95.02	94.79	0.23	1.39	0.53	0.86	3.59	1.26	2.34	4.98	1.79	3.19
5	7.5	1	7.5	2	25	95.02	94.64	0.38	1.39	0.68	0.71	3.59	1.64	1.96	4.98	2.32	2.67
5	10	1	10	2	25	95.02	94.46	0.56	1.39	0.78	0.61	3.59	1.91	1.68	4.98	2.69	2.29
5	12.5	1	12.5	2	25	95.02	94.25	0.77	1.39	0.86	0.53	3.59	2.11	1.48	4.98	2.97	2.01
5	15	1	15	2	25	95.02	94.01	1	1.39	0.91	0.48	3.59	2.26	1.33	4.98	3.17	1.81
5	17.5	1	17.5	2	25	95.02	93.74	1.27	1.39	0.95	0.44	3.59	2.38	1.22	4.98	3.33	1.65
5	20	1	20	2	25	95.02	93.44	1.57	1.39	0.98	0.41	3.59	2.47	1.13	4.98	3.45	1.53
5	22.5	1	22.5	2	25	95.02	93.11	1.9	1.39	1.01	0.38	3.59	2.54	1.06	4.98	3.54	1.44
5	25	1	25	2	25	95.02	92.76	2.26	1.39	1.03	0.36	3.59	2.59	1	4.98	3.62	1.36
5	1	1	1	1.5	33.33	86.2	86.13	0.07	4.95	0.32	4.63	8.85	0.5	8.35	13.8	0.82	12.98
5	1.5	1	1.5	1.5	33.33	86.2	86.1	0.1	4.95	0.47	4.48	8.85	0.74	8.11	13.8	1.21	12.59
5	2	1	2	1.5	33.33	86.2	86.06	0.14	4.95	0.61	4.34	8.85	0.97	7.88	13.8	1.58	12.22
5	2.5	1	2.5	1.5	33.33	86.2	86.03	0.17	4.95	0.75	4.2	8.85	1.19	7.66	13.8	1.94	11.86
5	5	1	5	1.5	33.33	86.2	85.84	0.37	4.95	1.34	3.61	8.85	2.17	6.68	13.8	3.51	10.29
5	7.5	1	7.5	1.5	33.33	86.2	85.62	0.58	4.95	1.81	3.14	8.85	2.98	5.87	13.8	4.79	9.01
5	10	1	10	1.5	33.33	86.2	85.39	0.81	4.95	2.18	2.77	8.85	3.64	5.22	13.8	5.82	7.98
5	12.5	1	12.5	1.5	33.33	86.2	85.14	1.06	4.95	2.48	2.47	8.85	4.17	4.68	13.8	6.65	7.15
5	15	1	15	1.5	33.33	86.2	84.87	1.33	4.95	2.71	2.23	8.85	4.61	4.25	13.8	7.32	6.48
5	17.5	1	17.5	1.5	33.33	86.2	84.58	1.62	4.95	2.91	2.04	8.85	4.96	3.89	13.8	7.87	5.93
5	20	1	20	1.5	33.33	86.2	84.27	1.93	4.95	3.06	1.89	8.85	5.26	3.59	13.8	8.32	5.48
5	22.5	1	22.5	1.5	33.33	86.2	83.94	2.26	4.95	3.19	1.76	8.85	5.5	3.35	13.8	8.69	5.11
5	25	1	25	1.5	33.33	86.2	83.6	2.6	4.95	3.29	1.65	8.85	5.71	3.14	13.8	9	4.8
10	1	1	1	3	16.67	99.16	99.15	0.01	0.02	0	0.01	0.82	0.13	0.69	0.84	0.13	0.7
10	1.5	1	1.5	3	16.67	99.16	99.15	0.02	0.02	0	0.01	0.82	0.18	0.64	0.84	0.19	0.65
10	2	1	2	3	16.67	99.16	99.14	0.03	0.02	0.01	0.01	0.82	0.23	0.59	0.84	0.23	0.6
10	2.5	1	2.5	3	16.67	99.16	99.13	0.03	0.02	0.01	0.01	0.82	0.27	0.55	0.84	0.28	0.56
10	5	1	5	3	16.67	99.16	99.08	0.08	0.02	0.01	0.01	0.82	0.41	0.4	0.84	0.42	0.41
10	7.5	1	7.5	3	16.67	99.16	99.01	0.15	0.02	0.01	0.01	0.82	0.5	0.32	0.84	0.51	0.33
10	10	1	10	3	16.67	99.16	98.91	0.25	0.02	0.01	0	0.82	0.55	0.27	0.84	0.56	0.28
10	12.5	1	12.5	3	16.67	99.16	98.78	0.38	0.02	0.01	0	0.82	0.58	0.24	0.84	0.59	0.25
10	15	1	15	3	16.67	99.16	98.61	0.56	0.02	0.01	0	0.82	0.6	0.22	0.84	0.61	0.22
10	17.5	1	17.5	3	16.67	99.16	98.39	0.77	0.02	0.01	0	0.82	0.62	0.2	0.84	0.63	0.21
10	20	1	20	3	16.67	99.16	98.13	1.03	0.02	0.01	0	0.82	0.63	0.19	0.84	0.64	0.2
10	22.5	1	22.5	3	16.67	99.16	97.83	1.33	0.02	0.01	0	0.82	0.63	0.18	0.84	0.65	0.19
10	25	1	25	3	16.67	99.16	97.49	1.67	0.02	0.01	0	0.82	0.64	0.18	0.84	0.65	0.18
10	1	1	1	2.5	20	97.59	97.56	0.03	0.13	0.02	0.11	2.28	0.27	2.01	2.41	0.29	2.12
10	1.5	1	1.5	2.5	20	97.59	97.55	0.04	0.13	0.03	0.11	2.28	0.39	1.89	2.41	0.41	2
10	2	1	2	2.5	20	97.59	97.54	0.05	0.13	0.04	0.1	2.28	0.49	1.78	2.41	0.53	1.88
10	2.5	1	2.5	2.5	20	97.59	97.52	0.07	0.13	0.04	0.09	2.28	0.59	1.69	2.41	0.63	1.78
10	5	1	5	2.5	20	97.59	97.43	0.16	0.13	0.07	0.07	2.28	0.97	1.31	2.41	1.03	1.38
10	7.5	1	7.5	2.5	20	97.59	97.32	0.27	0.13	0.08	0.05	2.28	1.21	1.06	2.41	1.29	1.12
10	10	1	10	2.5	20	97.59	97.18	0.41	0.13	0.09	0.05	2.28	1.37	0.9	2.41	1.46	0.95
10	12.5	1	12.5	2.5	20	97.59	97	0.59	0.13	0.09	0.04	2.28	1.48	0.79	2.41	1.58	0.83
10	15	1	15	2.5	20	97.59	96.79	0.8	0.13	0.1	0.04	2.28	1.56	0.71	2.41	1.66	0.75
10	17.5	1	17.5	2.5	20	97.59	96.54	1.05	0.13	0.1	0.03	2.28	1.62	0.66	2.41	1.72	0.69
10	20	1	20	2.5	20	97.59	96.26	1.33	0.13	0.1	0.03	2.28	1.66	0.61	2.41	1.76	0.65
10	22.5	1	22.5	2.5	20	97.59	95.94	1.65	0.13	0.1	0.03	2.28	1.7	0.58	2.41	1.8	0.61
10	25	1	25	2.5	20	97.59	95.58	2.01	0.13	0.11	0.03	2.28	1.72	0.55	2.41	1.83	0.58

a	$-i=$ $k_m s_m$	k_m	s_m	k_p	s_p	C	C_A	C_R	U	U_A	U_R	O	O_A	O_R	W	W_A	W_R
10	1	1	1	2	25	93.7	93.65	0.05	0.82	0.09	0.73	5.48	0.45	5.03	6.3	0.54	5.76
10	1.5	1	1.5	2	25	93.7	93.63	0.07	0.82	0.13	0.69	5.48	0.66	4.82	6.3	0.79	5.51
10	2	1	2	2	25	93.7	93.6	0.1	0.82	0.16	0.66	5.48	0.86	4.62	6.3	1.02	5.28
10	2.5	1	2.5	2	25	93.7	93.58	0.12	0.82	0.2	0.62	5.48	1.04	4.44	6.3	1.24	5.06
10	5	1	5	2	25	93.7	93.44	0.26	0.82	0.33	0.49	5.48	1.82	3.66	6.3	2.15	4.15
10	7.5	1	7.5	2	25	93.7	93.27	0.43	0.82	0.41	0.4	5.48	2.39	3.09	6.3	2.8	3.49
10	10	1	10	2	25	93.7	93.08	0.62	0.82	0.47	0.35	5.48	2.81	2.67	6.3	3.29	3.01
10	12.5	1	12.5	2	25	93.7	92.86	0.84	0.82	0.52	0.3	5.48	3.13	2.35	6.3	3.65	2.65
10	15	1	15	2	25	93.7	92.61	1.09	0.82	0.55	0.27	5.48	3.37	2.11	6.3	3.92	2.38
10	17.5	1	17.5	2	25	93.7	92.33	1.37	0.82	0.57	0.25	5.48	3.55	1.93	6.3	4.12	2.18
10	20	1	20	2	25	93.7	92.02	1.68	0.82	0.59	0.23	5.48	3.7	1.78	6.3	4.28	2.02
10	22.5	1	22.5	2	25	93.7	91.69	2.01	0.82	0.6	0.22	5.48	3.81	1.67	6.3	4.41	1.89
10	25	1	25	2	25	93.7	91.33	2.37	0.82	0.61	0.21	5.48	3.9	1.58	6.3	4.51	1.78
10	1	1	1	1.5	33.33	84.9	84.83	0.07	3.59	0.24	3.35	11.51	0.61	10.9	15.1	0.86	14.24
10	1.5	1	1.5	1.5	33.33	84.9	84.79	0.11	3.59	0.36	3.23	11.51	0.9	10.6	15.1	1.26	13.84
10	2	1	2	1.5	33.33	84.9	84.76	0.14	3.59	0.47	3.13	11.51	1.18	10.32	15.1	1.65	13.45
10	2.5	1	2.5	1.5	33.33	84.9	84.72	0.18	3.59	0.57	3.02	11.51	1.46	10.05	15.1	2.03	13.07
10	5	1	5	1.5	33.33	84.9	84.52	0.38	3.59	1.02	2.58	11.51	2.68	8.83	15.1	3.7	11.4
10	7.5	1	7.5	1.5	33.33	84.9	84.31	0.59	3.59	1.36	2.23	11.51	3.7	7.81	15.1	5.06	10.04
10	10	1	10	1.5	33.33	84.9	84.07	0.83	3.59	1.64	1.96	11.51	4.54	6.97	15.1	6.18	8.92
10	12.5	1	12.5	1.5	33.33	84.9	83.82	1.08	3.59	1.85	1.74	11.51	5.24	6.27	15.1	7.09	8.01
10	15	1	15	1.5	33.33	84.9	83.54	1.36	3.59	2.02	1.57	11.51	5.81	5.7	15.1	7.83	7.27
10	17.5	1	17.5	1.5	33.33	84.9	83.25	1.65	3.59	2.15	1.44	11.51	6.28	5.22	15.1	8.44	6.66
10	20	1	20	1.5	33.33	84.9	82.94	1.96	3.59	2.26	1.33	11.51	6.68	4.83	15.1	8.94	6.16
10	22.5	1	22.5	1.5	33.33	84.9	82.62	2.28	3.59	2.35	1.24	11.51	7.01	4.5	15.1	9.36	5.74
10	25	1	25	1.5	33.33	84.9	82.28	2.62	3.59	2.43	1.17	11.51	7.28	4.22	15.1	9.71	5.39
15	1	1	1	3	16.67	98.21	98.19	0.02	0	0	0	1.79	0.26	1.53	1.79	0.26	1.53
15	1.5	1	1.5	3	16.67	98.21	98.17	0.04	0	0	0	1.79	0.36	1.42	1.79	0.37	1.43
15	2	1	2	3	16.67	98.21	98.16	0.05	0	0	0	1.79	0.46	1.33	1.79	0.46	1.33
15	2.5	1	2.5	3	16.67	98.21	98.14	0.06	0	0	0	1.79	0.55	1.24	1.79	0.55	1.24
15	5	1	5	3	16.67	98.21	98.06	0.15	0	0	0	1.79	0.86	0.93	1.79	0.86	0.93
15	7.5	1	7.5	3	16.67	98.21	97.94	0.27	0	0	0	1.79	1.05	0.74	1.79	1.05	0.74
15	10	1	10	3	16.67	98.21	97.79	0.42	0	0	0	1.79	1.16	0.63	1.79	1.16	0.63
15	12.5	1	12.5	3	16.67	98.21	97.6	0.61	0	0	0	1.79	1.24	0.55	1.79	1.24	0.55
15	15	1	15	3	16.67	98.21	97.38	0.83	0	0	0	1.79	1.29	0.5	1.79	1.29	0.5
15	17.5	1	17.5	3	16.67	98.21	97.11	1.1	0	0	0	1.79	1.32	0.46	1.79	1.33	0.46
15	20	1	20	3	16.67	98.21	96.81	1.4	0	0	0	1.79	1.35	0.44	1.79	1.35	0.44
15	22.5	1	22.5	3	16.67	98.21	96.47	1.74	0	0	0	1.79	1.37	0.42	1.79	1.37	0.42
15	25	1	25	3	16.67	98.21	96.1	2.11	0	0	0	1.79	1.39	0.4	1.79	1.39	0.4
15	1	1	1	2.5	20	95.94	95.9	0.04	0.06	0.01	0.05	4.01	0.43	3.57	4.06	0.44	3.62
15	1.5	1	1.5	2.5	20	95.94	95.88	0.06	0.06	0.01	0.04	4.01	0.62	3.38	4.06	0.64	3.43
15	2	1	2	2.5	20	95.94	95.86	0.08	0.06	0.02	0.04	4.01	0.8	3.2	4.06	0.82	3.24
15	2.5	1	2.5	2.5	20	95.94	95.83	0.1	0.06	0.02	0.04	4.01	0.97	3.04	4.06	0.99	3.08
15	5	1	5	2.5	20	95.94	95.71	0.23	0.06	0.03	0.03	4.01	1.61	2.39	4.06	1.64	2.42
15	7.5	1	7.5	2.5	20	95.94	95.55	0.38	0.06	0.04	0.02	4.01	2.04	1.96	4.06	2.08	1.98
15	10	1	10	2.5	20	95.94	95.37	0.57	0.06	0.04	0.02	4.01	2.34	1.67	4.06	2.38	1.69
15	12.5	1	12.5	2.5	20	95.94	95.15	0.78	0.06	0.04	0.02	4.01	2.54	1.46	4.06	2.59	1.48
15	15	1	15	2.5	20	95.94	94.9	1.03	0.06	0.04	0.02	4.01	2.69	1.31	4.06	2.73	1.33
15	17.5	1	17.5	2.5	20	95.94	94.62	1.31	0.06	0.04	0.01	4.01	2.8	1.21	4.06	2.84	1.22
15	20	1	20	2.5	20	95.94	94.31	1.62	0.06	0.04	0.01	4.01	2.88	1.12	4.06	2.93	1.14
15	22.5	1	22.5	2.5	20	95.94	93.97	1.97	0.06	0.04	0.01	4.01	2.95	1.06	4.06	2.99	1.07
15	25	1	25	2.5	20	95.94	93.61	2.33	0.06	0.05	0.01	4.01	3	1.01	4.06	3.04	1.02
15	1	1	1	2	25	91.46	91.4	0.06	0.47	0.05	0.41	8.08	0.62	7.46	8.54	0.67	7.87
15	1.5	1	1.5	2	25	91.46	91.37	0.09	0.47	0.08	0.39	8.08	0.9	7.17	8.54	0.98	7.56
15	2	1	2	2	25	91.46	91.34	0.12	0.47	0.1	0.37	8.08	1.17	6.9	8.54	1.27	7.27
15	2.5	1	2.5	2	25	91.46	91.31	0.15	0.47	0.12	0.35	8.08	1.43	6.64	8.54	1.55	6.99
15	5	1	5	2	25	91.46	91.14	0.32	0.47	0.19	0.27	8.08	2.53	5.55	8.54	2.72	5.82
15	7.5	1	7.5	2	25	91.46	90.95	0.51	0.47	0.24	0.22	8.08	3.36	4.72	8.54	3.6	4.94
15	10	1	10	2	25	91.46	90.73	0.72	0.47	0.28	0.19	8.08	3.99	4.09	8.54	4.26	4.28
15	12.5	1	12.5	2	25	91.46	90.49	0.97	0.47	0.3	0.17	8.08	4.47	3.61	8.54	4.76	3.78
15	15	1	15	2	25	91.46	90.22	1.23	0.47	0.32	0.15	8.08	4.83	3.24	8.54	5.15	3.39
15	17.5	1	17.5	2	25	91.46	89.93	1.53	0.47	0.33	0.14	8.08	5.12	2.96	8.54	5.45	3.1
15	20	1	20	2	25	91.46	89.62	1.84	0.47	0.34	0.13	8.08	5.34	2.73	8.54	5.68	2.86
15	22.5	1	22.5	2	25	91.46	89.28	2.18	0.47	0.34	0.12	8.08	5.52	2.55	8.54	5.87	2.68
15	25	1	25	2	25	91.46	88.92	2.53	0.47	0.35	0.12	8.08	5.67	2.41	8.54	6.02	2.52

a	$-i=$ $k_m s_m$	k_m	s_m	k_p	s_p	C	C_A	C_R	U	U_A	U_R	O	O_A	O_R	W	W_A	W_R
15	1	1	1	1.5	33.33	82.76	82.68	0.07	2.56	0.18	2.37	14.69	0.73	13.96	17.24	0.91	16.33
15	1.5	1	1.5	1.5	33.33	82.76	82.64	0.11	2.56	0.27	2.29	14.69	1.07	13.61	17.24	1.34	15.9
15	2	1	2	1.5	33.33	82.76	82.6	0.15	2.56	0.35	2.21	14.69	1.41	13.27	17.24	1.76	15.48
15	2.5	1	2.5	1.5	33.33	82.76	82.57	0.19	2.56	0.43	2.13	14.69	1.74	12.95	17.24	2.17	15.08
15	5	1	5	1.5	33.33	82.76	82.36	0.4	2.56	0.75	1.8	14.69	3.23	11.46	17.24	3.98	13.26
15	7.5	1	7.5	1.5	33.33	82.76	82.13	0.62	2.56	1.01	1.55	14.69	4.49	10.19	17.24	5.5	11.75
15	10	1	10	1.5	33.33	82.76	81.89	0.86	2.56	1.2	1.36	14.69	5.55	9.14	17.24	6.75	10.49
15	12.5	1	12.5	1.5	33.33	82.76	81.64	1.12	2.56	1.35	1.21	14.69	6.44	8.25	17.24	7.79	9.46
15	15	1	15	1.5	33.33	82.76	81.36	1.39	2.56	1.47	1.09	14.69	7.17	7.51	17.24	8.64	8.6
15	17.5	1	17.5	1.5	33.33	82.76	81.07	1.69	2.56	1.56	1	14.69	7.79	6.9	17.24	9.35	7.89
15	20	1	20	1.5	33.33	82.76	80.76	1.99	2.56	1.64	0.92	14.69	8.31	6.38	17.24	9.94	7.3
15	22.5	1	22.5	1.5	33.33	82.76	80.44	2.31	2.56	1.7	0.86	14.69	8.74	5.94	17.24	10.44	6.8
15	25	1	25	1.5	33.33	82.76	80.11	2.64	2.56	1.75	0.81	14.69	9.11	5.58	17.24	10.86	6.39
20	1	1	1	3	16.67	96.41	96.36	0.04	0	0	0	3.59	0.47	3.13	3.59	0.47	3.13
20	1.5	1	1.5	3	16.67	96.41	96.34	0.06	0	0	0	3.59	0.67	2.92	3.59	0.67	2.93
20	2	1	2	3	16.67	96.41	96.32	0.09	0	0	0	3.59	0.85	2.74	3.59	0.85	2.74
20	2.5	1	2.5	3	16.67	96.41	96.29	0.11	0	0	0	3.59	1.02	2.58	3.59	1.02	2.58
20	5	1	5	3	16.67	96.41	96.15	0.26	0	0	0	3.59	1.64	1.96	3.59	1.64	1.96
20	7.5	1	7.5	3	16.67	96.41	95.97	0.44	0	0	0	3.59	2.02	1.57	3.59	2.02	1.57
20	10	1	10	3	16.67	96.41	95.75	0.66	0	0	0	3.59	2.26	1.33	3.59	2.26	1.33
20	12.5	1	12.5	3	16.67	96.41	95.49	0.92	0	0	0	3.59	2.43	1.17	3.59	2.43	1.17
20	15	1	15	3	16.67	96.41	95.19	1.21	0	0	0	3.59	2.54	1.06	3.59	2.54	1.06
20	17.5	1	17.5	3	16.67	96.41	94.87	1.54	0	0	0	3.59	2.62	0.97	3.59	2.62	0.98
20	20	1	20	3	16.67	96.41	94.51	1.89	0	0	0	3.59	2.68	0.91	3.59	2.68	0.92
20	22.5	1	22.5	3	16.67	96.41	94.14	2.27	0	0	0	3.59	2.72	0.87	3.59	2.73	0.87
20	25	1	25	3	16.67	96.41	93.75	2.66	0	0	0	3.59	2.76	0.83	3.59	2.76	0.83
20	1	1	1	2.5	20	93.3	93.24	0.06	0.02	0	0.02	6.68	0.66	6.02	6.7	0.66	6.04
20	1.5	1	1.5	2.5	20	93.3	93.21	0.09	0.02	0.01	0.02	6.68	0.95	5.73	6.7	0.96	5.75
20	2	1	2	2.5	20	93.3	93.18	0.12	0.02	0.01	0.02	6.68	1.23	5.45	6.7	1.24	5.47
20	2.5	1	2.5	2.5	20	93.3	93.15	0.15	0.02	0.01	0.01	6.68	1.49	5.19	6.7	1.49	5.21
20	5	1	5	2.5	20	93.3	92.97	0.33	0.02	0.01	0.01	6.68	2.53	4.15	6.7	2.54	4.17
20	7.5	1	7.5	2.5	20	93.3	92.76	0.53	0.02	0.01	0.01	6.68	3.25	3.43	6.7	3.26	3.44
20	10	1	10	2.5	20	93.3	92.53	0.77	0.02	0.02	0.01	6.68	3.76	2.92	6.7	3.77	2.93
20	12.5	1	12.5	2.5	20	93.3	92.26	1.04	0.02	0.02	0.01	6.68	4.12	2.56	6.7	4.13	2.57
20	15	1	15	2.5	20	93.3	91.97	1.33	0.02	0.02	0.01	6.68	4.38	2.3	6.7	4.4	2.31
20	17.5	1	17.5	2.5	20	93.3	91.65	1.65	0.02	0.02	0.01	6.68	4.58	2.11	6.7	4.59	2.11
20	20	1	20	2.5	20	93.3	91.31	1.99	0.02	0.02	0.01	6.68	4.72	1.96	6.7	4.74	1.96
20	22.5	1	22.5	2.5	20	93.3	90.95	2.35	0.02	0.02	0.01	6.68	4.84	1.84	6.7	4.86	1.84
20	25	1	25	2.5	20	93.3	90.57	2.72	0.02	0.02	0	6.68	4.93	1.75	6.7	4.95	1.75
20	1	1	1	2	25	88.24	88.17	0.07	0.26	0.03	0.22	11.51	0.81	10.7	11.76	0.84	10.92
20	1.5	1	1.5	2	25	88.24	88.13	0.11	0.26	0.04	0.21	11.51	1.18	10.32	11.76	1.23	10.53
20	2	1	2	2	25	88.24	88.1	0.14	0.26	0.06	0.2	11.51	1.54	9.96	11.76	1.6	10.16
20	2.5	1	2.5	2	25	88.24	88.06	0.18	0.26	0.07	0.19	11.51	1.89	9.62	11.76	1.96	9.81
20	5	1	5	2	25	88.24	87.86	0.38	0.26	0.11	0.15	11.51	3.38	8.13	11.76	3.49	8.27
20	7.5	1	7.5	2	25	88.24	87.63	0.6	0.26	0.14	0.12	11.51	4.54	6.97	11.76	4.68	7.08
20	10	1	10	2	25	88.24	87.39	0.85	0.26	0.15	0.1	11.51	5.44	6.07	11.76	5.59	6.17
20	12.5	1	12.5	2	25	88.24	87.12	1.12	0.26	0.17	0.09	11.51	6.14	5.37	11.76	6.3	5.46
20	15	1	15	2	25	88.24	86.83	1.4	0.26	0.18	0.08	11.51	6.68	4.83	11.76	6.85	4.91
20	17.5	1	17.5	2	25	88.24	86.53	1.71	0.26	0.18	0.07	11.51	7.1	4.4	11.76	7.29	4.48
20	20	1	20	2	25	88.24	86.2	2.03	0.26	0.19	0.07	11.51	7.44	4.06	11.76	7.63	4.13
20	22.5	1	22.5	2	25	88.24	85.86	2.37	0.26	0.19	0.07	11.51	7.72	3.79	11.76	7.91	3.86
20	25	1	25	2	25	88.24	85.51	2.73	0.26	0.19	0.06	11.51	7.94	3.57	11.76	8.13	3.63
20	1	1	1	1.5	33.33	79.81	79.73	0.08	1.79	0.14	1.65	18.41	0.84	17.56	20.19	0.98	19.21
20	1.5	1	1.5	1.5	33.33	79.81	79.69	0.12	1.79	0.2	1.59	18.41	1.25	17.16	20.19	1.45	18.74
20	2	1	2	1.5	33.33	79.81	79.65	0.16	1.79	0.26	1.53	18.41	1.65	16.76	20.19	1.9	18.29
20	2.5	1	2.5	1.5	33.33	79.81	79.61	0.2	1.79	0.31	1.47	18.41	2.03	16.37	20.19	2.34	17.85
20	5	1	5	1.5	33.33	79.81	79.39	0.42	1.79	0.55	1.24	18.41	3.81	14.6	20.19	4.35	15.84
20	7.5	1	7.5	1.5	33.33	79.81	79.16	0.65	1.79	0.73	1.06	18.41	5.33	13.07	20.19	6.06	14.13
20	10	1	10	1.5	33.33	79.81	78.91	0.9	1.79	0.86	0.93	18.41	6.64	11.77	20.19	7.5	12.69
20	12.5	1	12.5	1.5	33.33	79.81	78.65	1.16	1.79	0.97	0.82	18.41	7.74	10.67	20.19	8.71	11.49
20	15	1	15	1.5	33.33	79.81	78.37	1.44	1.79	1.05	0.74	18.41	8.67	9.73	20.19	9.72	10.47
20	17.5	1	17.5	1.5	33.33	79.81	78.08	1.73	1.79	1.11	0.68	18.41	9.46	8.95	20.19	10.57	9.62
20	20	1	20	1.5	33.33	79.81	77.78	2.03	1.79	1.16	0.63	18.41	10.12	8.28	20.19	11.28	8.91
20	22.5	1	22.5	1.5	33.33	79.81	77.46	2.34	1.79	1.2	0.58	18.41	10.68	7.72	20.19	11.89	8.31
20	25	1	25	1.5	33.33	79.81	77.14	2.67	1.79	1.24	0.55	18.41	11.16	7.24	20.19	12.4	7.8

a	i= $k_m s_m$	k_m	s_m	k_p	s_p	C	C_A	C_R	U	U_A	U_R	O	O_A	O_R	W	W_A	W_R
0	0	0	2.5	3	16.67	99.73	99.66	0.07	0.13	0.02	0.11	0.13	0.02	0.11	0.27	0.04	0.23
0	0	0	5	3	16.67	99.73	99.53	0.2	0.13	0.03	0.1	0.13	0.03	0.1	0.27	0.06	0.21
0	0	0	7.5	3	16.67	99.73	99.29	0.44	0.13	0.04	0.1	0.13	0.04	0.1	0.27	0.08	0.19
0	0	0	10	3	16.67	99.73	98.89	0.84	0.13	0.05	0.09	0.13	0.05	0.09	0.27	0.09	0.18
0	0	0	12.5	3	16.67	99.73	98.23	1.5	0.13	0.05	0.09	0.13	0.05	0.09	0.27	0.1	0.17
0	0	0	15	3	16.67	99.73	97.27	2.46	0.13	0.05	0.08	0.13	0.05	0.08	0.27	0.1	0.17
0	0	0	17.5	3	16.67	99.73	95.98	3.75	0.13	0.05	0.08	0.13	0.05	0.08	0.27	0.11	0.16
0	0	0	20	3	16.67	99.73	94.35	5.38	0.13	0.05	0.08	0.13	0.05	0.08	0.27	0.11	0.16
0	0	0	22.5	3	16.67	99.73	92.43	7.3	0.13	0.06	0.08	0.13	0.06	0.08	0.27	0.11	0.16
0	0	0	25	3	16.67	99.73	90.27	9.46	0.13	0.06	0.08	0.13	0.06	0.08	0.27	0.11	0.16
0	0	0	2.5	2.5	20	98.76	98.55	0.21	0.62	0.07	0.55	0.62	0.07	0.55	1.24	0.14	1.1
0	0	0	5	2.5	20	98.76	98.23	0.53	0.62	0.12	0.5	0.62	0.12	0.5	1.24	0.24	1
0	0	0	7.5	2.5	20	98.76	97.76	1	0.62	0.16	0.47	0.62	0.16	0.47	1.24	0.31	0.93
0	0	0	10	2.5	20	98.76	97.09	1.66	0.62	0.18	0.44	0.62	0.18	0.44	1.24	0.36	0.88
0	0	0	12.5	2.5	20	98.76	96.18	2.58	0.62	0.2	0.42	0.62	0.2	0.42	1.24	0.4	0.84
0	0	0	15	2.5	20	98.76	95	3.76	0.62	0.21	0.41	0.62	0.21	0.41	1.24	0.43	0.81
0	0	0	17.5	2.5	20	98.76	93.53	5.23	0.62	0.23	0.4	0.62	0.23	0.4	1.24	0.45	0.79
0	0	0	20	2.5	20	98.76	91.8	6.96	0.62	0.23	0.39	0.62	0.23	0.39	1.24	0.47	0.77
0	0	0	22.5	2.5	20	98.76	89.85	8.91	0.62	0.24	0.38	0.62	0.24	0.38	1.24	0.48	0.76
0	0	0	25	2.5	20	98.76	87.71	11.05	0.62	0.25	0.37	0.62	0.25	0.37	1.24	0.5	0.75
0	0	0	2.5	2	25	95.45	94.96	0.49	2.28	0.19	2.09	2.28	0.19	2.09	4.55	0.38	4.17
0	0	0	5	2	25	95.45	94.34	1.11	2.28	0.33	1.94	2.28	0.33	1.94	4.55	0.67	3.88
0	0	0	7.5	2	25	95.45	93.56	1.89	2.28	0.45	1.83	2.28	0.45	1.83	4.55	0.9	3.65
0	0	0	10	2	25	95.45	92.59	2.86	2.28	0.54	1.73	2.28	0.54	1.73	4.55	1.08	3.47
0	0	0	12.5	2	25	95.45	91.41	4.04	2.28	0.61	1.66	2.28	0.61	1.66	4.55	1.22	3.33
0	0	0	15	2	25	95.45	90.02	5.43	2.28	0.67	1.6	2.28	0.67	1.6	4.55	1.34	3.21
0	0	0	17.5	2	25	95.45	88.44	7.01	2.28	0.72	1.56	2.28	0.72	1.56	4.55	1.44	3.11
0	0	0	20	2	25	95.45	86.67	8.78	2.28	0.76	1.52	2.28	0.76	1.52	4.55	1.52	3.03
0	0	0	22.5	2	25	95.45	84.74	10.71	2.28	0.79	1.48	2.28	0.79	1.48	4.55	1.58	2.97
0	0	0	25	2	25	95.45	82.7	12.75	2.28	0.82	1.45	2.28	0.82	1.45	4.55	1.64	2.91
0	0	0	2.5	1.5	33.33	86.64	85.81	0.82	6.68	0.36	6.32	6.68	0.36	6.32	13.36	0.71	12.65
0	0	0	5	1.5	33.33	86.64	84.87	1.77	6.68	0.67	6.01	6.68	0.67	6.01	13.36	1.33	12.03
0	0	0	7.5	1.5	33.33	86.64	83.8	2.84	6.68	0.93	5.75	6.68	0.93	5.75	13.36	1.86	11.5
0	0	0	10	1.5	33.33	86.64	82.6	4.04	6.68	1.16	5.52	6.68	1.16	5.52	13.36	2.32	11.04
0	0	0	12.5	1.5	33.33	86.64	81.27	5.37	6.68	1.36	5.32	6.68	1.36	5.32	13.36	2.72	10.64
0	0	0	15	1.5	33.33	86.64	79.82	6.82	6.68	1.53	5.15	6.68	1.53	5.15	13.36	3.06	10.3
0	0	0	17.5	1.5	33.33	86.64	78.25	8.39	6.68	1.68	5	6.68	1.68	5	13.36	3.35	10.01
0	0	0	20	1.5	33.33	86.64	76.58	10.05	6.68	1.81	4.87	6.68	1.81	4.87	13.36	3.61	9.75
0	0	0	22.5	1.5	33.33	86.64	74.84	11.8	6.68	1.92	4.76	6.68	1.92	4.76	13.36	3.84	9.52
0	0	0	25	1.5	33.33	86.64	73.02	13.62	6.68	2.02	4.66	6.68	2.02	4.66	13.36	4.04	9.33
0	0	0	2.5	1	50	68.27	67.28	0.99	15.87	0.46	15.4	15.87	0.46	15.4	31.73	0.93	30.81
0	0	0	5	1	50	68.27	66.24	2.03	15.87	0.9	14.97	15.87	0.9	14.97	31.73	1.79	29.94
0	0	0	7.5	1	50	68.27	65.13	3.14	15.87	1.3	14.57	15.87	1.3	14.57	31.73	2.6	29.13
0	0	0	10	1	50	68.27	63.97	4.3	15.87	1.67	14.19	15.87	1.67	14.19	31.73	3.35	28.38
0	0	0	12.5	1	50	68.27	62.76	5.51	15.87	2.02	13.84	15.87	2.02	13.84	31.73	4.05	27.69
0	0	0	15	1	50	68.27	61.5	6.77	15.87	2.35	13.52	15.87	2.35	13.52	31.73	4.69	27.04
0	0	0	17.5	1	50	68.27	60.2	8.07	15.87	2.64	13.22	15.87	2.64	13.22	31.73	5.29	26.44
0	0	0	20	1	50	68.27	58.86	9.41	15.87	2.92	12.95	15.87	2.92	12.95	31.73	5.84	25.89
0	0	0	22.5	1	50	68.27	57.49	10.78	15.87	3.17	12.69	15.87	3.17	12.69	31.73	6.35	25.38
0	0	0	25	1	50	68.27	56.1	12.17	15.87	3.41	12.46	15.87	3.41	12.46	31.73	6.82	24.91
5	0	0	2.5	3	16.67	99.6	99.51	0.1	0.05	0.01	0.04	0.35	0.05	0.3	0.4	0.06	0.34
5	0	0	5	3	16.67	99.6	99.34	0.26	0.05	0.01	0.04	0.35	0.08	0.27	0.4	0.09	0.3
5	0	0	7.5	3	16.67	99.6	99.06	0.54	0.05	0.01	0.03	0.35	0.1	0.25	0.4	0.11	0.28
5	0	0	10	3	16.67	99.6	98.59	1.01	0.05	0.02	0.03	0.35	0.11	0.23	0.4	0.13	0.27
5	0	0	12.5	3	16.67	99.6	97.88	1.73	0.05	0.02	0.03	0.35	0.12	0.22	0.4	0.14	0.26
5	0	0	15	3	16.67	99.6	96.86	2.74	0.05	0.02	0.03	0.35	0.13	0.22	0.4	0.15	0.25
5	0	0	17.5	3	16.67	99.6	95.52	4.09	0.05	0.02	0.03	0.35	0.13	0.21	0.4	0.15	0.24
5	0	0	20	3	16.67	99.6	93.86	5.74	0.05	0.02	0.03	0.35	0.14	0.21	0.4	0.16	0.24
5	0	0	22.5	3	16.67	99.6	91.93	7.68	0.05	0.02	0.03	0.35	0.14	0.2	0.4	0.16	0.23
5	0	0	25	3	16.67	99.6	89.77	9.84	0.05	0.02	0.03	0.35	0.15	0.2	0.4	0.17	0.23

a	k_m	s_m	k_m	s_m	k_p	s_p	C	C_A	C_R	U	U_A	U_R	O	O_A	O_R	W	W_A	W_R
										$i=$								
5	0	0		2.5	2.5	20	98.48	98.23	0.24	0.3	0.04	0.26	1.22	0.13	1.09	1.52	0.17	1.35
5	0	0		5	2.5	20	98.48	97.88	0.6	0.3	0.06	0.24	1.22	0.23	1	1.52	0.29	1.23
5	0	0		7.5	2.5	20	98.48	97.37	1.11	0.3	0.08	0.22	1.22	0.29	0.93	1.52	0.37	1.15
5	0	0		10	2.5	20	98.48	96.66	1.82	0.3	0.09	0.21	1.22	0.34	0.88	1.52	0.43	1.09
5	0	0		12.5	2.5	20	98.48	95.7	2.78	0.3	0.1	0.2	1.22	0.38	0.84	1.52	0.48	1.04
5	0	0		15	2.5	20	98.48	94.48	4	0.3	0.11	0.19	1.22	0.41	0.81	1.52	0.51	1.01
5	0	0		17.5	2.5	20	98.48	92.99	5.49	0.3	0.11	0.19	1.22	0.43	0.79	1.52	0.54	0.98
5	0	0		20	2.5	20	98.48	91.25	7.23	0.3	0.11	0.18	1.22	0.45	0.77	1.52	0.57	0.95
5	0	0		22.5	2.5	20	98.48	89.29	9.19	0.3	0.12	0.18	1.22	0.47	0.76	1.52	0.58	0.94
5	0	0		25	2.5	20	98.48	87.16	11.32	0.3	0.12	0.18	1.22	0.48	0.74	1.52	0.6	0.92
5	0	0		2.5	2	25	95.02	94.51	0.51	1.39	0.12	1.27	3.59	0.28	3.31	4.98	0.4	4.58
5	0	0		5	2	25	95.02	93.86	1.16	1.39	0.22	1.17	3.59	0.5	3.09	4.98	0.72	4.27
5	0	0		7.5	2	25	95.02	93.05	1.97	1.39	0.29	1.1	3.59	0.68	2.92	4.98	0.96	4.02
5	0	0		10	2	25	95.02	92.05	2.97	1.39	0.34	1.05	3.59	0.82	2.77	4.98	1.16	3.82
5	0	0		12.5	2	25	95.02	90.85	4.16	1.39	0.39	1	3.59	0.93	2.66	4.98	1.32	3.66
5	0	0		15	2	25	95.02	89.46	5.56	1.39	0.42	0.97	3.59	1.03	2.57	4.98	1.45	3.54
5	0	0		17.5	2	25	95.02	87.86	7.15	1.39	0.45	0.94	3.59	1.1	2.49	4.98	1.55	3.43
5	0	0		20	2	25	95.02	86.1	8.92	1.39	0.47	0.92	3.59	1.17	2.43	4.98	1.64	3.34
5	0	0		22.5	2	25	95.02	84.18	10.84	1.39	0.49	0.9	3.59	1.22	2.37	4.98	1.72	3.27
5	0	0		25	2	25	95.02	82.15	12.87	1.39	0.51	0.88	3.59	1.27	2.32	4.98	1.78	3.2
5	0	0		2.5	1.5	33.33	86.2	85.37	0.83	4.95	0.28	4.67	8.85	0.45	8.41	13.8	0.73	13.07
5	0	0		5	1.5	33.33	86.2	84.42	1.79	4.95	0.52	4.43	8.85	0.84	8.02	13.8	1.35	12.44
5	0	0		7.5	1.5	33.33	86.2	83.34	2.86	4.95	0.72	4.22	8.85	1.18	7.68	13.8	1.9	11.9
5	0	0		10	1.5	33.33	86.2	82.13	4.07	4.95	0.9	4.05	8.85	1.47	7.38	13.8	2.37	11.43
5	0	0		12.5	1.5	33.33	86.2	80.8	5.4	4.95	1.04	3.9	8.85	1.73	7.12	13.8	2.77	11.03
5	0	0		15	1.5	33.33	86.2	79.35	6.85	4.95	1.17	3.78	8.85	1.95	6.9	13.8	3.12	10.67
5	0	0		17.5	1.5	33.33	86.2	77.79	8.41	4.95	1.28	3.67	8.85	2.15	6.7	13.8	3.43	10.37
5	0	0		20	1.5	33.33	86.2	76.13	10.07	4.95	1.38	3.57	8.85	2.32	6.53	13.8	3.7	10.1
5	0	0		22.5	1.5	33.33	86.2	74.39	11.81	4.95	1.46	3.49	8.85	2.47	6.38	13.8	3.93	9.87
5	0	0		25	1.5	33.33	86.2	72.59	13.61	4.95	1.53	3.42	8.85	2.6	6.25	13.8	4.13	9.66
5	0	0		2.5	1	50	68.03	67.04	0.99	13.57	0.42	13.15	18.41	0.51	17.9	31.97	0.93	31.05
5	0	0		5	1	50	68.03	66	2.03	13.57	0.8	12.76	18.41	0.99	17.42	31.97	1.79	30.18
5	0	0		7.5	1	50	68.03	64.89	3.13	13.57	1.16	12.41	18.41	1.44	16.96	31.97	2.6	29.37
5	0	0		10	1	50	68.03	63.74	4.29	13.57	1.49	12.08	18.41	1.86	16.54	31.97	3.35	28.62
5	0	0		12.5	1	50	68.03	62.53	5.5	13.57	1.8	11.77	18.41	2.26	16.15	31.97	4.05	27.92
5	0	0		15	1	50	68.03	61.27	6.76	13.57	2.08	11.49	18.41	2.62	15.79	31.97	4.7	27.27
5	0	0		17.5	1	50	68.03	59.97	8.05	13.57	2.34	11.23	18.41	2.96	15.45	31.97	5.3	26.67
5	0	0		20	1	50	68.03	58.64	9.39	13.57	2.58	10.99	18.41	3.28	15.13	31.97	5.85	26.12
5	0	0		22.5	1	50	68.03	57.28	10.75	13.57	2.8	10.77	18.41	3.57	14.84	31.97	6.37	25.61
5	0	0		25	1	50	68.03	55.89	12.13	13.57	3	10.57	18.41	3.84	14.57	31.97	6.84	25.13
10	0	0		2.5	3	16.67	99.16	98.99	0.17	0.02	0	0.01	0.82	0.11	0.71	0.84	0.11	0.73
10	0	0		5	3	16.67	99.16	98.71	0.45	0.02	0	0.01	0.82	0.18	0.64	0.84	0.18	0.66
10	0	0		7.5	3	16.67	99.16	98.29	0.88	0.02	0.01	0.01	0.82	0.22	0.6	0.84	0.23	0.61
10	0	0		10	3	16.67	99.16	97.65	1.52	0.02	0.01	0.01	0.82	0.26	0.56	0.84	0.26	0.57
10	0	0		12.5	3	16.67	99.16	96.75	2.41	0.02	0.01	0.01	0.82	0.28	0.54	0.84	0.29	0.55
10	0	0		15	3	16.67	99.16	95.57	3.59	0.02	0.01	0.01	0.82	0.3	0.52	0.84	0.3	0.53
10	0	0		17.5	3	16.67	99.16	94.11	5.06	0.02	0.01	0.01	0.82	0.31	0.51	0.84	0.32	0.52
10	0	0		20	3	16.67	99.16	92.37	6.79	0.02	0.01	0.01	0.82	0.32	0.5	0.84	0.33	0.51
10	0	0		22.5	3	16.67	99.16	90.4	8.76	0.02	0.01	0.01	0.82	0.33	0.49	0.84	0.34	0.5
10	0	0		25	3	16.67	99.16	88.25	10.92	0.02	0.01	0.01	0.82	0.34	0.48	0.84	0.34	0.49
10	0	0		2.5	2.5	20	97.59	97.25	0.34	0.13	0.02	0.12	2.28	0.23	2.05	2.41	0.25	2.16
10	0	0		5	2.5	20	97.59	96.78	0.81	0.13	0.03	0.11	2.28	0.4	1.88	2.41	0.42	1.99
10	0	0		7.5	2.5	20	97.59	96.14	1.45	0.13	0.04	0.1	2.28	0.52	1.76	2.41	0.56	1.85
10	0	0		10	2.5	20	97.59	95.29	2.3	0.13	0.04	0.09	2.28	0.61	1.66	2.41	0.65	1.76
10	0	0		12.5	2.5	20	97.59	94.22	3.37	0.13	0.05	0.09	2.28	0.68	1.59	2.41	0.73	1.68
10	0	0		15	2.5	20	97.59	92.9	4.69	0.13	0.05	0.09	2.28	0.74	1.54	2.41	0.79	1.62
10	0	0		17.5	2.5	20	97.59	91.35	6.24	0.13	0.05	0.08	2.28	0.78	1.49	2.41	0.84	1.57
10	0	0		20	2.5	20	97.59	89.57	8.02	0.13	0.05	0.08	2.28	0.82	1.45	2.41	0.87	1.54
10	0	0		22.5	2.5	20	97.59	87.61	9.98	0.13	0.05	0.08	2.28	0.85	1.42	2.41	0.9	1.51
10	0	0		25	2.5	20	97.59	85.51	12.08	0.13	0.06	0.08	2.28	0.88	1.4	2.41	0.93	1.48

a	$k_m s_m$	k_m	s_m	k_p	s_p	C	C_A	C_R	U	U_A	U_R	O	O_A	O_R	W	W_A	W_R
10	0	0	2.5	2	25	93.7	93.11	0.59	0.82	0.08	0.74	5.48	0.4	5.08	6.3	0.47	5.83
10	0	0	5	2	25	93.7	92.39	1.32	0.82	0.13	0.69	5.48	0.72	4.76	6.3	0.85	5.45
10	0	0	7.5	2	25	93.7	91.5	2.2	0.82	0.18	0.64	5.48	0.98	4.5	6.3	1.16	5.14
10	0	0	10	2	25	93.7	90.44	3.26	0.82	0.21	0.61	5.48	1.19	4.29	6.3	1.4	4.9
10	0	0	12.5	2	25	93.7	89.19	4.51	0.82	0.24	0.58	5.48	1.37	4.11	6.3	1.6	4.7
10	0	0	15	2	25	93.7	87.76	5.94	0.82	0.26	0.56	5.48	1.51	3.97	6.3	1.77	4.53
10	0	0	17.5	2	25	93.7	86.15	7.55	0.82	0.27	0.55	5.48	1.63	3.85	6.3	1.9	4.4
10	0	0	20	2	25	93.7	84.39	9.31	0.82	0.29	0.53	5.48	1.73	3.75	6.3	2.02	4.28
10	0	0	22.5	2	25	93.7	82.5	11.2	0.82	0.3	0.52	5.48	1.82	3.66	6.3	2.11	4.19
10	0	0	25	2	25	93.7	80.5	13.2	0.82	0.31	0.51	5.48	1.89	3.59	6.3	2.2	4.1
10	0	0	2.5	1.5	33.33	84.9	84.04	0.86	3.59	0.21	3.38	11.51	0.54	10.96	15.1	0.76	14.34
10	0	0	5	1.5	33.33	84.9	83.06	1.84	3.59	0.4	3.2	11.51	1.02	10.48	15.1	1.42	13.68
10	0	0	7.5	1.5	33.33	84.9	81.96	2.94	3.59	0.55	3.04	11.51	1.45	10.06	15.1	2	13.1
10	0	0	10	1.5	33.33	84.9	80.75	4.15	3.59	0.68	2.92	11.51	1.82	9.68	15.1	2.5	12.6
10	0	0	12.5	1.5	33.33	84.9	79.41	5.49	3.59	0.79	2.81	11.51	2.15	9.36	15.1	2.94	12.16
10	0	0	15	1.5	33.33	84.9	77.97	6.93	3.59	0.88	2.71	11.51	2.44	9.07	15.1	3.32	11.78
10	0	0	17.5	1.5	33.33	84.9	76.42	8.48	3.59	0.96	2.64	11.51	2.69	8.81	15.1	3.65	11.45
10	0	0	20	1.5	33.33	84.9	74.78	10.12	3.59	1.03	2.57	11.51	2.92	8.59	15.1	3.94	11.16
10	0	0	22.5	1.5	33.33	84.9	73.07	11.83	3.59	1.09	2.51	11.51	3.12	8.39	15.1	4.2	10.9
10	0	0	25	1.5	33.33	84.9	71.3	13.6	3.59	1.14	2.46	11.51	3.29	8.22	15.1	4.43	10.67
10	0	0	2.5	1	50	67.31	66.32	0.98	11.51	0.37	11.14	21.19	0.56	20.63	32.69	0.93	31.77
10	0	0	5	1	50	67.31	65.28	2.03	11.51	0.71	10.8	21.19	1.09	20.1	32.69	1.8	30.9
10	0	0	7.5	1	50	67.31	64.18	3.12	11.51	1.02	10.48	21.19	1.58	19.6	32.69	2.61	30.08
10	0	0	10	1	50	67.31	63.03	4.28	11.51	1.31	10.19	21.19	2.05	19.13	32.69	3.37	29.33
10	0	0	12.5	1	50	67.31	61.83	5.48	11.51	1.58	9.93	21.19	2.49	18.7	32.69	4.07	28.62
10	0	0	15	1	50	67.31	60.59	6.72	11.51	1.82	9.68	21.19	2.9	18.28	32.69	4.72	27.97
10	0	0	17.5	1	50	67.31	59.3	8.01	11.51	2.05	9.46	21.19	3.28	17.9	32.69	5.33	27.36
10	0	0	20	1	50	67.31	57.98	9.33	11.51	2.25	9.26	21.19	3.64	17.54	32.69	5.89	26.8
10	0	0	22.5	1	50	67.31	56.63	10.67	11.51	2.44	9.07	21.19	3.97	17.21	32.69	6.41	26.28
10	0	0	25	1	50	67.31	55.26	12.04	11.51	2.61	8.89	21.19	4.28	16.9	32.69	6.9	25.8
15	0	0	2.5	3	16.67	98.21	97.89	0.32	0	0	0	1.79	0.22	1.57	1.79	0.22	1.58
15	0	0	5	3	16.67	98.21	97.41	0.8	0	0	0	1.79	0.36	1.43	1.79	0.36	1.43
15	0	0	7.5	3	16.67	98.21	96.74	1.47	0	0	0	1.79	0.46	1.32	1.79	0.46	1.33
15	0	0	10	3	16.67	98.21	95.83	2.38	0	0	0	1.79	0.53	1.25	1.79	0.54	1.25
15	0	0	12.5	3	16.67	98.21	94.67	3.54	0	0	0	1.79	0.59	1.2	1.79	0.59	1.2
15	0	0	15	3	16.67	98.21	93.25	4.96	0	0	0	1.79	0.63	1.16	1.79	0.63	1.16
15	0	0	17.5	3	16.67	98.21	91.61	6.6	0	0	0	1.79	0.66	1.13	1.79	0.66	1.13
15	0	0	20	3	16.67	98.21	89.77	8.44	0	0	0	1.79	0.69	1.1	1.79	0.69	1.1
15	0	0	22.5	3	16.67	98.21	87.76	10.45	0	0	0	1.79	0.71	1.08	1.79	0.71	1.08
15	0	0	25	3	16.67	98.21	85.63	12.58	0	0	0	1.79	0.72	1.06	1.79	0.73	1.07
15	0	0	2.5	2.5	20	95.94	95.43	0.5	0.06	0.01	0.05	4.01	0.37	3.63	4.06	0.38	3.68
15	0	0	5	2.5	20	95.94	94.78	1.16	0.06	0.01	0.04	4.01	0.65	3.35	4.06	0.67	3.4
15	0	0	7.5	2.5	20	95.94	93.94	2	0.06	0.02	0.04	4.01	0.87	3.14	4.06	0.88	3.18
15	0	0	10	2.5	20	95.94	92.89	3.04	0.06	0.02	0.04	4.01	1.03	2.98	4.06	1.05	3.02
15	0	0	12.5	2.5	20	95.94	91.64	4.29	0.06	0.02	0.04	4.01	1.16	2.85	4.06	1.18	2.89
15	0	0	15	2.5	20	95.94	90.19	5.75	0.06	0.02	0.04	4.01	1.26	2.75	4.06	1.28	2.78
15	0	0	17.5	2.5	20	95.94	88.54	7.4	0.06	0.02	0.04	4.01	1.34	2.67	4.06	1.36	2.7
15	0	0	20	2.5	20	95.94	86.73	9.21	0.06	0.02	0.03	4.01	1.41	2.6	4.06	1.43	2.63
15	0	0	22.5	2.5	20	95.94	84.78	11.16	0.06	0.02	0.03	4.01	1.46	2.54	4.06	1.49	2.58
15	0	0	25	2.5	20	95.94	82.72	13.21	0.06	0.02	0.03	4.01	1.51	2.5	4.06	1.53	2.53
15	0	0	2.5	2	25	91.46	90.75	0.71	0.47	0.05	0.42	8.08	0.54	7.53	8.54	0.59	7.95
15	0	0	5	2	25	91.46	89.9	1.56	0.47	0.08	0.39	8.08	0.99	7.08	8.54	1.07	7.47
15	0	0	7.5	2	25	91.46	88.9	2.56	0.47	0.1	0.36	8.08	1.36	6.71	8.54	1.47	7.08
15	0	0	10	2	25	91.46	87.74	3.72	0.47	0.12	0.34	8.08	1.67	6.41	8.54	1.79	6.75
15	0	0	12.5	2	25	91.46	86.41	5.04	0.47	0.14	0.33	8.08	1.93	6.15	8.54	2.06	6.48
15	0	0	15	2	25	91.46	84.94	6.52	0.47	0.15	0.32	8.08	2.14	5.94	8.54	2.29	6.25
15	0	0	17.5	2	25	91.46	83.32	8.14	0.47	0.16	0.31	8.08	2.32	5.76	8.54	2.48	6.06
15	0	0	20	2	25	91.46	81.57	9.89	0.47	0.17	0.3	8.08	2.47	5.61	8.54	2.64	5.91
15	0	0	22.5	2	25	91.46	79.72	11.74	0.47	0.17	0.29	8.08	2.6	5.48	8.54	2.77	5.77
15	0	0	25	2	25	91.46	77.79	13.67	0.47	0.18	0.29	8.08	2.71	5.36	8.54	2.89	5.65

a	$k_m s_m$ $i=$	k_m	s_m	k_p	s_p	C	C_A	C_R	U	U_A	U_R	O	O_A	O_R	W	W_A	W_R
15	0	0	2.5	1.5	33.33	82.76	81.85	0.91	2.56	0.16	2.4	14.69	0.65	14.04	17.24	0.81	16.44
15	0	0	5	1.5	33.33	82.76	80.83	1.92	2.56	0.3	2.26	14.69	1.23	13.46	17.24	1.52	15.72
15	0	0	7.5	1.5	33.33	82.76	79.71	3.05	2.56	0.41	2.15	14.69	1.75	12.94	17.24	2.15	15.09
15	0	0	10	1.5	33.33	82.76	78.47	4.28	2.56	0.5	2.06	14.69	2.21	12.48	17.24	2.71	14.53
15	0	0	12.5	1.5	33.33	82.76	77.14	5.62	2.56	0.58	1.98	14.69	2.62	12.07	17.24	3.2	14.05
15	0	0	15	1.5	33.33	82.76	75.7	7.06	2.56	0.64	1.91	14.69	2.98	11.7	17.24	3.63	13.62
15	0	0	17.5	1.5	33.33	82.76	74.17	8.58	2.56	0.7	1.86	14.69	3.3	11.38	17.24	4.01	13.24
15	0	0	20	1.5	33.33	82.76	72.57	10.19	2.56	0.75	1.81	14.69	3.59	11.1	17.24	4.34	12.91
15	0	0	22.5	1.5	33.33	82.76	70.9	11.85	2.56	0.79	1.77	14.69	3.84	10.84	17.24	4.64	12.61
15	0	0	25	1.5	33.33	82.76	69.18	13.57	2.56	0.83	1.73	14.69	4.07	10.61	17.24	4.9	12.35
15	0	0	2.5	1	50	66.12	65.14	0.98	9.68	0.32	9.36	24.2	0.6	23.59	33.88	0.93	32.95
15	0	0	5	1	50	66.12	64.1	2.02	9.68	0.62	9.06	24.2	1.18	23.02	33.88	1.8	32.08
15	0	0	7.5	1	50	66.12	63.01	3.11	9.68	0.9	8.78	24.2	1.72	22.47	33.88	2.62	31.26
15	0	0	10	1	50	66.12	61.87	4.25	9.68	1.15	8.53	24.2	2.24	21.96	33.88	3.38	30.49
15	0	0	12.5	1	50	66.12	60.69	5.43	9.68	1.38	8.3	24.2	2.72	21.47	33.88	4.1	29.78
15	0	0	15	1	50	66.12	59.46	6.66	9.68	1.58	8.1	24.2	3.18	21.02	33.88	4.76	29.11
15	0	0	17.5	1	50	66.12	58.2	7.93	9.68	1.77	7.91	24.2	3.61	20.59	33.88	5.38	28.49
15	0	0	20	1	50	66.12	56.9	9.23	9.68	1.95	7.73	24.2	4.01	20.19	33.88	5.96	27.92
15	0	0	22.5	1	50	66.12	55.58	10.55	9.68	2.11	7.57	24.2	4.38	19.81	33.88	6.49	27.39
15	0	0	25	1	50	66.12	54.23	11.89	9.68	2.25	7.43	24.2	4.73	19.46	33.88	6.99	26.89
20	0	0	2.5	3	16.67	96.41	95.85	0.56	0	0	0	3.59	0.4	3.2	3.59	0.4	3.2
20	0	0	5	3	16.67	96.41	95.09	1.32	0	0	0	3.59	0.68	2.92	3.59	0.68	2.92
20	0	0	7.5	3	16.67	96.41	94.08	2.33	0	0	0	3.59	0.88	2.71	3.59	0.88	2.72
20	0	0	10	3	16.67	96.41	92.83	3.58	0	0	0	3.59	1.03	2.57	3.59	1.03	2.57
20	0	0	12.5	3	16.67	96.41	91.34	5.06	0	0	0	3.59	1.14	2.46	3.59	1.14	2.46
20	0	0	15	3	16.67	96.41	89.67	6.73	0	0	0	3.59	1.22	2.37	3.59	1.22	2.37
20	0	0	17.5	3	16.67	96.41	87.85	8.56	0	0	0	3.59	1.29	2.3	3.59	1.29	2.3
20	0	0	20	3	16.67	96.41	85.91	10.49	0	0	0	3.59	1.34	2.25	3.59	1.34	2.25
20	0	0	22.5	3	16.67	96.41	83.9	12.51	0	0	0	3.59	1.39	2.21	3.59	1.39	2.21
20	0	0	25	3	16.67	96.41	81.82	14.59	0	0	0	3.59	1.42	2.17	3.59	1.42	2.17
20	0	0	2.5	2.5	20	93.3	92.57	0.73	0.02	0	0.02	6.68	0.57	6.11	6.7	0.57	6.13
20	0	0	5	2.5	20	93.3	91.67	1.63	0.02	0.01	0.02	6.68	1.01	5.67	6.7	1.02	5.69
20	0	0	7.5	2.5	20	93.3	90.58	2.72	0.02	0.01	0.02	6.68	1.36	5.32	6.7	1.36	5.34
20	0	0	10	2.5	20	93.3	89.3	4	0.02	0.01	0.02	6.68	1.63	5.05	6.7	1.64	5.07
20	0	0	12.5	2.5	20	93.3	87.84	5.45	0.02	0.01	0.02	6.68	1.85	4.84	6.7	1.85	4.85
20	0	0	15	2.5	20	93.3	86.24	7.06	0.02	0.01	0.01	6.68	2.02	4.66	6.7	2.03	4.68
20	0	0	17.5	2.5	20	93.3	84.5	8.79	0.02	0.01	0.01	6.68	2.16	4.52	6.7	2.17	4.54
20	0	0	20	2.5	20	93.3	82.67	10.63	0.02	0.01	0.01	6.68	2.27	4.41	6.7	2.28	4.42
20	0	0	22.5	2.5	20	93.3	80.75	12.54	0.02	0.01	0.01	6.68	2.37	4.31	6.7	2.38	4.32
20	0	0	25	2.5	20	93.3	78.78	14.52	0.02	0.01	0.01	6.68	2.45	4.23	6.7	2.46	4.24
20	0	0	2.5˚	2	25	88.24	87.37	0.86	0.26	0.03	0.23	11.51	0.71	10.8	11.76	0.74	11.03
20	0	0	5	2	25	88.24	86.37	1.86	0.26	0.05	0.21	11.51	1.31	10.19	11.76	1.36	10.4
20	0	0	7.5	2	25	88.24	85.23	3	0.26	0.06	0.2	11.51	1.82	9.68	11.76	1.88	9.88
20	0	0	10	2	25	88.24	83.96	4.28	0.26	0.07	0.19	11.51	2.25	9.26	11.76	2.32	9.44
20	0	0	12.5	2	25	88.24	82.55	5.69	0.26	0.08	0.18	11.51	2.61	8.89	11.76	2.69	9.07
20	0	0	15	2	25	88.24	81.03	7.21	0.26	0.08	0.17	11.51	2.92	8.59	11.76	3	8.76
20	0	0	17.5	2	25	88.24	79.4	8.84	0.26	0.09	0.17	11.51	3.18	8.33	11.76	3.26	8.5
20	0	0	20	2	25	88.24	77.68	10.56	0.26	0.09	0.16	11.51	3.4	8.11	11.76	3.49	8.27
20	0	0	22.5	2	25	88.24	75.89	12.35	0.26	0.1	0.16	11.51	3.59	7.92	11.76	3.68	8.08
20	0	0	25	2	25	88.24	74.04	14.19	0.26	0.1	0.16	11.51	3.75	7.76	11.76	3.85	7.91
20	0	0	2.5	1.5	33.33	79.81	78.84	0.97	1.79	0.12	1.67	18.41	0.75	17.65	20.19	0.87	19.32
20	0	0	5	1.5	33.33	79.81	77.78	2.03	1.79	0.22	1.57	18.41	1.44	16.96	20.19	1.66	18.54
20	0	0	7.5	1.5	33.33	79.81	76.62	3.18	1.79	0.29	1.49	18.41	2.06	16.34	20.19	2.36	17.84
20	0	0	10	1.5	33.33	79.81	75.37	4.43	1.79	0.36	1.43	18.41	2.62	15.79	20.19	2.98	17.21
20	0	0	12.5	1.5	33.33	79.81	74.03	5.77	1.79	0.42	1.37	18.41	3.12	15.28	20.19	3.54	16.66
20	0	0	15	1.5	33.33	79.81	72.61	7.19	1.79	0.46	1.32	18.41	3.57	14.84	20.19	4.03	16.16
20	0	0	17.5	1.5	33.33	79.81	71.12	8.69	1.79	0.5	1.28	18.41	3.97	14.44	20.19	4.47	15.72
20	0	0	20	1.5	33.33	79.81	69.57	10.24	1.79	0.53	1.25	18.41	4.33	14.08	20.19	4.86	15.33
20	0	0	22.5	1.5	33.33	79.81	67.96	11.85	1.79	0.56	1.22	18.41	4.65	13.76	20.19	5.21	14.98
20	0	0	25	1.5	33.33	79.81	66.31	13.49	1.79	0.59	1.2	18.41	4.93	13.47	20.19	5.52	14.67

a	i= $k_m s_m$	k_m	s_m	k_p	s_p	C	C_A	C_R	U	U_A	U_R	O	O_A	O_R	W	W_A	W_R
20	0	0	2.5	1	50	64.5	63.52	0.98	8.08	0.28	7.79	27.43	0.65	26.78	35.5	0.93	34.57
20	0	0	5	1	50	64.5	62.49	2.01	8.08	0.54	7.53	27.43	1.26	26.16	35.5	1.81	33.7
20	0	0	7.5	1	50	64.5	61.41	3.09	8.08	0.78	7.3	27.43	1.85	25.57	35.5	2.63	32.87
20	0	0	10	1	50	64.5	60.29	4.21	8.08	0.99	7.08	27.43	2.42	25.01	35.5	3.41	32.09
20	0	0	12.5	1	50	64.5	59.12	5.38	8.08	1.19	6.89	27.43	2.95	24.48	35.5	4.13	31.37
20	0	0	15	1	50	64.5	57.92	6.58	8.08	1.36	6.71	27.43	3.45	23.98	35.5	4.81	30.69
20	0	0	17.5	1	50	64.5	56.68	7.82	8.08	1.52	6.55	27.43	3.92	23.5	35.5	5.45	30.05
20	0	0	20	1	50	64.5	55.42	9.08	8.08	1.67	6.41	27.43	4.37	23.06	35.5	6.04	29.46
20	0	0	22.5	1	50	64.5	54.13	10.37	8.08	1.8	6.27	27.43	4.79	22.64	35.5	6.59	28.91
20	0	0	25	1	50	64.5	52.82	11.68	8.08	1.93	6.15	27.43	5.18	22.24	35.5	7.11	28.39
25	0	0	2.5	3	16.67	93.32	92.44	0.88	0	0	0	6.68	0.67	6.01	6.68	0.67	6.01
25	0	0	5	3	16.67	93.32	91.3	2.02	0	0	0	6.68	1.16	5.52	6.68	1.16	5.52
25	0	0	7.5	3	16.67	93.32	89.91	3.41	0	0	0	6.68	1.53	5.15	6.68	1.53	5.15
25	0	0	10	3	16.67	93.32	88.29	5.03	0	0	0	6.68	1.81	4.87	6.68	1.81	4.87
25	0	0	12.5	3	16.67	93.32	86.49	6.83	0	0	0	6.68	2.02	4.66	6.68	2.02	4.66
25	0	0	15	3	16.67	93.32	84.59	8.73	0	0	0	6.68	2.18	4.5	6.68	2.18	4.5
25	0	0	17.5	3	16.67	93.32	82.63	10.68	0	0	0	6.68	2.32	4.37	6.68	2.32	4.37
25	0	0	20	3	16.67	93.32	80.66	12.66	0	0	0	6.68	2.42	4.26	6.68	2.42	4.26
25	0	0	22.5	3	16.67	93.32	78.67	14.65	0	0	0	6.68	2.51	4.17	6.68	2.51	4.17
25	0	0	25	3	16.67	93.32	76.7	16.62	0	0	0	6.68	2.58	4.1	6.68	2.58	4.1
25	0	0	2.5	2.5	20	89.43	88.43	1	0.01	0	0.01	10.56	0.82	9.75	10.57	0.82	9.76
25	0	0	5	2.5	20	89.43	87.25	2.18	0.01	0	0.01	10.56	1.48	9.09	10.57	1.48	9.1
25	0	0	7.5	2.5	20	89.43	85.88	3.54	0.01	0	0.01	10.56	2	8.56	10.57	2.01	8.57
25	0	0	10	2.5	20	89.43	84.36	5.06	0.01	0	0.01	10.56	2.43	8.14	10.57	2.43	8.14
25	0	0	12.5	2.5	20	89.43	82.72	6.71	0.01	0	0.01	10.56	2.77	7.79	10.57	2.78	7.8
25	0	0	15	2.5	20	89.43	80.99	8.44	0.01	0	0.01	10.56	3.05	7.51	10.57	3.05	7.52
25	0	0	17.5	2.5	20	89.43	79.2	10.23	0.01	0	0.01	10.56	3.28	7.28	10.57	3.28	7.29
25	0	0	20	2.5	20	89.43	77.37	12.05	0.01	0	0.01	10.56	3.47	7.09	10.57	3.47	7.1
25	0	0	22.5	2.5	20	89.43	75.52	13.9	0.01	0	0.01	10.56	3.63	6.93	10.57	3.64	6.94
25	0	0	25	2.5	20	89.43	73.66	15.77	0.01	0	0.01	10.56	3.77	6.8	10.57	3.77	6.8
25	0	0	2.5	2	25	84	82.96	1.04	0.13	0.01	0.12	15.87	0.9	14.97	16	0.91	15.09
25	0	0	5	2	25	84	81.8	2.2	0.13	0.02	0.11	15.87	1.67	14.19	16	1.7	14.3
25	0	0	7.5	2	25	84	80.51	3.49	0.13	0.03	0.1	15.87	2.35	13.52	16	2.38	13.62
25	0	0	10	2	25	84	79.12	4.88	0.13	0.04	0.1	15.87	2.92	12.95	16	2.96	13.04
25	0	0	12.5	2	25	84	77.64	6.36	0.13	0.04	0.09	15.87	3.41	12.46	16	3.45	12.55
25	0	0	15	2	25	84	76.08	7.92	0.13	0.05	0.09	15.87	3.83	12.04	16	3.87	12.13
25	0	0	17.5	2	25	84	74.47	9.53	0.13	0.05	0.09	15.87	4.19	11.68	16	4.24	11.76
25	0	0	20	2	25	84	72.8	11.2	0.13	0.05	0.09	15.87	4.5	11.37	16	4.55	11.45
25	0	0	22.5	2	25	84	71.09	12.91	0.13	0.05	0.08	15.87	4.77	11.1	16	4.82	11.18
25	0	0	25	2	25	84	69.36	14.64	0.13	0.05	0.08	15.87	5	10.86	16	5.06	10.95
25	0	0	2.5	1.5	33.33	76.11	75.09	1.03	1.22	0.08	1.14	22.66	0.86	21.8	23.89	0.94	22.94
25	0	0	5	1.5	33.33	76.11	73.98	2.14	1.22	0.15	1.07	22.66	1.65	21.01	23.89	1.81	22.08
25	0	0	7.5	1.5	33.33	76.11	72.79	3.33	1.22	0.21	1.01	22.66	2.38	20.28	23.89	2.59	21.3
25	0	0	10	1.5	33.33	76.11	71.52	4.59	1.22	0.26	0.97	22.66	3.04	19.62	23.89	3.3	20.59
25	0	0	12.5	1.5	33.33	76.11	70.19	5.92	1.22	0.29	0.93	22.66	3.64	19.02	23.89	3.93	19.95
25	0	0	15	1.5	33.33	76.11	68.8	7.32	1.22	0.32	0.9	22.66	4.18	18.48	23.89	4.5	19.38
25	0	0	17.5	1.5	33.33	76.11	67.35	8.77	1.22	0.35	0.87	22.66	4.67	18	23.89	5.02	18.87
25	0	0	20	1.5	33.33	76.11	65.86	10.26	1.22	0.37	0.85	22.66	5.1	17.56	23.89	5.48	18.41
25	0	0	22.5	1.5	33.33	76.11	64.33	11.79	1.22	0.39	0.83	22.66	5.5	17.16	23.89	5.89	17.99
25	0	0	25	1.5	33.33	76.11	62.77	13.35	1.22	0.41	0.81	22.66	5.85	16.81	23.89	6.26	17.62
25	0	0	2.5	1	50	62.47	61.49	0.97	6.68	0.24	6.44	30.85	0.68	30.17	37.53	0.93	36.61
25	0	0	5	1	50	62.47	60.47	1.99	6.68	0.47	6.22	30.85	1.34	29.51	37.53	1.81	35.73
25	0	0	7.5	1	50	62.47	59.41	3.06	6.68	0.67	6.01	30.85	1.98	28.88	37.53	2.64	34.89
25	0	0	10	1	50	62.47	58.31	4.16	6.68	0.85	5.83	30.85	2.58	28.27	37.53	3.43	34.1
25	0	0	12.5	1	50	62.47	57.17	5.3	6.68	1.01	5.67	30.85	3.16	27.69	37.53	4.17	33.36
25	0	0	15	1	50	62.47	56	6.47	6.68	1.16	5.52	30.85	3.71	27.15	37.53	4.87	32.67
25	0	0	17.5	1	50	62.47	54.79	7.67	6.68	1.3	5.39	30.85	4.23	26.63	37.53	5.52	32.01
25	0	0	20	1	50	62.47	53.57	8.9	6.68	1.42	5.26	30.85	4.72	26.14	37.53	6.14	31.4
25	0	0	22.5	1	50	62.47	52.32	10.15	6.68	1.53	5.15	30.85	5.18	25.67	37.53	6.71	30.83
25	0	0	25	1	50	62.47	51.05	11.41	6.68	1.63	5.05	30.85	5.62	25.24	37.53	7.25	30.29

a	$i=$ $k_m s_m$	k_m	s_m	k_p	s_p	C	C_A	C_R	U	U_A	U_R	O	O_A	O_R	W	W_A	W_R
0	2.5	1	2.5	3	16.67	99.73	99.51	0.22	0.13	0	0.13	0.13	0	0.13	0.27	0.01	0.26
0	2.5	1.5	1.67	3	16.67	99.73	99.54	0.19	0.13	0	0.13	0.13	0	0.13	0.27	0	0.27
0	2.5	2	1.25	3	16.67	99.73	99.55	0.18	0.13	0	0.13	0.13	0	0.13	0.27	0	0.27
0	2.5	2.5	1	3	16.67	99.73	99.56	0.17	0.13	0	0.13	0.13	0	0.13	0.27	0	0.27
0	2.5	3	0.83	3	16.67	99.73	99.56	0.17	0.13	0	0.13	0.13	0	0.13	0.27	0	0.27
0	5	1	5	3	16.67	99.73	99.01	0.72	0.13	0.01	0.13	0.13	0.01	0.13	0.27	0.02	0.25
0	5	1.5	3.33	3	16.67	99.73	99.18	0.55	0.13	0	0.13	0.13	0	0.13	0.27	0	0.27
0	5	2	2.5	3	16.67	99.73	99.24	0.49	0.13	0	0.13	0.13	0	0.13	0.27	0	0.27
0	5	2.5	2	3	16.67	99.73	99.27	0.46	0.13	0	0.13	0.13	0	0.13	0.27	0	0.27
0	5	3	1.67	3	16.67	99.73	99.28	0.45	0.13	0	0.13	0.13	0	0.13	0.27	0	0.27
0	7.5	1	7.5	3	16.67	99.73	97.97	1.76	0.13	0.01	0.13	0.13	0.01	0.13	0.27	0.02	0.25
0	7.5	1.5	5	3	16.67	99.73	98.53	1.2	0.13	0	0.13	0.13	0	0.13	0.27	0.01	0.26
0	7.5	2	3.75	3	16.67	99.73	98.71	1.02	0.13	0	0.13	0.13	0	0.13	0.27	0	0.27
0	7.5	2.5	3	3	16.67	99.73	98.79	0.94	0.13	0	0.13	0.13	0	0.13	0.27	0	0.27
0	7.5	3	2.5	3	16.67	99.73	98.83	0.9	0.13	0	0.13	0.13	0	0.13	0.27	0	0.27
0	10	1	10	3	16.67	99.73	96	3.73	0.13	0.01	0.12	0.13	0.01	0.12	0.27	0.02	0.25
0	10	1.5	6.67	3	16.67	99.73	97.4	2.33	0.13	0	0.13	0.13	0	0.13	0.27	0.01	0.26
0	10	2	5	3	16.67	99.73	97.84	1.89	0.13	0	0.13	0.13	0	0.13	0.27	0	0.27
0	10	2.5	4	3	16.67	99.73	98.04	1.69	0.13	0	0.13	0.13	0	0.13	0.27	0	0.27
0	10	3	3.33	3	16.67	99.73	98.14	1.59	0.13	0	0.13	0.13	0	0.13	0.27	0	0.27
0	15	1	15	3	16.67	99.73	88.14	11.59	0.13	0.01	0.12	0.13	0.01	0.12	0.27	0.03	0.24
0	15	1.5	10	3	16.67	99.73	92.81	6.92	0.13	0	0.13	0.13	0	0.13	0.27	0.01	0.26
0	15	2	7.5	3	16.67	99.73	94.44	5.29	0.13	0	0.13	0.13	0	0.13	0.27	0	0.27
0	15	2.5	6	3	16.67	99.73	95.18	4.55	0.13	0	0.13	0.13	0	0.13	0.27	0	0.27
0	15	3	5	3	16.67	99.73	95.57	4.16	0.13	0	0.13	0.13	0	0.13	0.27	0	0.27
0	20	1	20	3	16.67	99.73	75.27	24.46	0.13	0.01	0.12	0.13	0.01	0.12	0.27	0.03	0.24
0	20	1.5	13.33	3	16.67	99.73	84.05	15.68	0.13	0.01	0.13	0.13	0.01	0.13	0.27	0.01	0.26
0	20	2	10	3	16.67	99.73	87.74	11.99	0.13	0	0.13	0.13	0	0.13	0.27	0	0.27
0	20	2.5	8	3	16.67	99.73	89.54	10.19	0.13	0	0.13	0.13	0	0.13	0.27	0	0.27
0	20	3	6.67	3	16.67	99.73	90.53	9.2	0.13	0	0.13	0.13	0	0.13	0.27	0	0.27
0	25	1	25	3	16.67	99.73	59.89	39.84	0.13	0.02	0.12	0.13	0.02	0.12	0.27	0.03	0.24
0	25	1.5	16.67	3	16.67	99.73	71.29	28.44	0.13	0.01	0.13	0.13	0.01	0.13	0.27	0.01	0.26
0	25	2	12.5	3	16.67	99.73	77.06	22.67	0.13	0	0.13	0.13	0	0.13	0.27	0	0.27
0	25	2.5	10	3	16.67	99.73	80.2	19.53	0.13	0	0.13	0.13	0	0.13	0.27	0	0.27
0	25	3	8.33	3	16.67	99.73	82.04	17.69	0.13	0	0.13	0.13	0	0.13	0.27	0	0.27
0	2.5	1	2.5	2.5	20	98.76	98.12	0.63	0.62	0.02	0.61	0.62	0.02	0.61	1.24	0.03	1.21
0	2.5	1.5	1.67	2.5	20	98.76	98.2	0.56	0.62	0	0.62	0.62	0	0.62	1.24	0.01	1.23
0	2.5	2	1.25	2.5	20	98.76	98.22	0.54	0.62	0	0.62	0.62	0	0.62	1.24	0	1.24
0	2.5	2.5	1	2.5	20	98.76	98.23	0.53	0.62	0	0.62	0.62	0	0.62	1.24	0	1.24
0	2.5	3	0.83	2.5	20	98.76	98.24	0.52	0.62	0	0.62	0.62	0	0.62	1.24	0	1.24
0	5	.1	5	2.5	20	98.76	97.04	1.72	0.62	0.03	0.59	0.62	0.03	0.59	1.24	0.06	1.19
0	5	1.5	3.33	2.5	20	98.76	97.34	1.42	0.62	0.01	0.61	0.62	0.01	0.61	1.24	0.02	1.23
0	5	2	2.5	2.5	20	98.76	97.44	1.32	0.62	0	0.62	0.62	0	0.62	1.24	0	1.24
0	5	2.5	2	2.5	20	98.76	97.48	1.28	0.62	0	0.62	0.62	0	0.62	1.24	0	1.24
0	5	3	1.67	2.5	20	98.76	97.51	1.25	0.62	0	0.62	0.62	0	0.62	1.24	0	1.24
0	7.5	1	7.5	2.5	20	98.76	95.26	3.5	0.62	0.04	0.58	0.62	0.04	0.58	1.24	0.07	1.17
0	7.5	1.5	5	2.5	20	98.76	96.05	2.71	0.62	0.01	0.61	0.62	0.01	0.61	1.24	0.02	1.22
0	7.5	2	3.75	2.5	20	98.76	96.32	2.44	0.62	0	0.62	0.62	0	0.62	1.24	0.01	1.24
0	7.5	2.5	3	2.5	20	98.76	96.44	2.32	0.62	0	0.62	0.62	0	0.62	1.24	0	1.24
0	7.5	3	2.5	2.5	20	98.76	96.5	2.26	0.62	0	0.62	0.62	0	0.62	1.24	0	1.24

a	$i= k_m s_m$	k_m	s_m	k_p	s_p	C	C_A	C_R	U	U_A	U_R	O	O_A	O_R	W	W_A	W_R
0	10	1	10	2.5	20	98.76	92.54	6.21	0.62	0.04	0.58	0.62	0.04	0.58	1.24	0.09	1.15
0	10	1.5	6.67	2.5	20	98.76	94.19	4.56	0.62	0.01	0.61	0.62	0.01	0.61	1.24	0.03	1.22
0	10	2	5	2.5	20	98.76	94.76	4	0.62	0	0.62	0.62	0	0.62	1.24	0.01	1.24
0	10	2.5	4	2.5	20	98.76	95.01	3.75	0.62	0	0.62	0.62	0	0.62	1.24	0	1.24
0	10	3	3.33	2.5	20	98.76	95.15	3.61	0.62	0	0.62	0.62	0	0.62	1.24	0	1.24
0	15	1	15	2.5	20	98.76	83.78	14.98	0.62	0.05	0.57	0.62	0.05	0.57	1.24	0.11	1.13
0	15	1.5	10	2.5	20	98.76	88.22	10.54	0.62	0.02	0.6	0.62	0.02	0.6	1.24	0.03	1.21
0	15	2	7.5	2.5	20	98.76	89.86	8.9	0.62	0	0.62	0.62	0	0.62	1.24	0.01	1.23
0	15	2.5	6	2.5	20	98.76	90.63	8.13	0.62	0	0.62	0.62	0	0.62	1.24	0	1.24
0	15	3	5	2.5	20	98.76	91.04	7.71	0.62	0	0.62	0.62	0	0.62	1.24	0	1.24
0	20	1	20	2.5	20	98.76	71.17	27.58	0.62	0.06	0.56	0.62	0.06	0.56	1.24	0.12	1.12
0	20	1.5	13.33	2.5	20	98.76	78.81	19.95	0.62	0.02	0.6	0.62	0.02	0.6	1.24	0.04	1.2
0	20	2	10	2.5	20	98.76	82.03	16.73	0.62	0.01	0.62	0.62	0.01	0.62	1.24	0.01	1.23
0	20	2.5	8	2.5	20	98.76	83.63	15.13	0.62	0	0.62	0.62	0	0.62	1.24	0	1.24
0	20	3	6.67	2.5	20	98.76	84.53	14.23	0.62	0	0.62	0.62	0	0.62	1.24	0	1.24
0	25	1	25	2.5	20	98.76	56.74	42.02	0.62	0.07	0.55	0.62	0.07	0.55	1.24	0.13	1.11
0	25	1.5	16.67	2.5	20	98.76	66.4	32.36	0.62	0.02	0.6	0.62	0.02	0.6	1.24	0.05	1.2
0	25	2	12.5	2.5	20	98.76	71.13	27.63	0.62	0.01	0.61	0.62	0.01	0.61	1.24	0.01	1.23
0	25	2.5	10	2.5	20	98.76	73.67	25.09	0.62	0	0.62	0.62	0	0.62	1.24	0	1.24
0	25	3	8.33	2.5	20	98.76	75.16	23.6	0.62	0	0.62	0.62	0	0.62	1.24	0	1.24
0	2.5	1	2.5	2	25	95.45	94.05	1.4	2.28	0.04	2.23	2.28	0.04	2.23	4.55	0.08	4.47
0	2.5	1.5	1.67	2	25	95.45	94.18	1.27	2.28	0.01	2.26	2.28	0.01	2.26	4.55	0.02	4.53
0	2.5	2	1.25	2	25	95.45	94.22	1.23	2.28	0	2.27	2.28	0	2.27	4.55	0.01	4.54
0	2.5	2.5	1	2	25	95.45	94.24	1.21	2.28	0	2.27	2.28	0	2.27	4.55	0	4.55
0	2.5	3	0.83	2	25	95.45	94.24	1.21	2.28	0	2.27	2.28	0	2.27	4.55	0	4.55
0	5	1	5	2	25	95.45	92.09	3.36	2.28	0.08	2.2	2.28	0.08	2.2	4.55	0.15	4.4
0	5	1.5	3.33	2	25	95.45	92.52	2.93	2.28	0.02	2.26	2.28	0.02	2.26	4.55	0.04	4.51
0	5	2	2.5	2	25	95.45	92.66	2.79	2.28	0	2.27	2.28	0	2.27	4.55	0.01	4.54
0	5	2.5	2	2	25	95.45	92.72	2.73	2.28	0	2.27	2.28	0	2.27	4.55	0	4.55
0	5	3	1.67	2	25	95.45	92.75	2.7	2.28	0	2.27	2.28	0	2.27	4.55	0	4.55
0	7.5	1	7.5	2	25	95.45	89.45	6	2.28	0.1	2.17	2.28	0.1	2.17	4.55	0.21	4.34
0	7.5	1.5	5	2	25	95.45	90.39	5.06	2.28	0.03	2.25	2.28	0.03	2.25	4.55	0.06	4.49
0	7.5	2	3.75	2	25	95.45	90.71	4.74	2.28	0.01	2.27	2.28	0.01	2.27	4.55	0.01	4.54
0	7.5	2.5	3	2	25	95.45	90.85	4.6	2.28	0	2.27	2.28	0	2.27	4.55	0	4.55
0	7.5	3	2.5	2	25	95.45	90.93	4.52	2.28	0	2.27	2.28	0	2.27	4.55	0	4.55
0	10	1	10	2	25	95.45	86.01	9.44	2.28	0.13	2.15	2.28	0.13	2.15	4.55	0.26	4.29
0	10	1.5	6.67	2	25	95.45	87.72	7.73	2.28	0.04	2.24	2.28	0.04	2.24	4.55	0.07	4.48
0	10	2	5	2	25	95.45	88.32	7.13	2.28	0.01	2.27	2.28	0.01	2.27	4.55	0.02	4.53
0	10	2.5	4	2	25	95.45	88.58	6.87	2.28	0	2.27	2.28	0	2.27	4.55	0	4.55
0	10	3	3.33	2	25	95.45	88.72	6.73	2.28	0	2.27	2.28	0	2.27	4.55	0	4.55
0	15	1	15	2	25	95.45	76.71	18.74	2.28	0.16	2.11	2.28	0.16	2.11	4.55	0.33	4.22
0	15	1.5	10	2	25	95.45	80.55	14.9	2.28	0.05	2.23	2.28	0.05	2.23	4.55	0.1	4.45
0	15	2	7.5	2	25	95.45	81.98	13.47	2.28	0.01	2.26	2.28	0.01	2.26	4.55	0.03	4.52
0	15	2.5	6	2	25	95.45	82.65	12.8	2.28	0	2.27	2.28	0	2.27	4.55	0.01	4.54
0	15	3	5	2	25	95.45	83.02	12.43	2.28	0	2.27	2.28	0	2.27	4.55	0	4.55
0	20	1	20	2	25	95.45	64.87	30.58	2.28	0.19	2.08	2.28	0.19	2.08	4.55	0.38	4.17
0	20	1.5	13.33	2	25	95.45	70.95	24.5	2.28	0.06	2.22	2.28	0.06	2.22	4.55	0.12	4.43
0	20	2	10	2	25	95.45	73.46	21.99	2.28	0.02	2.26	2.28	0.02	2.26	4.55	0.03	4.52
0	20	2.5	8	2	25	95.45	74.69	20.76	2.28	0	2.27	2.28	0	2.27	4.55	0.01	4.54
0	20	3	6.67	2	25	95.45	75.38	20.07	2.28	0	2.27	2.28	0	2.27	4.55	0	4.55
0	25	1	25	2	25	95.45	51.86	43.59	2.28	0.21	2.06	2.28	0.21	2.06	4.55	0.42	4.13
0	25	1.5	16.67	2	25	95.45	59.41	36.04	2.28	0.07	2.21	2.28	0.07	2.21	4.55	0.14	4.41
0	25	2	12.5	2	25	95.45	62.89	32.56	2.28	0.02	2.26	2.28	0.02	2.26	4.55	0.04	4.51
0	25	2.5	10	2	25	95.45	64.69	30.76	2.28	0	2.27	2.28	0	2.27	4.55	0.01	4.54
0	25	3	8.33	2	25	95.45	65.73	29.72	2.28	0	2.27	2.28	0	2.27	4.55	0	4.55

a	$i=$ $k_m s_m$	k_m	s_m	k_p	s_p	C	C_A	C_R	U	U_A	U_R	O	O_A	O_R	W	W_A	W_R
0	2.5	1	2.5	1.5	33.33	86.64	84.31	2.32	6.68	0.08	6.6	6.68	0.08	6.6	13.36	0.15	13.21
0	2.5	1.5	1.67	1.5	33.33	86.64	84.49	2.14	6.68	0.02	6.66	6.68	0.02	6.66	13.36	0.04	13.32
0	2.5	2	1.25	1.5	33.33	86.64	84.55	2.09	6.68	0	6.68	6.68	0	6.68	13.36	0.01	13.35
0	2.5	2.5	1	1.5	33.33	86.64	84.56	2.07	6.68	0	6.68	6.68	0	6.68	13.36	0	13.36
0	2.5	3	0.83	1.5	33.33	86.64	84.57	2.07	6.68	0	6.68	6.68	0	6.68	13.36	0	13.36
0	5	1	5	1.5	33.33	86.64	81.52	5.12	6.68	0.15	6.53	6.68	0.15	6.53	13.36	0.29	13.07
0	5	1.5	3.33	1.5	33.33	86.64	82.01	4.63	6.68	0.04	6.64	6.68	0.04	6.64	13.36	0.08	13.29
0	5	2	2.5	1.5	33.33	86.64	82.16	4.48	6.68	0.01	6.67	6.68	0.01	6.67	13.36	0.02	13.34
0	5	2.5	2	1.5	33.33	86.64	82.22	4.42	6.68	0	6.68	6.68	0	6.68	13.36	0	13.36
0	5	3	1.67	1.5	33.33	86.64	82.24	4.4	6.68	0	6.68	6.68	0	6.68	13.36	0	13.36
0	7.5	1	7.5	1.5	33.33	86.64	78.23	8.41	6.68	0.21	6.47	6.68	0.21	6.47	13.36	0.42	12.94
0	7.5	1.5	5	1.5	33.33	86.64	79.16	7.48	6.68	0.05	6.63	6.68	0.05	6.63	13.36	0.11	13.25
0	7.5	2	3.75	1.5	33.33	86.64	79.46	7.18	6.68	0.01	6.67	6.68	0.01	6.67	13.36	0.03	13.33
0	7.5	2.5	3	1.5	33.33	86.64	79.58	7.06	6.68	0	6.68	6.68	0	6.68	13.36	0.01	13.36
0	7.5	3	2.5	1.5	33.33	86.64	79.64	7	6.68	0	6.68	6.68	0	6.68	13.36	0	13.36
0	10	1	10	1.5	33.33	86.64	74.44	12.2	6.68	0.26	6.42	6.68	0.26	6.42	13.36	0.53	12.83
0	10	1.5	6.67	1.5	33.33	86.64	75.93	10.71	6.68	0.07	6.61	6.68	0.07	6.61	13.36	0.14	13.22
0	10	2	5	1.5	33.33	86.64	76.43	10.21	6.68	0.02	6.66	6.68	0.02	6.66	13.36	0.03	13.33
0	10	2.5	4	1.5	33.33	86.64	76.64	9.99	6.68	0	6.68	6.68	0	6.68	13.36	0.01	13.35
0	10	3	3.33	1.5	33.33	86.64	76.75	9.89	6.68	0	6.68	6.68	0	6.68	13.36	0	13.36
0	15	1	15	1.5	33.33	86.64	65.47	21.16	6.68	0.36	6.32	6.68	0.36	6.32	13.36	0.72	12.64
0	15	1.5	10	1.5	33.33	86.64	68.35	18.29	6.68	0.1	6.58	6.68	0.1	6.58	13.36	0.2	13.16
0	15	2	7.5	1.5	33.33	86.64	69.39	17.25	6.68	0.03	6.66	6.68	0.03	6.66	13.36	0.05	13.31
0	15	2.5	6	1.5	33.33	86.64	69.85	16.79	6.68	0.01	6.68	6.68	0.01	6.68	13.36	0.01	13.35
0	15	3	5	1.5	33.33	86.64	70.09	16.55	6.68	0	6.68	6.68	0	6.68	13.36	0	13.36
0	20	1	20	1.5	33.33	86.64	55.17	31.47	6.68	0.43	6.25	6.68	0.43	6.25	13.36	0.87	12.49
0	20	1.5	13.33	1.5	33.33	86.64	59.43	27.21	6.68	0.13	6.56	6.68	0.13	6.56	13.36	0.25	13.11
0	20	2	10	1.5	33.33	86.64	61.07	25.56	6.68	0.03	6.65	6.68	0.03	6.65	13.36	0.06	13.3
0	20	2.5	8	1.5	33.33	86.64	61.84	24.8	6.68	0.01	6.67	6.68	0.01	6.67	13.36	0.01	13.35
0	20	3	6.67	1.5	33.33	86.64	62.25	24.39	6.68	0	6.68	6.68	0	6.68	13.36	0	13.36
0	25	1	25	1.5	33.33	86.64	44.28	42.36	6.68	0.49	6.19	6.68	0.49	6.19	13.36	0.99	12.38
0	25	1.5	16.67	1.5	33.33	86.64	49.51	37.13	6.68	0.15	6.53	6.68	0.15	6.53	13.36	0.3	13.07
0	25	2	12.5	1.5	33.33	86.64	51.68	34.96	6.68	0.04	6.64	6.68	0.04	6.64	13.36	0.08	13.28
0	25	2.5	10	1.5	33.33	86.64	52.74	33.9	6.68	0.01	6.67	6.68	0.01	6.67	13.36	0.02	13.34
0	25	3	8.33	1.5	33.33	86.64	53.31	33.32	6.68	0	6.68	6.68	0	6.68	13.36	0	13.36
0	2.5	1	2.5	1	50	68.27	65.53	2.74	15.87	0.1	15.77	15.87	0.1	15.77	31.73	0.2	31.53
0	2.5	1.5	1.67	1	50	68.27	65.71	2.56	15.87	0.02	15.84	15.87	0.02	15.84	31.73	0.05	31.68
0	2.5	2	1.25	1	50	68.27	65.76	2.51	15.87	0.01	15.86	15.87	0.01	15.86	31.73	0.01	31.72
0	2.5	2.5	1	1	50	68.27	65.78	2.49	15.87	0	15.86	15.87	0	15.86	31.73	0	31.73
0	2.5	3	0.83	1	50	68.27	65.78	2.49	15.87	0	15.87	15.87	0	15.87	31.73	0	31.73
0	5	1	5	1	50	68.27	62.56	5.71	15.87	0.19	15.67	15.87	0.19	15.67	31.73	0.39	31.34
0	5	1.5	3.33	1	50	68.27	62.98	5.28	15.87	0.05	15.82	15.87	0.05	15.82	31.73	0.1	31.63
0	5	2	2.5	1	50	68.27	63.11	5.16	15.87	0.01	15.85	15.87	0.01	15.85	31.73	0.02	31.71
0	5	2.5	2	1	50	68.27	63.14	5.12	15.87	0	15.86	15.87	0	15.86	31.73	0.01	31.73
0	5	3	1.67	1	50	68.27	63.16	5.11	15.87	0	15.87	15.87	0	15.87	31.73	0	31.73
0	7.5	1	7.5	1	50	68.27	59.38	8.89	15.87	0.28	15.58	15.87	0.28	15.58	31.73	0.57	31.17
0	7.5	1.5	5	1	50	68.27	60.09	8.18	15.87	0.07	15.79	15.87	0.07	15.79	31.73	0.14	31.59
0	7.5	2	3.75	1	50	68.27	60.3	7.97	15.87	0.02	15.85	15.87	0.02	15.85	31.73	0.03	31.7
0	7.5	2.5	3	1	50	68.27	60.38	7.89	15.87	0	15.86	15.87	0	15.86	31.73	0.01	31.72
0	7.5	3	2.5	1	50	68.27	60.41	7.86	15.87	0	15.86	15.87	0	15.86	31.73	0	31.73
0	10	1	10	1	50	68.27	55.99	12.28	15.87	0.37	15.5	15.87	0.37	15.5	31.73	0.73	31
0	10	1.5	6.67	1	50	68.27	57.03	11.24	15.87	0.09	15.77	15.87	0.09	15.77	31.73	0.19	31.54
0	10	2	5	1	50	68.27	57.35	10.92	15.87	0.02	15.84	15.87	0.02	15.84	31.73	0.05	31.69
0	10	2.5	4	1	50	68.27	57.47	10.8	15.87	0	15.86	15.87	0	15.86	31.73	0.01	31.72
0	10	3	3.33	1	50	68.27	57.52	10.74	15.87	0	15.86	15.87	0	15.86	31.73	0	31.73

a	$i =$ $k_m s_m$	k_m	s_m	k_p	s_p	C	C_A	C_R	U	U_A	U_R	O	O_A	O_R	W	W_A	W_R
0	15	1	15	1	50	68.27	48.7	19.57	15.87	0.52	15.34	15.87	0.52	15.34	31.73	1.05	30.68
0	15	1.5	10	1	50	68.27	50.48	17.79	15.87	0.14	15.73	15.87	0.14	15.73	31.73	0.27	31.46
0	15	2	7.5	1	50	68.27	51.06	17.21	15.87	0.03	15.83	15.87	0.03	15.83	31.73	0.07	31.66
0	15	2.5	6	1	50	68.27	51.28	16.99	15.87	0.01	15.86	15.87	0.01	15.86	31.73	0.01	31.72
0	15	3	5	1	50	68.27	51.39	16.88	15.87	0	15.86	15.87	0	15.86	31.73	0	31.73
0	20	1	20	1	50	68.27	40.94	27.33	15.87	0.66	15.2	15.87	0.66	15.2	31.73	1.33	30.4
0	20	1.5	13.33	1	50	68.27	43.44	24.83	15.87	0.18	15.69	15.87	0.18	15.69	31.73	0.35	31.38
0	20	2	10	1	50	68.27	44.28	23.98	15.87	0.04	15.82	15.87	0.04	15.82	31.73	0.09	31.64
0	20	2.5	8	1	50	68.27	44.63	23.64	15.87	0.01	15.86	15.87	0.01	15.86	31.73	0.02	31.71
0	20	3	6.67	1	50	68.27	44.79	23.47	15.87	0	15.86	15.87	0	15.86	31.73	0	31.73
0	25	1	25	1	50	68.27	32.99	35.28	15.87	0.78	15.08	15.87	0.78	15.08	31.73	1.56	30.17
0	25	1.5	16.67	1	50	68.27	36.05	32.22	15.87	0.22	15.65	15.87	0.22	15.65	31.73	0.43	31.3
0	25	2	12.5	1	50	68.27	37.13	31.14	15.87	0.05	15.81	15.87	0.05	15.81	31.73	0.11	31.62
0	25	2.5	10	1	50	68.27	37.58	30.68	15.87	0.01	15.85	15.87	0.01	15.85	31.73	0.02	31.71
0	25	3	8.33	1	50	68.27	37.81	30.46	15.87	0	15.86	15.87	0	15.86	31.73	0	31.73

a	$i=$ $k_m s_m$	k_m	s_m	k_p	s_p	C	C_A	C_R	U	U_A	U_R	O	O_A	O_R	W	W_A	W_R
5	2.5	1	2.5	3	16.67	99.6	99.31	0.29	0.05	0	0.05	0.35	0.01	0.34	0.4	0.01	0.38
5	2.5	1.5	1.67	3	16.67	99.6	99.35	0.25	0.05	0	0.05	0.35	0	0.34	0.4	0	0.39
5	2.5	2	1.25	3	16.67	99.6	99.37	0.24	0.05	0	0.05	0.35	0	0.35	0.4	0	0.39
5	2.5	2.5	1	3	16.67	99.6	99.37	0.23	0.05	0	0.05	0.35	0	0.35	0.4	0	0.39
5	2.5	3	0.83	3	16.67	99.6	99.37	0.23	0.05	0	0.05	0.35	0	0.35	0.4	0	0.39
5	5	1	5	3	16.67	99.6	98.7	0.91	0.05	0	0.05	0.35	0.02	0.33	0.4	0.02	0.37
5	5	1.5	3.33	3	16.67	99.6	98.9	0.7	0.05	0	0.05	0.35	0.01	0.34	0.4	0.01	0.39
5	5	2	2.5	3	16.67	99.6	98.97	0.64	0.05	0	0.05	0.35	0	0.35	0.4	0	0.39
5	5	2.5	2	3	16.67	99.6	99	0.61	0.05	0	0.05	0.35	0	0.35	0.4	0	0.39
5	5	3	1.67	3	16.67	99.6	99.01	0.59	0.05	0	0.05	0.35	0	0.35	0.4	0	0.39
5	7.5	1	7.5	3	16.67	99.6	97.49	2.12	0.05	0	0.04	0.35	0.02	0.32	0.4	0.03	0.37
5	7.5	1.5	5	3	16.67	99.6	98.12	1.49	0.05	0	0.05	0.35	0.01	0.34	0.4	0.01	0.39
5	7.5	2	3.75	3	16.67	99.6	98.32	1.29	0.05	0	0.05	0.35	0	0.34	0.4	0	0.39
5	7.5	2.5	3	3	16.67	99.6	98.41	1.2	0.05	0	0.05	0.35	0	0.35	0.4	0	0.39
5	7.5	3	2.5	3	16.67	99.6	98.45	1.15	0.05	0	0.05	0.35	0	0.35	0.4	0	0.39
5	10	1	10	3	16.67	99.6	95.34	4.27	0.05	0	0.04	0.35	0.03	0.32	0.4	0.03	0.36
5	10	1.5	6.67	3	16.67	99.6	96.82	2.79	0.05	0	0.05	0.35	0.01	0.34	0.4	0.01	0.39
5	10	2	5	3	16.67	99.6	97.3	2.31	0.05	0	0.05	0.35	0	0.34	0.4	0	0.39
5	10	2.5	4	3	16.67	99.6	97.51	2.1	0.05	0	0.05	0.35	0	0.35	0.4	0	0.39
5	10	3	3.33	3	16.67	99.6	97.62	1.98	0.05	0	0.05	0.35	0	0.35	0.4	0	0.39
5	15	1	15	3	16.67	99.6	87.22	12.38	0.05	0.01	0.04	0.35	0.03	0.31	0.4	0.04	0.36
5	15	1.5	10	3	16.67	99.6	91.87	7.73	0.05	0	0.05	0.35	0.01	0.34	0.4	0.01	0.38
5	15	2	7.5	3	16.67	99.6	93.53	6.08	0.05	0	0.05	0.35	0	0.34	0.4	0	0.39
5	15	2.5	6	3	16.67	99.6	94.28	5.32	0.05	0	0.05	0.35	0	0.35	0.4	0	0.39
5	15	3	5	3	16.67	99.6	94.69	4.92	0.05	0	0.05	0.35	0	0.35	0.4	0	0.39
5	20	1	20	3	16.67	99.6	74.38	25.22	0.05	0.01	0.04	0.35	0.04	0.31	0.4	0.04	0.35
5	20	1.5	13.33	3	16.67	99.6	82.91	16.7	0.05	0	0.05	0.35	0.01	0.33	0.4	0.01	0.38
5	20	2	10	3	16.67	99.6	86.5	13.1	0.05	0	0.05	0.35	0	0.34	0.4	0	0.39
5	20	2.5	8	3	16.67	99.6	88.27	11.33	0.05	0	0.05	0.35	0	0.35	0.4	0	0.39
5	20	3	6.67	3	16.67	99.6	89.26	10.35	0.05	0	0.05	0.35	0	0.35	0.4	0	0.39
5	25	1	25	3	16.67	99.6	59.2	40.4	0.05	0.01	0.04	0.35	0.04	0.31	0.4	0.05	0.35
5	25	1.5	16.67	3	16.67	99.6	70.19	29.41	0.05	0	0.05	0.35	0.01	0.33	0.4	0.02	0.38
5	25	2	12.5	3	16.67	99.6	75.72	23.89	0.05	0	0.05	0.35	0	0.34	0.4	0	0.39
5	25	2.5	10	3	16.67	99.6	78.72	20.88	0.05	0	0.05	0.35	0	0.35	0.4	0	0.39
5	25	3	8.33	3	16.67	99.6	80.49	19.12	0.05	0	0.05	0.35	0	0.35	0.4	0	0.39
5	2.5	1	2.5	2.5	20	98.48	97.75	0.73	0.3	0.01	0.29	1.22	0.03	1.19	1.52	0.04	1.48
5	2.5	1.5	1.67	2.5	20	98.48	97.84	0.64	0.3	0	0.3	1.22	0.01	1.21	1.52	0.01	1.51
5	2.5	2	1.25	2.5	20	98.48	97.86	0.62	0.3	0	0.3	1.22	0	1.22	1.52	0	1.52
5	2.5	2.5	1	2.5	20	98.48	97.87	0.61	0.3	0	0.3	1.22	0	1.22	1.52	0	1.52
5	2.5	3	0.83	2.5	20	98.48	97.88	0.6	0.3	0	0.3	1.22	0	1.22	1.52	0	1.52
5	5	1	5	2.5	20	98.48	96.55	1.93	0.3	0.01	0.28	1.22	0.05	1.17	1.52	0.07	1.45
5	5	1.5	3.33	2.5	20	98.48	96.87	1.61	0.3	0	0.29	1.22	0.01	1.21	1.52	0.02	1.5
5	5	2	2.5	2.5	20	98.48	96.98	1.5	0.3	0	0.3	1.22	0	1.22	1.52	0	1.52
5	5	2.5	2	2.5	20	98.48	97.03	1.45	0.3	0	0.3	1.22	0	1.22	1.52	0	1.52
5	5	3	1.67	2.5	20	98.48	97.05	1.43	0.3	0	0.3	1.22	0	1.22	1.52	0	1.52
5	7.5	1	7.5	2.5	20	98.48	94.64	3.84	0.3	0.02	0.28	1.22	0.07	1.15	1.52	0.09	1.43
5	7.5	1.5	5	2.5	20	98.48	95.47	3.01	0.3	0.01	0.29	1.22	0.02	1.2	1.52	0.03	1.5
5	7.5	2	3.75	2.5	20	98.48	95.75	2.73	0.3	0	0.3	1.22	0	1.22	1.52	0.01	1.51
5	7.5	2.5	3	2.5	20	98.48	95.87	2.61	0.3	0	0.3	1.22	0	1.22	1.52	0	1.52
5	7.5	3	2.5	2.5	20	98.48	95.94	2.54	0.3	0	0.3	1.22	0	1.22	1.52	0	1.52

a	$i=$ $k_m s_m$	k_m	s_m	k_p	s_p	C	C_A	C_R	U	U_A	U_R	O	O_A	O_R	W	W_A	W_R
5	10	1	10	2.5	20	98.48	91.81	6.67	0.3	0.02	0.28	1.22	0.08	1.14	1.52	0.11	1.41
5	10	1.5	6.67	2.5	20	98.48	93.48	5	0.3	0.01	0.29	1.22	0.02	1.2	1.52	0.03	1.49
5	10	2	5	2.5	20	98.48	94.06	4.42	0.3	0	0.3	1.22	0.01	1.22	1.52	0.01	1.51
5	10	2.5	4	2.5	20	98.48	94.32	4.16	0.3	0	0.3	1.22	0	1.22	1.52	0	1.52
5	10	3	3.33	2.5	20	98.48	94.46	4.02	0.3	0	0.3	1.22	0	1.22	1.52	0	1.52
5	15	1	15	2.5	20	98.48	82.92	15.56	0.3	0.03	0.27	1.22	0.1	1.12	1.52	0.13	1.39
5	15	1.5	10	2.5	20	98.48	87.3	11.18	0.3	0.01	0.29	1.22	0.03	1.19	1.52	0.04	1.48
5	15	2	7.5	2.5	20	98.48	88.93	9.55	0.3	0	0.3	1.22	0.01	1.21	1.52	0.01	1.51
5	15	2.5	6	2.5	20	98.48	89.69	8.79	0.3	0	0.3	1.22	0	1.22	1.52	0	1.52
5	15	3	5	2.5	20	98.48	90.1	8.38	0.3	0	0.3	1.22	0	1.22	1.52	0	1.52
5	20	1	20	2.5	20	98.48	70.39	28.09	0.3	0.03	0.27	1.22	0.12	1.1	1.52	0.15	1.37
5	20	1.5	13.33	2.5	20	98.48	77.81	20.67	0.3	0.01	0.29	1.22	0.04	1.18	1.52	0.05	1.47
5	20	2	10	2.5	20	98.48	80.95	17.53	0.3	0	0.3	1.22	0.01	1.21	1.52	0.01	1.51
5	20	2.5	8	2.5	20	98.48	82.5	15.98	0.3	0	0.3	1.22	0	1.22	1.52	0	1.52
5	20	3	6.67	2.5	20	98.48	83.37	15.1	0.3	0	0.3	1.22	0	1.22	1.52	0	1.52
5	25	1	25	2.5	20	98.48	56.13	42.35	0.3	0.03	0.27	1.22	0.13	1.1	1.52	0.16	1.36
5	25	1.5	16.67	2.5	20	98.48	65.5	32.98	0.3	0.01	0.29	1.22	0.04	1.18	1.52	0.05	1.47
5	25	2	12.5	2.5	20	98.48	70.05	28.43	0.3	0	0.29	1.22	0.01	1.21	1.52	0.02	1.5
5	25	2.5	10	2.5	20	98.48	72.48	26	0.3	0	0.3	1.22	0	1.22	1.52	0	1.52
5	25	3	8.33	2.5	20	98.48	73.9	24.57	0.3	0	0.3	1.22	0	1.22	1.52	0	1.52
5	2.5	1	2.5	2	25	95.02	93.54	1.47	1.39	0.03	1.36	3.59	0.06	3.53	4.98	0.09	4.9
5	2.5	1.5	1.67	2	25	95.02	93.68	1.34	1.39	0.01	1.38	3.59	0.02	3.58	4.98	0.02	4.96
5	2.5	2	1.25	2	25	95.02	93.72	1.3	1.39	0	1.39	3.59	0	3.59	4.98	0.01	4.98
5	2.5	2.5	1	2	25	95.02	93.74	1.28	1.39	0	1.39	3.59	0	3.59	4.98	0	4.98
5	2.5	3	0.83	2	25	95.02	93.74	1.27	1.39	0	1.39	3.59	0	3.59	4.98	0	4.98
5	5	1	5	2	25	95.02	91.51	3.5	1.39	0.05	1.34	3.59	0.11	3.48	4.98	0.16	4.82
5	5	1.5	3.33	2	25	95.02	91.95	3.07	1.39	0.01	1.38	3.59	0.03	3.56	4.98	0.04	4.94
5	5	2	2.5	2	25	95.02	92.09	2.92	1.39	0	1.39	3.59	0.01	3.59	4.98	0.01	4.97
5	5	2.5	2	2	25	95.02	92.15	2.87	1.39	0	1.39	3.59	0	3.59	4.98	0	4.98
5	5	3	1.67	2	25	95.02	92.18	2.84	1.39	0	1.39	3.59	0	3.59	4.98	0	4.98
5	7.5	1	7.5	2	25	95.02	88.8	6.22	1.39	0.07	1.32	3.59	0.16	3.44	4.98	0.22	4.76
5	7.5	1.5	5	2	25	95.02	89.75	5.27	1.39	0.02	1.37	3.59	0.04	3.55	4.98	0.06	4.92
5	7.5	2	3.75	2	25	95.02	90.07	4.94	1.39	0	1.39	3.59	0.01	3.58	4.98	0.02	4.97
5	7.5	2.5	3	2	25	95.02	90.21	4.8	1.39	0	1.39	3.59	0	3.59	4.98	0	4.98
5	7.5	3	2.5	2	25	95.02	90.29	4.73	1.39	0	1.39	3.59	0	3.59	4.98	0	4.98
5	10	1	10	2	25	95.02	85.32	9.7	1.39	0.08	1.31	3.59	0.19	3.4	4.98	0.27	4.71
5	10	1.5	6.67	2	25	95.02	87.02	8	1.39	0.02	1.37	3.59	0.05	3.54	4.98	0.08	4.91
5	10	2	5	2	25	95.02	87.61	7.4	1.39	0.01	1.38	3.59	0.01	3.58	4.98	0.02	4.96
5	10	2.5	4	2	25	95.02	87.88	7.14	1.39	0	1.39	3.59	0	3.59	4.98	0	4.98
5	10	3	3.33	2	25	95.02	88.02	7	1.39	0	1.39	3.59	0	3.59	4.98	0	4.98
5	15	1	15	2	25	95.02	76.01	19.01	1.39	0.1	1.29	3.59	0.25	3.34	4.98	0.35	4.63
5	15	1.5	10	2	25	95.02	79.77	15.24	1.39	0.03	1.36	3.59	0.07	3.52	4.98	0.1	4.88
5	15	2	7.5	2	25	95.02	81.19	13.83	1.39	0.01	1.38	3.59	0.02	3.57	4.98	0.03	4.96
5	15	2.5	6	2	25	95.02	81.84	13.17	1.39	0	1.39	3.59	0	3.59	4.98	0.01	4.98
5	15	3	5	2	25	95.02	82.2	12.82	1.39	0	1.39	3.59	0	3.59	4.98	0	4.98
5	20	1	20	2	25	95.02	64.25	30.77	1.39	0.12	1.27	3.59	0.29	3.3	4.98	0.41	4.57
5	20	1.5	13.33	2	25	95.02	70.19	24.83	1.39	0.04	1.35	3.59	0.09	3.5	4.98	0.13	4.86
5	20	2	10	2	25	95.02	72.64	22.38	1.39	0.01	1.38	3.59	0.02	3.57	4.98	0.03	4.95
5	20	2.5	8	2	25	95.02	73.83	21.18	1.39	0	1.39	3.59	0.01	3.59	4.98	0.01	4.98
5	20	3	6.67	2	25	95.02	74.5	20.52	1.39	0	1.39	3.59	0	3.59	4.98	0	4.98
5	25	1	25	2	25	95.02	51.38	43.64	1.39	0.13	1.26	3.59	0.32	3.27	4.98	0.46	4.53
5	25	1.5	16.67	2	25	95.02	58.75	36.27	1.39	0.04	1.35	3.59	0.1	3.49	4.98	0.15	4.84
5	25	2	12.5	2	25	95.02	62.12	32.9	1.39	0.01	1.38	3.59	0.03	3.56	4.98	0.04	4.94
5	25	2.5	10	2	25	95.02	63.86	31.15	1.39	0	1.39	3.59	0.01	3.59	4.98	0.01	4.97
5	25	3	8.33	2	25	95.02	64.86	30.15	1.39	0	1.39	3.59	0	3.59	4.98	0	4.98

a	$i=$ $k_m s_m$	k_m	s_m	k_p	s_p	C	C_A	C_R	U	U_A	U_R	O	O_A	O_R	W	W_A	W_R
5	2.5	1	2.5	1.5	33.33	86.2	83.85	2.35	4.95	0.06	4.89	8.85	0.1	8.75	13.8	0.16	13.64
5	2.5	1.5	1.67	1.5	33.33	86.2	84.03	2.17	4.95	0.02	4.93	8.85	0.02	8.83	13.8	0.04	13.76
5	2.5	2	1.25	1.5	33.33	86.2	84.08	2.12	4.95	0	4.94	8.85	0.01	8.85	13.8	0.01	13.79
5	2.5	2.5	1	1.5	33.33	86.2	84.1	2.1	4.95	0	4.95	8.85	0	8.85	13.8	0	13.8
5	2.5	3	0.83	1.5	33.33	86.2	84.11	2.09	4.95	0	4.95	8.85	0	8.85	13.8	0	13.8
5	5	1	5	1.5	33.33	86.2	81.04	5.16	4.95	0.11	4.83	8.85	0.18	8.67	13.8	0.3	13.5
5	5	1.5	3.33	1.5	33.33	86.2	81.52	4.68	4.95	0.03	4.92	8.85	0.05	8.8	13.8	0.08	13.72
5	5	2	2.5	1.5	33.33	86.2	81.67	4.53	4.95	0.01	4.94	8.85	0.01	8.84	13.8	0.02	13.78
5	5	2.5	2	1.5	33.33	86.2	81.73	4.47	4.95	0	4.95	8.85	0	8.85	13.8	0	13.79
5	5	3	1.67	1.5	33.33	86.2	81.76	4.44	4.95	0	4.95	8.85	0	8.85	13.8	0	13.8
5	7.5	1	7.5	1.5	33.33	86.2	77.73	8.47	4.95	0.16	4.78	8.85	0.26	8.59	13.8	0.42	13.37
5	7.5	1.5	5	1.5	33.33	86.2	78.66	7.55	4.95	0.04	4.9	8.85	0.07	8.78	13.8	0.11	13.69
5	7.5	2	3.75	1.5	33.33	86.2	78.96	7.25	4.95	0.01	4.94	8.85	0.02	8.83	13.8	0.03	13.77
5	7.5	2.5	3	1.5	33.33	86.2	79.08	7.12	4.95	0	4.94	8.85	0	8.85	13.8	0.01	13.79
5	7.5	3	2.5	1.5	33.33	86.2	79.14	7.07	4.95	0	4.95	8.85	0	8.85	13.8	0	13.8
5	10	1	10	1.5	33.33	86.2	73.94	12.26	4.95	0.2	4.74	8.85	0.33	8.52	13.8	0.54	13.26
5	10	1.5	6.67	1.5	33.33	86.2	75.42	10.78	4.95	0.06	4.89	8.85	0.09	8.76	13.8	0.14	13.65
5	10	2	5	1.5	33.33	86.2	75.92	10.28	4.95	0.01	4.93	8.85	0.02	8.83	13.8	0.04	13.76
5	10	2.5	4	1.5	33.33	86.2	76.13	10.07	4.95	0	4.94	8.85	0	8.85	13.8	0.01	13.79
5	10	3	3.33	1.5	33.33	86.2	76.23	9.97	4.95	0	4.95	8.85	0	8.85	13.8	0	13.8
5	15	1	15	1.5	33.33	86.2	65.01	21.19	4.95	0.28	4.67	8.85	0.45	8.4	13.8	0.73	13.07
5	15	1.5	10	1.5	33.33	86.2	67.85	18.35	4.95	0.08	4.87	8.85	0.13	8.73	13.8	0.2	13.6
5	15	2	7.5	1.5	33.33	86.2	68.87	17.33	4.95	0.02	4.93	8.85	0.03	8.82	13.8	0.05	13.75
5	15	2.5	6	1.5	33.33	86.2	69.32	16.88	4.95	0	4.94	8.85	0.01	8.84	13.8	0.01	13.79
5	15	3	5	1.5	33.33	86.2	69.56	16.64	4.95	0	4.95	8.85	0	8.85	13.8	0	13.8
5	20	1	20	1.5	33.33	86.2	54.77	31.43	4.95	0.33	4.61	8.85	0.55	8.3	13.8	0.89	12.91
5	20	1.5	13.33	1.5	33.33	86.2	58.97	27.23	4.95	0.1	4.85	8.85	0.16	8.69	13.8	0.26	13.54
5	20	2	10	1.5	33.33	86.2	60.59	25.62	4.95	0.02	4.92	8.85	0.04	8.81	13.8	0.07	13.73
5	20	2.5	8	1.5	33.33	86.2	61.33	24.87	4.95	0.01	4.94	8.85	0.01	8.84	13.8	0.01	13.78
5	20	3	6.67	1.5	33.33	86.2	61.73	24.47	4.95	0	4.95	8.85	0	8.85	13.8	0	13.8
5	25	1	25	1.5	33.33	86.2	43.96	42.24	4.95	0.38	4.57	8.85	0.63	8.22	13.8	1.01	12.79
5	25	1.5	16.67	1.5	33.33	86.2	49.11	37.09	4.95	0.11	4.83	8.85	0.19	8.66	13.8	0.3	13.5
5	25	2	12.5	1.5	33.33	86.2	51.25	34.95	4.95	0.03	4.92	8.85	0.05	8.8	13.8	0.08	13.72
5	25	2.5	10	1.5	33.33	86.2	52.28	33.92	4.95	0.01	4.94	8.85	0.01	8.84	13.8	0.02	13.78
5	25	3	8.33	1.5	33.33	86.2	52.84	33.36	4.95	0	4.95	8.85	0	8.85	13.8	0	13.79
5	2.5	1	2.5	1	50	68.03	65.29	2.74	13.57	0.09	13.48	18.41	0.11	18.3	31.97	0.2	31.77
5	2.5	1.5	1.67	1	50	68.03	65.47	2.56	13.57	0.02	13.54	18.41	0.03	18.38	31.97	0.05	31.92
5	2.5	2	1.25	1	50	68.03	65.52	2.51	13.57	0.01	13.56	18.41	0.01	18.4	31.97	0.01	31.96
5	2.5	2.5	1	1	50	68.03	65.54	2.49	13.57	0	13.57	18.41	0	18.4	31.97	0	31.97
5	2.5	3	0.83	1	50	68.03	65.54	2.49	13.57	0	13.57	18.41	0	18.41	31.97	0	31.97
5	5	1	5	1	50	68.03	62.33	5.7	13.57	0.17	13.39	18.41	0.21	18.19	31.97	0.39	31.59
5	5	1.5	3.33	1	50	68.03	62.75	5.28	13.57	0.04	13.52	18.41	0.05	18.35	31.97	0.1	31.88
5	5	2	2.5	1	50	68.03	62.87	5.16	13.57	0.01	13.56	18.41	0.01	18.39	31.97	0.02	31.95
5	5	2.5	2	1	50	68.03	62.91	5.12	13.57	0	13.56	18.41	0	18.4	31.97	0.01	31.97
5	5	3	1.67	1	50	68.03	62.92	5.11	13.57	0	13.57	18.41	0	18.41	31.97	0	31.97
5	7.5	1	7.5	1	50	68.03	59.15	8.88	13.57	0.25	13.31	18.41	0.31	18.09	31.97	0.57	31.41
5	7.5	1.5	5	1	50	68.03	59.86	8.17	13.57	0.06	13.5	18.41	0.08	18.33	31.97	0.14	31.83
5	7.5	2	3.75	1	50	68.03	60.07	7.96	13.57	0.02	13.55	18.41	0.02	18.39	31.97	0.03	31.94
5	7.5	2.5	3	1	50	68.03	60.14	7.89	13.57	0	13.56	18.41	0	18.4	31.97	0.01	31.97
5	7.5	3	2.5	1	50	68.03	60.17	7.86	13.57	0	13.57	18.41	0	18.41	31.97	0	31.97
5	10	1	10	1	50	68.03	55.77	12.26	13.57	0.33	13.24	18.41	0.41	18	31.97	0.74	31.24
5	10	1.5	6.67	1	50	68.03	56.81	11.22	13.57	0.08	13.48	18.41	0.1	18.3	31.97	0.19	31.78
5	10	2	5	1	50	68.03	57.12	10.9	13.57	0.02	13.55	18.41	0.03	18.38	31.97	0.05	31.93
5	10	2.5	4	1	50	68.03	57.24	10.79	13.57	0	13.56	18.41	0.01	18.4	31.97	0.01	31.96
5	10	3	3.33	1	50	68.03	57.29	10.73	13.57	0	13.57	18.41	0	18.4	31.97	0	31.97

a	$k_m s_m$	k_m	s_m	k_p	s_p	C	C_A	C_R	U	U_A	U_R	O	O_A	O_R	W	W_A	W_R
5	15	1	15	1	50	68.03	48.51	19.52	13.57	0.47	13.1	18.41	0.58	17.82	31.97	1.05	30.92
5	15	1.5	10	1	50	68.03	50.27	17.75	13.57	0.12	13.44	18.41	0.15	18.25	31.97	0.27	31.7
5	15	2	7.5	1	50	68.03	50.84	17.18	13.57	0.03	13.54	18.41	0.04	18.37	31.97	0.07	31.91
5	15	2.5	6	1	50	68.03	51.07	16.96	13.57	0.01	13.56	18.41	0.01	18.4	31.97	0.01	31.96
5	15	3	5	1	50	68.03	51.17	16.86	13.57	0	13.57	18.41	0	18.4	31.97	0	31.97
5	20	1	20	1	50	68.03	40.77	27.25	13.57	0.59	12.98	18.41	0.74	17.66	31.97	1.33	30.64
5	20	1.5	13.33	1	50	68.03	43.26	24.77	13.57	0.16	13.41	18.41	0.2	18.21	31.97	0.36	31.62
5	20	2	10	1	50	68.03	44.09	23.93	13.57	0.04	13.53	18.41	0.05	18.36	31.97	0.09	31.89
5	20	2.5	8	1	50	68.03	44.43	23.59	13.57	0.01	13.56	18.41	0.01	18.4	31.97	0.02	31.95
5	20	3	6.67	1	50	68.03	44.6	23.43	13.57	0	13.56	18.41	0	18.4	31.97	0	31.97
5	25	1	25	1	50	68.03	32.86	35.17	13.57	0.69	12.88	18.41	0.88	17.53	31.97	1.57	30.4
5	25	1.5	16.67	1	50	68.03	35.9	32.13	13.57	0.19	13.38	18.41	0.24	18.17	31.97	0.43	31.54
5	25	2	12.5	1	50	68.03	36.97	31.06	13.57	0.05	13.52	18.41	0.06	18.35	31.97	0.11	31.87
5	25	2.5	10	1	50	68.03	37.42	30.61	13.57	0.01	13.56	18.41	0.01	18.39	31.97	0.02	31.95
5	25	3	8.33	1	50	68.03	37.64	30.39	13.57	0	13.56	18.41	0	18.4	31.97	0	31.97

a	$k_m s_m$	k_m	s_m	k_p	s_p	C	C_A	C_R	U	U_A	U_R	O	O_A	O_R	W	W_A	W_R
10	2.5	1	2.5	3	16.67	99.16	98.64	0.52	0.02	0	0.02	0.82	0.02	0.8	0.84	0.02	0.81
10	2.5	1.5	1.67	3	16.67	99.16	98.71	0.46	0.02	0	0.02	0.82	0.01	0.81	0.84	0.01	0.83
10	2.5	2	1.25	3	16.67	99.16	98.73	0.44	0.02	0	0.02	0.82	0	0.82	0.84	0	0.83
10	2.5	2.5	1	3	16.67	99.16	98.74	0.43	0.02	0	0.02	0.82	0	0.82	0.84	0	0.84
10	2.5	3	0.83	3	16.67	99.16	98.74	0.42	0.02	0	0.02	0.82	0	0.82	0.84	0	0.84
10	5	1	5	3	16.67	99.16	97.66	1.5	0.02	0	0.01	0.82	0.04	0.78	0.84	0.04	0.79
10	5	1.5	3.33	3	16.67	99.16	97.95	1.21	0.02	0	0.02	0.82	0.01	0.81	0.84	0.01	0.82
10	5	2	2.5	3	16.67	99.16	98.05	1.11	0.02	0	0.02	0.82	0	0.82	0.84	0	0.83
10	5	2.5	2	3	16.67	99.16	98.09	1.07	0.02	0	0.02	0.82	0	0.82	0.84	0	0.84
10	5	3	1.67	3	16.67	99.16	98.12	1.05	0.02	0	0.02	0.82	0	0.82	0.84	0	0.84
10	7.5	1	7.5	3	16.67	99.16	95.97	3.19	0.02	0	0.01	0.82	0.05	0.77	0.84	0.06	0.78
10	7.5	1.5	5	3	16.67	99.16	96.77	2.4	0.02	0	0.02	0.82	0.02	0.8	0.84	0.02	0.82
10	7.5	2	3.75	3	16.67	99.16	97.03	2.13	0.02	0	0.02	0.82	0	0.82	0.84	0	0.83
10	7.5	2.5	3	3	16.67	99.16	97.15	2.01	0.02	0	0.02	0.82	0	0.82	0.84	0	0.83
10	7.5	3	2.5	3	16.67	99.16	97.22	1.95	0.02	0	0.02	0.82	0	0.82	0.84	0	0.84
10	10	1	10	3	16.67	99.16	93.29	5.87	0.02	0	0.01	0.82	0.06	0.76	0.84	0.07	0.77
10	10	1.5	6.67	3	16.67	99.16	94.98	4.19	0.02	0	0.02	0.82	0.02	0.8	0.84	0.02	0.82
10	10	2	5	3	16.67	99.16	95.56	3.61	0.02	0	0.02	0.82	0.01	0.81	0.84	0.01	0.83
10	10	2.5	4	3	16.67	99.16	95.82	3.35	0.02	0	0.02	0.82	0	0.82	0.84	0	0.83
10	10	3	3.33	3	16.67	99.16	95.96	3.21	0.02	0	0.02	0.82	0	0.82	0.84	0	0.84
10	15	1	15	3	16.67	99.16	84.49	14.68	0.02	0	0.01	0.82	0.08	0.74	0.84	0.08	0.76
10	15	1.5	10	3	16.67	99.16	89.03	10.13	0.02	0	0.02	0.82	0.02	0.79	0.84	0.03	0.81
10	15	2	7.5	3	16.67	99.16	90.74	8.43	0.02	0	0.02	0.82	0.01	0.81	0.84	0.01	0.83
10	15	2.5	6	3	16.67	99.16	91.54	7.63	0.02	0	0.02	0.82	0	0.82	0.84	0	0.83
10	15	3	5	3	16.67	99.16	91.98	7.19	0.02	0	0.02	0.82	0	0.82	0.84	0	0.84
10	20	1	20	3	16.67	99.16	71.76	27.4	0.02	0	0.01	0.82	0.09	0.73	0.84	0.09	0.75
10	20	1.5	13.33	3	16.67	99.16	79.54	19.62	0.02	0	0.02	0.82	0.03	0.79	0.84	0.03	0.81
10	20	2	10	3	16.67	99.16	82.86	16.31	0.02	0	0.02	0.82	0.01	0.81	0.84	0.01	0.83
10	20	2.5	8	3	16.67	99.16	84.51	14.65	0.02	0	0.02	0.82	0	0.82	0.84	0	0.83
10	20	3	6.67	3	16.67	99.16	85.45	13.71	0.02	0	0.02	0.82	0	0.82	0.84	0	0.84
10	25	1	25	3	16.67	99.16	57.19	41.98	0.02	0	0.01	0.82	0.09	0.73	0.84	0.09	0.74
10	25	1.5	16.67	3	16.67	99.16	67	32.16	0.02	0	0.02	0.82	0.03	0.79	0.84	0.03	0.8
10	25	2	12.5	3	16.67	99.16	71.83	27.34	0.02	0	0.02	0.82	0.01	0.81	0.84	0.01	0.83
10	25	2.5	10	3	16.67	99.16	74.42	24.74	0.02	0	0.02	0.82	0	0.82	0.84	0	0.83
10	25	3	8.33	3	16.67	99.16	75.96	23.21	0.02	0	0.02	0.82	0	0.82	0.84	0	0.84
10	2.5	1	2.5	2.5	20	97.59	96.59	1	0.13	0	0.13	2.28	0.05	2.22	2.41	0.05	2.36
10	2.5	1.5	1.67	2.5	20	97.59	96.69	0.9	0.13	0	0.13	2.28	0.01	2.26	2.41	0.01	2.4
10	2.5	2	1.25	2.5	20	97.59	96.73	0.86	0.13	0	0.13	2.28	0	2.27	2.41	0	2.41
10	2.5	2.5	1	2.5	20	97.59	96.74	0.85	0.13	0	0.13	2.28	0	2.27	2.41	0	2.41
10	2.5	3	0.83	2.5	20	97.59	96.75	0.84	0.13	0	0.13	2.28	0	2.27	2.41	0	2.41
10	5	1	5	2.5	20	97.59	95.04	2.55	0.13	0.01	0.13	2.28	0.09	2.18	2.41	0.1	2.31
10	5	1.5	3.33	2.5	20	97.59	95.42	2.17	0.13	0	0.13	2.28	0.02	2.25	2.41	0.03	2.38
10	5	2	2.5	2.5	20	97.59	95.55	2.04	0.13	0	0.13	2.28	0.01	2.27	2.41	0.01	2.4
10	5	2.5	2	2.5	20	97.59	95.61	1.98	0.13	0	0.13	2.28	0	2.27	2.41	0	2.41
10	5	3	1.67	2.5	20	97.59	95.63	1.96	0.13	0	0.13	2.28	0	2.27	2.41	0	2.41
10	7.5	1	7.5	2.5	20	97.59	92.76	4.83	0.13	0.01	0.13	2.28	0.12	2.15	2.41	0.13	2.28
10	7.5	1.5	5	2.5	20	97.59	93.67	3.92	0.13	0	0.13	2.28	0.03	2.24	2.41	0.04	2.37
10	7.5	2	3.75	2.5	20	97.59	93.99	3.6	0.13	0	0.13	2.28	0.01	2.27	2.41	0.01	2.4
10	7.5	2.5	3	2.5	20	97.59	94.12	3.47	0.13	0	0.13	2.28	0	2.27	2.41	0	2.41
10	7.5	3	2.5	2.5	20	97.59	94.2	3.39	0.13	0	0.13	2.28	0	2.27	2.41	0	2.41

a	$i=$ $k_m s_m$	k_m	s_m	k_p	s_p	C	C_A	C_R	U	U_A	U_R	O	O_A	O_R	W	W_A	W_R
10	10	1	10	2.5	20	97.59	89.59	8	0.13	0.01	0.12	2.28	0.15	2.13	2.41	0.16	2.25
10	10	1.5	6.67	2.5	20	97.59	91.33	6.26	0.13	0	0.13	2.28	0.04	2.23	2.41	0.05	2.36
10	10	2	5	2.5	20	97.59	91.94	5.65	0.13	0	0.13	2.28	0.01	2.26	2.41	0.01	2.4
10	10	2.5	4	2.5	20	97.59	92.22	5.37	0.13	0	0.13	2.28	0	2.27	2.41	0	2.41
10	10	3	3.33	2.5	20	97.59	92.37	5.22	0.13	0	0.13	2.28	0	2.27	2.41	0	2.41
10	15	1	15	2.5	20	97.59	80.39	17.2	0.13	0.01	0.12	2.28	0.19	2.09	2.41	0.2	2.21
10	15	1.5	10	2.5	20	97.59	84.56	13.03	0.13	0	0.13	2.28	0.06	2.22	2.41	0.06	2.35
10	15	2	7.5	2.5	20	97.59	86.14	11.45	0.13	0	0.13	2.28	0.02	2.26	2.41	0.02	2.39
10	15	2.5	6	2.5	20	97.59	86.88	10.71	0.13	0	0.13	2.28	0	2.27	2.41	0	2.41
10	15	3	5	2.5	20	97.59	87.28	10.31	0.13	0	0.13	2.28	0	2.27	2.41	0	2.41
10	20	1	20	2.5	20	97.59	68.09	29.5	0.13	0.01	0.12	2.28	0.21	2.06	2.41	0.23	2.18
10	20	1.5	13.33	2.5	20	97.59	74.9	22.69	0.13	0	0.13	2.28	0.07	2.21	2.41	0.07	2.34
10	20	2	10	2.5	20	97.59	77.76	19.83	0.13	0	0.13	2.28	0.02	2.26	2.41	0.02	2.39
10	20	2.5	8	2.5	20	97.59	79.18	18.41	0.13	0	0.13	2.28	0	2.27	2.41	0	2.41
10	20	3	6.67	2.5	20	97.59	79.97	17.62	0.13	0	0.13	2.28	0	2.27	2.41	0	2.41
10	25	1	25	2.5	20	97.59	54.36	43.23	0.13	0.02	0.12	2.28	0.23	2.04	2.41	0.25	2.16
10	25	1.5	16.67	2.5	20	97.59	62.86	34.73	0.13	0.01	0.13	2.28	0.08	2.2	2.41	0.08	2.33
10	25	2	12.5	2.5	20	97.59	66.89	30.7	0.13	0	0.13	2.28	0.02	2.25	2.41	0.02	2.39
10	25	2.5	10	2.5	20	97.59	69.02	28.57	0.13	0	0.13	2.28	0.01	2.27	2.41	0.01	2.4
10	25	3	8.33	2.5	20	97.59	70.26	27.33	0.13	0	0.13	2.28	0	2.27	2.41	0	2.41
10	2.5	1	2.5	2	25	93.7	92.01	1.69	0.82	0.02	0.8	5.48	0.09	5.39	6.3	0.1	6.2
10	2.5	1.5	1.67	2	25	93.7	92.16	1.54	0.82	0	0.82	5.48	0.02	5.46	6.3	0.03	6.27
10	2.5	2	1.25	2	25	93.7	92.21	1.49	0.82	0	0.82	5.48	0.01	5.47	6.3	0.01	6.29
10	2.5	2.5	1	2	25	93.7	92.22	1.48	0.82	0	0.82	5.48	0	5.48	6.3	0	6.3
10	2.5	3	0.83	2	25	93.7	92.23	1.47	0.82	0	0.82	5.48	0	5.48	6.3	0	6.3
10	5	1	5	2	25	93.7	89.77	3.93	0.82	0.03	0.79	5.48	0.16	5.32	6.3	0.19	6.11
10	5	1.5	3.33	2	25	93.7	90.23	3.47	0.82	0.01	0.81	5.48	0.04	5.44	6.3	0.05	6.25
10	5	2	2.5	2	25	93.7	90.38	3.32	0.82	0	0.82	5.48	0.01	5.47	6.3	0.01	6.29
10	5	2.5	2	2	25	93.7	90.44	3.26	0.82	0	0.82	5.48	0	5.48	6.3	0	6.3
10	5	3	1.67	2	25	93.7	90.47	3.23	0.82	0	0.82	5.48	0	5.48	6.3	0	6.3
10	7.5	1	7.5	2	25	93.7	86.87	6.83	0.82	0.04	0.78	5.48	0.22	5.26	6.3	0.26	6.04
10	7.5	1.5	5	2	25	93.7	87.84	5.87	0.82	0.01	0.81	5.48	0.06	5.42	6.3	0.07	6.23
10	7.5	2	3.75	2	25	93.7	88.16	5.54	0.82	0	0.82	5.48	0.01	5.47	6.3	0.02	6.28
10	7.5	2.5	3	2	25	93.7	88.3	5.4	0.82	0	0.82	5.48	0	5.48	6.3	0	6.3
10	7.5	3	2.5	2	25	93.7	88.38	5.32	0.82	0	0.82	5.48	0	5.48	6.3	0	6.3
10	10	1	10	2	25	93.7	83.25	10.45	0.82	0.05	0.77	5.48	0.28	5.2	6.3	0.33	5.97
10	10	1.5	6.67	2	25	93.7	84.93	8.77	0.82	0.01	0.81	5.48	0.08	5.4	6.3	0.09	6.21
10	10	2	5	2	25	93.7	85.52	8.18	0.82	0	0.82	5.48	0.02	5.46	6.3	0.02	6.28
10	10	2.5	4	2	25	93.7	85.78	7.92	0.82	0	0.82	5.48	0	5.48	6.3	0.01	6.29
10	10	3	3.33	2	25	93.7	85.91	7.79	0.82	0	0.82	5.48	0	5.48	6.3	0	6.3
10	15	1	15	2	25	93.7	73.91	19.79	0.82	0.06	0.76	5.48	0.37	5.11	6.3	0.43	5.87
10	15	1.5	10	2	25	93.7	77.49	16.21	0.82	0.02	0.8	5.48	0.11	5.37	6.3	0.13	6.17
10	15	2	7.5	2	25	93.7	78.83	14.87	0.82	0.01	0.81	5.48	0.03	5.45	6.3	0.03	6.27
10	15	2.5	6	2	25	93.7	79.45	14.25	0.82	0	0.82	5.48	0.01	5.47	6.3	0.01	6.29
10	15	3	5	2	25	93.7	79.78	13.92	0.82	0	0.82	5.48	0	5.48	6.3	0	6.3
10	20	1	20	2	25	93.7	62.42	31.28	0.82	0.07	0.75	5.48	0.43	5.05	6.3	0.5	5.79
10	20	1.5	13.33	2	25	93.7	67.96	25.74	0.82	0.02	0.8	5.48	0.13	5.35	6.3	0.15	6.15
10	20	2	10	2	25	93.7	70.22	23.48	0.82	0.01	0.81	5.48	0.03	5.45	6.3	0.04	6.26
10	20	2.5	8	2	25	93.7	71.31	22.39	0.82	0	0.82	5.48	0.01	5.47	6.3	0.01	6.29
10	20	3	6.67	2	25	93.7	71.92	21.78	0.82	0	0.82	5.48	0	5.48	6.3	0	6.3
10	25	1	25	2	25	93.7	49.96	43.74	0.82	0.08	0.74	5.48	0.48	5	6.3	0.56	5.74
10	25	1.5	16.67	2	25	93.7	56.8	36.9	0.82	0.03	0.79	5.48	0.15	5.33	6.3	0.18	6.12
10	25	2	12.5	2	25	93.7	59.88	33.82	0.82	0.01	0.81	5.48	0.04	5.44	6.3	0.05	6.25
10	25	2.5	10	2	25	93.7	61.44	32.26	0.82	0	0.82	5.48	0.01	5.47	6.3	0.01	6.29
10	25	3	8.33	2	25	93.7	62.34	31.36	0.82	0	0.82	5.48	0	5.48	6.3	0	6.3

a	$i=$ $k_m s_m$	k_m	s_m	k_p	s_p	C	C_A	C_R	U	U_A	U_R	O	O_A	O_R	W	W_A	W_R
10	2.5	1	2.5	1.5	33.33	84.9	82.47	2.43	3.59	0.05	3.55	11.51	0.12	11.39	15.1	0.16	14.94
10	2.5	1.5	1.67	1.5	33.33	84.9	82.65	2.25	3.59	0.01	3.58	11.51	0.03	11.48	15.1	0.04	15.06
10	2.5	2	1.25	1.5	33.33	84.9	82.71	2.19	3.59	0	3.59	11.51	0.01	11.5	15.1	0.01	15.09
10	2.5	2.5	1	1.5	33.33	84.9	82.72	2.18	3.59	0	3.59	11.51	0	11.51	15.1	0	15.1
10	2.5	3	0.83	1.5	33.33	84.9	82.73	2.17	3.59	0	3.59	11.51	0	11.51	15.1	0	15.1
10	5	1	5	1.5	33.33	84.9	79.6	5.3	3.59	0.09	3.51	11.51	0.22	11.28	15.1	0.31	14.79
10	5	1.5	3.33	1.5	33.33	84.9	80.08	4.82	3.59	0.02	3.57	11.51	0.06	11.45	15.1	0.08	15.02
10	5	2	2.5	1.5	33.33	84.9	80.23	4.67	3.59	0.01	3.59	11.51	0.01	11.49	15.1	0.02	15.08
10	5	2.5	2	1.5	33.33	84.9	80.29	4.61	3.59	0	3.59	11.51	0	11.5	15.1	0	15.1
10	5	3	1.67	1.5	33.33	84.9	80.31	4.59	3.59	0	3.59	11.51	0	11.51	15.1	0	15.1
10	7.5	1	7.5	1.5	33.33	84.9	76.26	8.64	3.59	0.12	3.47	11.51	0.32	11.19	15.1	0.44	14.66
10	7.5	1.5	5	1.5	33.33	84.9	77.17	7.73	3.59	0.03	3.56	11.51	0.08	11.42	15.1	0.12	14.98
10	7.5	2	3.75	1.5	33.33	84.9	77.46	7.44	3.59	0.01	3.59	11.51	0.02	11.49	15.1	0.03	15.07
10	7.5	2.5	3	1.5	33.33	84.9	77.58	7.32	3.59	0	3.59	11.51	0	11.5	15.1	0.01	15.09
10	7.5	3	2.5	1.5	33.33	84.9	77.64	7.26	3.59	0	3.59	11.51	0	11.51	15.1	0	15.1
10	10	1	10	1.5	33.33	84.9	72.47	12.43	3.59	0.16	3.44	11.51	0.41	11.1	15.1	0.57	14.53
10	10	1.5	6.67	1.5	33.33	84.9	73.91	10.99	3.59	0.04	3.55	11.51	0.11	11.4	15.1	0.15	14.95
10	10	2	5	1.5	33.33	84.9	74.39	10.51	3.59	0.01	3.58	11.51	0.03	11.48	15.1	0.04	15.06
10	10	2.5	4	1.5	33.33	84.9	74.59	10.31	3.59	0	3.59	11.51	0.01	11.5	15.1	0.01	15.09
10	10	3	3.33	1.5	33.33	84.9	74.7	10.2	3.59	0	3.59	11.51	0	11.51	15.1	0	15.1
10	15	1	15	1.5	33.33	84.9	63.63	21.27	3.59	0.21	3.38	11.51	0.56	10.94	15.1	0.77	14.33
10	15	1.5	10	1.5	33.33	84.9	66.37	18.53	3.59	0.06	3.53	11.51	0.15	11.35	15.1	0.21	14.89
10	15	2	7.5	1.5	33.33	84.9	67.34	17.56	3.59	0.01	3.58	11.51	0.04	11.47	15.1	0.05	15.05
10	15	2.5	6	1.5	33.33	84.9	67.77	17.13	3.59	0	3.59	11.51	0.01	11.5	15.1	0.01	15.09
10	15	3	5	1.5	33.33	84.9	67.99	16.91	3.59	0	3.59	11.51	0	11.51	15.1	0	15.1
10	20	1	20	1.5	33.33	84.9	53.59	31.31	3.59	0.25	3.34	11.51	0.69	10.81	15.1	0.94	14.16
10	20	1.5	13.33	1.5	33.33	84.9	57.61	27.29	3.59	0.07	3.52	11.51	0.2	11.31	15.1	0.27	14.83
10	20	2	10	1.5	33.33	84.9	59.14	25.76	3.59	0.02	3.57	11.51	0.05	11.46	15.1	0.07	15.03
10	20	2.5	8	1.5	33.33	84.9	59.84	25.06	3.59	0	3.59	11.51	0.01	11.5	15.1	0.02	15.08
10	20	3	6.67	1.5	33.33	84.9	60.22	24.68	3.59	0	3.59	11.51	0	11.5	15.1	0	15.1
10	25	1	25	1.5	33.33	84.9	43.04	41.86	3.59	0.28	3.31	11.51	0.79	10.71	15.1	1.07	14.03
10	25	1.5	16.67	1.5	33.33	84.9	47.96	36.94	3.59	0.09	3.51	11.51	0.23	11.27	15.1	0.32	14.78
10	25	2	12.5	1.5	33.33	84.9	49.98	34.92	3.59	0.02	3.57	11.51	0.06	11.45	15.1	0.08	15.02
10	25	2.5	10	1.5	33.33	84.9	50.94	33.96	3.59	0.01	3.59	11.51	0.01	11.49	15.1	0.02	15.08
10	25	3	8.33	1.5	33.33	84.9	51.46	33.44	3.59	0	3.59	11.51	0	11.5	15.1	0	15.1
10	2.5	1	2.5	1	50	67.31	64.57	2.73	11.51	0.08	11.43	21.19	0.12	21.07	32.69	0.2	32.49
10	2.5	1.5	1.67	1	50	67.31	64.76	2.55	11.51	0.02	11.49	21.19	0.03	21.16	32.69	0.05	32.64
10	2.5	2	1.25	1	50	67.31	64.8	2.5	11.51	0	11.5	21.19	0.01	21.18	32.69	0.01	32.68
10	2.5	2.5	1	1	50	67.31	64.82	2.49	11.51	0	11.51	21.19	0	21.18	32.69	0	32.69
10	2.5	3	0.83	1	50	67.31	64.82	2.48	11.51	0	11.51	21.19	0	21.19	32.69	0	32.69
10	5	1	5	1	50	67.31	61.62	5.69	11.51	0.15	11.35	21.19	0.23	20.95	32.69	0.39	32.31
10	5	1.5	3.33	1	50	67.31	62.04	5.27	11.51	0.04	11.47	21.19	0.06	21.13	32.69	0.1	32.6
10	5	2	2.5	1	50	67.31	62.16	5.15	11.51	0.01	11.5	21.19	0.01	21.17	32.69	0.02	32.67
10	5	2.5	2	1	50	67.31	62.2	5.11	11.51	0	11.5	21.19	0	21.18	32.69	0.01	32.69
10	5	3	1.67	1	50	67.31	62.21	5.1	11.51	0	11.51	21.19	0	21.18	32.69	0	32.69
10	7.5	1	7.5	1	50	67.31	58.46	8.85	11.51	0.22	11.28	21.19	0.34	20.84	32.69	0.57	32.13
10	7.5	1.5	5	1	50	67.31	59.16	8.15	11.51	0.06	11.45	21.19	0.09	21.1	32.69	0.14	32.55
10	7.5	2	3.75	1	50	67.31	59.37	7.94	11.51	0.01	11.49	21.19	0.02	21.16	32.69	0.03	32.66
10	7.5	2.5	3	1	50	67.31	59.44	7.87	11.51	0	11.5	21.19	0	21.18	32.69	0.01	32.69
10	7.5	3	2.5	1	50	67.31	59.47	7.84	11.51	0	11.51	21.19	0	21.18	32.69	0	32.69
10	10	1	10	1	50	67.31	55.11	12.2	11.51	0.29	11.22	21.19	0.45	20.74	32.69	0.74	31.96
10	10	1.5	6.67	1	50	67.31	56.13	11.18	11.51	0.07	11.43	21.19	0.11	21.07	32.69	0.19	32.5
10	10	2	5	1	50	67.31	56.44	10.86	11.51	0.02	11.49	21.19	0.03	21.16	32.69	0.05	32.65
10	10	2.5	4	1	50	67.31	56.56	10.75	11.51	0	11.5	21.19	0.01	21.18	32.69	0.01	32.68
10	10	3	3.33	1	50	67.31	56.61	10.7	11.51	0	11.51	21.19	0	21.18	32.69	0	32.69

a	$i=$ $k_m s_m$	k_m	s_m	k_p	s_p	C	C_A	C_R	U	U_A	U_R	O	O_A	O_R	W	W_A	W_R
10	15	1	15	1	50	67.31	47.92	19.39	11.51	0.41	11.1	21.19	0.64	20.54	32.69	1.05	31.64
10	15	1.5	10	1	50	67.31	49.66	17.65	11.51	0.11	11.4	21.19	0.17	21.02	32.69	0.27	32.42
10	15	2	7.5	1	50	67.31	50.21	17.09	11.51	0.03	11.48	21.19	0.04	21.15	32.69	0.07	32.63
10	15	2.5	6	1	50	67.31	50.43	16.88	11.51	0.01	11.5	21.19	0.01	21.18	32.69	0.01	32.68
10	15	3	5	1	50	67.31	50.53	16.78	11.51	0	11.51	21.19	0	21.18	32.69	0	32.69
10	20	1	20	1	50	67.31	40.28	27.03	11.51	0.52	10.99	21.19	0.82	20.37	32.69	1.34	31.36
10	20	1.5	13.33	1	50	67.31	42.71	24.59	11.51	0.14	11.37	21.19	0.22	20.97	32.69	0.36	32.34
10	20	2	10	1	50	67.31	43.53	23.78	11.51	0.03	11.47	21.19	0.05	21.13	32.69	0.09	32.6
10	20	2.5	8	1	50	67.31	43.86	23.45	11.51	0.01	11.5	21.19	0.01	21.17	32.69	0.02	32.67
10	20	3	6.67	1	50	67.31	44.02	23.29	11.51	0	11.51	21.19	0	21.18	32.69	0	32.69
10	25	1	25	1	50	67.31	32.46	34.84	11.51	0.61	10.9	21.19	0.97	20.21	32.69	1.58	31.11
10	25	1.5	16.67	1	50	67.31	35.44	31.86	11.51	0.17	11.34	21.19	0.26	20.92	32.69	0.43	32.26
10	25	2	12.5	1	50	67.31	36.49	30.82	11.51	0.04	11.46	21.19	0.07	21.12	32.69	0.11	32.59
10	25	2.5	10	1	50	67.31	36.92	30.38	11.51	0.01	11.5	21.19	0.01	21.17	32.69	0.02	32.67
10	25	3	8.33	1	50	67.31	37.14	30.17	11.51	0	11.51	21.19	0	21.18	32.69	0	32.69

a	$i=$ $k_m s_m$	k_m	s_m	k_p	s_p	C	C_A	C_R	U	U_A	U_R	O	O_A	O_R	W	W_A	W_R
15	2.5	1	2.5	3	16.67	98.21	97.25	0.96	0	0	0	1.79	0.05	1.74	1.79	0.05	1.74
15	2.5	1.5	1.67	3	16.67	98.21	97.36	0.85	0	0	0	1.79	0.01	1.77	1.79	0.01	1.78
15	2.5	2	1.25	3	16.67	98.21	97.4	0.81	0	0	0	1.79	0	1.78	1.79	0	1.79
15	2.5	2.5	1	3	16.67	98.21	97.41	0.8	0	0	0	1.79	0	1.79	1.79	0	1.79
15	2.5	3	0.83	3	16.67	98.21	97.42	0.79	0	0	0	1.79	0	1.79	1.79	0	1.79
15	5	1	5	3	16.67	98.21	95.65	2.56	0	0	0	1.79	0.08	1.7	1.79	0.08	1.71
15	5	1.5	3.33	3	16.67	98.21	96.08	2.13	0	0	0	1.79	0.02	1.76	1.79	0.02	1.77
15	5	2	2.5	3	16.67	98.21	96.22	1.99	0	0	0	1.79	0.01	1.78	1.79	0.01	1.79
15	5	2.5	2	3	16.67	98.21	96.29	1.92	0	0	0	1.79	0	1.79	1.79	0	1.79
15	5	3	1.67	3	16.67	98.21	96.32	1.89	0	0	0	1.79	0	1.79	1.79	0	1.79
15	7.5	1	7.5	3	16.67	98.21	93.19	5.02	0	0	0	1.79	0.11	1.68	1.79	0.11	1.68
15	7.5	1.5	5	3	16.67	98.21	94.22	3.99	0	0	0	1.79	0.03	1.75	1.79	0.03	1.76
15	7.5	2	3.75	3	16.67	98.21	94.58	3.63	0	0	0	1.79	0.01	1.78	1.79	0.01	1.78
15	7.5	2.5	3	3	16.67	98.21	94.74	3.46	0	0	0	1.79	0	1.78	1.79	0	1.79
15	7.5	3	2.5	3	16.67	98.21	94.83	3.38	0	0	0	1.79	0	1.79	1.79	0	1.79
15	10	1	10	3	16.67	98.21	89.73	8.48	0	0	0	1.79	0.13	1.65	1.79	0.13	1.66
15	10	1.5	6.67	3	16.67	98.21	91.67	6.54	0	0	0	1.79	0.04	1.75	1.79	0.04	1.75
15	10	2	5	3	16.67	98.21	92.37	5.84	0	0	0	1.79	0.01	1.78	1.79	0.01	1.78
15	10	2.5	4	3	16.67	98.21	92.7	5.51	0	0	0	1.79	0	1.78	1.79	0	1.79
15	10	3	3.33	3	16.67	98.21	92.87	5.34	0	0	0	1.79	0	1.79	1.79	0	1.79
15	15	1	15	3	16.67	98.21	79.99	18.22	0	0	0	1.79	0.16	1.62	1.79	0.16	1.63
15	15	1.5	10	3	16.67	98.21	84.29	13.92	0	0	0	1.79	0.05	1.74	1.79	0.05	1.74
15	15	2	7.5	3	16.67	98.21	85.99	12.22	0	0	0	1.79	0.01	1.77	1.79	0.01	1.78
15	15	2.5	6	3	16.67	98.21	86.82	11.39	0	0	0	1.79	0	1.78	1.79	0	1.79
15	15	3	5	3	16.67	98.21	87.28	10.93	0	0	0	1.79	0	1.79	1.79	0	1.79
15	20	1	20	3	16.67	98.21	67.55	30.66	0	0	0	1.79	0.18	1.6	1.79	0.18	1.61
15	20	1.5	13.33	3	16.67	98.21	74.13	24.08	0	0	0	1.79	0.06	1.73	1.79	0.06	1.73
15	20	2	10	3	16.67	98.21	76.96	21.25	0	0	0	1.79	0.02	1.77	1.79	0.02	1.77
15	20	2.5	8	3	16.67	98.21	78.4	19.81	0	0	0	1.79	0	1.78	1.79	0	1.79
15	20	3	6.67	3	16.67	98.21	79.23	18.98	0	0	0	1.79	0	1.79	1.79	0	1.79
15	25	1	25	3	16.67	98.21	53.94	44.26	0	0	0	1.79	0.19	1.59	1.79	0.2	1.6
15	25	1.5	16.67	3	16.67	98.21	61.97	36.24	0	0	0	1.79	0.07	1.72	1.79	0.07	1.72
15	25	2	12.5	3	16.67	98.21	65.72	32.49	0	0	0	1.79	0.02	1.77	1.79	0.02	1.77
15	25	2.5	10	3	16.67	98.21	67.69	30.52	0	0	0	1.79	0	1.78	1.79	0	1.79
15	25	3	8.33	3	16.67	98.21	68.84	29.36	0	0	0	1.79	0	1.79	1.79	0	1.79
15	2.5	1	2.5	2.5	20	95.94	94.48	1.46	0.06	0	0.06	4.01	0.08	3.92	4.06	0.08	3.98
15	2.5	1.5	1.67	2.5	20	95.94	94.62	1.32	0.06	0	0.06	4.01	0.02	3.99	4.06	0.02	4.04
15	2.5	2	1.25	2.5	20	95.94	94.66	1.27	0.06	0	0.06	4.01	0.01	4	4.06	0.01	4.06
15	2.5	2.5	1	2.5	20	95.94	94.68	1.26	0.06	0	0.06	4.01	0	4	4.06	0	4.06
15	2.5	3	0.83	2.5	20	95.94	94.69	1.25	0.06	0	0.06	4.01	0	4.01	4.06	0	4.06
15	5	1	5	2.5	20	95.94	92.39	3.55	0.06	0	0.05	4.01	0.15	3.86	4.06	0.15	3.91
15	5	1.5	3.33	2.5	20	95.94	92.86	3.08	0.06	0	0.06	4.01	0.04	3.97	4.06	0.04	4.02
15	5	2	2.5	2.5	20	95.94	93.01	2.92	0.06	0	0.06	4.01	0.01	4	4.06	0.01	4.05
15	5	2.5	2	2.5	20	95.94	93.08	2.86	0.06	0	0.06	4.01	0	4	4.06	0	4.06
15	5	3	1.67	2.5	20	95.94	93.11	2.82	0.06	0	0.06	4.01	0	4.01	4.06	0	4.06
15	7.5	1	7.5	2.5	20	95.94	89.54	6.39	0.06	0	0.05	4.01	0.2	3.8	4.06	0.21	3.86
15	7.5	1.5	5	2.5	20	95.94	90.57	5.37	0.06	0	0.06	4.01	0.06	3.95	4.06	0.06	4.01
15	7.5	2	3.75	2.5	20	95.94	90.92	5.01	0.06	0	0.06	4.01	0.01	3.99	4.06	0.01	4.05
15	7.5	2.5	3	2.5	20	95.94	91.08	4.86	0.06	0	0.06	4.01	0	4	4.06	0	4.06
15	7.5	3	2.5	2.5	20	95.94	91.16	4.78	0.06	0	0.06	4.01	0	4.01	4.06	0	4.06

a	$k_m s_m$	k_m	s_m	k_p	s_p	C	C_A	C_R	U	U_A	U_R	O	O_A	O_R	W	W_A	W_R
15	10	1	10	2.5	20	95.94	85.88	10.05	0.06	0	0.05	4.01	0.25	3.76	4.06	0.25	3.81
15	10	1.5	6.67	2.5	20	95.94	87.69	8.24	0.06	0	0.06	4.01	0.07	3.94	4.06	0.07	3.99
15	10	2	5	2.5	20	95.94	88.34	7.6	0.06	0	0.06	4.01	0.02	3.99	4.06	0.02	4.05
15	10	2.5	4	2.5	20	95.94	88.63	7.31	0.06	0	0.06	4.01	0	4	4.06	0	4.06
15	10	3	3.33	2.5	20	95.94	88.79	7.15	0.06	0	0.06	4.01	0	4.01	4.06	0	4.06
15	15	1	15	2.5	20	95.94	76.27	19.67	0.06	0.01	0.05	4.01	0.31	3.69	4.06	0.32	3.74
15	15	1.5	10	2.5	20	95.94	80.09	15.84	0.06	0	0.06	4.01	0.09	3.91	4.06	0.1	3.97
15	15	2	7.5	2.5	20	95.94	81.56	14.37	0.06	0	0.06	4.01	0.02	3.98	4.06	0.03	4.04
15	15	2.5	6	2.5	20	95.94	82.26	13.68	0.06	0	0.06	4.01	0.01	4	4.06	0.01	4.06
15	15	3	5	2.5	20	95.94	82.64	13.3	0.06	0	0.06	4.01	0	4	4.06	0	4.06
15	20	1	20	2.5	20	95.94	64.38	31.56	0.06	0.01	0.05	4.01	0.36	3.64	4.06	0.37	3.7
15	20	1.5	13.33	2.5	20	95.94	70.23	25.71	0.06	0	0.06	4.01	0.11	3.89	4.06	0.12	3.95
15	20	2	10	2.5	20	95.94	72.66	23.28	0.06	0	0.06	4.01	0.03	3.97	4.06	0.03	4.03
15	20	2.5	8	2.5	20	95.94	73.85	22.08	0.06	0	0.06	4.01	0.01	4	4.06	0.01	4.06
15	20	3	6.67	2.5	20	95.94	74.52	21.41	0.06	0	0.06	4.01	0	4	4.06	0	4.06
15	25	1	25	2.5	20	95.94	51.5	44.44	0.06	0.01	0.05	4.01	0.39	3.61	4.06	0.4	3.66
15	25	1.5	16.67	2.5	20	95.94	58.67	37.26	0.06	0	0.06	4.01	0.13	3.88	4.06	0.13	3.93
15	25	2	12.5	2.5	20	95.94	61.92	34.02	0.06	0	0.06	4.01	0.04	3.97	4.06	0.04	4.03
15	25	2.5	10	2.5	20	95.94	63.59	32.35	0.06	0	0.06	4.01	0.01	4	4.06	0.01	4.05
15	25	3	8.33	2.5	20	95.94	64.54	31.4	0.06	0	0.06	4.01	0	4	4.06	0	4.06
15	2.5	1	2.5	2	25	91.46	89.44	2.02	0.47	0.01	0.46	8.08	0.12	7.96	8.54	0.13	8.41
15	2.5	1.5	1.67	2	25	91.46	89.61	1.85	0.47	0	0.46	8.08	0.03	8.05	8.54	0.03	8.51
15	2.5	2	1.25	2	25	91.46	89.66	1.8	0.47	0	0.47	8.08	0.01	8.07	8.54	0.01	8.53
15	2.5	2.5	1	2	25	91.46	89.68	1.78	0.47	0	0.47	8.08	0	8.07	8.54	0	8.54
15	2.5	3	0.83	2	25	91.46	89.68	1.77	0.47	0	0.47	8.08	0	8.08	8.54	0	8.54
15	5	1	5	2	25	91.46	86.87	4.59	0.47	0.02	0.45	8.08	0.22	7.86	8.54	0.24	8.3
15	5	1.5	3.33	2	25	91.46	87.36	4.1	0.47	0	0.46	8.08	0.06	8.02	8.54	0.06	8.48
15	5	2	2.5	2	25	91.46	87.52	3.94	0.47	0	0.46	8.08	0.01	8.06	8.54	0.02	8.53
15	5	2.5	2	2	25	91.46	87.58	3.88	0.47	0	0.47	8.08	0	8.07	8.54	0	8.54
15	5	3	1.67	2	25	91.46	87.61	3.85	0.47	0	0.47	8.08	0	8.08	8.54	0	8.54
15	7.5	1	7.5	2	25	91.46	83.69	7.77	0.47	0.02	0.44	8.08	0.31	7.77	8.54	0.33	8.21
15	7.5	1.5	5	2	25	91.46	84.67	6.79	0.47	0.01	0.46	8.08	0.08	7.99	8.54	0.09	8.45
15	7.5	2	3.75	2	25	91.46	85	6.46	0.47	0	0.46	8.08	0.02	8.06	8.54	0.02	8.52
15	7.5	2.5	3	2	25	91.46	85.14	6.32	0.47	0	0.47	8.08	0	8.07	8.54	0	8.54
15	7.5	3	2.5	2	25	91.46	85.21	6.25	0.47	0	0.47	8.08	0	8.07	8.54	0	8.54
15	10	1	10	2	25	91.46	79.88	11.58	0.47	0.03	0.44	8.08	0.39	7.69	8.54	0.42	8.13
15	10	1.5	6.67	2	25	91.46	81.51	9.94	0.47	0.01	0.46	8.08	0.11	7.97	8.54	0.11	8.43
15	10	2	5	2	25	91.46	82.08	9.38	0.47	0	0.46	8.08	0.03	8.05	8.54	0.03	8.51
15	10	2.5	4	2	25	91.46	82.33	9.13	0.47	0	0.47	8.08	0.01	8.07	8.54	0.01	8.54
15	10	3	3.33	2	25	91.46	82.46	9	0.47	0	0.47	8.08	0	8.07	8.54	0	8.54
15	15	1	15	2	25	91.46	70.53	20.93	0.47	0.04	0.43	8.08	0.51	7.56	8.54	0.55	7.99
15	15	1.5	10	2	25	91.46	73.8	17.66	0.47	0.01	0.45	8.08	0.15	7.93	8.54	0.16	8.38
15	15	2	7.5	2	25	91.46	75.02	16.44	0.47	0	0.46	8.08	0.04	8.04	8.54	0.04	8.5
15	15	2.5	6	2	25	91.46	75.57	15.89	0.47	0	0.47	8.08	0.01	8.07	8.54	0.01	8.53
15	15	3	5	2	25	91.46	75.87	15.59	0.47	0	0.47	8.08	0	8.07	8.54	0	8.54
15	20	1	20	2	25	91.46	59.47	31.99	0.47	0.04	0.42	8.08	0.61	7.47	8.54	0.65	7.89
15	20	1.5	13.33	2	25	91.46	64.39	27.07	0.47	0.01	0.45	8.08	0.18	7.89	8.54	0.2	8.35
15	20	2	10	2	25	91.46	66.35	25.11	0.47	0	0.46	8.08	0.05	8.03	8.54	0.05	8.49
15	20	2.5	8	2	25	91.46	67.28	24.17	0.47	0	0.47	8.08	0.01	8.06	8.54	0.01	8.53
15	20	3	6.67	2	25	91.46	67.8	23.66	0.47	0	0.47	8.08	0	8.07	8.54	0	8.54
15	25	1	25	2	25	91.46	47.66	43.79	0.47	0.05	0.42	8.08	0.68	7.39	8.54	0.73	7.81
15	25	1.5	16.67	2	25	91.46	53.7	37.76	0.47	0.02	0.45	8.08	0.21	7.86	8.54	0.23	8.31
15	25	2	12.5	2	25	91.46	56.31	35.15	0.47	0	0.46	8.08	0.06	8.02	8.54	0.06	8.48
15	25	2.5	10	2	25	91.46	57.61	33.85	0.47	0	0.47	8.08	0.01	8.06	8.54	0.01	8.53
15	25	3	8.33	2	25	91.46	58.33	33.13	0.47	0	0.47	8.08	0	8.07	8.54	0	8.54

a	i= $k_m s_m$	k_m	s_m	k_p	s_p	C	C_A	C_R	U	U_A	U_R	O	O_A	O_R	W	W_A	W_R
15	2.5	1	2.5	1.5	33.33	82.76	80.2	2.55	2.56	0.03	2.52	14.69	0.14	14.55	17.24	0.17	17.07
15	2.5	1.5	1.67	1.5	33.33	82.76	80.39	2.36	2.56	0.01	2.55	14.69	0.03	14.65	17.24	0.04	17.2
15	2.5	2	1.25	1.5	33.33	82.76	80.45	2.31	2.56	0	2.56	14.69	0.01	14.68	17.24	0.01	17.23
15	2.5	2.5	1	1.5	33.33	82.76	80.46	2.29	2.56	0	2.56	14.69	0	14.68	17.24	0	17.24
15	2.5	3	0.83	1.5	33.33	82.76	80.47	2.28	2.56	0	2.56	14.69	0	14.69	17.24	0	17.24
15	5	1	5	1.5	33.33	82.76	77.24	5.51	2.56	0.07	2.49	14.69	0.27	14.42	17.24	0.33	16.91
15	5	1.5	3.33	1.5	33.33	82.76	77.73	5.03	2.56	0.02	2.54	14.69	0.07	14.62	17.24	0.08	17.16
15	5	2	2.5	1.5	33.33	82.76	77.87	4.88	2.56	0	2.55	14.69	0.02	14.67	17.24	0.02	17.22
15	5	2.5	2	1.5	33.33	82.76	77.93	4.83	2.56	0	2.56	14.69	0	14.68	17.24	0	17.24
15	5	3	1.67	1.5	33.33	82.76	77.95	4.8	2.56	0	2.56	14.69	0	14.69	17.24	0	17.24
15	7.5	1	7.5	1.5	33.33	82.76	73.86	8.9	2.56	0.09	2.47	14.69	0.39	14.3	17.24	0.48	16.77
15	7.5	1.5	5	1.5	33.33	82.76	74.75	8.01	2.56	0.02	2.53	14.69	0.1	14.59	17.24	0.12	17.12
15	7.5	2	3.75	1.5	33.33	82.76	75.03	7.73	2.56	0.01	2.55	14.69	0.02	14.66	17.24	0.03	17.21
15	7.5	2.5	3	1.5	33.33	82.76	75.14	7.61	2.56	0	2.56	14.69	0.01	14.68	17.24	0.01	17.24
15	7.5	3	2.5	1.5	33.33	82.76	75.19	7.56	2.56	0	2.56	14.69	0	14.68	17.24	0	17.24
15	10	1	10	1.5	33.33	82.76	70.07	12.69	2.56	0.12	2.44	14.69	0.49	14.19	17.24	0.61	16.63
15	10	1.5	6.67	1.5	33.33	82.76	71.45	11.3	2.56	0.03	2.53	14.69	0.13	14.56	17.24	0.16	17.08
15	10	2	5	1.5	33.33	82.76	71.91	10.84	2.56	0.01	2.55	14.69	0.04	14.65	17.24	0.04	17.21
15	10	2.5	4	1.5	33.33	82.76	72.1	10.66	2.56	0	2.56	14.69	0.01	14.68	17.24	0.01	17.24
15	10	3	3.33	1.5	33.33	82.76	72.19	10.56	2.56	0	2.56	14.69	0	14.68	17.24	0	17.24
15	15	1	15	1.5	33.33	82.76	61.4	21.36	2.56	0.15	2.41	14.69	0.69	14	17.24	0.84	16.41
15	15	1.5	10	1.5	33.33	82.76	63.96	18.8	2.56	0.04	2.52	14.69	0.19	14.5	17.24	0.23	17.02
15	15	2	7.5	1.5	33.33	82.76	64.86	17.89	2.56	0.01	2.55	14.69	0.05	14.64	17.24	0.06	17.19
15	15	2.5	6	1.5	33.33	82.76	65.25	17.5	2.56	0	2.56	14.69	0.01	14.68	17.24	0.01	17.23
15	15	3	5	1.5	33.33	82.76	65.45	17.3	2.56	0	2.56	14.69	0	14.68	17.24	0	17.24
15	20	1	20	1.5	33.33	82.76	51.68	31.07	2.56	0.18	2.38	14.69	0.85	13.84	17.24	1.03	16.22
15	20	1.5	13.33	1.5	33.33	82.76	55.41	27.34	2.56	0.05	2.5	14.69	0.24	14.45	17.24	0.29	16.95
15	20	2	10	1.5	33.33	82.76	56.81	25.95	2.56	0.01	2.54	14.69	0.06	14.63	17.24	0.07	17.17
15	20	2.5	8	1.5	33.33	82.76	57.44	25.32	2.56	0	2.56	14.69	0.01	14.67	17.24	0.02	17.23
15	20	3	6.67	1.5	33.33	82.76	57.77	24.99	2.56	0	2.56	14.69	0	14.68	17.24	0	17.24
15	25	1	25	1.5	33.33	82.76	41.53	41.22	2.56	0.21	2.35	14.69	0.97	13.71	17.24	1.18	16.07
15	25	1.5	16.67	1.5	33.33	82.76	46.1	36.66	2.56	0.06	2.5	14.69	0.28	14.4	17.24	0.35	16.9
15	25	2	12.5	1.5	33.33	82.76	47.92	34.83	2.56	0.02	2.54	14.69	0.07	14.61	17.24	0.09	17.16
15	25	2.5	10	1.5	33.33	82.76	48.77	33.98	2.56	0	2.55	14.69	0.02	14.67	17.24	0.02	17.22
15	25	3	8.33	1.5	33.33	82.76	49.23	33.53	2.56	0	2.56	14.69	0	14.68	17.24	0	17.24
15	2.5	1	2.5	1	50	66.12	63.4	2.73	9.68	0.07	9.61	24.2	0.13	24.07	33.88	0.2	33.68
15	2.5	1.5	1.67	1	50	66.12	63.58	2.55	9.68	0.02	9.66	24.2	0.03	24.16	33.88	0.05	33.83
15	2.5	2	1.25	1	50	66.12	63.63	2.5	9.68	0.01	9.68	24.2	0.01	24.19	33.88	0.01	33.86
15	2.5	2.5	1	1	50	66.12	63.64	2.48	9.68	0	9.68	24.2	0	24.19	33.88	0	33.87
15	2.5	3	0.83	1	50	66.12	63.64	2.48	9.68	0	9.68	24.2	0	24.2	33.88	0	33.88
15	5	1	5	1	50	66.12	60.46	5.66	9.68	0.13	9.55	24.2	0.25	23.94	33.88	0.39	33.49
15	5	1.5	3.33	1	50	66.12	60.87	5.25	9.68	0.03	9.65	24.2	0.06	24.13	33.88	0.1	33.78
15	5	2	2.5	1	50	66.12	60.99	5.13	9.68	0.01	9.67	24.2	0.02	24.18	33.88	0.02	33.85
15	5	2.5	2	1	50	66.12	61.03	5.1	9.68	0	9.68	24.2	0	24.19	33.88	0.01	33.87
15	5	3	1.67	1	50	66.12	61.04	5.08	9.68	0	9.68	24.2	0	24.2	33.88	0	33.88
15	7.5	1	7.5	1	50	66.12	57.34	8.79	9.68	0.2	9.48	24.2	0.37	23.82	33.88	0.57	33.31
15	7.5	1.5	5	1	50	66.12	58.02	8.1	9.68	0.05	9.63	24.2	0.09	24.1	33.88	0.14	33.73
15	7.5	2	3.75	1	50	66.12	58.22	7.9	9.68	0.01	9.67	24.2	0.02	24.17	33.88	0.03	33.84
15	7.5	2.5	3	1	50	66.12	58.29	7.83	9.68	0	9.68	24.2	0	24.19	33.88	0.01	33.87
15	7.5	3	2.5	1	50	66.12	58.32	7.8	9.68	0	9.68	24.2	0	24.2	33.88	0	33.87
15	10	1	10	1	50	66.12	54.03	12.1	9.68	0.25	9.43	24.2	0.49	23.71	33.88	0.74	33.14
15	10	1.5	6.67	1	50	66.12	55.03	11.1	9.68	0.07	9.61	24.2	0.12	24.07	33.88	0.19	33.69
15	10	2	5	1	50	66.12	55.33	10.8	9.68	0.02	9.66	24.2	0.03	24.17	33.88	0.05	33.83
15	10	2.5	4	1	50	66.12	55.44	10.69	9.68	0	9.68	24.2	0.01	24.19	33.88	0.01	33.87
15	10	3	3.33	1	50	66.12	55.49	10.64	9.68	0	9.68	24.2	0	24.2	33.88	0	33.87

| | i= | | | | | | | | | | | | | | | | |
a	$k_m s_m$	k_m	s_m	k_p	s_p	C	C_A	C_R	U	U_A	U_R	O	O_A	O_R	W	W_A	W_R
15	15	1	15	1	50	66.12	46.96	19.17	9.68	0.36	9.32	24.2	0.7	23.49	33.88	1.06	32.82
15	15	1.5	10	1	50	66.12	48.64	17.48	9.68	0.09	9.59	24.2	0.18	24.02	33.88	0.28	33.6
15	15	2	7.5	1	50	66.12	49.18	16.95	9.68	0.02	9.66	24.2	0.04	24.15	33.88	0.07	33.81
15	15	2.5	6	1	50	66.12	49.38	16.74	9.68	0.01	9.67	24.2	0.01	24.19	33.88	0.01	33.86
15	15	3	5	1	50	66.12	49.48	16.64	9.68	0	9.68	24.2	0	24.19	33.88	0	33.87
15	20	1	20	1	50	66.12	39.47	26.66	9.68	0.45	9.23	24.2	0.9	23.3	33.88	1.35	32.53
15	20	1.5	13.33	1	50	66.12	41.82	24.3	9.68	0.12	9.56	24.2	0.24	23.96	33.88	0.36	33.52
15	20	2	10	1	50	66.12	42.61	23.52	9.68	0.03	9.65	24.2	0.06	24.14	33.88	0.09	33.79
15	20	2.5	8	1	50	66.12	42.92	23.2	9.68	0.01	9.67	24.2	0.01	24.18	33.88	0.02	33.86
15	20	3	6.67	1	50	66.12	43.07	23.05	9.68	0	9.68	24.2	0	24.19	33.88	0	33.87
15	25	1	25	1	50	66.12	31.81	34.31	9.68	0.52	9.16	24.2	1.07	23.13	33.88	1.59	32.28
15	25	1.5	16.67	1	50	66.12	34.7	31.42	9.68	0.15	9.53	24.2	0.29	23.91	33.88	0.44	33.44
15	25	2	12.5	1	50	66.12	35.7	30.42	9.68	0.04	9.64	24.2	0.07	24.13	33.88	0.11	33.77
15	25	2.5	10	1	50	66.12	36.11	30.01	9.68	0.01	9.67	24.2	0.02	24.18	33.88	0.02	33.85
15	25	3	8.33	1	50	66.12	36.31	29.81	9.68	0	9.68	24.2	0	24.19	33.88	0	33.87

a	$k_m s_m$ (i=)	k_m	s_m	k_p	s_p	C	C_A	C_R	U	U_A	U_R	O	O_A	O_R	W	W_A	W_R
20	2.5	1	2.5	3	16.67	96.41	94.77	1.63	0	0	0	3.59	0.09	3.51	3.59	0.09	3.51
20	2.5	1.5	1.67	3	16.67	96.41	94.94	1.46	0	0	0	3.59	0.02	3.57	3.59	0.02	3.57
20	2.5	2	1.25	3	16.67	96.41	95	1.41	0	0	0	3.59	0.01	3.59	3.59	0.01	3.59
20	2.5	2.5	1	3	16.67	96.41	95.02	1.39	0	0	0	3.59	0	3.59	3.59	0	3.59
20	2.5	3	0.83	3	16.67	96.41	95.03	1.38	0	0	0	3.59	0	3.59	3.59	0	3.59
20	5	1	5	3	16.67	96.41	92.3	4.11	0	0	0	3.59	0.16	3.44	3.59	0.16	3.44
20	5	1.5	3.33	3	16.67	96.41	92.89	3.52	0	0	0	3.59	0.04	3.55	3.59	0.04	3.55
20	5	2	2.5	3	16.67	96.41	93.09	3.32	0	0	0	3.59	0.01	3.58	3.59	0.01	3.58
20	5	2.5	2	3	16.67	96.41	93.17	3.23	0	0	0	3.59	0	3.59	3.59	0	3.59
20	5	3	1.67	3	16.67	96.41	93.22	3.19	0	0	0	3.59	0	3.59	3.59	0	3.59
20	7.5	1	7.5	3	16.67	96.41	88.85	7.55	0	0	0	3.59	0.21	3.38	3.59	0.21	3.39
20	7.5	1.5	5	3	16.67	96.41	90.13	6.28	0	0	0	3.59	0.06	3.53	3.59	0.06	3.54
20	7.5	2	3.75	3	16.67	96.41	90.58	5.82	0	0	0	3.59	0.01	3.58	3.59	0.01	3.58
20	7.5	2.5	3	3	16.67	96.41	90.79	5.62	0	0	0	3.59	0	3.59	3.59	0	3.59
20	7.5	3	2.5	3	16.67	96.41	90.9	5.51	0	0	0	3.59	0	3.59	3.59	0	3.59
20	10	1	10	3	16.67	96.41	84.5	11.91	0	0	0	3.59	0.25	3.34	3.59	0.25	3.34
20	10	1.5	6.67	3	16.67	96.41	86.63	9.78	0	0	0	3.59	0.07	3.52	3.59	0.07	3.52
20	10	2	5	3	16.67	96.41	87.43	8.97	0	0	0	3.59	0.02	3.57	3.59	0.02	3.58
20	10	2.5	4	3	16.67	96.41	87.81	8.6	0	0	0	3.59	0	3.59	3.59	0	3.59
20	10	3	3.33	3	16.67	96.41	88.01	8.39	0	0	0	3.59	0	3.59	3.59	0	3.59
20	15	1	15	3	16.67	96.41	73.87	22.53	0	0	0	3.59	0.31	3.28	3.59	0.31	3.28
20	15	1.5	10	3	16.67	96.41	77.68	18.72	0	0	0	3.59	0.1	3.5	3.59	0.1	3.5
20	15	2	7.5	3	16.67	96.41	79.26	17.14	0	0	0	3.59	0.03	3.57	3.59	0.03	3.57
20	15	2.5	6	3	16.67	96.41	80.05	16.36	0	0	0	3.59	0.01	3.59	3.59	0.01	3.59
20	15	3	5	3	16.67	96.41	80.49	15.92	0	0	0	3.59	0	3.59	3.59	0	3.59
20	20	1	20	3	16.67	96.41	61.96	34.45	0	0	0	3.59	0.35	3.24	3.59	0.35	3.24
20	20	1.5	13.33	3	16.67	96.41	67	29.41	0	0	0	3.59	0.12	3.48	3.59	0.12	3.48
20	20	2	10	3	16.67	96.41	69.14	27.27	0	0	0	3.59	0.03	3.56	3.59	0.03	3.56
20	20	2.5	8	3	16.67	96.41	70.23	26.17	0	0	0	3.59	0.01	3.59	3.59	0.01	3.59
20	20	3	6.67	3	16.67	96.41	70.86	25.54	0	0	0	3.59	0	3.59	3.59	0	3.59
20	25	1	25	3	16.67	96.41	49.65	46.76	0	0	0	3.59	0.38	3.21	3.59	0.38	3.21
20	25	1.5	16.67	3	16.67	96.41	55.49	40.91	0	0	0	3.59	0.13	3.46	3.59	0.13	3.46
20	25	2	12.5	3	16.67	96.41	57.92	38.48	0	0	0	3.59	0.04	3.56	3.59	0.04	3.56
20	25	2.5	10	3	16.67	96.41	59.12	37.29	0	0	0	3.59	0.01	3.58	3.59	0.01	3.59
20	25	3	8.33	3	16.67	96.41	59.79	36.61	0	0	0	3.59	0	3.59	3.59	0	3.59
20	2.5	1	2.5	2.5	20	93.3	91.21	2.08	0.02	0	0.02	6.68	0.12	6.56	6.7	0.12	6.58
20	2.5	1.5	1.67	2.5	20	93.3	91.4	1.9	0.02	0	0.02	6.68	0.03	6.65	6.7	0.03	6.67
20	2.5	2	1.25	2.5	20	93.3	91.46	1.84	0.02	0	0.02	6.68	0.01	6.67	6.7	0.01	6.7
20	2.5	2.5	1	2.5	20	93.3	91.48	1.82	0.02	0	0.02	6.68	0	6.68	6.7	0	6.7
20	2.5	3	0.83	2.5	20	93.3	91.49	1.81	0.02	0	0.02	6.68	0	6.68	6.7	0	6.7
20	5	1	5	2.5	20	93.3	88.43	4.87	0.02	0	0.02	6.68	0.23	6.45	6.7	0.23	6.48
20	5	1.5	3.33	2.5	20	93.3	88.99	4.3	0.02	0	0.02	6.68	0.06	6.62	6.7	0.06	6.64
20	5	2	2.5	2.5	20	93.3	89.18	4.12	0.02	0	0.02	6.68	0.01	6.67	6.7	0.01	6.69
20	5	2.5	2	2.5	20	93.3	89.26	4.04	0.02	0	0.02	6.68	0	6.68	6.7	0	6.7
20	5	3	1.67	2.5	20	93.3	89.3	4	0.02	0	0.02	6.68	0	6.68	6.7	0	6.7
20	7.5	1	7.5	2.5	20	93.3	84.91	8.39	0.02	0	0.02	6.68	0.31	6.37	6.7	0.32	6.39
20	7.5	1.5	5	2.5	20	93.3	86.04	7.26	0.02	0	0.02	6.68	0.09	6.6	6.7	0.09	6.62
20	7.5	2	3.75	2.5	20	93.3	86.43	6.87	0.02	0	0.02	6.68	0.02	6.66	6.7	0.02	6.68
20	7.5	2.5	3	2.5	20	93.3	86.6	6.7	0.02	0	0.02	6.68	0	6.68	6.7	0	6.7
20	7.5	3	2.5	2.5	20	93.3	86.69	6.61	0.02	0	0.02	6.68	0	6.68	6.7	0	6.7

a	$i=$ $k_m s_m$	k_m	s_m	k_p	s_p	C	C_A	C_R	U	U_A	U_R	O	O_A	O_R	W	W_A	W_R
20	10	1	10	2.5	20	93.3	80.71	12.59	0.02	0	0.02	6.68	0.39	6.3	6.7	0.39	6.32
20	10	1.5	6.67	2.5	20	93.3	82.53	10.76	0.02	0	0.02	6.68	0.11	6.57	6.7	0.11	6.59
20	10	2	5	2.5	20	93.3	83.19	10.1	0.02	0	0.02	6.68	0.03	6.65	6.7	0.03	6.68
20	10	2.5	4	2.5	20	93.3	83.49	9.8	0.02	0	0.02	6.68	0.01	6.67	6.7	0.01	6.7
20	10	3	3.33	2.5	20	93.3	83.65	9.65	0.02	0	0.02	6.68	0	6.68	6.7	0	6.7
20	15	1	15	2.5	20	93.3	70.73	22.57	0.02	0	0.02	6.68	0.5	6.18	6.7	0.5	6.2
20	15	1.5	10	2.5	20	93.3	74.05	19.24	0.02	0	0.02	6.68	0.15	6.53	6.7	0.15	6.56
20	15	2	7.5	2.5	20	93.3	75.34	17.96	0.02	0	0.02	6.68	0.04	6.64	6.7	0.04	6.67
20	15	2.5	6	2.5	20	93.3	75.94	17.35	0.02	0	0.02	6.68	0.01	6.67	6.7	0.01	6.7
20	15	3	5	2.5	20	93.3	76.27	17.02	0.02	0	0.02	6.68	0	6.68	6.7	0	6.7
20	20	1	20	2.5	20	93.3	59.45	33.84	0.02	0	0.02	6.68	0.58	6.1	6.7	0.58	6.12
20	20	1.5	13.33	2.5	20	93.3	64.1	29.2	0.02	0	0.02	6.68	0.18	6.5	6.7	0.18	6.52
20	20	2	10	2.5	20	93.3	65.96	27.34	0.02	0	0.02	6.68	0.05	6.63	6.7	0.05	6.66
20	20	2.5	8	2.5	20	93.3	66.86	26.44	0.02	0	0.02	6.68	0.01	6.67	6.7	0.01	6.69
20	20	3	6.67	2.5	20	93.3	67.35	25.94	0.02	0	0.02	6.68	0	6.68	6.7	0	6.7
20	25	1	25	2.5	20	93.3	47.69	45.6	0.02	0	0.02	6.68	0.63	6.05	6.7	0.64	6.07
20	25	1.5	16.67	2.5	20	93.3	53.24	40.05	0.02	0	0.02	6.68	0.21	6.47	6.7	0.21	6.5
20	25	2	12.5	2.5	20	93.3	55.53	37.77	0.02	0	0.02	6.68	0.06	6.62	6.7	0.06	6.65
20	25	2.5	10	2.5	20	93.3	56.63	36.67	0.02	0	0.02	6.68	0.01	6.67	6.7	0.01	6.69
20	25	3	8.33	2.5	20	93.3	57.23	36.06	0.02	0	0.02	6.68	0	6.68	6.7	0	6.7
20	2.5	1	2.5	2	25	88.24	85.79	2.44	0.26	0.01	0.25	11.51	0.15	11.35	11.76	0.16	11.6
20	2.5	1.5	1.67	2	25	88.24	85.99	2.25	0.26	0	0.25	11.51	0.04	11.47	11.76	0.04	11.72
20	2.5	2	1.25	2	25	88.24	86.04	2.19	0.26	0	0.26	11.51	0.01	11.5	11.76	0.01	11.75
20	2.5	2.5	1	2	25	88.24	86.06	2.17	0.26	0	0.26	11.51	0	11.5	11.76	0	11.76
20	2.5	3	0.83	2	25	88.24	86.07	2.17	0.26	0	0.26	11.51	0	11.51	11.76	0	11.76
20	5	1	5	2	25	88.24	82.82	5.42	0.26	0.01	0.25	11.51	0.29	11.22	11.76	0.3	11.46
20	5	1.5	3.33	2	25	88.24	83.35	4.89	0.26	0	0.25	11.51	0.07	11.43	11.76	0.08	11.68
20	5	2	2.5	2	25	88.24	83.51	4.73	0.26	0	0.25	11.51	0.02	11.49	11.76	0.02	11.74
20	5	2.5	2	2	25	88.24	83.58	4.66	0.26	0	0.26	11.51	0	11.5	11.76	0	11.76
20	5	3	1.67	2	25	88.24	83.61	4.63	0.26	0	0.26	11.51	0	11.51	11.76	0	11.76
20	7.5	1	7.5	2	25	88.24	79.31	8.92	0.26	0.01	0.24	11.51	0.41	11.1	11.76	0.42	11.34
20	7.5	1.5	5	2	25	88.24	80.3	7.94	0.26	0	0.25	11.51	0.11	11.4	11.76	0.11	11.65
20	7.5	2	3.75	2	25	88.24	80.63	7.61	0.26	0	0.25	11.51	0.03	11.48	11.76	0.03	11.73
20	7.5	2.5	3	2	25	88.24	80.76	7.47	0.26	0	0.26	11.51	0.01	11.5	11.76	0.01	11.76
20	7.5	3	2.5	2	25	88.24	80.83	7.41	0.26	0	0.26	11.51	0	11.51	11.76	0	11.76
20	10	1	10	2	25	88.24	75.3	12.94	0.26	0.02	0.24	11.51	0.52	10.99	11.76	0.53	11.23
20	10	1.5	6.67	2	25	88.24	76.86	11.38	0.26	0	0.25	11.51	0.14	11.37	11.76	0.14	11.62
20	10	2	5	2	25	88.24	77.4	10.84	0.26	0	0.25	11.51	0.03	11.47	11.76	0.04	11.73
20	10	2.5	4	2	25	88.24	77.62	10.61	0.26	0	0.26	11.51	0.01	11.5	11.76	0.01	11.75
20	10	3	3.33	2	25	88.24	77.74	10.5	0.26	0	0.26	11.51	0	11.51	11.76	0	11.76
20	15	1	15	2	25	88.24	66	22.23	0.26	0.02	0.23	11.51	0.69	10.81	11.76	0.71	11.05
20	15	1.5	10	2	25	88.24	68.88	19.36	0.26	0.01	0.25	11.51	0.2	11.31	11.76	0.2	11.56
20	15	2	7.5	2	25	88.24	69.92	18.32	0.26	0	0.25	11.51	0.05	11.46	11.76	0.05	11.71
20	15	2.5	6	2	25	88.24	70.39	17.85	0.26	0	0.26	11.51	0.01	11.5	11.76	0.01	11.75
20	15	3	5	2	25	88.24	70.63	17.6	0.26	0	0.26	11.51	0	11.5	11.76	0	11.76
20	20	1	20	2	25	88.24	55.54	32.69	0.26	0.02	0.23	11.51	0.83	10.68	11.76	0.85	10.91
20	20	1.5	13.33	2	25	88.24	59.67	28.56	0.26	0.01	0.25	11.51	0.24	11.26	11.76	0.25	11.51
20	20	2	10	2	25	88.24	61.26	26.98	0.26	0	0.25	11.51	0.06	11.44	11.76	0.07	11.7
20	20	2.5	8	2	25	88.24	61.99	26.25	0.26	0	0.25	11.51	0.01	11.49	11.76	0.01	11.75
20	20	3	6.67	2	25	88.24	62.38	25.86	0.26	0	0.26	11.51	0	11.5	11.76	0	11.76
20	25	1	25	2	25	88.24	44.61	43.63	0.26	0.03	0.23	11.51	0.93	10.57	11.76	0.96	10.8
20	25	1.5	16.67	2	25	88.24	49.63	38.61	0.26	0.01	0.25	11.51	0.29	11.22	11.76	0.29	11.47
20	25	2	12.5	2	25	88.24	51.66	36.58	0.26	0	0.25	11.51	0.08	11.43	11.76	0.08	11.68
20	25	2.5	10	2	25	88.24	52.62	35.62	0.26	0	0.25	11.51	0.02	11.49	11.76	0.02	11.74
20	25	3	8.33	2	25	88.24	53.14	35.1	0.26	0	0.26	11.51	0	11.5	11.76	0	11.76

a	$i=$ $k_m s_m$	k_m	s_m	k_p	s_p	C	C_A	C_R	U	U_A	U_R	O	O_A	O_R	W	W_A	W_R
20	2.5	1	2.5	1.5	33.33	79.81	77.11	2.7	1.79	0.03	1.76	18.41	0.16	18.24	20.19	0.19	20.01
20	2.5	1.5	1.67	1.5	33.33	79.81	77.3	2.51	1.79	0.01	1.78	18.41	0.04	18.37	20.19	0.05	20.15
20	2.5	2	1.25	1.5	33.33	79.81	77.35	2.45	1.79	0	1.78	18.41	0.01	18.4	20.19	0.01	20.18
20	2.5	2.5	1	1.5	33.33	79.81	77.37	2.44	1.79	0	1.79	18.41	0	18.4	20.19	0	20.19
20	2.5	3	0.83	1.5	33.33	79.81	77.38	2.43	1.79	0	1.79	18.41	0	18.41	20.19	0	20.19
20	5	1	5	1.5	33.33	79.81	74.04	5.77	1.79	0.05	1.74	18.41	0.31	18.09	20.19	0.36	19.83
20	5	1.5	3.33	1.5	33.33	79.81	74.52	5.29	1.79	0.01	1.77	18.41	0.08	18.33	20.19	0.09	20.1
20	5	2	2.5	1.5	33.33	79.81	74.66	5.15	1.79	0	1.78	18.41	0.02	18.39	20.19	0.02	20.17
20	5	2.5	2	1.5	33.33	79.81	74.71	5.1	1.79	0	1.79	18.41	0	18.4	20.19	0	20.19
20	5	3	1.67	1.5	33.33	79.81	74.73	5.07	1.79	0	1.79	18.41	0	18.41	20.19	0	20.19
20	7.5	1	7.5	1.5	33.33	79.81	70.6	9.2	1.79	0.07	1.72	18.41	0.45	17.95	20.19	0.52	19.67
20	7.5	1.5	5	1.5	33.33	79.81	71.46	8.35	1.79	0.02	1.77	18.41	0.12	18.29	20.19	0.13	20.06
20	7.5	2	3.75	1.5	33.33	79.81	71.73	8.08	1.79	0	1.78	18.41	0.03	18.38	20.19	0.03	20.16
20	7.5	2.5	3	1.5	33.33	79.81	71.83	7.98	1.79	0	1.79	18.41	0.01	18.4	20.19	0.01	20.19
20	7.5	3	2.5	1.5	33.33	79.81	71.88	7.93	1.79	0	1.79	18.41	0	18.4	20.19	0	20.19
20	10	1	10	1.5	33.33	79.81	66.83	12.98	1.79	0.08	1.7	18.41	0.58	17.82	20.19	0.67	19.53
20	10	1.5	6.67	1.5	33.33	79.81	68.13	11.67	1.79	0.02	1.76	18.41	0.15	18.25	20.19	0.17	20.02
20	10	2	5	1.5	33.33	79.81	68.56	11.25	1.79	0.01	1.78	18.41	0.04	18.37	20.19	0.04	20.15
20	10	2.5	4	1.5	33.33	79.81	68.73	11.08	1.79	0	1.79	18.41	0.01	18.4	20.19	0.01	20.19
20	10	3	3.33	1.5	33.33	79.81	68.81	11	1.79	0	1.79	18.41	0	18.4	20.19	0	20.19
20	15	1	15	1.5	33.33	79.81	58.39	21.42	1.79	0.11	1.68	18.41	0.81	17.59	20.19	0.93	19.27
20	15	1.5	10	1.5	33.33	79.81	60.73	19.08	1.79	0.03	1.75	18.41	0.22	18.19	20.19	0.25	19.94
20	15	2	7.5	1.5	33.33	79.81	61.53	18.27	1.79	0.01	1.78	18.41	0.05	18.35	20.19	0.06	20.13
20	15	2.5	6	1.5	33.33	79.81	61.87	17.93	1.79	0	1.78	18.41	0.01	18.39	20.19	0.01	20.18
20	15	3	5	1.5	33.33	79.81	62.05	17.76	1.79	0	1.79	18.41	0	18.4	20.19	0	20.19
20	20	1	20	1.5	33.33	79.81	49.12	30.69	1.79	0.13	1.65	18.41	1.01	17.39	20.19	1.14	19.05
20	20	1.5	13.33	1.5	33.33	79.81	52.47	27.33	1.79	0.04	1.75	18.41	0.28	18.13	20.19	0.32	19.87
20	20	2	10	1.5	33.33	79.81	53.69	26.12	1.79	0.01	1.78	18.41	0.07	18.34	20.19	0.08	20.11
20	20	2.5	8	1.5	33.33	79.81	54.22	25.58	1.79	0	1.78	18.41	0.02	18.39	20.19	0.02	20.17
20	20	3	6.67	1.5	33.33	79.81	54.5	25.31	1.79	0	1.79	18.41	0	18.4	20.19	0	20.19
20	25	1	25	1.5	33.33	79.81	39.51	40.3	1.79	0.15	1.64	18.41	1.17	17.24	20.19	1.32	18.88
20	25	1.5	16.67	1.5	33.33	79.81	43.61	36.2	1.79	0.05	1.74	18.41	0.34	18.07	20.19	0.38	19.81
20	25	2	12.5	1.5	33.33	79.81	45.18	34.63	1.79	0.01	1.77	18.41	0.09	18.32	20.19	0.1	20.1
20	25	2.5	10	1.5	33.33	79.81	45.89	33.91	1.79	0	1.78	18.41	0.02	18.39	20.19	0.02	20.17
20	25	3	8.33	1.5	33.33	79.81	46.27	33.54	1.79	0	1.79	18.41	0	18.4	20.19	0	20.19
20	2.5	1	2.5	1	50	64.5	61.78	2.71	8.08	0.06	8.01	27.43	0.14	27.29	35.5	0.2	35.3
20	2.5	1.5	1.67	1	50	64.5	61.96	2.54	8.08	0.02	8.06	27.43	0.03	27.39	35.5	0.05	35.45
20	2.5	2	1.25	1	50	64.5	62.01	2.49	8.08	0	8.07	27.43	0.01	27.42	35.5	0.01	35.49
20	2.5	2.5	1	1	50	64.5	62.02	2.48	8.08	0	8.07	27.43	0	27.42	35.5	0	35.5
20	2.5	3	0.83	1	50	64.5	62.03	2.47	8.08	0	8.08	27.43	0	27.42	35.5	0	35.5
20	5	1	5	1	50	64.5	58.88	5.62	8.08	0.12	7.96	27.43	0.27	27.15	35.5	0.39	35.11
20	5	1.5	3.33	1	50	64.5	59.28	5.22	8.08	0.03	8.05	27.43	0.07	27.36	35.5	0.1	35.4
20	5	2	2.5	1	50	64.5	59.39	5.11	8.08	0.01	8.07	27.43	0.02	27.41	35.5	0.02	35.48
20	5	2.5	2	1	50	64.5	59.43	5.07	8.08	0	8.07	27.43	0	27.42	35.5	0.01	35.5
20	5	3	1.67	1	50	64.5	59.44	5.06	8.08	0	8.08	27.43	0	27.42	35.5	0	35.5
20	7.5	1	7.5	1	50	64.5	55.8	8.7	8.08	0.17	7.91	27.43	0.4	27.03	35.5	0.57	34.93
20	7.5	1.5	5	1	50	64.5	56.46	8.04	8.08	0.04	8.03	27.43	0.1	27.33	35.5	0.14	35.36
20	7.5	2	3.75	1	50	64.5	56.66	7.84	8.08	0.01	8.07	27.43	0.02	27.4	35.5	0.03	35.47
20	7.5	2.5	3	1	50	64.5	56.72	7.78	8.08	0	8.07	27.43	0.01	27.42	35.5	0.01	35.49
20	7.5	3	2.5	1	50	64.5	56.75	7.75	8.08	0	8.08	27.43	0	27.42	35.5	0	35.5
20	10	1	10	1	50	64.5	52.55	11.95	8.08	0.22	7.86	27.43	0.52	26.9	35.5	0.74	34.76
20	10	1.5	6.67	1	50	64.5	53.51	10.99	8.08	0.06	8.02	27.43	0.13	27.3	35.5	0.19	35.31
20	10	2	5	1	50	64.5	53.8	10.7	8.08	0.01	8.06	27.43	0.03	27.39	35.5	0.05	35.46
20	10	2.5	4	1	50	64.5	53.9	10.6	8.08	0	8.07	27.43	0.01	27.42	35.5	0.01	35.49
20	10	3	3.33	1	50	64.5	53.95	10.55	8.08	0	8.08	27.43	0	27.42	35.5	0	35.5

a	$i=$ $k_m s_m$	k_m	s_m	k_p	s_p	C	C_A	C_R	U	U_A	U_R	O	O_A	O_R	W	W_A	W_R
20	15	1	15	1	50	64.5	45.64	18.86	8.08	0.31	7.77	27.43	0.76	26.67	35.5	1.07	34.43
20	15	1.5	10	1	50	64.5	47.26	17.24	8.08	0.08	7.99	27.43	0.19	27.23	35.5	0.28	35.22
20	15	2	7.5	1	50	64.5	47.76	16.73	8.08	0.02	8.06	27.43	0.05	27.38	35.5	0.07	35.43
20	15	2.5	6	1	50	64.5	47.96	16.54	8.08	0	8.07	27.43	0.01	27.42	35.5	0.01	35.49
20	15	3	5	1	50	64.5	48.05	16.45	8.08	0	8.07	27.43	0	27.42	35.5	0	35.5
20	20	1	20	1	50	64.5	38.36	26.14	8.08	0.39	7.69	27.43	0.98	26.45	35.5	1.36	34.14
20	20	1.5	13.33	1	50	64.5	40.61	23.89	8.08	0.11	7.97	27.43	0.25	27.17	35.5	0.36	35.14
20	20	2	10	1	50	64.5	41.35	23.15	8.08	0.03	8.05	27.43	0.06	27.36	35.5	0.09	35.41
20	20	2.5	8	1	50	64.5	41.64	22.86	8.08	0.01	8.07	27.43	0.01	27.41	35.5	0.02	35.48
20	20	3	6.67	1	50	64.5	41.78	22.72	8.08	0	8.07	27.43	0	27.42	35.5	0	35.5
20	25	1	25	1	50	64.5	30.93	33.57	8.08	0.45	7.63	27.43	1.16	26.26	35.5	1.61	33.89
20	25	1.5	16.67	1	50	64.5	33.69	30.81	8.08	0.13	7.95	27.43	0.31	27.11	35.5	0.44	35.06
20	25	2	12.5	1	50	64.5	34.63	29.87	8.08	0.03	8.04	27.43	0.08	27.35	35.5	0.11	35.39
20	25	2.5	10	1	50	64.5	35.01	29.49	8.08	0.01	8.07	27.43	0.02	27.41	35.5	0.02	35.48
20	25	3	8.33	1	50	64.5	35.19	29.31	8.08	0	8.07	27.43	0	27.42	35.5	0	35.5

a	i= k_m, s_m	k_m	s_m	k_p	s_p	C	C_A	C_R	U	U_A	U_R	O	O_A	O_R	W	W_A	W_R
25	2.5	1	2.5	3	16.67	93.32	90.76	2.56	0	0	0	6.68	0.15	6.53	6.68	0.15	6.53
25	2.5	1.5	1.67	3	16.67	93.32	91	2.32	0	0	0	6.68	0.04	6.64	6.68	0.04	6.64
25	2.5	2	1.25	3	16.67	93.32	91.08	2.24	0	0	0	6.68	0.01	6.67	6.68	0.01	6.67
25	2.5	2.5	1	3	16.67	93.32	91.11	2.21	0	0	0	6.68	0	6.68	6.68	0	6.68
25	2.5	3	0.83	3	16.67	93.32	91.12	2.2	0	0	0	6.68	0	6.68	6.68	0	6.68
25	5	1	5	3	16.67	93.32	87.22	6.1	0	0	0	6.68	0.26	6.42	6.68	0.26	6.42
25	5	1.5	3.33	3	16.67	93.32	87.96	5.36	0	0	0	6.68	0.07	6.61	6.68	0.07	6.61
25	5	2	2.5	3	16.67	93.32	88.21	5.1	0	0	0	6.68	0.02	6.66	6.68	0.02	6.66
25	5	2.5	2	3	16.67	93.32	88.32	5	0	0	0	6.68	0	6.68	6.68	0	6.68
25	5	3	1.67	3	16.67	93.32	88.37	4.94	0	0	0	6.68	0	6.68	6.68	0	6.68
25	7.5	1	7.5	3	16.67	93.32	82.73	10.59	0	0	0	6.68	0.36	6.32	6.68	0.36	6.32
25	7.5	1.5	5	3	16.67	93.32	84.17	9.15	0	0	0	6.68	0.1	6.58	6.68	0.1	6.58
25	7.5	2	3.75	3	16.67	93.32	84.69	8.63	0	0	0	6.68	0.03	6.66	6.68	0.03	6.66
25	7.5	2.5	3	3	16.67	93.32	84.92	8.4	0	0	0	6.68	0.01	6.68	6.68	0.01	6.68
25	7.5	3	2.5	3	16.67	93.32	85.04	8.28	0	0	0	6.68	0	6.68	6.68	0	6.68
25	10	1	10	3	16.67	93.32	77.54	15.78	0	0	0	6.68	0.43	6.25	6.68	0.43	6.25
25	10	1.5	6.67	3	16.67	93.32	79.7	13.62	0	0	0	6.68	0.13	6.56	6.68	0.13	6.56
25	10	2	5	3	16.67	93.32	80.53	12.79	0	0	0	6.68	0.03	6.65	6.68	0.03	6.65
25	10	2.5	4	3	16.67	93.32	80.91	12.41	0	0	0	6.68	0.01	6.67	6.68	0.01	6.67
25	10	3	3.33	3	16.67	93.32	81.12	12.2	0	0	0	6.68	0	6.68	6.68	0	6.68
25	15	1	15	3	16.67	93.32	66.36	26.96	0	0	0	6.68	0.55	6.13	6.68	0.55	6.13
25	15	1.5	10	3	16.67	93.32	69.41	23.91	0	0	0	6.68	0.17	6.51	6.68	0.17	6.51
25	15	2	7.5	3	16.67	93.32	70.7	22.62	0	0	0	6.68	0.04	6.64	6.68	0.04	6.64
25	15	2.5	6	3	16.67	93.32	71.34	21.98	0	0	0	6.68	0.01	6.67	6.68	0.01	6.67
25	15	3	5	3	16.67	93.32	71.7	21.62	0	0	0	6.68	0	6.68	6.68	0	6.68
25	20	1	20	3	16.67	93.32	55.28	38.03	0	0	0	6.68	0.63	6.05	6.68	0.63	6.05
25	20	1.5	13.33	3	16.67	93.32	58.58	34.74	0	0	0	6.68	0.2	6.48	6.68	0.2	6.48
25	20	2	10	3	16.67	93.32	59.87	33.45	0	0	0	6.68	0.05	6.63	6.68	0.05	6.63
25	20	2.5	8	3	16.67	93.32	60.51	32.81	0	0	0	6.68	0.01	6.67	6.68	0.01	6.67
25	20	3	6.67	3	16.67	93.32	60.86	32.46	0	0	0	6.68	0	6.68	6.68	0	6.68
25	25	1	25	3	16.67	93.32	44.51	48.81	0	0	0	6.68	0.68	6	6.68	0.68	6
25	25	1.5	16.67	3	16.67	93.32	48.05	45.27	0	0	0	6.68	0.23	6.45	6.68	0.23	6.45
25	25	2	12.5	3	16.67	93.32	49.1	44.22	0	0	0	6.68	0.06	6.62	6.68	0.06	6.62
25	25	2.5	10	3	16.67	93.32	49.47	43.85	0	0	0	6.68	0.02	6.67	6.68	0.02	6.67
25	25	3	8.33	3	16.67	93.32	49.63	43.69	0	0	0	6.68	0	6.68	6.68	0	6.68
25	2.5	1	2.5	2.5	20	89.43	86.59	2.83	0.01	0	0.01	10.56	0.18	10.39	10.57	0.18	10.4
25	2.5	1.5	1.67	2.5	20	89.43	86.83	2.6	0.01	0	0.01	10.56	0.04	10.52	10.57	0.05	10.53
25	2.5	2	1.25	2.5	20	89.43	86.9	2.53	0.01	0	0.01	10.56	0.01	10.55	10.57	0.01	10.56
25	2.5	2.5	1	2.5	20	89.43	86.92	2.5	0.01	0	0.01	10.56	0	10.56	10.57	0	10.57
25	2.5	3	0.83	2.5	20	89.43	86.93	2.49	0.01	0	0.01	10.56	0	10.56	10.57	0	10.57
25	5	1	5	2.5	20	89.43	83.04	6.39	0.01	0	0.01	10.56	0.33	10.24	10.57	0.33	10.24
25	5	1.5	3.33	2.5	20	89.43	83.69	5.74	0.01	0	0.01	10.56	0.09	10.48	10.57	0.09	10.49
25	5	2	2.5	2.5	20	89.43	83.9	5.53	0.01	0	0.01	10.56	0.02	10.54	10.57	0.02	10.55
25	5	2.5	2	2.5	20	89.43	83.98	5.44	0.01	0	0.01	10.56	0	10.56	10.57	0	10.57
25	5	3	1.67	2.5	20	89.43	84.03	5.4	0.01	0	0.01	10.56	0	10.56	10.57	0	10.57
25	7.5	1	7.5	2.5	20	89.43	78.84	10.59	0.01	0	0.01	10.56	0.46	10.11	10.57	0.46	10.11
25	7.5	1.5	5	2.5	20	89.43	80.03	9.4	0.01	0	0.01	10.56	0.12	10.44	10.57	0.12	10.45
25	7.5	2	3.75	2.5	20	89.43	80.44	8.99	0.01	0	0.01	10.56	0.03	10.53	10.57	0.03	10.54
25	7.5	2.5	3	2.5	20	89.43	80.61	8.82	0.01	0	0.01	10.56	0.01	10.56	10.57	0.01	10.57
25	7.5	3	2.5	2.5	20	89.43	80.7	8.73	0.01	0	0.01	10.56	0	10.56	10.57	0	10.57

a	k_m	s_m	k_m/s_m (i=)	k_p	s_p	C	C_A	C_R	U	U_A	U_R	O	O_A	O_R	W	W_A	W_R
25	10	1	10	2.5	20	89.43	74.15	15.28	0.01	0	0.01	10.56	0.57	10	10.57	0.57	10
25	10	1.5	6.67	2.5	20	89.43	75.91	13.52	0.01	0	0.01	10.56	0.16	10.41	10.57	0.16	10.42
25	10	2	5	2.5	20	89.43	76.54	12.89	0.01	0	0.01	10.56	0.04	10.53	10.57	0.04	10.53
25	10	2.5	4	2.5	20	89.43	76.82	12.61	0.01	0	0.01	10.56	0.01	10.56	10.57	0.01	10.56
25	10	3	3.33	2.5	20	89.43	76.96	12.47	0.01	0	0.01	10.56	0	10.56	10.57	0	10.57
25	15	1	15	2.5	20	89.43	64.01	25.42	0.01	0	0.01	10.56	0.74	9.82	10.57	0.74	9.83
25	15	1.5	10	2.5	20	89.43	66.69	22.73	0.01	0	0.01	10.56	0.22	10.35	10.57	0.22	10.36
25	15	2	7.5	2.5	20	89.43	67.72	21.71	0.01	0	0.01	10.56	0.06	10.51	10.57	0.06	10.52
25	15	2.5	6	2.5	20	89.43	68.19	21.24	0.01	0	0.01	10.56	0.01	10.55	10.57	0.01	10.56
25	15	3	5	2.5	20	89.43	68.44	20.99	0.01	0	0.01	10.56	0	10.56	10.57	0	10.57
25	20	1	20	2.5	20	89.43	53.57	35.85	0.01	0	0.01	10.56	0.87	9.69	10.57	0.87	9.7
25	20	1.5	13.33	2.5	20	89.43	56.87	32.55	0.01	0	0.01	10.56	0.27	10.3	10.57	0.27	10.31
25	20	2	10	2.5	20	89.43	58.08	31.34	0.01	0	0.01	10.56	0.07	10.49	10.57	0.07	10.5
25	20	2.5	8	2.5	20	89.43	58.63	30.8	0.01	0	0.01	10.56	0.02	10.55	10.57	0.02	10.56
25	20	3	6.67	2.5	20	89.43	58.92	30.51	0.01	0	0.01	10.56	0	10.56	10.57	0	10.57
25	25	1	25	2.5	20	89.43	43.14	46.29	0.01	0	0.01	10.56	0.96	9.6	10.57	0.96	9.61
25	25	1.5	16.67	2.5	20	89.43	46.94	42.49	0.01	0	0.01	10.56	0.31	10.26	10.57	0.31	10.27
25	25	2	12.5	2.5	20	89.43	48.21	41.22	0.01	0	0.01	10.56	0.08	10.48	10.57	0.08	10.49
25	25	2.5	10	2.5	20	89.43	48.71	40.72	0.01	0	0.01	10.56	0.02	10.55	10.57	0.02	10.55
25	25	3	8.33	2.5	20	89.43	48.94	40.48	0.01	0	0.01	10.56	0	10.56	10.57	0	10.57
25	2.5	1	2.5	2	25	84	81.08	2.92	0.13	0	0.13	15.87	0.19	15.67	16	0.2	15.8
25	2.5	1.5	1.67	2	25	84	81.3	2.7	0.13	0	0.13	15.87	0.05	15.82	16	0.05	15.95
25	2.5	2	1.25	2	25	84	81.36	2.64	0.13	0	0.13	15.87	0.01	15.85	16	0.01	15.99
25	2.5	2.5	1	2	25	84	81.38	2.62	0.13	0	0.13	15.87	0	15.86	16	0	16
25	2.5	3	0.83	2	25	84	81.39	2.61	0.13	0	0.13	15.87	0	15.87	16	0	16
25	5	1	5	2	25	84	77.69	6.31	0.13	0.01	0.13	15.87	0.37	15.5	16	0.37	15.63
25	5	1.5	3.33	2	25	84	78.24	5.76	0.13	0	0.13	15.87	0.09	15.77	16	0.1	15.9
25	5	2	2.5	2	25	84	78.41	5.59	0.13	0	0.13	15.87	0.02	15.84	16	0.02	15.98
25	5	2.5	2	2	25	84	78.47	5.53	0.13	0	0.13	15.87	0	15.86	16	0.01	16
25	5	3	1.67	2	25	84	78.5	5.5	0.13	0	0.13	15.87	0	15.86	16	0	16
25	7.5	1	7.5	2	25	84	73.86	10.14	0.13	0.01	0.13	15.87	0.52	15.34	16	0.53	15.47
25	7.5	1.5	5	2	25	84	74.83	9.17	0.13	0	0.13	15.87	0.14	15.73	16	0.14	15.86
25	7.5	2	3.75	2	25	84	75.15	8.85	0.13	0	0.13	15.87	0.03	15.83	16	0.03	15.97
25	7.5	2.5	3	2	25	84	75.27	8.73	0.13	0	0.13	15.87	0.01	15.86	16	0.01	15.99
25	7.5	3	2.5	2	25	84	75.33	8.67	0.13	0	0.13	15.87	0	15.86	16	0	16
25	10	1	10	2	25	84	69.67	14.33	0.13	0.01	0.13	15.87	0.66	15.2	16	0.67	15.33
25	10	1.5	6.67	2	25	84	71.12	12.88	0.13	0	0.13	15.87	0.18	15.69	16	0.18	15.82
25	10	2	5	2	25	84	71.6	12.4	0.13	0	0.13	15.87	0.04	15.82	16	0.04	15.96
25	10	2.5	4	2	25	84	71.8	12.2	0.13	0	0.13	15.87	0.01	15.86	16	0.01	15.99
25	10	3	3.33	2	25	84	71.9	12.1	0.13	0	0.13	15.87	0	15.86	16	0	16
25	15	1	15	2	25	84	60.54	23.46	0.13	0.01	0.12	15.87	0.9	14.97	16	0.91	15.09
25	15	1.5	10	2	25	84	62.94	21.06	0.13	0	0.13	15.87	0.25	15.61	16	0.25	15.75
25	15	2	7.5	2	25	84	63.78	20.22	0.13	0	0.13	15.87	0.06	15.8	16	0.06	15.94
25	15	2.5	6	2	25	84	64.14	19.86	0.13	0	0.13	15.87	0.01	15.85	16	0.01	15.99
25	15	3	5	2	25	84	64.32	19.68	0.13	0	0.13	15.87	0	15.86	16	0	16
25	20	1	20	2	25	84	50.83	33.16	0.13	0.01	0.12	15.87	1.09	14.78	16	1.1	14.9
25	20	1.5	13.33	2	25	84	54.08	29.92	0.13	0	0.13	15.87	0.31	15.55	16	0.32	15.68
25	20	2	10	2	25	84	55.23	28.77	0.13	0	0.13	15.87	0.08	15.78	16	0.08	15.92
25	20	2.5	8	2	25	84	55.73	28.27	0.13	0	0.13	15.87	0.02	15.85	16	0.02	15.98
25	20	3	6.67	2	25	84	55.98	28.02	0.13	0	0.13	15.87	0	15.86	16	0	16
25	25	1	25	2	25	84	40.93	43.07	0.13	0.01	0.12	15.87	1.23	14.63	16	1.24	14.76
25	25	1.5	16.67	2	25	84	44.82	39.18	0.13	0	0.13	15.87	0.37	15.49	16	0.38	15.62
25	25	2	12.5	2	25	84	46.22	37.78	0.13	0	0.13	15.87	0.1	15.77	16	0.1	15.9
25	25	2.5	10	2	25	84	46.81	37.19	0.13	0	0.13	15.87	0.02	15.84	16	0.02	15.98
25	25	3	8.33	2	25	84	47.11	36.89	0.13	0	0.13	15.87	0	15.86	16	0	16

a	i= $k_m s_m$	k_m	s_m	k_p	s_p	C	C_A	C_R	U	U_A	U_R	O	O_A	O_R	W	W_A	W_R
25	2.5	1	2.5	1.5	33.33	76.11	73.25	2.87	1.22	0.02	1.2	22.66	0.18	22.48	23.89	0.2	23.68
25	2.5	1.5	1.67	1.5	33.33	76.11	73.45	2.67	1.22	0	1.22	22.66	0.05	22.62	23.89	0.05	23.83
25	2.5	2	1.25	1.5	33.33	76.11	73.5	2.61	1.22	0	1.22	22.66	0.01	22.65	23.89	0.01	23.87
25	2.5	2.5	1	1.5	33.33	76.11	73.52	2.6	1.22	0	1.22	22.66	0	22.66	23.89	0	23.88
25	2.5	3	0.83	1.5	33.33	76.11	73.52	2.59	1.22	0	1.22	22.66	0	22.66	23.89	0	23.88
25	5	1	5	1.5	33.33	76.11	70.07	6.04	1.22	0.03	1.19	22.66	0.36	22.31	23.89	0.39	23.49
25	5	1.5	3.33	1.5	33.33	76.11	70.54	5.57	1.22	0.01	1.21	22.66	0.09	22.57	23.89	0.1	23.79
25	5	2	2.5	1.5	33.33	76.11	70.68	5.43	1.22	0	1.22	22.66	0.02	22.64	23.89	0.02	23.86
25	5	2.5	2	1.5	33.33	76.11	70.73	5.38	1.22	0	1.22	22.66	0	22.66	23.89	0.01	23.88
25	5	3	1.67	1.5	33.33	76.11	70.75	5.36	1.22	0	1.22	22.66	0	22.66	23.89	0	23.88
25	7.5	1	7.5	1.5	33.33	76.11	66.6	9.52	1.22	0.05	1.17	22.66	0.52	22.14	23.89	0.57	23.32
25	7.5	1.5	5	1.5	33.33	76.11	67.41	8.7	1.22	0.01	1.21	22.66	0.13	22.53	23.89	0.15	23.74
25	7.5	2	3.75	1.5	33.33	76.11	67.66	8.45	1.22	0	1.22	22.66	0.03	22.63	23.89	0.04	23.85
25	7.5	2.5	3	1.5	33.33	76.11	67.75	8.36	1.22	0	1.22	22.66	0.01	22.66	23.89	0.01	23.88
25	7.5	3	2.5	1.5	33.33	76.11	67.8	8.32	1.22	0	1.22	22.66	0	22.66	23.89	0	23.88
25	10	1	10	1.5	33.33	76.11	62.86	13.26	1.22	0.06	1.16	22.66	0.67	21.99	23.89	0.73	23.15
25	10	1.5	6.67	1.5	33.33	76.11	64.07	12.05	1.22	0.02	1.21	22.66	0.17	22.49	23.89	0.19	23.69
25	10	2	5	1.5	33.33	76.11	64.45	11.67	1.22	0	1.22	22.66	0.04	22.62	23.89	0.05	23.84
25	10	2.5	4	1.5	33.33	76.11	64.6	11.52	1.22	0	1.22	22.66	0.01	22.65	23.89	0.01	23.88
25	10	3	3.33	1.5	33.33	76.11	64.66	11.45	1.22	0	1.22	22.66	0	22.66	23.89	0	23.88
25	15	1	15	1.5	33.33	76.11	54.73	21.39	1.22	0.08	1.14	22.66	0.95	21.72	23.89	1.03	22.86
25	15	1.5	10	1.5	33.33	76.11	56.8	19.31	1.22	0.02	1.2	22.66	0.25	22.41	23.89	0.27	23.61
25	15	2	7.5	1.5	33.33	76.11	57.49	18.62	1.22	0.01	1.22	22.66	0.06	22.6	23.89	0.07	23.82
25	15	2.5	6	1.5	33.33	76.11	57.77	18.34	1.22	0	1.22	22.66	0.01	22.65	23.89	0.01	23.87
25	15	3	5	1.5	33.33	76.11	57.91	18.21	1.22	0	1.22	22.66	0	22.66	23.89	0	23.88
25	20	1	20	1.5	33.33	76.11	46.01	30.11	1.22	0.09	1.13	22.66	1.18	21.48	23.89	1.28	22.61
25	20	1.5	13.33	1.5	33.33	76.11	48.91	27.2	1.22	0.03	1.19	22.66	0.32	22.34	23.89	0.35	23.53
25	20	2	10	1.5	33.33	76.11	49.93	26.19	1.22	0.01	1.22	22.66	0.08	22.58	23.89	0.09	23.8
25	20	2.5	8	1.5	33.33	76.11	50.35	25.77	1.22	0	1.22	22.66	0.02	22.64	23.89	0.02	23.87
25	20	3	6.67	1.5	33.33	76.11	50.56	25.56	1.22	0	1.22	22.66	0	22.66	23.89	0	23.88
25	25	1	25	1.5	33.33	76.11	37.05	39.06	1.22	0.1	1.12	22.66	1.38	21.29	23.89	1.48	22.41
25	25	1.5	16.67	1.5	33.33	76.11	40.6	35.51	1.22	0.03	1.19	22.66	0.39	22.27	23.89	0.42	23.46
25	25	2	12.5	1.5	33.33	76.11	41.89	34.23	1.22	0.01	1.21	22.66	0.1	22.56	23.89	0.11	23.78
25	25	2.5	10	1.5	33.33	76.11	42.44	33.68	1.22	0	1.22	22.66	0.02	22.64	23.89	0.02	23.86
25	25	3	8.33	1.5	33.33	76.11	42.72	33.4	1.22	0	1.22	22.66	0	22.66	23.89	0	23.88
25	2.5	1	2.5	1	50	62.47	59.77	2.7	6.68	0.05	6.63	30.85	0.15	30.71	37.53	0.2	37.34
25	2.5	1.5	1.67	1	50	62.47	59.94	2.52	6.68	0.01	6.67	30.85	0.04	30.82	37.53	0.05	37.49
25	2.5	2	1.25	1	50	62.47	59.99	2.48	6.68	0	6.68	30.85	0.01	30.85	37.53	0.01	37.52
25	2.5	2.5	1	1	50	62.47	60	2.46	6.68	0	6.68	30.85	0	30.85	37.53	0	37.53
25	2.5	3	0.83	1	50	62.47	60.01	2.46	6.68	0	6.68	30.85	0	30.85	37.53	0	37.53
25	5	1	5	1	50	62.47	56.9	5.57	6.68	0.1	6.58	30.85	0.29	30.57	37.53	0.39	37.15
25	5	1.5	3.33	1	50	62.47	57.29	5.18	6.68	0.03	6.66	30.85	0.07	30.78	37.53	0.1	37.44
25	5	2	2.5	1	50	62.47	57.4	5.07	6.68	0.01	6.67	30.85	0.02	30.84	37.53	0.02	37.51
25	5	2.5	2	1	50	62.47	57.43	5.03	6.68	0	6.68	30.85	0	30.85	37.53	0.01	37.53
25	5	3	1.67	1	50	62.47	57.45	5.02	6.68	0	6.68	30.85	0	30.85	37.53	0	37.53
25	7.5	1	7.5	1	50	62.47	53.87	8.59	6.68	0.15	6.53	30.85	0.43	30.43	37.53	0.57	36.96
25	7.5	1.5	5	1	50	62.47	54.52	7.95	6.68	0.04	6.64	30.85	0.11	30.75	37.53	0.14	37.39
25	7.5	2	3.75	1	50	62.47	54.7	7.77	6.68	0.01	6.67	30.85	0.03	30.83	37.53	0.03	37.5
25	7.5	2.5	3	1	50	62.47	54.76	7.7	6.68	0	6.68	30.85	0.01	30.85	37.53	0.01	37.53
25	7.5	3	2.5	1	50	62.47	54.79	7.68	6.68	0	6.68	30.85	0	30.85	37.53	0	37.53
25	10	1	10	1	50	62.47	50.7	11.76	6.68	0.19	6.49	30.85	0.56	30.29	37.53	0.75	36.79
25	10	1.5	6.67	1	50	62.47	51.62	10.84	6.68	0.05	6.63	30.85	0.14	30.71	37.53	0.19	37.34
25	10	2	5	1	50	62.47	51.9	10.57	6.68	0.01	6.67	30.85	0.03	30.82	37.53	0.05	37.49
25	10	2.5	4	1	50	62.47	51.99	10.47	6.68	0	6.68	30.85	0.01	30.85	37.53	0.01	37.52
25	10	3	3.33	1	50	62.47	52.03	10.43	6.68	0	6.68	30.85	0	30.85	37.53	0	37.53

a	$\begin{matrix}i=\\k_m s_m\end{matrix}$	k_m	s_m	k_p	s_p	C	C_A	C_R	U	U_A	U_R	O	O_A	O_R	W	W_A	W_R
25	15	1	15	1	50	62.47	44	18.46	6.68	0.26	6.42	30.85	0.81	30.04	37.53	1.08	36.46
25	15	1.5	10	1	50	62.47	45.53	16.93	6.68	0.07	6.61	30.85	0.21	30.65	37.53	0.28	37.26
25	15	2	7.5	1	50	62.47	46.01	16.46	6.68	0.02	6.66	30.85	0.05	30.8	37.53	0.07	37.47
25	15	2.5	6	1	50	62.47	46.18	16.28	6.68	0	6.68	30.85	0.01	30.84	37.53	0.01	37.52
25	15	3	5	1	50	62.47	46.26	16.2	6.68	0	6.68	30.85	0	30.85	37.53	0	37.53
25	20	1	20	1	50	62.47	36.98	25.49	6.68	0.33	6.35	30.85	1.05	29.81	37.53	1.38	36.16
25	20	1.5	13.33	1	50	62.47	39.1	23.37	6.68	0.09	6.59	30.85	0.27	30.58	37.53	0.36	37.17
25	20	2	10	1	50	62.47	39.78	22.69	6.68	0.02	6.66	30.85	0.07	30.79	37.53	0.09	37.45
25	20	2.5	8	1	50	62.47	40.04	22.42	6.68	0.01	6.68	30.85	0.01	30.84	37.53	0.02	37.52
25	20	3	6.67	1	50	62.47	40.17	22.3	6.68	0	6.68	30.85	0	30.85	37.53	0	37.53
25	25	1	25	1	50	62.47	29.82	32.64	6.68	0.38	6.3	30.85	1.26	29.6	37.53	1.64	35.9
25	25	1.5	16.67	1	50	62.47	32.42	30.04	6.68	0.11	6.57	30.85	0.33	30.52	37.53	0.44	37.09
25	25	2	12.5	1	50	62.47	33.29	29.17	6.68	0.03	6.65	30.85	0.08	30.77	37.53	0.11	37.43
25	25	2.5	10	1	50	62.47	33.64	28.83	6.68	0.01	6.67	30.85	0.02	30.84	37.53	0.02	37.51
25	25	3	8.33	1	50	62.47	33.8	28.67	6.68	0	6.68	30.85	0	30.85	37.53	0	37.53

Tables XXIII a and b. Tables of the Convolution of Gaussian and Rectangular Probability Density Distributions

Introduction to part (a)

I.XXIII(1) This table gives the values of the tolerance factor k for indexed values of η and P, where η is the ratio of the standard deviation σ_G, of the Gaussian distribution, to the standard deviation σ_R, of the rectangular distribution, and P is the probability of an event occurring in the tolerance interval $\mu - k\sigma$ to $\mu + k\sigma$, where μ is the mean of the convoluted distribution and σ is its standard deviation.

$$\sigma = \left[\sigma_G^2 + \sigma_R^2 \right]^{1/2}$$

and

$$\eta = \frac{\sigma_G}{\sigma_R}$$

I.XXIII(2) The tabulated values of k were calculated from the formula for the probability P of an event lying within the limits $\mu - k\sigma$ to $\mu + k\sigma$ for the combined distributions, where

$$P = \left[\frac{1 + \eta^2}{6\pi} \right]^{1/2} \int_0^k dq \int_{\frac{\{q\sqrt{(1 + \eta^2)} - \sqrt{3}\}}{\eta}}^{\frac{\{q\sqrt{(1 + \eta^2)} + \sqrt{3}\}}{\eta}} e^{-y^2/2} \, dy$$

by finding each value of k which satisfied the expression for given values of P and η.

I.XXIII(3) Equation 4.53(2) gives

$$P_{RG}(x) = \frac{1}{2\sqrt{\{3(1 - \alpha^2)\}}\sigma} \int_{\frac{q - \sqrt{\{3(1 - \alpha^2)\}}}{\alpha}}^{\frac{q + \sqrt{\{3(1 - \alpha^2)\}}}{\alpha}} \frac{e^{-y^2/2}}{\sqrt{(2\pi)}} \, dy$$

where

$$\alpha = \frac{\sigma_G}{\sigma}$$

It is easily shown that

$$\alpha = \frac{\eta}{\sqrt{(1 + \eta^2)}}$$

whence 4.53(2) becomes

$$P_{RG}(x) = \frac{(1 + \eta^2)^{1/2}}{2(6\pi)^{1/2}\sigma} \int_{\{q\sqrt{(1+\eta^2)} - \sqrt{3}\}/\eta}^{\{q\sqrt{(1+\eta^2)} + \sqrt{3}\}/\eta} e^{-y^2/2}\, dy$$

If we put $x = q\sigma$ and integrate for q between $-k$ and k, we have

$$P = \frac{(1 + \eta^2)^{1/2}}{2(6\pi)^{1/2}} \int_{-k}^{k} dq \int_{\{q\sqrt{(1+\eta^2)} - \sqrt{3}\}/\eta}^{\{q\sqrt{(1+\eta^2)} + \sqrt{3}\}/\eta} e^{-y^2/2}\, dy$$

where $dx = \sigma dq$. It remains to be proved that the above expression is identical with the initial expression for P. Putting

$$\frac{q\sqrt{(1 + \eta^2)} + \sqrt{3}}{\eta} = qa + b$$

and

$$\frac{q\sqrt{(1 + \eta^2)} - \sqrt{3}}{\eta} = qa - b$$

the first integral becomes

$$\int_{qa-\eta}^{qa+\eta} e^{-y^2/2}\, dy = f(q)$$

Therefore the double integral becomes

$$\int_{-k}^{k} dq\, f(q) = \int_{-k}^{0} dq\, f(q) + \int_{0}^{k} dq\, f(q)$$

$$= -\int_{0}^{-k} dq\, f(q) + \int_{0}^{k} dq\, f(q)$$

The first term of the above equation may be written $\int_{0}^{k} dx\, f(-x)$ by putting $q = -x$.

We must now prove that $f(q)$ is an even function, i.e. that $f(q) = f(-q)$. Now

$$f(q) = \int_{qa-\eta}^{qa+\eta} e^{-y^2/2}\, dy$$

and

$$f(-q) = \int_{qa-\eta}^{-qa+\eta} e^{-y^2/2}\, dy$$

Putting $y = -x$

$$f(-q) = -\int_{qa+\eta}^{qa-\eta} e^{-x^2/2}\,dx = \int_{qa-\eta}^{qa+\eta} e^{-x^2/2}\,dx = f(q)$$

Therefore $\int_{-k}^{k} dq\, f(q) = 2\int_{0}^{k} dq\, f(q)$, and the two expressions for P are seen to be identical.

η \ P	0.999 000	0.997 300	0.990 000	0.954 500	0.950 000	0.916 733	0.900 000
0.00	1.730 32	1.727 37	1.714 73	1.653 24	1.645 45	1.587 83	1.558 85
0.01	1.737 80	1.730 66	1.714 82	1.653 16	1.645 37	1.587 75	1.558 77
0.02	1.751 29	1.739 51	1.717 10	1.652 91	1.645 12	1.587 51	1.558 53
0.03	1.766 84	1.750 71	1.721 76	1.652 54	1.644 73	1.587 11	1.558 14
0.04	1.783 56	1.763 27	1.728 01	1.652 30	1.644 35	1.586 56	1.557 60
0.05	1.801 09	1.776 76	1.735 38	1.652 48	1.644 28	1.585 88	1.556 90
0.06	1.819 18	1.790 92	1.743 57	1.653 21	1.644 66	1.585 14	1.556 08
0.07	1.837 69	1.805 58	1.752 40	1.654 49	1.645 52	1.584 47	1.555 19
0.08	1.856 51	1.820 63	1.761 73	1.656 27	1.646 86	1.583 94	1.554 32
0.09	1.875 58	1.835 98	1.771 47	1.658 49	1.648 63	1.583 63	1.553 53
0.10	1.894 82	1.851 56	1.781 53	1.661 11	1.650 78	1.583 55	1.552 87
0.11	1.914 19	1.867 31	1.791 86	1.664 07	1.653 27	1.583 70	1.552 37
0.12	1.933 65	1.883 21	1.802 40	1.667 33	1.656 06	1.584 09	1.552 04
0.13	1.953 16	1.899 20	1.813 12	1.670 85	1.659 11	1.584 70	1.551 89
0.14	1.972 70	1.915 26	1.823 97	1.674 59	1.662 38	1.585 52	1.551 92
0.15	1.992 24	1.931 36	1.834 93	1.678 53	1.665 86	1.586 52	1.552 11
0.16	2.011 75	1.947 47	1.845 97	1.682 64	1.669 50	1.587 69	1.552 45
0.17	2.031 22	1.963 58	1.857 07	1.686 90	1.673 30	1.589 02	1.552 95
0.18	2.050 63	1.979 67	1.868 21	1.691 28	1.677 23	1.590 49	1.553 57
0.19	2.069 96	1.995 71	1.879 37	1.695 77	1.681 27	1.592 09	1.554 32
0.20	2.089 20	2.011 71	1.890 53	1.700 34	1.685 40	1.593 80	1.555 18
0.21	2.108 33	2.027 63	1.901 68	1.704 99	1.689 61	1.595 61	1.556 15
0.22	2.127 35	2.043 47	1.912 81	1.709 70	1.693 88	1.597 50	1.557 20
0.23	2.146 24	2.059 22	1.923 91	1.714 46	1.698 21	1.599 48	1.558 33
0.24	2.165 00	2.074 86	1.934 95	1.719 26	1.702 58	1.601 52	1.559 54
0.25	2.183 61	2.090 40	1.945 95	1.724 08	1.706 98	1.603 62	1.560 82
0.26	2.202 06	2.105 81	1.956 87	1.728 92	1.711 41	1.605 77	1.562 15
0.27	2.220 35	2.121 10	1.967 73	1.733 76	1.715 85	1.607 96	1.563 53
0.28	2.238 47	2.136 26	1.978 50	1.738 61	1.720 29	1.610 18	1.564 95
0.29	2.256 42	2.151 27	1.989 19	1.743 45	1.724 74	1.612 44	1.566 41
0.30	2.274 18	2.166 14	1.999 79	1.748 28	1.729 18	1.614 71	1.567 90
0.31	2.291 76	2.180 85	2.010 29	1.753 09	1.733 60	1.617 00	1.569 42
0.32	2.309 15	2.195 41	2.020 69	1.757 88	1.73 801	1.619 30	1.570 95
0.33	2.326 34	2.209 81	2.030 98	1.762 64	1.742 39	1.621 61	1.572 51
0.34	2.343.33	2.224 04	2.041 16	1.767 36	1.746 74	1.623 92	1.574 07
0.35	2.360 12	2.238 10	2.051 22	1.772 05	1.751 07	1.626 23	1.575 65
0.36	2.376 70	2.251 99	2.061 16	1.776 69	1.755 35	1.628 53	1.577 22
0.37	2.393 08	2.265 71	2.070 99	1.781 29	1.759 60	1.630 82	1.578 80
0.38	2.409 24	2.279 25	2.080 68	1.785 85	1.763 81	1.633 10	1.580 38
0.39	2.425 19	2.292 61	2.090 26	1.790 35	1.767 97	1.635 37	1.581 95

P η	0.999 000	0.997 300	0.990 000	0.954 500	0 950 000	0.916 73	0.900 000
0.40	2.440 92	2.305 78	2.099 70	1.794 80	1.772 08	1.637 61	1.583 51
0.41	2.456 44	2.318 78	2.109 01	1.799 19	1.776 14	1.639 84	1.585 07
0.42	2.471 74	2.331 59	2.118 19	1.803 53	1.780 15	1.642 04	1 586 61
0.43	2.486 82	2.344 22	2.127 24	1.807 81	1.784 11	1 644 22	1.588 14
0.44	2.501 68	2.356 66	2.136 16	1.812 03	1.788 01	1.646 37	1.589 65
0.45	2.516 32	2.368 92	2.144 93	1.816 19	1.791 86	1.648 50	1.591 15
0.46	2.530 74	2.380 99	2.153 58	1.820 28	1.795 65	1.650 59	1.592 63
0.47	2.544 94	2.392 87	2.162 08	1.824 31	1.799 38	1.652 66	1.594 09
0.48	2.558 92	2.404 56	2.170 46	1.828 27	1.803 05	1.654 69	1.595 52
0.49	2.572 68	2.416 07	2.178 69	1.832 17	1.806 66	1.656 69	1.596 94
0.50	2.586 22	2.427 40	2.186 79	1.836 01	1.810 20	1.658 66	1.598 33
0.51	2.599 55	2.438 53	2.194 75	1.839 78	1.813 69	1.660 60	1.599 70
0.52	2.612 65	2.449 49	2.202 58	1.843 48	1.817 12	1.662 50	1.601 04
0.53	2.625 54	2.460 25	2.210 27	1.847 11	1.820 48	1.664 36	1.602 36
0.54	2.638 22	2.470 84	2.217 82	1.850 68	1 823 78	1.666 19	1.603 66
0.55	2.650 68	2.481 24	2.225 24	1.854 18	1.827 02	1.667 98	1.604 92
0.56	2.662 93	2.491 47	2.232 53	1.857 61	1.830 19	1.669 74	1.606 17
0.57	2.674 97	2.501 51	2.239 69	1.860 98	1.833 31	1.671 46	1.607 38
0.58	2.686 80	2.511 38	2.246 71	1.864 28	1.836 36	1.673 15	1.608 57
0.59	2.698 43	2.521 07	2.253 60	1.867 51	1.839 35	1.674 80	1.609 73
0.60	2.709 85	2.530 58	2.260 37	1.870 68	1.842 28	1.676 41	1.610 87
0.61	2.721 06	2.539 92	2.267 01	1.873 79	1.845 15	1.677 99	1.611 97
0.62	2.732 08	2.549 09	2.273 51	1.876 82	1.847 96	1.679 53	1.613 06
0.63	2.742 89	2.558 09	2.279 90	1.879 80	1.850 71	1.681 04	1.614 11
0.64	2.753 51	2.566 92	2.286 16	1.882 71	1.853 39	1.682 51	1.615 14
0.65	2.763 93	2.575 59	2.292 30	1.885 56	1.856 03	1.683 94	1.616 14
0.66	2.774 17	2.584 09	2.298 31	1.888 34	1.858 60	1.685 34	1.617 12
0.67	2.784 21	2.592 44	2.304 21	1.891 07	1.861 11	1.686 71	1.618 07
0.68	2.794 06	2.600 62	2.309 99	1.893 73	1.863 57	1.688 04	1.618 99
0.69	2.803 73	2.608 64	2.315 65	1.896 34	1.865 98	1.689 34	1.619 89
0.70	2.813 22	2.616 51	2.321 19	1.898 88	1.868 32	1.690 60	1.620 76
0.71	2.822 53	2.624 23	2.326 63	1.901 37	1.870 62	1.691 83	1.621 61
0.72	2.831 65	2.631 79	2.331 95	1.903 80	1.872 86	1.693 03	1.622 44
0.73	2.840 61	2.639 21	2.337 16	1.906 18	1.875 05	1.694 20	1.623 24
0.74	2.849 39	2.646 48	2.342 27	1.908 50	1.877 18	1.695 34	1.624 01
0.75	2.858 00	2.653 61	2.347 26	1.910 76	1.879 27	1.696 45	1.624 77
0.76	2.866 45	2.660 59	2.352 16	1.912 97	1.881 30	1.697 52	1.625 50
0.77	2.874 73	2.667 43	2.356 95	1.915 13	1.883 29	1.698 57	1.626 21
0.78	2.882 85	2.674 14	2.361 63	1.917 24	1.885 23	1.699 59	1.626 90
0.79	2.890 81	2.680 71	2.336 22	1.919 29	1.887 12	1.700 58	1.627 56

Uncertainty, Calibration and Probability

η \ P	0.999 000	0.997 300	0.990 000	0.954 500	0.950 000	0.916 73	0.900 000
0.80	2.898 61	2.687 15	2.370 71	1.921 30	1.888 96	1.701 54	1.628 21
0.81	2.906 25	2.693 45	2.375 11	1.923 25	1.890 76	1.702 47	1.628 83
0.82	2.913 75	2.669 63	2.379 41	1.925 16	1.892 51	1.703 38	1.629 44
0.83	2.921 10	2.705 68	2.383 61	1.927 02	1.894 22	1.704 26	1.630 02
0.84	2.928 30	2.711 61	2.387 73	1.928 84	1.895 89	1.705 12	1.630 59
0.85	2.935 35	2.717 42	2.391 76	1.930 61	1.897 51	1.705 95	1.631 14
0.86	2.942 27	2.723 10	2.395 70	1.932 34	1.899 10	1.706 75	1.631 67
0.87	2.949 04	2.728 70	2.399 55	1.934 02	1.900 64	1.707 54	1.632 18
0.88	2.955 68	2.734 12	2.403 32	1.935 66	1.902 14	1.708 30	1.632 67
0.89	2.962 19	2.739 46	2.407 00	1.937 26	1.903 61	1.709 03	1.633 15
0.90	2.968 56	2.744 69	2.410 61	1.938 82	1.905 04	1.709 75	1.633 61
0.91	2.974 81	2.749 81	2.414 13	1.940 34	1.906 43	1.710 44	1.634 06
0.92	2.980 93	2.754 82	2.417 58	1.941 82	1.907 78	1.711 11	1.634 49
0.93	2.986 92	2.759 73	2.420 95	1.943 27	1.909 10	1.711 76	1.634 90
0.94	2.992 79	2.764 53	2.424 25	1.944 67	1.910 39	1.712 40	1.635 30
0.95	2.998 54	2.769 24	2.427 47	1.946 04	1.911 64	1.713 01	1.635 69
0.96	3.004 18	2.773 84	2.430 62	1.947 38	1.912 86	1.713 60	1.636 06
0.97	3.009 70	2.778 35	2.433 70	1.948 68	1.914 05	1.714 18	1.636 42
0.98	3.015 10	2.782 76	2.436 71	1.949 95	1.915 20	1.714 73	1.636 77
0.99	3.020 40	2.787 08	2.439 66	1.951 18	1.916 33	1.715 27	1.637 10
1.00	3.025 58	2.791 31	2.442 54	1.952 38	1.917 42	1.715 80	1.637 42
1.05	3.049 96	2.811 15	2.455 99	1.957 94	1.922 48	1.718 18	1.638 86
1.10	3.071 94	2.828 97	2.467 99	1.962 81	1.926 90	1.720 20	1.640 06
1.15	3.091 74	2.844 97	2.470 70	1.967 08	1.930 77	1.721 93	1.641 04
0.20	3.109 59	2.859 35	2.488 26	1.970 81	1.934 14	1.723 39	1.641 84
1.25	3.125 69	2.872 27	2.496 79	1.974 08	1.937 09	1.724 63	1.642 50
1.30	3.140 20	2.883 89	2.504 40	1.976 95	1.939 66	1.725 68	1.643 04
1.35	3.153 30	2.894 33	2.511 20	1.979 46	1.941 92	1.726 58	1.643 47
1.40	3.165 12	2.903 72	2.517 28	1.981 67	1.943 89	1.727 34	1.643 82
1.45	3.175 81	2.912 18	2.522 71	1.983 61	1.945 62	1.727 99	1.644 11
1.50	3.185 47	2.919 81	2.527 58	1.985 32	1.947 14	1.728 54	1.644 34
1.55	3.194 21	2.926 68	2.531 94	1.986 82	1.948 48	1.729 01	1.644 52
1.60	3.202 12	2.932 89	2.535 85	1.988 15	1.949 65	1.729 41	1.644 67
1.65	3.209 29	2.938 49	2.539 37	1.989 32	1.950 69	1.729 76	1.644 78
1.70	3.215 80	2.943 56	2.542 53	1.990 36	1.951 61	1.730 05	1.644 88
1.75	3.221 71	2.948 15	2.545 37	1.991 28	1.952 42	1.730 30	1.644 95
1.80	3.227 08	2.952 31	2.547 93	1.992 10	1.953 14	1.730 52	1.645 00
1.85	3.231 97	2.956 09	2.550 24	1.992 83	1.953 78	1.730 71	1.645 04
1.90	3.236 42	2.959 52	2.552 33	1.993 48	1.954 34	1.730 87	1.645 07
1.95	3.240 48	2.962 63	2.554 22	1.994 06	1.954 86	1.731 01	1.645 10

η	0.999 000	0.997 300	0.990 000	0.954 500	0.950 000	0.916 73	0.900 000
2.0	3.244 18	2.965 47	2.555 94	1.994 58	1.955 31	1.731 13	1.645 11
2.1	3.250 67	2.970 42	2.558 90	1.995 46	1.956 08	1.731 33	1.645 13
2.2	3.256 10	2.974 54	2.561 35	1.996 18	1.956 70	1.731 48	1.645 13
2.3	3.260 67	2.978 00	2.563 39	1.996 76	1.957 20	1.731 59	1.645 12
2.4	3.264 54	2.980 91	2.565 09	1.997 24	1.957 62	1.731 68	1.645 11
2.5	3.267 82	2.983 36	2.566 52	1.997 64	1.957 96	1.731 75	1.645 09
2.6	3.270 61	2.985 45	2.567 72	1.997 96	1.958 24	1.731 81	1.645 08
2.7	3.273 00	2.987 23	2.568 74	1.998 24	1.958 47	1.731 85	1.645 06
2.8	3.275 06	2.988 75	2.569 61	1.998 47	1.958 67	1.731 88	1.645 04
2.9	3.276 82	2.990 06	2.570 35	1.998 66	1.958 83	1.731 91	1.645 03
3.0	3.278 35	2.991 18	2.570 99	1.998 82	1.958 97	1.731 93	1.645 01
3.1	3.279 68	2.992 16	2.571 54	1.998 97	1.959 09	1.731 95	1.645 00
3.2	3.280 83	2.993 01	2.572 01	1.999 09	1.959 19	1.731 96	1.644 99
3.3	3.281 84	2.993 74	2.572 42	1.999 19	1.959 28	1.731 97	1.644 98
3.4	3.282 72	2.994 39	2.572 78	1.999 28	1.959 36	1.731 98	1.644 97
3.5	3.283 50	2.994 95	2.573 09	1.999 36	1.959 42	1.731 99	1.644 96
4.0	3.286 22	2.996 92	2.574 18	1.999 62	1.959 65	1.732 02	1.644 92
4.5	3.287 75	2.998 03	2.574 77	1.999 76	1.959 76	1.732 03	1.644 90
5.0	3.288 66	2.998 68	2.575 13	1.999 84	1.959 83	1.732 03	1.644 89
6.0	3.289 60	2.999 35	2.575 48	1.999 92	1.959 90	1.732 03	1.644 87
7.0	3.290 02	2.999 64	2.575 64	1.999 96	1.959 93	1.732 04	1.644 86
8.0	3.290 22	2.999 79	2.575 72	1.999 98	1.959 94	1.732 04	1.644 86
9.0	3.29 034	2.999 87	2.575 76	1.999 98	1.959 95	1.732 04	1.644 86
10.0	3.290 40	2.999 91	2.575 78	1.999 99	1.959 96	1.732 04	1.644 86
20.0	3.290 52	2.999 99	2.575 83	2.000 00	1.959 96	1.732 04	1.644 85
30.0	3.290 53	3.000 00	2.575 83	2.000 00	1.959 96	1.732 04	1.644 85
40.0	3.290 53	3.000 00	2.575 83	2.000 00	1.959 96	1.732 04	1.644 85
50.0	3.290 53	3.000 00	2.575 83	2.000 00	1.959 96	1.732 04	1.644 85
∞	3.290 53	3.000 00	2.575 83	2.000 00	1.959 96	1.732 04	1.644 85

Introduction to part (b)

I.XXIII(4) This table gives the values of P, the probability of an event occurring between the limits $\mu - k\sigma$ to $\mu + k\sigma$, where μ is the mean of the convoluted distribution and σ is its standard deviation for indexed values of η and k. η is the ratio σ_G/σ_R, where σ_G is the standard deviation of the Gaussian distribution, and σ_R is the standard deviation of the rectangular distribution, while k is the tolerance factor defining the range of the event as given above. The expression for P given below is the same as that given in

$$P = \left[\frac{1 + \eta^2}{6\pi} \right]^{1/2} \int_0^k dq \int_{\frac{\{q\sqrt{(1 + \eta^2)} - \sqrt{3}\}}{\eta}}^{\frac{\{q\sqrt{(1 + \eta^2)} + \sqrt{3}\}}{\eta}} e^{-y^2/2} \, dy$$

paragraph I.XXIII(2). P is calculated for indexed values of η and k.

η \ k	0.700	0.800	0.900	1.000	1.100	1.200	1.300	1.400
0.00	0.404 15	0.461 88	0.519 62	0.577 35	0.635 08	0.692 82	0.750 56	0.808 29
0.01	0.404 17	0.461 90	0.519 64	0.577 38	0.635 12	0.692 85	0.750 59	0.808 33
0.02	0.404 23	0.461 97	0.519 72	0.577 47	0.635 21	0.692 96	0.750 71	0.808 45
0.03	0.404 33	0.462 09	0.519 85	0.577 61	0.635 37	0.693 13	0.750 89	0.808 65
0.04	0.404 47	0.462 25	0.520 03	0.577 81	0.635 59	0.693 37	0.751 16	0.808 94
0.05	0.404 65	0.462 46	0.520 26	0.578 07	0.635 88	0.693 69	0.751 49	0.809 30
0.06	0.404 87	0.462 71	0.520 55	0.578 39	0.636 23	0.694 07	0.751 91	0.809 74
0.07	0.405 13	0.463 01	0.520 89	0.578 76	0.636 64	0.694 52	0.752 39	0.810 27
0.08	0.405 44	0.463 36	0.521 28	0.579 19	0.637 11	0.695 03	0.752 95	0.810 87
0.09	0.405 78	0.463 75	0.521 72	0.579 68	0.637 65	0.695 62	0.753 59	0.811 56
0.10	0.406 16	0.464 18	0.522 21	0.580 23	0.638 25	0.696 28	0.754 30	0.812 31
0.11	0.406 58	0.464 67	0.522 75	0.580 83	0.638 92	0.697 00	0.755 08	0.813 14
0.12	0.407 04	0.465 19	0.523 34	0.581 49	0.639 64	0.697 79	0.755 94	0.814 01
0.13	0.407 55	0.465 77	0.523 99	0.582 21	0.640 43	0.698 65	0.756 86	0.814 92
0.14	0.408 09	0.466 38	0.524 68	0.582 98	0.641 28	0.699 58	0.757 84	0.815 85
0.15	0.408 67	0.467 05	0.525 43	0.583 81	0.642 19	0 700 57	0.758 88	0.816 79
0.16	0.409 29	0.467 75	0.526 22	0.584 69	0.643 16	0.701 62	0.759 97	0.817 71
0.17	0.409 94	0.468 51	0.527 07	0.585 63	0.644 19	0.702 73	0.761 08	0.818 60
0.18	0.410 64	0.469 30	0.527 97	0.586 63	0.645 28	0.703 89	0.762 22	0.819 47
0.19	0.411 38	0.470 14	0.528 91	0.587 68	0.646 43	0.705 09	0.763 37	0.820 29
0.20	0.412 15	0.471 03	0.529 91	0.588 78	0.647 63	0.706 34	0.764 52	0.821 08
0.21	0.412 96	0.471 95	0.530 95	0.589 93	0.648 87	0.707 61	0.765 66	0.821 8?
0.22	0.413 81	0.472 93	0.532 04	0.591 14	0.650 16	0.708 91	0.766 79	0.822 50
0.23	0.414 70	0.473 94	0.533 17	0.592 38	0.651 49	0.710 22	0.767 90	0.823 15
0.24	0.415 62	0.474 99	0.534 36	0.593 68	0.652 85	0.711 53	0.768 99	0.823 75
0.25	0.416 58	0.476 09	0.535 58	0.595 01	0.654 23	0.712 85	0.770 04	0.824 31
0.26	0.417 58	0.477 23	0.536 85	0.596 38	0.655 63	0.714 16	0.771 06	0.824 82
0.27	0.418 61	0.478 40	0.538 15	0.597 78	0.657 05	0.715 47	0.772 06	0.825 30
0.28	0.419 68	0.479 62	0.539 50	0.599 21	0.658 48	0.716 76	0.773 01	0.825 74
0.29	0.420 78	0.480 87	0.540 87	0.600 66	0.659 92	0.178 03	0.773 94	0.826 15
0.30	0.421 92	0.482 15	0.542 27	0.602 12	0.661 35	0.719 28	0.774 83	0.826 53
0.31	0.423 09	0.483 46	0.543 70	0.603 61	0.662 78	0.720 50	0.775 69	0.826 88
0.32	0.424 29	0.484 81	0.545 15	0.605 10	0.664 19	0.721 70	0.776 51	0.827 20
0.33	0.425 51	0.486 18	0.546 62	0.606 59	0.665 60	0.722 88	0.777 30	0.827 49
0.34	0.426 77	0.487 57	0.548 11	0.608 09	0.666 99	0.724 02	0.778 06	0.827 77
0.35	0.428 05	0.488 98	0.549 60	0.609 58	0.668 37	0.725 14	0.778 79	0.838 02
0.36	0.429 35	0.490 41	0.551 10	0.611 07	0.669 73	0.726 23	0.779 49	0.828 26
0.37	0.430 67	0.491 86	0.552 61	0.612 55	0.671 06	0.727 30	0.780 17	0.828 47
0.38	0.432 01	0.493 31	0.554 12	0.614 02	0.672 37	0.728 33	0.780 82	0.828 68
0.39	0.433 36	0.494 78	0.555 63	0.615 47	0.673 66	0.729 33	0.781 44	0.828 86

η \ k	1.500	1.600	1.700	1.800	1.900	2.00	2.100	2.200
0.00	0.866 03	0.923 76	0.981 50	1.000 00	1.000 00	1.000 00	1.000 00	1.000 00
0.01	0.866 07	0.923 81	0.981 54					
0.02	0.866 20	0.923 95	0.981 41	1.000 00				
0.03	0.866 42	0.924 18	0.980 61	0.999 94				
0.04	0.866 72	0.924 50	0.979 34	0.999 61				
0.05	0.867 11	0.924 87	0.977 84	0.998 95	1.000 00			
0.06	0.867 58	0.925 23	0.976 23	0.998 00	0.999 98			
0.07	0.868 14	0.925 49	0.974 56	0.996 84	0.999 91			
0.08	0.868 76	0.925 60	0.972 88	0.995 54	0.999 76	1.000 00		
0.09	0.869 43	0.925 57	0.971 20	0.994 14	0.999 50	0.999 98		
0.10	0.870 10	0.925 39	0.969 53	0.992 68	0.999 12	0.999 95		
0.11	0.870 75	0.925 08	0.967 88	0.991 18	0.998 63	0.999 89	1.000 00	
0.12	0.871 35	0.924 67	0.966 25	0.989 66	0.998 03	0.999 78	0.999 99	
0.13	0.871 90	0.924 16	0.964 66	0.988 13	0.997 34	0.999 62	0.999 97	
0.14	0.872 37	0.923 59	0.963 09	0.986 60	0.996 56	0.999 41	0.999 93	1.000 00
0.15	0.872 78	0.922 95	0.961 56	0.985 09	0.995 72	0.999 13	0.999 88	0.999 99
0.16	0.873 11	0.922 27	0.960 06	0.983 59	0.994 83	0.998 79	0.999 79	0.999 98
0.17	0.873 36	0.921 55	0.958 59	0.982 10	0.993 88	0.998 39	0.999 68	0.999 95
0.18	0.873 56	0.920 81	0.957 16	0.980 64	0.992 91	0.997 94	0.999 54	0.999 92
0.19	0.873 69	0.920 05	0.955 76	0.979 21	0.991 90	0.997 44	0.999 35	0.999 87
0.20	0.873 77	0.919 28	0.954 40	0.977 79	0.990 88	0.996 90	0.999 14	0.999 81
0.21	0873 80	0.918 50	0.953 07	0.976 41	0.989 84	0.996 31	0.998 89	0.999 72
0.22	0.873 78	0.917 72	0.951 77	0.975 05	0.988 80	0.995 69	0.998 60	0.999 62
0.23	0.873 73	0.916 93	0.950 51	0.973 72	0.987 75	0.995 04	0.998 28	0.999 49
0.24	0.873 65	0.916 16	0.949 28	0.972 43	0.986 70	0.994 37	0.997 93	0.999 34
0.25	0.873 53	0.915 38	0.948 08	0.971 15	0.985 65	0.993 68	0.997 55	0.999 17
0.26	0.873 40	0.914 62	0.946 91	0.969 91	0.984 62	0.992 96	0.997 14	0.998 98
0.27	0.873 24	0.913 86	0.945 78	0.968 70	0.983 58	0.992 24	0.996 72	0.998 76
0.28	0.873 06	0.913 12	0.944 67	0.967 52	0.982 56	0.991 50	0.996 27	0.998 53
0.29	0.872 87	0.912 39	0.943 60	0.966 37	0.981 56	0.990 76	0.995 80	0.998 27
0.30	0.872 66	0.911 67	0.942 55	0.965 24	0.980 56	0.990 01	0.995 31	0.998 00
0.31	0.872 45	0.910 96	0.941 54	0.964 14	0.979 58	0.989 27	0.994 81	0.997 71
0.32	0.872 23	0.910 27	0.940 55	0.963 07	0.978 62	0.988 52	0.994 30	0.997 40
0.33	0.872 00	0.909 60	0.939 59	0.962 03	0.977 67	0.987 77	0.993 78	0.997 08
0.34	0.871 77	0.908 94	0.938 66	0.961 02	0.976 74	0.987 02	0.993 26	0.996 75
0.35	0.871 53	0.908 29	0.937 76	0.960 03	0.975 83	0.986 28	0.992 72	0.996 40
0.36	0.871 30	0.907 66	0.936 88	0.959 07	0.974 93	0.985 55	0.992 18	0.996 05
0.37	0.871 06	0.907 05	0.936 03	0.958 14	0.974 06	0.984 82	0.991 64	0.995 68
0.38	0.870 82	0.906 45	0.935 20	0.957 23	0.973 20	0.984 10	0.991 10	0.995 31
0.39	0.870 59	0.905 87	0.934 40	0.956 35	0.972 36	0.983 39	0.990 56	0.994 93

k	2.300	2.400	2.500	2.600	2.700	2.800	2.900	3.000
0.00	1.000 00	1.000 00	1.000 00	1.000 00	1.000 00	1.000 00	1.000 00	1.000 00
0.01								
0.02								
0.03								
0.04								
0.05								
0.06								
0.07								
0.08								
0.09								
0.10								
0.11								
0.12								
0.13								
0.14								
0.15								
0.16								
0.17	1.000 00							
0.18	0.999 99							
0.19	0.999 98							
0.20	0.999 97	1.000 00						
0.21	0.999 94	0.999 99						
0.22	0.999 91	0.999 98						
0.23	0.999 87	0.999 97	1.000 00					
0.24	0.99 82	0.999 96	0.999 99					
0.25	0.99 76	0.999 94	0.999 99					
0.26	0.999 68	0.999 91	0.999 98	1.000 00				
0.27	0.999 59	0.999 88	0.999 97	0.999 99				
0.28	0.999 48	0.999 84	0.999 95	0.999 99				
0.29	0.999 36	0.999 79	0.999 94	0.999 98	1.000 00			
0.30	0.999 22	0.999 73	0.999 91	0.999 98	0.999 99			
0.31	0.999 08	0.999 66	0.999 89	0.999 97	0.999 99			
0.32	0.998 91	0.999 58	0.999 85	0.999 95	0.999 99	1.000 00		
0.33	0.998 74	0.999 50	0.999 82	0.999 94	0.999 98	0.999 99		
0.34	0.998 55	0.999 40	0.999 77	0.999 92	0.999 97	0.999 99		
0.35	0.998 35	0.999 29	0.999 72	0.999 90	0.999 97	0.999 99		
0.36	0.998 13	0.999 18	0.999 66	0.999 87	0.999 96	0.999 99	1.000 00	
0.37	0.997 91	0.999 06	0.999 60	0.999 84	0.999 94	0.999 98	0.999 99	
0.38	0.997 68	0.998 92	0.999 53	0.999 81	0.999 93	0.999 97	0.999 99	
0.39	0.997 44	0.998 78	0.999 46	0.999 77	0.999 91	0.999 97	0.999 99	1.000 00

η \ k	0.700	0.800	0.900	1.000	1.100	1.200	1.300	1.400
0.40	0.434 73	0.496 25	0.557 12	0.616 92	0.674 93	0.730 31	0.782 04	0.829 04
0.41	0.436 11	0.497 73	0.558 63	0.618 34	0.676 17	0.731 26	0.782 62	0.829 21
0.42	0.437 50	0.499 20	0.560 12	0.619 74	0.677 39	0.732 19	0.783 17	0.829 36
0.43	0.438 89	0.500 68	0.561 60	0.621 13	0.678 58	0.733 09	0.783 71	0.829 51
0.44	0.440 29	0.502 15	0.563 06	0.622 50	0.679 74	0.733 96	0.784 22	0.829 64
0.45	0.441 69	0.503 62	0.564 52	0.623 84	0.680 88	0.734 81	0.784 72	0.829 77
0.46	0.443 09	0.505 07	0.565 95	0.625 16	0.681 99	0.735 63	0.785 21	0.829 89
0.47	0.444 49	0.506 52	0.567 37	0.626 46	0.683 08	0.736 43	0.785 67	0.830 01
0.48	0.445 88	0.507 96	0.568 77	0.627 73	0.684 15	0.737 21	0.786 12	0.830 12
0.49	0.447 27	0.509 38	0.570 15	0.628 98	0.685 18	0.737 97	0.786 56	0.830 23
0.50	0.448 65	0.510 79	0.571 51	0.630 21	0.686 20	0.738 71	0.786 98	0.830 33
0.51	0.450 02	0.512 19	0.572 85	0.631 41	0.687 19	0.739 42	0.787 39	0.830 43
0.52	0.451 38	0.513 56	0.574 17	0.632 59	0.688 15	0.740 12	0.787 78	0.830 53
0.53	0.452 73	0.514 93	0.575 47	0.633 75	0.689 10	0.740 80	0.788 17	0.830 62
0.54	0.454 06	0.516 27	0.576 74	0.634 88	0.690 02	0.741 46	0.788 54	0.830 71
0.55	0.455 38	0.517 59	0.577 99	0.635 99	0.690 92	0.742 10	0.788 91	0.830 80
0.56	0.456 69	0.518 90	0.579 22	0.637 07	0.691 80	0.742 73	0.789 26	0.830 89
0.57	0.457 98	0.520 18	0.580 43	0.638 13	0.692 65	0.743 34	0.789 61	0.830 97
0.58	0.459 26	0.521 44	0.581 61	0.639 17	0.693 48	0.743 93	0.789 94	0.831 06
0.59	0.460 51	0.522 69	0.582 77	0.640 18	0.694 30	0.744 51	0.790 27	0.831 14
0.60	0.461 75	0.523 91	0.583 91	0.641 17	0.695 09	0.745 07	0.790 59	0.831 22
0.61	0.462 97	0.525 11	0.585 02	0.642 14	0.695 87	0.745 62	0.790 90	0.831 30
0.62	0.464 17	0.526 28	0.586 11	0.643 09	0.696 63	0.746 16	0.791 20	0.831 38
0.63	0.465 35	0.527 44	0.587 18	0.644 02	0.697 37	0.746 68	0.791 50	0.831 46
0.64	0.466 51	0.528 57	0.588 23	0.644 92	0.698 09	0.747 19	0.791 79	0.831 54
0.65	0.467 65	0.529 69	0.589 25	0.645 81	0.698 79	0.747 69	0.792 07	0.831 62
0.66	0.468 77	0.530 78	0.590 26	0.646 67	0.699 48	0.748 18	0.792 35	0.831 70
0.67	0.469.87	0.531 84	0.591 24	0.647 52	0.700 15	0.748 65	0.792 62	0.831 77
0.68	0.470 95	0.532 89	0.592 20	0.648 34	0.700 80	0.749 11	0.792 88	0.831 85
0.69	0.472 01	0.533 91	0.593 13	0.649 15	0.701 44	0.749 56	0.793 14	0.831 93
0.70	0.473 05	0.534 92	0.594 05	0.649 93	0.702 07	0.750 01	0.793 40	0.832 00
0.71	0.474 06	0.535 90	0.594 95	0.650 70	0.702 67	0.750 44	0.793 65	0.832 08
0.72	0.475 06	0.536 86	0.595 82	0.651 45	0.703 27	0.750 86	0.793 89	0.832 16
0.73	0.476 03	0.537 80	0.596 68	0.652 18	0.703 85	0.751 27	0.794 13	0.832 23
0.74	0.476 99	0.538 72	0.597 52	0.652 90	0.704 42	0.751 67	0.794 36	0.832 31
0.75	0.477 92	0.539 61	0.598 33	0.653 60	0.704 97	0.752 06	0.794 59	0.832 39
0.76	0.478 84	0.540 49	0.599 13	0.654 28	0.705 51	0.752 45	0.794 82	0.832 46
0.77	0.479 73	0.541 35	0.599 91	0.654 95	0.706 04	0.752 82	0.795 04	0.832 54
0.78	0.480 60	0.542 19	0.600 67	0.655 60	0.706 55	0.753 19	0.795 26	0.832 61
0.79	0.481 46	0.543 00	0.601 41	0. 656 23	0.707 05	0.753 55	0.795 47	0.832 69

η \ k	1.500	1.600	1.700	1.800	1.900	2.000	2.100	2.200
0.40	0.870 36	0.905 31	0.933 62	0.955 49	0.971 54	0.982 69	0.990 02	0.994 55
0.41	0.870 13	0.904 76	0.932 86	0.954 65	0.970 73	0.982 00	0.989 48	0.994 16
0.42	0.869 90	0.904 22	0.932 13	0.953 84	0.969 95	0.981 32	0.988 94	0.993 77
0.43	0.869 68	0.903 70	0.931 42	0.953 05	0.969 19	0.980 65	0.988 41	0.993 38
0.44	0.869 46	0.903 20	0.930 73	0.952 29	0.968 44	0.980 00	0.987 88	0.992 99
0.45	0.869 25	0.902 71	0.930 06	0.951 54	0.967 71	0.978 35	0.987 35	0.992 59
0.46	0.869 04	0.902 24	0.929 41	0.950 82	0.967 00	0.978 72	0.986 84	0.992 20
0.47	0.868 84	0.901 78	0.928 79	0.950 12	0.966 31	0.978 10	0.986 32	0.991 81
0.48	0.868 64	0.901 34	0.928 18	0.949 44	0.965 63	0.977 49	0.985 82	0.991 41
0.49	0.868 45	0.900 91	0.927 59	0.948 77	0.964 98	0.976 90	0.985 32	0.991 03
0.50	0.868 26	0.900 49	0.927 02	0.948 13	0.964 34	0.976 31	0.984 83	0.990 64
0.51	0.868 08	0.900 09	0.926 47	0.947 51	0.963 71	0.975 74	0.984 34	0.990 25
0.52	0.867 91	0.899 70	0.925 94	0.946 91	0.963 11	0.975 19	0.983 87	0.989 87
0.53	0.867 74	0.899 32	0.925 42	0.946 32	0.962 52	0.974 64	0.983 40	0.989 50
0.54	0.867 58	0.898 96	0.924 92	0.945 75	0.961 94	0.974 11	0.982 94	0.989 13
0.55	0.867 42	0.898 60	0.924 43	0.945 20	0.961 38	0.973 59	0.982 49	0.988 76
0.56	0.867 27	0.898 26	0.923 96	0.944 66	0.960 84	0.973 08	0.982 05	0.988 40
0.57	0.867 12	0.897 94	0.923 51	0.944 15	0.960 31	0.972 58	0.981 61	0.988 04
0.58	0.866 98	0.897 62	0.923 07	0.943 64	0.959 80	0.972 10	0.981 19	0.987 68
0.59	0.866 85	0.897 31	0.922 65	0.943 16	0.959 30	0.971 63	0.980 77	0.987 34
0.60	0.866 72	0.897 02	0.922 24	0.942 68	0.958 81	0.971 17	0.980 36	0.986 99
0.61	0.866 60	0.896 73	0.921 84	0.942 23	0.958 34	0.970 72	0.979 96	0.986 66
0.62	0.866 48	0.896 46	0.921 46	0.941 78	0.957 88	0.970 28	0.979 57	0.986 33
0.63	0.866 37	0.896 19	0.921 09	0.941 35	0.957 44	0.969 86	0.979 19	0.986 00
0.64	0.866 26	0.895 94	0.920 73	0.940 94	0.957 00	0.969 44	0.978 81	0.985 68
0.65	0.866 16	0.895 69	0.920 38	0.940 54	0.956 58	0.969 04	0.978 45	0.985 37
0.66	0.866 06	0.895 46	0.920 05	0.940 14	0.956 17	0.968 64	0.978 09	0.985 06
0.67	0.865 97	0.895 23	0.919 72	0.939 77	0.955 78	0.968 26	0.977 74	0.984 76
0.68	0.865 88	0.895 01	0.919 41	0.939 40	0.955 39	0.967 89	0.977 40	0.984 47
0.69	0.865 80	0.894 80	0.919 11	0.939 04	0.955 02	0.967 52	0.977 07	0.984 18
0.70	0.865 72	0.894 60	0.918 82	0.938 70	0.954 66	0.967 17	0.976 74	0.983 90
0.71	0.865 65	0.894 40	0.918 54	0.938 37	0.954 31	0.966 82	0.976 43	0.983 62
072	0.865 58	0.894 21	0.918 27	0.938 05	0.953 96	0.966 49	0.976 12	0.983 35
0.73	0.865 51	0.894 03	0.918 00	0.937 73	0.953 63	0.966 16	0.975 82	0.983 08
0.74	0.865 45	0.893 85	0.917 75	0.937 43	0.953 31	0.965 85	0.975 52	0.982 82
0.75	0.865 39	0.893 69	0.917 51	0.937 14	0.953 00	0.965 54	0.975 23	0.982 57
0.76	0.865 34	0.893 53	0.917 27	0.936 86	0.952 70	0.965 24	0.974 96	0.982 32
0.77	0.865 28	0.893 38	0.917 04	0.936 58	0.952 40	0.964 95	0.974 68	0.982 08
0.78	0.865 24	0.893 23	0.916 82	0.936 32	0.952 12	0.964 66	0.974 42	0.981 84
0.79	0.865 19	0.893 09	0.916 61	0.936 06	0.951 84	0.964 39	0.974 16	0.981 61

η \ k	2.300	2.400	2.500	2.600	2.700	2.800	2.900	3.000
0.40	0.997 19	0.998 63	0.999 37	0.999 73	0.999 89	0.999 96	0.999 99	1.000 00
0.41	0.996 93	0.998 48	0.999 29	0.999 68	0.999 87	0.999 95	0.999 98	0.999 99
0.42	0.996 67	0.998 31	0.999 19	0.999 63	0.999 84	0.999 94	0.999 98	0.999 99
0.43	0.996 41	0.998 15	0.999 09	0.999 58	0.999 82	0.999 92	0.999 97	0.999 99
0.44	0.996 13	0.997 97	0.998 99	0.999 52	0.999 79	0.999 91	0.999 96	0.999 99
0.45	0.995 86	0.997 79	0.998 88	0.999 46	0.999 75	0.999 89	0.999 95	0.999 98
0.46	0.995 58	0.997 61	0.998 76	0.999 39	0.999 71	0.999 87	0.999 95	0.999 98
0.47	0.995 30	0.997 42	0.998 64	0.999 32	0.999 67	0.999 85	0.999 94	0.999 97
0.48	0.995 01	0.997 22	0.998 52	0.999 25	0.999 63	0.999 83	0.999 92	0.999 97
0.49	0.994 73	0.997 03	0.998 39	0.999 17	0.999 59	0.999 80	0.999 91	0.999 96
0.50	0.994 44	0.996 83	0.998 26	0.999 09	0.999 54	0.999 78	0.999 90	0.999 95
0.51	0.994 16	0.996 63	0.998 13	0.999 00	0.999 49	0.999 75	0.999 88	0.999 95
0.52	0.993 87	0.996 42	0.997 99	0.998 92	0.999 44	0.999 72	0.999 87	0.999 94
0.53	0.993 59	0.996 22	0.997 85	0.998 83	0.999 38	0.999 69	0.999 85	0.999 93
0.54	0.993 30	0.996 01	0.997 71	0.998 73	0.999 32	0.999 65	0.999 83	0.999 92
0.55	0.993 02	0.995 81	0.997 57	0.998 64	0.999 27	0.999 62	0.999 81	0.999 91
0.56	0.992 74	0.995 60	0.997 42	0.998 54	0.999 20	0.999 58	0.999 79	0.999 90
0.57	0.992 46	0.995 39	0.997 28	0.998 44	0.999 14	0.999 54	0.999 76	0.999 88
0.58	0.992 18	0.995 19	0.997 13	0.998 34	0.999 08	0.999 50	0.999 74	0.999 87
0.59	0.991 90	0.994 98	0.996 98	0.998 24	0.999 01	0.999 46	0.999 71	0.999 85
0.60	0.991 63	0.994 77	0.996 83	0.998 14	0.998 94	0.999 42	0.999 69	0.999 84
0.61	0.991 36	0.994 57	0.996 68	0.998 04	0.998 87	0.999 37	0.999 66	0.999 82
0.62	0.991 10	0.994 37	0.996 54	0.997 93	0.998 80	0.999 33	0.999 63	0.999 81
0.63	0.990 84	0.994 16	0.996 39	0.997 83	0.998 73	0.999 28	0.999 60	0.999 79
0.64	0.990 58	0.993 96	0.996 24	0.997 72	0.998 66	0.999 23	0.999 57	0.999 77
0.65	0.990 32	0.993 76	0.996 09	0.997 61	0.998 58	0.999 18	0.999 54	0.999 75
0.66	0.990 07	0.993 57	0.995 94	0.997 51	0.998 51	0.999 13	0.999 51	0.999 73
0.67	0.989 82	0.993 37	0.995 79	0.997 40	0.998 43	0.999 08	0.999 48	0.999 71
0.68	0.989 58	0.993 18	0.995 65	0.997 29	0.998 36	0.999 03	0.999 44	0.999 69
0.69	0.989 34	0.992 99	0.995 50	0.997 19	0.998 28	0.998 98	0.999 41	0.999 67
0.70	0.989 10	0.992 80	0.995 36	0.997 08	0.998 21	0.998 93	0.999 37	0.999 64
0.71	0.998 87	0.992 61	0.995 21	0.996 97	0.998 13	0.998 87	0.999 34	0.999 62
0.72	0.988 64	0.992 43	0.995 07	0.996 87	0.998 05	0.998 82	0.999 30	0.999 60
0.73	0.988 42	0.992 25	0.994 93	0.996 76	0.997 98	0.998 77	0.999 27	0.999 58
0.74	0.988 20	0.992 07	0.994 79	0.996 66	0.997 90	0.998 71	0.999 23	0.999 55
0.75	0.987 99	0.991 90	0.994 66	0.996 55	0.997 83	0.998 66	0.999 19	0.999 53
0.76	0.987 78	0.991 73	0.994 52	0.996 45	0.997 75	0.998 61	0.999 16	0.999 50
0.77	0.987 57	0.991 56	0.994 39	0.996 35	0.997 68	0.998 55	0.999 12	0.999 48
0.78	0.987 37	0.991 39	0.994 25	0.996 25	0.997 60	0.998 50	0.999 08	0.999 45
0.79	0.987 17	0.991 23	0.994 12	0.996 15	0.997 53	0.998 45	0.999 04	0.999 43

k / η	0.700	0.800	0.900	1.000	1.100	1.200	1.300	1.400
0.80	0.482 29	0.543 80	0.602 14	0.656 85	0.707 54	0.753 90	0.795 68	0.832 76
0.81	0.483 11	0.544 58	0.602 84	0.657 45	0.708 02	0.754 24	0.795 88	0.832 84
0.82	0.483 90	0.545 35	0.603 53	0.658 04	0.708 49	0.754 57	0.796 08	0.832 91
0.83	0.484 68	0.546 09	0.604 21	0.658 62	0.708 95	0.754 90	0.796 28	0.832 99
0.84	0.485 44	0.546 82	0.604 87	0.659 18	0.709 39	0.755 22	0.796 47	0.833 06
0.85	0.486 18	0.547 52	0.605 51	0.659 73	0.709 83	0.755 53	0.796 67	0.833 14
0.86	0.486 91	0.548 22	0.606 13	0.660 26	0.710 25	0.755 84	0.796 85	0.833 21
0.87	0.487 61	0.548 89	0.606 75	0.660 78	0.710 67	0.756 14	0.797 03	0.833 28
0.88	0.488 30	0.549 55	0.607 34	0.661 29	0.711 07	0.756 43	0.797 21	0.833 36
0.89	0.488 97	0.550 19	0.607 92	0.661 79	0.711 47	0.756 72	0.797 39	0.833 43
0.90	0.489 63	0.550 82	0.608 49	0.662 27	0.711 85	0.757 00	0.797 56	0.833 50
0.91	0.490 27	0.551 43	0.609 04	0.662 75	0.712 23	0.757 27	0.797 73	0.833 57
0.92	0.490 89	0.552 02	0.609 58	0.663 21	0.712 60	0.757 54	0.797 89	0.833 64
0.93	0.491 50	0.552 60	0.610 11	0.663 66	0.712 96	0.757 80	0.798 06	0.833 71
0.94	0.492 09	0.553 17	0.610 62	0.664 10	0.713 31	0.758 05	0.798 22	0.833 78
0.95	0.492 67	0.553 72	0.611 12	0.664 52	0.713 65	0.758 30	0.798 37	0.833 85
0.96	0.493 23	0.554 26	0.611 61	0.664 94	0.713 99	0.758 55	0.798 53	0.833 92
0.97	0.493 78	0.554 78	0.612 08	0.665 35	0.714 31	0.758 79	0.798 68	0.833 99
0.98	0.494 32	0.555 29	0.612 55	0.665 75	0.714 63	0.759 02	0.798 83	0.834 06
0.99	0.494 84	0.555 79	0.613 00	0.666 13	0.714 94	0.759 25	0.798 97	0.834 13
1.00	0.495 35	0.556 28	0.613 44	0.666 51	0.715 25	0.759 47	0.799 12	0.834 19
1.05	0.497 70	0.558 52	0.615 48	0.668 27	0.716 66	0.760 52	0.799 79	0.834 52
1.10	0.499 77	0.560 50	0.617 28	0.669 82	0.717 91	0.761 45	0.800 40	0.834 82
1.15	0.501 58	0.562 24	0.618 86	0.671 19	0.719 03	0.762 29	0.800 95	0.835 11
1.20	0.503 17	0.563 76	0.620 25	0.672 39	0.720 01	0.763 03	0.801 46	0.835 38
1.25	0.504 57	0.565 10	0.621 48	0.673 46	0.720 89	0.763 70	0.801 91	0.835 63
1.30	0.505 79	0.566 28	0.622 56	0.674 41	0.721 67	0.764 29	0.802 32	0.835 86
1.35	0.506 87	0.567 32	0.623 52	0.675 24	0.722 36	0.764 82	0.802 69	0.836 08
1.40	0.507 82	0.568 24	0.624 36	0.675 98	0.722 97	0.765 30	0.803 02	0.836 27
1.45	0.508 66	0.569 05	0.625 11	0.676 64	0.723 52	0.765 72	0.803 32	0.836 45
1.50	0.509 40	0.569 77	0.625 77	0.677 22	0.724 00	0.766 10	0.803 59	0.836 62
1.55	0.510 05	0.570 40	0.626 36	0.677 74	0.724 44	0.766 44	0.803 84	0.836 77
1.60	0.510 63	0.570 96	0.626 88	0.678 20	0.724 83	0.766 75	0.804 06	0.836 91
1.65	0.511 15	0.571 46	0.627 35	0.678 61	0.725 17	0.767 02	0.804 26	0.837 03
1.70	0.511 61	0.571 91	0.627 76	0.678 98	0.725 48	0.767 27	0.804 44	0.837 15
1.75	0.512 02	0.572 31	0.628 13	0.679 31	0.725 76	0.767 49	0.804 60	0.837 25
1.80	0.512 38	0.572 66	0.628 46	0.679 61	0.726 01	0.767 69	0.804 75	0.837 35
1.85	0.512 71	0.572 98	0.628 76	0.679 87	0.726 24	0.767 87	0.804 88	0.837 44
1.90	0.513 00	0.573 26	0.629 03	0.680 11	0.726 44	0.768 04	0.805 00	0.837 52
1.95	0.513 26	0.573 52	0.629 27	0.680 33	0.726 62	0.768 18	0.805 11	0.837 59

η \ k	1.500	1.600	1.700	1.800	1.900	2.000	2.100	2.200
0.80	0.865 15	0.892 96	0.916 40	0.935 81	0.951 57	0.964 12	0.973 90	0.981 38
0.81	0.865 11	0.892 83	0.916 21	0.935 57	0.951 31	0.963 86	0.973 66	0.981 16
0.82	0.865 08	0.892 70	0.916 02	0.935 34	0.951 06	0.963 61	0.973 42	0.980 94
0.83	0.865 04	0.892 58	0.915 83	0.935 12	0.950 82	0.963 36	0.973 19	0.980 73
0.84	0.865 01	0.892 47	0.915 66	0.934 90	0.950 58	0.963 12	0.972 96	0.980 53
0.85	0.864 99	0.892 36	0.915 48	0.934 69	0.950 35	0.962 89	0.972 74	0.980 33
0.86	0.864 96	0.892 25	0.915 32	0.934 48	0.950 13	0.962 66	0.972 52	0.980 13
0.87	0.864 94	0.892 15	0.915 16	0.934 29	0.949 91	0.962 44	0.972 31	0.979 94
0.88	0.864 92	0.892 06	0.915 01	0.934 10	0.947 70	0.962 23	0.972 11	0.979 75
0.89	0.864 90	0.891 96	0.914 86	0.933 91	0.949 50	0.962 02	0.971 91	0.979 57
0.90	0 864 89	0.891 87	0.914 72	0.933 73	0.949 30	0.961 82	0.971 72	0.979 39
0.91	0 864 87	0.891 79	0.914 58	0.933 56	0.949 11	0.961 63	0.971 53	0.979 22
0.92	0.864 86	0.891 71	0.914 45	0.933 39	0.948 92	0.961 44	0.971 34	0.979 05
0.93	0.864 85	0.891 63	0.914 32	0.933 23	0.948 74	0.961 25	0.971 16	0.978 89
0.94	0.864 84	0.891 56	0.914 20	0.933 08	0.948 57	0.961 07	0.970 99	0.978 72
0.95	0.864 83	0.891 49	0.914 08	0.932 92	0.948 40	0.960 90	0.970 82	0.978 57
0.96	0.864 83	0.891 42	0.913 96	0.932 78	0.948 24	0.960 73	0.970 66	0.978 42
0.97	0.864 82	0.891 35	0.913 85	0.932 64	0.948 08	0.960 56	0.970 50	0.978 27
0.98	0.864 82	0.891 29	0.913 75	0.932 50	0.947 92	0.960 40	0.970 34	0.978 12
0.99	0.864 82	0.891 23	0.913 64	0.932 37	0.947 77	0.960 25	0.970 19	0.977 98
1.00	0.864 82	0.891 18	0.913 54	0.932 24	0.947 63	0.960 10	0.970 04	0.977 84
1.05	0.864 83	0.890 93	0.913 10	0.931 66	0.946 97	0.959 41	0.969 36	0.977 20
1.10	0.864 87	0.890 74	0.912 73	0.931 17	0.946 40	0.958 81	0.968 77	0.976 64
1.15	0.864 92	0.890 59	0.912 43	0.930 75	0.945 91	0.958 29	0.968 25	0.976 15
1.20	0.864 98	0.890 48	0.912 17	0.930 39	0.945 49	0.957 84	0.967 80	0.975 72
1.25	0.865 04	0.890 39	0.911 96	0.930 09	0.945 13	0.957 45	0.967 40	0.975 33
1.30	0.865 12	0.890 32	0.911 79	0.929 83	0.944 82	0.957 11	0.967 06	0.975 00
1.35	0.865 19	0.890 27	0.911 64	0.929 61	0.944 55	0.956 82	0.966 75	0.974 71
1.40	0.865 26	0.890 24	0.911 52	0.929 43	0.944 32	0.956 56	0.966 49	0.974 45
1.45	0.865 33	0.890 21	0.911 41	0.929 27	0.944 12	0.956 34	0.966 26	0.974 22
1.50	0.865 40	0.890 20	0.911 33	0 929 13	0.943 95	0.956 14	0.966 05	0.974 01
1.55	0.865 46	0.890 19	0.911 26	0.929 01	0.943 79	0.955 97	0.965 87	0.973 83
1.60	0.865 52	0.890 18	0.911 20	0.928 91	0.943 66	0.955 82	0.965 71	0.973 67
1.65	0.865 58	0.890 18	0.911 15	0.928 82	0.943 55	0.955 68	0.965 57	0.973 53
1.70	0.865 64	0.890 18	0.911 10	0.928 74	0.943 44	0.955 57	0.965 45	0.973 41
1.75	0.865 69	0.890 19	0.911 07	0.928 67	0.943 35	0.955 46	0.965 33	0.973 30
1.80	0.865 74	0.890 19	0.911 04	0.928 62	0.943 28	0.955 37	0.965 24	0.973 20
1.85	0.865 78	0.890 20	0.911 01	0.928 56	0.943 21	0.955 29	0.965 15	0.973 11
1.90	0.865 82	0.890 21	0.910 99	0.928 52	0.943 14	0.955 22	0.965 07	0.973 03
1.95	0.865 86	0.890 21	0.910 97	0.928 48	0.943 09	0.955 15	0.965 00	0.972 96

η \ k	2.300	2.400	2.500	2.600	2.700	2.800	2.900	3.000
0.80	0.986 97	0.991 06	0.993 99	0.996 05	0.997 45	0.998 39	0.999 01	0.999 40
0.81	0.986 78	0.990 91	0.993 87	0.995 95	0.997 38	0.998 34	0.998 97	0.999 37
0.82	0.986 59	0.990 75	0.993 74	0.995 85	0.997 30	0.998 28	0.998 93	0.999 35
0.83	0.986 41	0.990 60	0.993 62	0.995 76	0.997 23	0.998 23	0.998 89	0.999 32
0.84	0.986 23	0.990 45	0.993 50	0.995 66	0.997 16	0.998 18	0.998 86	0.999 30
0.85	0.986 06	0.990 30	0.993 38	0.995 57	0.997 09	0.998 13	0.998 82	0.999 27
0.86	0.985 89	0.990 16	0.993 26	0.995 48	0.997 02	0.998 07	0.998 78	0.999 24
0.87	0.985 72	0.990 02	0.993 15	0.995 38	0.996 95	0.998 02	0.998 74	0.999 22
0.88	0.985 56	0.989 88	0.993 03	0.995 29	0.996 88	0.997 97	0.998 70	0.999 19
0.89	0.985 40	0.989 74	0.992 92	0.995 21	0.996 81	0.997 92	0.998 67	0.999 16
0.90	0.985 24	0.989 61	0.992 81	0.995 12	0.996 75	0.997 87	0.998 63	0.999 14
0.91	0.985 09	0.989 48	0.992 71	0.995 03	0.996 68	0.997 82	0.998 59	0.999 11
0.92	0.984 94	0.989 35	0.992 60	0.994 95	0.996 61	0.997 77	0.998 56	0.999 08
0.93	0.984 79	0.989 23	0.992 50	0.994 87	0.996 55	0.997 72	0.998 52	0.999 06
0.94	0.984 65	0.989 11	0.992 40	0.994 79	0.996 48	0.997 67	0.998 49	0.999 03
0.95	0.984 51	0.988 99	0.992 30	0.994 71	0.996 42	0.997 62	0.998 45	0.999 01
0.96	0.984 37	0.988 87	0.992 20	0.994 63	0.996 36	0.997 58	0.998 41	0.998 98
0.97	0.984 24	0.988 75	0.992 11	0.994 55	0.996 30	0.997 53	0.998 38	0.998 96
0.98	0.984 11	0.988 64	0.992 01	0.994 47	0.996 24	0.997 48	0.998 35	0.998 93
0.99	0.983 98	0.988 53	0.991 92	0.994 40	0.996 18	0.997 44	0.998 31	0.998 91
1.00	0.983 86	0.988 43	0.991 83	0.994 33	0.996 12	0.997 39	0.998 28	0.998 88
1.05	0.983 28	0.987 92	0.991 41	0.993 98	0.995 85	0.997 18	0.998 11	0.998 76
1.10	0.982 77	0.987 48	0.991 03	0.993 66	0.995 59	0.996 98	0.997 96	0.998 65
1.15	0.982 32	0.987 08	0.990 68	0.993 38	0.995 36	0.996 80	0.997 82	0.998 54
1.20	0.981 92	0.986 72	0.990 38	0.993 12	0.995 15	0.996 63	0.997 69	0.998 44
1.25	0.981 57	0.986 41	0.990 10	0.992 89	0.994 96	0.996 47	0.997 57	0.998 34
1.30	0.981 26	0.986 13	0.989 86	0.992 68	0.994 79	0.996 33	0.997 46	0.998 26
1.35	0.980 98	0.985 88	0.989 64	0.992 49	0.994 63	0.996 21	0.997 36	0.998 18
1.40	0.980 74	0.985 65	0.989 44	0.992 33	0.994 49	0.996 09	0.997 26	0.998 11
1.45	0.980 52	0.985 46	0.989 27	0.992 17	0.994 36	0.995 99	0.997 18	0.998 04
1.50	0.980 33	0.985 28	0.989 11	0.992 04	0.994 25	0.995 89	0.997 10	0.997 98
1.55	0.980 16	0.985 12	0.988 97	0.991 92	0.994 14	0.995 81	0.997 03	0.997 93
1.60	0.980 01	0.984 98	0.988 84	0.991 81	0.994 05	0.995 73	0.996 97	0.997 88
1.65	0.979 87	0.984 86	0.988 73	0.991 71	0.993 97	0.995 66	0.996 91	0.997 83
1.70	0.979 75	0.984 74	0.988 63	0.991 62	0.993 89	0.995 60	0.996 86	0.997 79
1.75	0.979 64	0.984 64	0.988 54	0.991 54	0.993 82	0.995 54	0.996 81	0.997 75
1.80	0.979 55	0.984 55	0.988 46	0.991 47	0.993 76	0.995 48	0.996 77	0.997 71
1.85	0.979 46	0.984 47	0.988 38	0.991 40	0.993 70	0.995 44	0.996 73	0.997 68
1.90	0.979 38	0.984 40	0.988 32	0.991 34	0.993 65	0.995 39	0.996 69	0.997 65
1.95	0.979 31	0.984 33	0.988 26	0.991 29	0.993 61	0.995 35	0.996 66	0.997 63

η \ k	0.700	0.800	0.900	1.000	1.100	1.200	1.300	1.400
2.0	0.513 49	0.573 75	0.629 48	0.680 52	0.726 79	0.768 32	0.805 21	0.837 66
2.1	0.531 90	0.574 15	0.629 85	0.680 85	0.727 07	0.768 55	0.805 39	0.837 77
2.2	0.514 23	0.574 47	0.630 16	0.681 12	0.727 31	0.768 74	0.805 53	0.837 87
2.3	0.514 50	0.574 73	0.630 41	0.681 35	0.727 50	0.768 90	0.805 65	0.837 96
2.4	0.514 72	0.574 95	0.630 61	0.681 54	0.727 66	0.769 03	0.805 75	0.838 02
2.5	0.514 91	0.575 14	0.630 79	0.681 69	0.727 80	0.769 14	0.805 84	0.838 08
2.6	0.515 06	0.575 29	0.630 93	0.681 82	0.727 91	0.769 23	0.805 91	0.838 13
2.7	0.515 19	0.575 42	0.631 05	0.681 93	0.728 01	0.769 31	0.805 97	0.838 18
2.8	0.515 30	0.575 53	0.631 16	0.682 03	0.728 09	0.769 38	0.806 02	0.838 21
2.9	0.515 40	0.575 62	0.631 24	0.682 11	0.728 16	0.769 44	0.806 07	0.838 25
3.0	0.515 48	0.575 70	0.631 32	0.682 18	0.728 22	0.769 49	0.806 10	0.838 27
3.1	0.515 55	0.575 77	0.631 38	0.682 24	0.728 27	0.769 53	0.806 14	0.838 30
3.2	0.515 61	0.575 83	0.631 44	0.682 29	0.728 31	0.769 56	0.806 17	0.838 32
3.3	0.515 66	0.575 88	0.631 49	0.682 33	0.728 35	0.769 60	0.806 19	0.838 33
3.4	0.515 70	0.575 92	0.631 53	0.682 37	0.728 39	0.769 62	0.806 21	0.838 35
3.5	0.515 74	0.575 96	0.631 57	0.682 40	0.728 41	0.769 65	0.806 23	0.838 36
4.0	0.515 87	0.576 09	0.631 69	0.682 52	0.728 52	0.769 73	0.806 30	0.838 41
4.5	0.515 95	0.576 16	0.631 76	0.682 58	0.728 57	0.769 78	0.806 33	0.838 44
5.0	0.515 99	0.576 21	0.631 80	0.682 62	0.728 60	0.769 81	0.806 36	0.838 45
6.0	0.516 03	0.576 25	0.631 84	0.682 65	0.728 64	0.769 83	0.806 38	0.838 47
7.0	0.516 05	0.576 27	0.631 86	0.682 67	0.728 65	0.769 85	0.806 39	0.838 48
8.0	0.516 06	0.576 28	0.631 87	0.682 68	0.728 66	0.769 85	0.806 39	0.838 48
9.0	0.516 06	0.576 28	0.631 87	0.682 68	0.728 66	0.769 86	0.806 39	0.838 48
10.0	0.516 07	0.576 28	0.631 87	0.682 68	0.728 66	0.769 86	0.806 40	0.838 48
20.0	0.516 07	0.576 29	0.631 88	0.682 69	0.728 67	0.769 86	0.806 40	0.838 49
30.0	0.516 07	0.576 29	0.631 88	0.682 69	0.728 67	0.769 86	0.806 40	0.838 49
40.0	0.516 07	0.576 29	0.631 88	0.682 69	0.728 67	0.769 86	0.806 40	0.838 49
50.0	0.516 07	0.576 29	0.631 88	0.682 69	0.728 67	0.769 86	0.806 40	0.838 49
∞	0.516 07	0.576 29	0.631 88	0.682 69	0.728 67	0.769 86	0.806 40	0.838 49

η ⟍ k	1.500	1.600	1.700	1.800	1.900	2.000	2.100	2.200
2.0	0.865 90	0.890 22	0.910 96	0.928 45	0.943 04	0.955 09	0.964 94	0.972 89
2.1	0.865 96	0.890 24	0.910 93	0.928 39	0.942 96	0.955 00	0.964 83	0.972 78
2.2	0.866 02	0.890 25	0.910 91	0.928 34	0.942 90	0.954 92	0.964 74	0.972 69
2.3	0.866 06	0.890 27	0.910 90	0.928 31	0.942 84	0.954 85	0.964 67	0.972 62
2.4	0.866 10	0.890 28	0.910 89	0.928 28	0.942 80	0.954 80	0.964 61	0.972 56
2.5	0.866 14	0.890 29	0.910 89	0.928 26	0.942 77	0.954 76	0.964 57	0.972 51
2.6	0.866 17	0.890 30	0.910 88	0.928 24	0.942 74	0.954 72	0.964 53	0.972 46
2.7	0.866 19	0.890 31	0.910 88	0.928 22	0.942 71	0.954 69	0.964 49	0.972 43
2.8	0.866 21	0.890 32	0.910 87	0.928 21	0.942 69	0.954 67	0.964 46	0.972 40
2.9	0.866 23	0.890 33	0.910 87	0.928 20	0.942 68	0.954 64	0.964 44	0.972 37
3.0	0.866 25	0.890 34	0.910 87	0.928 19	0.942 66	0.954 63	0.964 42	0.972 35
3.1	0.866 26	0.890 34	0.910 87	0.928 19	0.942 65	0.954 61	0.964 40	0.972 33
3.2	0.866 28	0.890 35	0.910 87	0.928 18	0.942 64	0.954 60	0.964 39	0.972 32
3.3	0.866 29	0.890 36	0.910 87	0.928 18	0.942 63	0.954 59	0.964 37	0.972 30
3.4	0.866 30	0.890 36	0.910 87	0.928 17	0.942 63	0.954 58	0.964 36	0.972 29
3.5	0.866 31	0.890 36	0.910 87	0.928 17	0.942 62	0.954 57	0.964 35	0.972 28
4.0	0.866 34	0.890 38	0.910 87	0.928 16	0.942 60	0.954 54	0.964 32	0.972 25
4.5	0.866 35	0.890 39	0.910 87	0.928 15	0.942 59	0.954 53	0.964 30	0.972 23
5.0	0.866 36	0.890 39	0.910 87	0.928 15	0.942 58	0.954 52	0.964 29	0.972 22
6.0	0.866 38	0.890 40	0.910 87	0.928 14	0.942 57	0.954 51	0.964 28·	0.972 20
7.0	0.866 38	0.890 40	0.910 87	0.928 14	0.942 57	0.954 50	0.964 28	0.972 20
8.0	0.866 38	0.890 40	0.910 87	0.928 14	0.942 57	0.954 50	0.964 27	0.972 20
9.0	0.866 38	0.890 40	0.910 87	0.928 14	0.942 57	0.954 50	0.964 27	0.972 20
10.0	0.866 38	0.890 40	0.910 87	0.928 14	0.942 57	0.954 50	0.964 27	0.972 19
20.0	0.866 39	0.890 40	0.910 87	0.928 14	0.942 57	0.954 50	0.964 27	0.972 19
30.0	0.866 39	0.890 40	0.910 87	0.928 14	0.942 57	0.954 50	0.964 27	0.972 19
40.0	0.866 39	0.890 40	0.910 87	0.928 14	0.942 57	0.954 50	0.964 27	0.972 19
50.0	0.866 39	0.890 40	0.910 87	0.928 14	0.942 57	0.954 50	0.964 27	0.972 19
∞	0.866 39	0.890 40	0.910 87	0.928 14	0.942 57	0.954 50	0.964 27	0.972 19

η \ k	2.300	2.400	2.500	2.600	2.700	2.800	2.900	3.000
2.0	0.979 25	0.984 27	0.988 20	0.991 24	0.993 56	0.995 32	0.996 63	0.997 60
2.1	0.979 14	0.984 17	0.988 11	0.991 16	0.993 49	0.995 26	0.996 58	0.997 56
2.2	0.979 05	0.984 09	0.988 03	0.991 09	0.993 43	0.995 21	0.996 54	0.997 52
2.3	0.978 98	0.984 02	0.987 97	0.991 03	0.993 38	0.995 16	0.996 50	0.997 49
2.4	0.978 92	0.983 96	0.987 91	0.990 98	0.993 34	0.995 13	0.996 47	0.997 47
2.5	0.978 87	0.983 91	0.987 87	0.990 94	0.993 30	0.995 09	0.996 44	0.997 45
2.6	0.978 83	0.983 87	0.987 83	0.990 91	0.993 27	0.995 07	0.996 42	0.997 43
2.7	0.978 79	0.983 84	0.987 80	0.990 88	0.993 25	0.995 05	0.996 40	0.997 41
2.8	0.978 76	0.983 81	0.987 77	0.990 85	0.993 22	0.995 03	0.996 39	0.997 40
2.9	0.978 74	0.983 78	0.987 75	0.990 83	0.993 20	0.995 01	0.996 37	0.997 39
3.0	0.978 71	0.983 76	0.987 73	0.990 81	0.993 19	0.995 00	0.996 36	0.997 38
3.1	0.978 69	0.983 74	0.987 71	0.990 80	0.993 18	0.994 99	0.996 35	0.997 37
3.2	0.978 68	0.983 73	0.987 70	0.990 79	0.993 16	0.994 97	0.996 34	0.997 36
3.3	0.978 66	0.983 72	0.987 69	0.990 77	0.993 15	0.994 97	0.996 33	0.997 36
3.4	0.978 65	0.983 70	0.987 67	0.990 76	0.993 14	0.994 96	0.996 33	0.997 35
3.5	0.978 64	0.983 69	0.987 66	0.990 76	0.993 14	0.994 95	0.996 32	0.997 34
4.0	0.978 61	0.983 66	0.987 63	0.990 72	0.993 11	0.994 93	0.996 30	0.997 33
4.5	0.978 59	0.983 64	0.987 61	0.990 71	0.993 09	0.994 91	0.996 29	0.997 32
5.0	0.978 57	0.983 63	0.987 60	0.990 70	0.993 08	0.994 91	0.996.28	0.997 31
6.0	0.978 56	0.983 62	0.987 59	0.990 69	0.993 07	0.994 90	0.996 28	0.997 31
7.0	0.978 56	0.983 61	0.987 59	0.990 68	0.993 07	0.994 89	0.996 27	0.997 30
8.0	0.978 56	0.983 61	0.987 58	0.990 68	0.993 07	0.994 89	0.996 27	0.997 30
9.0	0.978 55	0.983 61	0.987 58	0.990 68	0.993 07	0.994 89	0.996 27	0.997 30
10.0	0.978 55	0.983 61	0.987 58	0.990 68	0.993 07	0.994 89	0.996 27	0.997 30
20.0	0.978 55	0.983 61	0.987 58	0.990 68	0.993 07	0.994 89	0.996 27	0.997 30
30.0	0.978 55	0.983 60	0.987 58	0.990 68	0.993 07	0.994 89	0.996 27	0.997 30
40.0	0.978 55	0.983 60	0.987 58	0.990 68	0.993 07	0.994 89	0.996 27	0.997 30
50.0	0.978 55	0.983 60	0.987 58	0.990 68	0.993 07	0.994 89	0.996 27	0.997 30
∞	0.978 55	0.983 60	0.987 58	0.990 68	0.993 07	0.994 89	0.996 27	0.997 30

Explanatory notes

(a) *Combination of a rectangular distribution with a Gaussian distribution*
I.XXIII(5)   This is the simplest use to which the tables can be put. The standard deviation σ_G of the Gaussian distribution is required and also either the standard deviation of the rectangular distribution σ_R or its semi-range 'a'. Calculate the ratio $\sigma_G/\sigma_R = \eta$, or

$$\frac{\sqrt{(3)}\sigma_G}{a} = \eta$$

if 'a' is used. (Note: $\sigma_R = a/\sqrt{3}$)
   (i) *Probability*
If the probability for a given tolerance range $\mu - k\sigma$ to $\mu + k\sigma$ is required, where

$$\sigma = \left[\sigma_G^2 + \sigma_R^2\right]^{1/2} \equiv \left[\sigma_G^2 + \frac{a^2}{3}\right]^{1/2}$$

then part (b) should be consulted, the entry under the appropriate η and k giving the required value for the probability P.

(ii) *Tolerance factor k*

If the value of k, the tolerance factor for given probability P and given value for η, is required, then part (a) should be consulted for k. The required tolerance range is given by

$$\mu - k\sigma \text{ to } \mu + k\sigma$$

where μ is the mean value and σ the standard deviation of the combination.

(b) *Combination of a rectangular distribution with several Gaussian distributions*

I.XXIII(6)   First calculate the total Gaussian standard deviation σ_G as

$$\left[\sum_1^n \sigma_r^2\right]^{1/2} \equiv \sigma_G$$

where σ_r is the standard deviation of each Gaussian component, then proceed as in paragraph I.XXIII(5) as required.

(c) *Combination of a rectangular distribution with several Gaussian distributions and several rectangular distributions of smaller range than the first*

I.XXIII(7)   First calculate the standard deviation of the rectangular distributions, with the exception of the one having the largest range, as

$$\sigma_{RS} = \left[\sum_1^n \frac{a_r^2}{3}\right]^{1/2}$$

where a_r is the semi-range of each of the small rectangular distributions, and calculate the standard deviation of the Gaussian distributions as

$$\sigma_G = \left[\sum_1^n \sigma_r^2\right]^{1/2}$$

where σ_r is the standard deviation of each Gaussian distribution. The total standard deviation is given by

$$\sigma = [\sigma_R^2 + \sigma_{RS}^2 + \sigma_G^2]^{1/2}$$

where σ_R is the standard deviation of the largest rectangular distribution. η should then be calculated as given below

$$\eta = \frac{[\sigma_{RS}^2 + \sigma_G^2]^{1/2}}{\sigma_R}$$

Then proceed as in paragraph I.XXIII(5) as required.

The answers obtained will not be the exact values for P and k as in paragraphs I.XXIII(5) and (6), since the combination of the rectangular distributions above has been assumed to be Gaussian, which is not quite true. However, where a value of k is sought as in paragraph I.XXIII(5(ii)), the correct value of k will in fact be slightly smaller than that obtained. Where a value of P is sought as in I.XXIII(5(i)), the value of P obtained will be smaller than the true value for given k.

Appendix II

Proof that

$$F_0(p; v_1; v_2) = \frac{1}{F_0(1 - p; v_2; v_1)} \quad †$$

We can write

$$f(F) = \frac{A F^{v_1/2 - 1}}{(v_2 + v_1 F)^{(v_1 + v_2)/2}}$$

where

$$A = \frac{\Gamma((v_1 + v_2)/2) v_1^{v_1/2} v_2^{v_2/2}}{\Gamma(v_1/2)\Gamma(v_2/2)}$$

Now the probability of

$$F \geqslant F_0 = \frac{s_1^2}{s_2^2} = \frac{\chi_1^2/v_1}{\chi_2^2/v_2}$$

is given by

$$\int_{F_0}^{\infty} \frac{A F^{v_1/2 - 1} \, dF}{(v_2 + v_1 F)^{(v_1 + v_2)/2}} = p \qquad \text{say} \qquad \text{II(1)}$$

Let us now write

$$F_0' = \frac{\chi_2^2/v_2}{\chi_1^2/v_1} = \frac{1}{F_0}$$

Then the probability that

$$F \geqslant F_0' \equiv \frac{1}{F_0}$$

is given by

$$\int_{F_0'}^{\infty} \frac{A F^{v_2/2} \, dF}{(v_1 + v_2 F)^{(v_1 + v_2)/2}} = \int_{\frac{1}{F_0}}^{\infty} = I \qquad \text{II(2)}$$

† See equation 6.20(1).

Let us now put $F = 1/Q$ in the integral of I, which gives

$$I = \int_{F_0}^{0} \frac{A(1/Q)^{v_2/2-1} d(1/Q)}{(v_1 + v_2/Q)^{(v_1 + v_2)/2}} = -\int_{F_0}^{0} \frac{AQ^{-v/2+1} Q^{(v_1 + v_2)/2} dQ}{(v_2 + v_1 Q)^{(v_1 + v_2)/2} \cdot Q^2}$$

giving

$$I = \int_{0}^{F_0} \frac{AQ^{v_1/2-1} dQ}{(v_2 + v_1 Q)^{(v_1 + v_2)/2}}$$

which, by equation II(1), is equal to $(1 - p)$. Thus if the probability of

$$F \geqslant F_0 = \frac{\chi_1^2/v_1}{\chi_2^2/v_2} = \frac{s_1^2}{s_2^2} \quad \text{is} \quad p \quad \text{say}$$

then the probability

$$F \geqslant \frac{1}{F_0} = \frac{s_2^2}{s_1^2} = \frac{\chi_2^2/v_2}{\chi_1^2/v_1} \quad \text{is} \quad (1 - p)$$

Appendix III

Proof that

$$\underbrace{\sum_{q=1}^{q=m} \sum_{r=1}^{r=n_q} \frac{w_{rq}(x_{rq} - \mu)^2}{\sigma^2}}_{\chi_0^2} = \underbrace{\sum_{q=1}^{q=m} \sum_{r=1}^{r=n_q} \frac{w_{rq}(x_{rq} - \bar{x}_q)^2}{\sigma^2}}_{\chi_1^2}$$

$$+ \underbrace{\sum_{q=1}^{q=m} \sum_{r=1}^{r=n_q} \frac{w_{rq}(x_{rq} - \bar{x})^2}{\sigma^2}}_{\chi_2^2} + \underbrace{\sum_{q=1}^{q=m} \sum_{r=1}^{r=n_q} \frac{w_{rq}(x_{rq} - \mu)^2\dagger}{\sigma^2}}_{\chi_3^2}$$

Let $\sum_{q=1}^{q=m} \sum_{r=1}^{r=n_q}$ be written as $\sum\sum$ in the interests of clarity. Now $\sum\sum w_{rq}(x_{rq} - \mu)^2$ can be written as

$$\sum\sum w_{rq}[(x_{rq} - \bar{x}_q) + (\bar{x}_q - \bar{x}) + (\bar{x} - \mu)]^2$$

$$= \sum\sum w_{rq}[(x_{rq} - \bar{x}_q)^2 + (\bar{x}_q - \bar{x})^2 + (\bar{x} - \mu)^2] + 2\sum\sum w_{rq}(x_{rq} - \bar{x}_q)(\bar{x}_q - x)$$

$$+ 2\sum\sum w_{rq}(x_{rq} - \bar{x}_q)(\bar{x} - \mu) + 2\sum\sum w_{rq}(\bar{x}_q - \bar{x})(\bar{x} - \mu) \qquad \text{III}(1)$$

Now

$$\sum\sum w_{rq}(x_{rq} - \bar{x}_q)(\bar{x}_q - \bar{x}) = \sum_{q=1}^{q=m} (\bar{x}_q - \bar{x}) \sum_{r=1}^{r=n_q} w_{rq}(x_{rq} - \bar{x}_q) = 0$$

since

$$\sum_{r=1}^{r=n_q} w_{rq}(x_{rq} - \bar{x}_q) = 0$$

that is

$$\bar{x}_q = \sum_{r=1}^{r=n_q} \frac{w_{rq}x_{rq}}{\sum_{r=1}^{r=n_q} w_{rq}} \qquad \text{(see 9.13(1))}$$

Similarly

$$\sum\sum w_{rq}(x_{rq} - \bar{x}_q)(\bar{x} - \mu) = \sum_{q=1}^{q=m} (\bar{x} - \mu) \sum_{r=1}^{r=n_q} w_{rq}(x_{rq} - \bar{x}_q) = 0$$

from above.

† See paragraph 9.14, equation 9.14(1).

Also

$$\sum\sum w_{rq}(\bar{x}_q - \bar{x})(\bar{x} - \mu) = (\bar{x} - \mu) \sum_{q=1}^{q=m}(\bar{x}_q - \bar{x}) \sum_{r=1}^{r=n_q} w_{rq}$$

$$= (\bar{x} - \mu) \sum_{q=1}^{q=m} w_q(\bar{x}_q - \bar{x}) = 0$$

since

$$\sum_{q=1}^{q=m} w_q(\bar{x}_q - \bar{x}) = 0$$

that is

$$\bar{x} = \sum_{q=1}^{q=m} \frac{w_q \bar{x}_q}{\sum_{q=1}^{q=m} w_q} \qquad \text{(see 2.58(2))}$$

and where

$$\sum_{r=1}^{r=n_q} w_{rq} = w_q \qquad \text{(see 2.58(3))}$$

Thus the last three terms of III(1) are zero, and so dividing both sides of this equation by σ^2 we have the required expression.

Appendix IV

The Uncertainty in the Difference in Height between Points on a Surface (viz. a Surface Plate)

In paragraphs 10.02 to 10.11 formulae have been derived for calculating the topography of a surface, where measurements have been made in a sequential manner along sight lines shown in Figure 10.03. Now at each measuring position there will be an uncertainty associated with the angle α_r, and since the calculated topography is a combination of many such measurements, the uncertainty in the difference in height between points on the surface will be a function of these measurements, that is to say it will involve the uncertainties in the measurements between the two points considered, and also the uncertainties in the data involved, since the latter are themselves a function of the measurements. The required uncertainty will thus be a somewhat complicated function of the position of the two points and of the measurement constants.

Before proceeding with the derivation of the required uncertainties we shall state the nomenclature to be used. Consider Figure IV.1. This is similar to Figure 10.03. The arrows show the directions of measurement, and the n at the head of each arrow defines the number of measurements for the sight line involved. The letters in brackets at each line intersection give the calculated height of each intersection above the datum plane defined by the two diagonals AC and BD. The angles α for each measurement position are

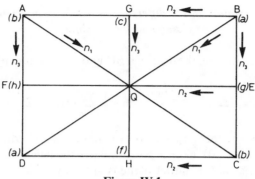

Figure IV.1

designated as follows:

$_L\alpha_r \equiv r$th measured angle from B on side AB   ($L \equiv$ long)
$_L\alpha'_r \equiv r$th measured angle from C on side CD
$_S\alpha_r \equiv r$th measured angle from B on side BC   ($S \equiv$ short)
$_S\alpha'_r \equiv r$th measured angle from A on side AD
$_D\alpha_r \equiv r$th measured angle from B on side BD   ($D \equiv$ diagonal)
$_D\alpha'_r \equiv r$th measured angle from A on side AD
$_{LM}\alpha_r \equiv r$th measured angle from E on side EF   ($LM \equiv$ long middle)
$_{SM}\alpha_r \equiv r$th measured angle from G on side GH   ($SM \equiv$ short middle)

$_1d_d =$ step distance for each measurement on sides AC or BD
$_2d_d =$ step distance for each measurement on sides BA, EF, and CD
$_3d_d =$ step distance for each measurement on sides BC, GH, and AD

$_D\Delta_L \equiv$ difference in height between a point on a long side (AC or CD) and a point on a diagonal (AC or BD)
$_L\Delta_L \equiv$ difference in height between two points on a long side
$_D\Delta_S \equiv$ difference in height between a point on a short side (AD or BC) and a point on a diagonal
$_D\Delta_{LM} \equiv$ difference in height between a point on the long middle side (EF) and a point on a diagonal (AC or BD)
$_D\Delta_{SM} \equiv$ difference in height between a point on the short middle side (GH) and a point on a diagonal
$_D\Delta_D \equiv$ difference in height between two points on a diagonal

$_D(\sigma_\Delta)_L$ is the standard deviation of $_D\Delta_L$
$_L(\sigma_\Delta)_L$ is the standard deviation of $_L\Delta_L$
$_D(\sigma_\Delta)_S$ is the standard deviation of $_D\Delta_S$
$_D(\sigma_\Delta)_{LM}$ is the standard deviation of $_D\Delta_{LM}$
$_D(\sigma_\Delta)_{SM}$ is the standard deviation of $_D\Delta_{SM}$
$_D(\sigma_\Delta)_D$ is the standard deviation of $_D\Delta_D$
σ_α is the common standard deviation of all the α

(1) *The uncertainty in the difference in height between two points on a diagonal line of sight*

The height of a point p on the diagonal BD above the datum plane is given by

$$_D z_p = {_1}d_d\left(\sum_{r=0}^{r=p} {_D}\alpha_r - \frac{p}{n}\sum_{r=0}^{r=n_1} {_D}\alpha_r + \tfrac{1}{2}\sum_{r=0}^{r=n_1} {_D}\alpha_r - \sum_{r=0}^{r=n_1/2} {_D}\alpha_r\right) \qquad \text{IV(1)}$$

using 10.06(2) (note that $_D\alpha_0$ is zero by definition), and where p is measured from the point B($p = 0$), $n_1 \geqslant p \geqslant 0$. Similarly the height of a point q on the same diagonal above the datum plane is given by

$$_D z_q = {_1}d_d\left(\sum_{r=0}^{r=q} {_D}\alpha_r - \frac{q}{n}\sum_{r=0}^{r=n_1} {_D}\alpha_r + \tfrac{1}{2}\sum_{r=0}^{r=n_1} {_D}\alpha_r - \sum_{r=1}^{r=n_1/2} {_D}\alpha_r\right) \qquad \text{IV(2)}$$

$$n_1 \geqslant q \geqslant 0$$

again using 10.06(2), and where once again q is measured from B.

Let us suppose that $p > q$, and thus the range 1 to n_1 can be divided into a number of parts, that is 1 to $q, q + 1$ to p and $p + 1$ to n_1. Thus $_D z_p$ can be written as

$$_D z_p = {_1}d_d\left\{\sum_{r=0}^{r=q} {_D}\alpha_r + \sum_{r=q+1}^{r=p} {_D}\alpha_r \right.$$

$$\left. + \left(\frac{1}{2} - \frac{p}{n_1}\right)\left(\sum_{r=0}^{r=q} {_D}\alpha_\alpha + \sum_{r=q+1}^{r=p} {_D}\alpha_\alpha + \sum_{r=p+1}^{r=n_1} {_D}\alpha_\alpha\right) - \sum_{r=0}^{r=n_1/2} {_D}\alpha_\alpha\right\}$$

$$\text{IV(3)}$$

Similarly

$$_D z_q = {_1}d_d\left\{\sum_{r=0}^{r=q} {_D}\alpha_r \right.$$

$$\left. + \left(\frac{1}{2} - \frac{q}{n_1}\right)\left(\sum_{r=0}^{r=q} {_D}\alpha_r + \sum_{r=q+1}^{r=p} {_D}\alpha_r + \sum_{r=p+1}^{r=n_1} {_D}\alpha_r\right) - \sum_{r=0}^{r=n_1/2} {_D}\alpha_r\right\}$$

The difference in height $_D z_p - {_D}z_q$ is thus given by

$$_D\Delta_D = {_1}d_d\left\{\sum_{r=q+1}^{r=p} {_D}\alpha_r + \left(\frac{q}{n_1} - \frac{p}{n_1}\right)\left(\sum_{r=0}^{r=q} {_D}\alpha_r + \sum_{r=q+1}^{r=p} {_D}\alpha_r + \sum_{r=p+1}^{r=n_1} {_D}\alpha_r\right)\right\}$$

$$= {_1}d_d\left\{\frac{q-p}{n_1}\left(\sum_{r=0}^{r=q} {_D}\alpha_r + \sum_{r=p+1}^{r=n_1} {_D}\alpha_r\right) + \left(1 + \frac{q-p}{n_1}\right)\sum_{r=q+1}^{r=p} {_D}\alpha_r\right\} \quad \text{IV(4)}$$

where $p > q$.

It is worth noting here that it is most important that terms which cancel should be removed before the standard deviation of $_D\Delta_D$ is found, because this is a squaring process and terms not so removed will add in the squaring process and produce a wrong answer. The procedure just stated applies equally well to all the subsequent uncertainties.

Now the $_D\alpha_r$ are subject to uncertainty, and can thus be considered as random variables. Further, since each $_D\alpha_r$ is obtained from using the same instrument, the standard deviation of each $_D\alpha_r$ will be the same. Thus we write the common standard deviation of each $_D\alpha_r$ as σ_α.

Now $_D\Delta_D$ is a linear function of the $_D\alpha_r$ and so

$$_D(\sigma_\Delta)_D^2 = \left\{ \frac{(q-p)^2}{n_1^2}(q+n_1-p) + \left(1 + \frac{q-p}{n_1}\right)^2 (p-q) \right\}_1 d_d^2 \sigma_\alpha^2$$

(see 3.40(4) and 3.40(5))

$$= \left(1 + \frac{q-p}{n_1}\right)(p-q)\sigma_\alpha^2 {}_1 d_d^2 \qquad\qquad \text{IV(5)}$$

after a little simplification. Note that $p > q$.

The uncertainty in the difference in height between two points on a diagonal is thus given by a tolerance range $\pm k_1 {}_D(\sigma_\Delta)_D$ where k_1 corresponds to the

selected tolerance probability $\underset{-k_1 \text{ to } k_1}{P}$

$$_D(\sigma_\Delta)_D = \sqrt{\left\{ \left(1 + \frac{q-p}{n_1}\right)(p-q) \right\}} \sigma_\alpha {}_1 d_d \qquad\qquad \text{IV(6)}$$

where $p > q$.

It is to be noted that if $p = q$, then $_D(\sigma_\Delta)_D$ is zero; also if $q = 0$ and $p = n_1$, then $_D(\sigma_\Delta)_n$ is again zero.

It is easily shown that the maximum value of $_D(\sigma_\Delta)_D$ is given when $p - q = n_1/2$, giving

$$_D(\sigma_\Delta)_D \text{max} = \frac{\sqrt{n_1}}{2} \cdot \sigma_\alpha {}_1 d_d \qquad\qquad \text{IV(7)}$$

(2)   *The uncertainty in the difference in height between two points on an outside line of sight*

(a)   We shall first consider the side BA
The distance of a point on BA above the datum plane is given by

$$_L Z_p = \left[\left(1 - \frac{p}{n_2}\right) \sum_{r=0}^{r=p} {}_L\alpha_r - \frac{p}{n_2} \sum_{r=p+1}^{r=n_2} {}_L\alpha_r \right] {}_2 d_d + a\left(1 - \frac{p}{n_2}\right) + \frac{bp}{n_2}$$

$$\text{IV(8)}$$

where p is measured from B, and where $n_2 \geqslant p \geqslant 0$ (see 10.08(1)). Similarly the distance of another point on BA above the datum plane is given by

$$_L Z_q = \left[\left(1 - \frac{q}{n_2} \right) \sum_{r=0}^{r=q} {}_L \alpha_r - \frac{q}{n_2} \sum_{r=q+1}^{r=n_2} {}_L \alpha_r \right] {}_2 d_d + a \left(1 - \frac{q}{n_2} \right) + \frac{bq}{n_2}$$

IV(9)

where q is measured from B and $n_2 \geqslant q \geqslant 0$.

Let us suppose that $p \geqslant q$. Thus the difference in height $_L \Delta_L$ between the two points is given by IV(8) minus IV(9)

$$_L \Delta_L = \left[\frac{q}{n_2} \sum_{r=q+1}^{r=n_2} {}_L \alpha_r - \frac{p}{n_2} \sum_{r=p+1}^{r=n_2} {}_L \alpha_r \right.$$

$$\left. + \left(1 - \frac{p}{n_2} \right) \sum_0^p {}_L \alpha_r - \left(1 - \frac{q}{n_2} \right) \sum_0^q {}_L \alpha_r \right] {}_2 d_d + \frac{(p-q)}{n_2} (b - a)$$

IV(10)

Now

$$a = \frac{1}{2} \left(\sum_{n_1/2+1}^{n_1} {}_D \alpha_r - \sum_0^{n_1/2} {}_D \alpha_r \right) {}_1 d_d$$

IV(11)

and

$$b = \frac{1}{2} \left(\sum_{n_1/2+1}^{n_1} {}_D \alpha'_r - \sum_0^{n_1/2} {}_D \alpha_r \right) {}_1 d_d$$

IV(12)

These values are obtained by putting $p = n_1$ or 0 in 10.06(2) and splitting the range 0 to n_1 into 0 to $n_1/2$ and $n_1/2 + 1$ to n_1. The a term is the end value for the side BA, whilst b is the end value for the side CA. The range 0 to n_2 along the side BA can be divided into the sub-ranges 0 to q, $q + 1$ to p, and $p + 1$ to n_2, and inserting these ranges in equation IV(10) yields

$$_L \Delta_L = \left[\left(\frac{q-p}{n_2} \right) \sum_0^q {}_L \alpha_r + \left(1 - \frac{p-q}{n_2} \right) \sum_{q+1}^q {}_L \alpha_r + \left(\frac{q-p}{n_2} \right) \sum_{p+1}^{n_2} {}_L \alpha_r \right] {}_2 d_d$$

$$+ \left(\frac{p-q}{n_2} \right) \frac{{}_1 d_d}{2} \left[\sum_{n_1/2+1}^{n_1} {}_D \alpha'_r - \sum_{n_1/2+1}^{n_1} {}_D \alpha_r + \sum_0^{n_1/2} {}_D \alpha_r - \sum_0^{n_1/2} {}_D \alpha'_r \right]$$ IV(13)

Once again $_L \Delta_L$ is a linear function of the α_r and so the standard deviation of $_L \Delta_L$ is given by

$$_L (\sigma_\Delta)_L^2 = \left[\frac{(p-q)^2}{n_2^2} (n_2 - p + q) + \left(1 - \frac{p-q}{n_2} \right)^2 (p - q) \right] \sigma_\alpha \, {}_2 d_d^2$$

$$+ \frac{(p-q)^2}{n_2^2} \frac{{}_1 d_d^2}{4} (2n_1) \sigma_\alpha^2$$

$$= \left[(p-q) \left(1 - \frac{p-q}{n_2} \right) {}_2 d_d^2 + \frac{(p-q)^2 n_1 \, {}_1 d_d^2}{2 n_2^2} \right] \sigma_\alpha^2$$

IV(14)

Note that the first term has the same form as that for a diagonal side. The second term represents the contribution to the uncertainty from the diagonal sides which arises from the uncertainty in a and b. Putting $p - q = h$ and differentiating IV(14) it can be shown that $_L(\sigma_\Delta)_L$ has its maximum value when

$$h = \frac{n_2}{(2 - (_1d_d^2/_2d_d^2) \cdot (n_1/n_2))} \qquad \text{IV(15)}$$

giving the maximum value of $_L(\sigma_\Delta)_L$ as

$$_L(\sigma_\Delta)_L \max = \frac{\sqrt{(n_2)_2d_d\sigma_\alpha}}{\sqrt{\{2(2 - _1d_d^2n_1/_2d_d^2n_2)\}}} \qquad \text{IV(16)}$$

It is to be noted that this maximum value is generally not attained in practical cases because $_1d_d \geqslant _2d_d$ and $n_1 > n_2$, and so $h > n_2$, and since $p - q$ cannot exceed n_2 this means that the uncertainty between two points on an outside line of sight increases with the distance apart of the two points; unlike the uncertainty for two points on a diagonal side which reaches a maximum value when the points are $n_1/2$ apart.

(*b*) Side CD

Formulae IV(14), IV(15) and IV(16) apply to the side CD where p and q are measured from C.

(*c*) Sides BC and AD

The formulae for the sides BC and AD are obtained by replacing n_2 by n_3 and $_2d_d$ by $_3d_d$, giving

$$_s(\sigma_\Delta)_S^2 = \left\{(p - q)\left(1 - \frac{p - q}{n_3}\right)_3d_d^2 + \frac{(p - q)^2 n_1 \,_1d_d^2}{2n_2^2}\right\}\sigma_\alpha^2 \qquad \text{IV(17)}$$

where p and q are both measured from A or D for AD, or B or C for BC

$$h = \frac{n_3}{(2 - _1d_d^2n_1/_3d_d^2n_3)} \qquad \text{IV(18)}$$

and

$$_s(\sigma_\Delta)_S = \frac{\sqrt{(n_3)_3d_d\sigma_\alpha}}{\sqrt{\{2(2 - _1d_d^2n_1/_3d_d^2n_3)\}}} \qquad \text{IV(19)}$$

(3) *The uncertainty in the difference in height between two points on side EF or side GH*

Let us consider GH

The distance of a point on GH above the datum plane is given by

$$_{SM}z_p = \left[\left(1 - \frac{p}{n_3}\right)\sum_{r=0}^{r=p} {}_{SM}\alpha_r - \frac{p}{n_3}\sum_{r=p+1}^{r=n_3} {}_{SM}\alpha_r\right]_3d_d + c\left(1 - \frac{p}{n_3}\right) + \frac{fp}{n_3}$$

$$\text{IV(20)}$$

where $n_3 \geqslant p \geqslant 0$ and p is measured from G $(p = 0)$.

Similarly the distance above the datum plane of a second point on GH is given by

$$_{SM}Z_q = \left[\left(1 - \frac{q}{n_3} \right) \sum_{r=0}^{r=q} {}_{SM}\alpha_r - \frac{q}{n_3} \sum_{r=q+1}^{n_3} {}_{SM}\alpha_r \right] {}_3d_d + c \left(1 - \frac{q}{n_3} \right) + \frac{fq}{n_3}$$

$$\text{IV(21)}$$

where $n_3 \geqslant q \geqslant 0$ and q is measured from G $(q = 0)$.

The difference in height between these two points is thus IV(20) minus IV(21), given by

$$_{SM}\Delta_{SM} = \left[\left(\frac{q-p}{n_3} \right) \sum_{r=0}^{r=q} {}_{SM}\alpha_r + \left(1 - \frac{p-q}{n_3} \right) \sum_{r=q+1}^{r=p} {}_{SM}\alpha_r \right.$$

$$\left. - \left(\frac{p-q}{n_3} \right) \sum_{r=p+1}^{r=n_3} {}_{SM}\alpha_r \right] {}_3d_d + \frac{(p-q)}{n_3}(f - c) \qquad \text{IV(22)}$$

where $p \geqslant q$.

c is the height of the mid-point of BA above the datum plane, whilst f is the height of the mid-point of CD above the datum plane.

$$c = \left(\frac{1}{2} \sum_{r=0}^{r=n_2/2} {}_L\alpha_r - \frac{1}{2} \sum_{r=n_2/2+1}^{r=n_2} {}_L\alpha_r \right) {}_2d_d + \frac{a+b}{2} \qquad \text{IV(23a)}$$

$$= \left(\frac{1}{2} \sum_{r=0}^{r=n_2/2} {}_L\alpha_r - \frac{1}{2} \sum_{r=n_2/2+1}^{r=n_2} {}_L\alpha_r \right) {}_2d_d + \frac{1}{4} \left(\sum_{r=n_1/2+1}^{r=n_1} {}_D\alpha_r - \sum_{r=0}^{r=n_1/2} {}_D\alpha_r \right) {}_1d_d$$

$$+ \frac{1}{4} \left(\sum_{r=n_1/2+1}^{r=n_1} {}_D\alpha'_r - \sum_{r=0}^{r=n_1/2} {}_D\alpha'_r \right) {}_1d_d \qquad \text{IV(23b)}$$

by putting $p = n_2/2$ in equation IV(8) and substituting for a and b from IV(11) and IV(12).

Similarly

$$f = \frac{1}{2} \left(\sum_{r=0}^{r=n_2/2} {}_L\alpha'_r - \sum_{r=n_2/2+1}^{r=n_2} {}_L\alpha'_r \right) {}_2d_d + \frac{a+b}{2} \qquad \text{IV(24a)}$$

Thus $f - c$ is given by

$$\frac{1}{2} \left(\sum_{r=0}^{r=n_2/2} {}_L\alpha'_r - \sum_{r=0}^{r=n_2/2} {}_L\alpha_r - \sum_{r=n_2/2+1}^{r=n_2} {}_L\alpha'_r + \sum_{r=n_2/2+1}^{r=n_2} {}_L\alpha_r \right) {}_2d_d \qquad \text{IV(24b)}$$

and so

$$
\begin{aligned}
_{SM}\Delta_{SM} = &\left[\left(\frac{q-p}{n_3}\right)\sum_{r=0}^{r=q}{}_{SM}\alpha_r + \left(1 - \frac{p-q}{n_3}\right)\sum_{r=q+1}^{r=p}{}_{SM}\alpha_r \right. \\
&- \left(\frac{p-q}{n_3}\right)\sum_{r=p+1}^{r=n_3}{}_{SM}\alpha_r \Bigg]{}_3 d_d \\
&+ \frac{(p-q)}{2n_3}\left(\sum_{r=0}^{r=n_2/2}{}_L\alpha'_r - \sum_{r=0}^{r=n_2/2}{}_L\alpha_r - \sum_{r=n_2/2+1}^{r=n_2}{}_L\alpha'_r + \sum_{r=n_2/2+1}^{r=n_2}{}_L\alpha_r\right){}_2 d_d
\end{aligned}
$$

$$\text{IV}(25)$$

The standard deviation of $_{SM}\Delta_{SM}$ is thus given by

$$
_{SM}(\sigma_\Delta)^2_{SM} = \left[(p-q)\left(1 - \frac{p-q}{n_3}\right){}_3 d_d^2 + \frac{(p-q)^2 n_2 \, _2 d_d^2}{2n_3^2}\right]\sigma_\alpha^2
$$

$$\text{IV}(26)$$

which is exactly analogous to the expressions under heading (2) for the outside sight lines. Similar expressions hold for the maximum of $_{SM}(\sigma_\Delta)_{SM}$, that is

$$
h = (p - q) = \frac{n_3}{2 - \, _2 d_d^2 n_2 / \, _3 d_d^2 n_3}
$$

$$\text{IV}(27)$$

and

$$
_{SM}(\sigma_\Delta)_{SM} \text{ max} = \frac{\sqrt{(n_3)}\, _3 d_d \sigma_\alpha}{\sqrt{\{2(2 - \, _2 d_d n_2 / \, _3 d_d n_3)\}}}
$$

$$\text{IV}(28)$$

Since $_2 d_d \geqslant _3 d_d$ and $n_2 > n_3$, $h > n_3$ and so the maximum of $_{SM}(\sigma_\Delta)_{SM}$ lies outside the range of $p - q$ which is limited to 0 to n_2. In the above formulae p and q can of course be measured from H.

Uncertainty for points on side EF

Analogous reasoning leads to

$$
_{LM}(\sigma_\Delta)^2_{LM} = \left[(p-q)\left(1 - \frac{p-q}{n_2}\right){}_2 d_d^2 + \frac{(p-q)^2}{2n_2^2}n_3 \, _3 d_d^2\right]\sigma_\alpha^2 \quad \text{IV}(29)
$$

where p and q are measured from E or F ($p = 0$, $q = 0$) and $p \geqslant q$, with

$$
h = p - q = \frac{n_2}{[2 - \, _3 d_d^2 n_3 / \, _2 d_d^2 n_2]}
$$

$$\text{IV}(30)$$

and $h < n_2$, since $_2 d_d > _3 d_d$ and $n_2 > n_3$ and thus the denominator of n_2 in equation IV(30) is greater than unity. Thus the maximum of $_{LM}(\sigma_\Delta)^2_{LM}$ lies within the range of points on EF and

$$_{LM}(\sigma_\Delta)_{LM} \max = \frac{\sqrt{(n_2)_2 d_a \sigma_\alpha}}{\sqrt{\{2(2 - {}_3 d_a^2 n_3 / {}_2 d_a^2 n_2)\}}} \qquad \text{IV(31)}$$

(4) *The uncertainty between a point on a diagonal and a point on an outside side*

(a) We shall consider first the sides BA and BD
The distance of a point on the diagonal side BD above the datum plane is given by

$$_D z_p = {}_1 d_d \left(\sum_{r=0}^{r=p} {}_D \alpha_r - \frac{p}{n} \sum_{r=0}^{r=n_1} {}_D \alpha_r + \frac{1}{2} \sum_{r=0}^{r=n_1} {}_D \alpha_r - \sum_{r=0}^{r=n_1/2} {}_D \alpha_r \right)$$

from equation IV(1)

$$= {}_1 d_d \left\{ \left(1 - \frac{p}{n_1}\right) \sum_{r=0}^{r=p} {}_D \alpha_r - \frac{p}{n} \sum_{r=p+1}^{r=n_1} {}_D \alpha_r + \frac{1}{2} \left(\underbrace{\sum_{r=n_1/2+1}^{r=n_1} {}_D \alpha_r - \sum_{r=0}^{r=n_1/2} {}_D \alpha_r}_{a} \right) \right\}$$

$$\text{IV(32)}$$

where p is measured from $B(p = 0)$.

The distance of a point on the outside side BA above the datum plane is given by

$$_L z_q = \left[\left(1 - \frac{q}{n_2}\right) \sum_{r=0}^{r=q} {}_L \alpha_r - \frac{q}{n_2} \sum_{r=q+1}^{r=n_2} {}_L \alpha_r \right] {}_2 d_d + a\left(1 + \frac{q}{n_2}\right) + \frac{bq}{n_2}$$

$$\text{IV(33)}$$

where a and b are given by IV(11) and IV(12), and q is measured from B $(q = 0)$.

The difference in height between a point on BA and one on BD is thus given by IV(33) minus IV(32) as $_L \Delta_D$

$$_L \Delta_D = \left[\left(1 - \frac{q}{n_2}\right) \sum_{r=0}^{r=q} {}_L \alpha_r - \frac{q}{n_2} \sum_{r=q+1}^{r=n_2} {}_L \alpha_r \right] {}_2 d_d$$

$$- \left[\left(1 - \frac{p}{n_1}\right) \sum_{r=0}^{r=p} {}_D \alpha_r - \frac{p}{n_1} \sum_{r=p+1}^{r=n_1} {}_D \alpha_r \right] {}_1 d_d + \frac{q}{n_2}(b - a)$$

where

$$b - a = \frac{1}{2} \left(\sum_{n_1/2+1}^{n_1} {}_D \alpha_r' - \sum_{n_1/2+1}^{n_1} {}_D \alpha_r + \sum_0^{n_1/2} {}_D \alpha_r - \sum_0^{n_1/2} {}_D \alpha_r' \right) {}_1 d_d$$

from IV(11) and IV(12).

$$_L\Delta_D = \left[\left(1 - \frac{q}{n_2}\right)\sum_{r=0}^{r=q}{_L\alpha_r} - \frac{q}{n_2}\sum_{r=q+1}^{r=n_2}{_L\alpha_r}\right]{_2d_d}$$

$$+ \frac{q}{2n_2}\left[\sum_{r=n_1/2+1}^{r=n_1}{_D\alpha'_r} - \sum_{r=0}^{r=n_1/2}{_D\alpha'_r}\right]{_1d_d} + \text{a term involving BD}$$

$$\text{IV(34)}$$

The term involving BD has two forms, depending on whether $p > n_1/2$ or $p < n_1/2$.

If $p > n_1/2$ the BD term is

$$\left[\left(\frac{p}{n_1} - \frac{q}{2n_2}\right)\sum_{r=p+1}^{r=n_1}{_D\alpha_r} - \left(1 - \frac{p}{n_1} - \frac{q}{2n_2}\right)\sum_{r=0}^{r=n_1/2}{_D\alpha_r}\right.$$

$$\left. - \left(1 - \frac{p}{n_1} + \frac{q}{2n_2}\right)\sum_{r=n_1/2+1}^{r=p}\right]{_1d_d} \quad \text{IV(35)}$$

If $p < n_1/2$ the BD term is

$$\left[\left(\frac{p}{n_1} + \frac{q}{2n_2}\right)\sum_{r=p+1}^{r=n_1/2}{_D\alpha_r} - \left(1 - \frac{p}{n_1} - \frac{q}{2n_2}\right)\sum_{r=0}^{r=p}{_D\alpha_r}\right.$$

$$\left. + \left(\frac{p}{n_1} - \frac{q}{2n_2}\right)\sum_{r=n_1/2+1}^{r=n_1}\right]{_1d_d} \quad \text{IV(36)}$$

Thus if $p > n_1/2$

$$_L(\sigma_\Delta)^2_D = \left[\left(1 - \frac{q}{n_2}\right)^2 q + \frac{q^2}{n_2^2}(n_2 - q)\right]{_2d_d^2}\sigma_\alpha^2 + \frac{q^2}{4n_2^2}n_1\,{_1d_d^2}\sigma_\alpha^2$$

$$+ \text{ last term}$$

$$= \left[q\left(1 + \frac{q}{n_2}\right){_2d_d^2} + \frac{q^2}{4n_2^2}n_1\,{_1d_d^2}\right]\sigma_\alpha^2$$

$$+ \left[\left(\frac{p}{n_1} - \frac{q}{2n_2}\right)^2(n_1 - p) + \left(1 - \frac{p}{n_1} - \frac{q}{2n_2}\right)^2\frac{n_1}{2}\right.$$

$$\left. + \left(1 - \frac{p}{n_1} + \frac{q}{2n_2}\right)^2\left(p - \frac{n_1}{2}\right)\right]{_1d_d^2}\sigma_\alpha^2$$

$$= \left[q\left(1 - \frac{q}{n_2}\right){_2d_d^2} + \frac{q^2}{4n_2^2}n_1\,{_1d_d^2}\right]\sigma_\alpha^2$$

$$+ \left[\left(1 - \frac{p}{n_1}\right)\left(p - \frac{qn_1}{n_2}\right) + \frac{q^2n_1}{4n_2^2}\right]{_1d_d^2}\sigma_\alpha^2 \quad \text{IV(37a)}$$

If we put $q = m + n_2/2$, IV(37a) can be written as

$$_L(\sigma_\Delta)_D^2 \left\{ \left(\frac{n_2}{4} - \frac{m^2}{n_2} \right)_2 d_d^2 + \left[\frac{n_1 m^2}{2n_2^2} + \frac{m}{n_2} \left(p - \frac{n_1}{2} \right) \right. \right.$$

$$\left. \left. + p \left(\frac{3}{2} - \frac{p}{n_1} \right) - \frac{3}{8} n_1 \right]_1 d_d^2 \right\} \sigma_\alpha^2 \qquad \text{IV(37b)}$$

which if $p = n_1/2$ is symmetrical in m, that is the standard deviation is then symmetrical about the centre of an outside side. $p = n_1/2$ corresponds to the centre.

If $p < n_1/2$

$$_L(\sigma_\Delta)_D^2 = \left[q \left(1 - \frac{q}{n_2} \right)_2 d_d^2 + \frac{q^2}{4n_2^2} n_1 \, _1d_d^2 \right] \sigma_\alpha^2$$

$$+ \left[\left(\frac{p}{n_1} + \frac{q}{2n_2} \right)^2 \left(\frac{n_1}{2} - p \right) + \left(1 - \frac{p}{n_1} - \frac{q}{2n_2} \right)^2 p \right.$$

$$\left. + \left(\frac{p}{n_1} - \frac{q}{2n_2} \right)^2 \frac{n_1}{2} \right]_1 d_d^2 \sigma_\alpha^2$$

$$= \left[q \left(1 - \frac{q}{n_2} \right)_2 d_d^2 + \frac{q^2}{4n_2^2} n_1 \, _1d_d^2 \right] \sigma_\alpha^2$$

$$+ \left[p \left(1 - \frac{q}{n_2} - \frac{p}{n_1} \right) + \frac{n_1 q^2}{4n_2^2} \right]_1 d_d^2 \sigma_\alpha^2 \qquad \text{IV(38a)}$$

In a similar manner it can be shown that IV(38a) can be written as

$$_L(\sigma_\Delta)_D^2 = \left[\left(\frac{n_2}{4} - \frac{m^2}{n_2} \right)_2 d_d^2 \right.$$

$$\left. + \left\{ \frac{n_1 m^2}{2n_2^2} + \frac{m}{n_2} \left(p - \frac{n_1}{2} \right) + p \left(\frac{1}{2} - \frac{p}{n_1} \right) + \frac{n_1}{8} \right\}_1 d_d^2 \right] \sigma_\alpha^2$$

$$\text{IV(38b)}$$

These formulae are also true for the sides CD and CA, where p and q are measured from C.

(b) The formulae for points on sides BD and BC, and AC and AD are obtained by replacing n_2 by n_3 and $_2d_d$ by $_3d_d$ in IV(37) and IV(38), where in the first case p and q are measured from B, and in the second case from A.

(5) *The uncertainty in height between a point on a diagonal side and a point on a non-diagonal side which passes through the centre, Q, of the sight line figure*

(*a*) Let us begin by considering a point on BD and a point on GH
The distance of a point on the diagonal side BD above the datum plane is given by

$$_D z_D = \left[\sum_{r=0}^{r=p} {}_D\alpha_r - \frac{p}{n_1} \sum_{r=0}^{r=n_1} {}_D\alpha_r + \frac{1}{2} \sum_{r=0}^{r=n_1} {}_D\alpha_r - \sum_{r=0}^{r=n_1/2} {}_D\alpha_r \right] {}_1 d_d$$

(see equation IV(1)). p is measured from B and $n_1 \geqslant p \geqslant 0$.

Similarly the distance of a point on the side GH above the datum plane is given by

$$_{SM} z_q = \left[\left(1 - \frac{q}{n_3} \right) \sum_{r=0}^{r=q} {}_{SM}\alpha_r - \frac{q}{n_3} \sum_{r=q+1}^{r=n_3} {}_{SM}\alpha_r \right] {}_3 d_d + c \left(1 - \frac{q}{n_3} \right) + \frac{fq}{n_3}$$

where q is measured from G and $n_3 \geqslant q \geqslant 0$ (see equation IV(20)).

The difference in height between the point on GH and the point on BD is thus given by

$$_{SM}\Delta_D = {}_{SM}z_q - {}_D z_p$$

$$= \left\{ \left(1 - \frac{q}{n_3} \right) \sum_{r=0}^{r=q} {}_{SM}\alpha_r - \frac{q}{n_3} \sum_{r=q+1}^{r=n_3} {}_{SM}\alpha_r \right\} {}_3 d_d + c \left(1 - \frac{q}{n_3} \right) + \frac{fq}{n_3}$$

$$- \left(\sum_{r=0}^{r=p} {}_D\alpha_r - \frac{p}{n_1} \sum_{r=0}^{r=n_1} {}_D\alpha_r + \frac{1}{2} \sum_{r=0}^{r=n_1} {}_D\alpha_r - \sum_{r=0}^{r=n_1/2} {}_D\alpha_r \right) {}_1 d_d \qquad \text{IV(39)}$$

Now c is given by IV(23) and f by IV(24), and so substituting for c and f and rearranging we have

$$_{SM}\Delta_D = \left\{ \left(1 - \frac{q}{n_3} \right) \sum_{r=0}^{r=q} {}_{SM}\alpha_r - \frac{q}{n_3} \sum_{r=q+1}^{r=n_3} {}_{SM}\alpha_r \right\} {}_3 d_d$$

$$+ \frac{1}{2} \left(1 - \frac{q}{n_3} \right) \left(\sum_{r=0}^{r=n_2/2} {}_L\alpha_r - \sum_{r=n_2/2+1}^{r=n_2} {}_L\alpha_r \right) {}_2 d_d$$

$$+ \frac{q}{2n_3} \left(\sum_{r=0}^{r=n_2/2} {}_L\alpha'_r - \sum_{r=n_2/2+1}^{r=n_2} {}_L\alpha'_r \cdot {}_2 d_d \right)$$

$$- \left(\sum_{r=0}^{r=p} {}_D\alpha_r - \frac{p}{n_1} \sum_{r=0}^{r=n_1} {}_D\alpha_r + \frac{1}{2} \sum_{r=0}^{r=n_1} {}_D\alpha_r - \sum_{r=0}^{r=n_1/2} {}_D\alpha_r \right) {}_1 d_d$$

$$+ \frac{1}{4} \left(\sum_{r=n_1/2+1}^{r=n_1} {}_D\alpha_r - \sum_{r=0}^{r=n_1/2} {}_D\alpha_r + \sum_{r=n_1/2+1}^{r=n_1} {}_D\alpha'_r - \sum_{r=0}^{r=n_1/2} {}_D\alpha'_r \right) {}_1 d_d$$

$$\text{IV(40)}$$

Before we can reduce the last two brackets, we must decide whether $n_1 \geqslant p \geqslant n_1/2$ or $0 \leqslant p \leqslant n_1/2$.

If $p \geqslant n_1/2$, the last two brackets reduce to

$$\left[\left(\frac{p}{n_1} - \frac{3}{4}\right) \sum_{r=0}^{r=n_1/2} {}_D\alpha_r + \left(\frac{p}{n_1} - \frac{5}{4}\right) \sum_{r=n_1/2+1}^{r=p} {}_D\alpha_r + \left(\frac{p}{n_1} - \frac{1}{4}\right) \sum_{r=p+1}^{r=n_1} {}_D\alpha_r\right] {}_1 d_d$$

$$+ \frac{1}{4}\left[\sum_{r=n_1/2+1}^{r=n_1} {}_D\alpha'_r - \sum_{r=0}^{r=n_1/2} {}_D\alpha'_r\right] {}_1 d_d \qquad \text{IV(41)}$$

If $p \leqslant n_1/2$, the last two brackets reduce to

$$\left[\left(\frac{p}{n_1} - \frac{3}{4}\right) \sum_{r=0}^{r=p} {}_D\alpha_r + \left(\frac{p}{n_1} + \frac{1}{4}\right) \sum_{r=p+1}^{r=n_1/2} {}_D\alpha_r + \left(\frac{p}{n_1} - \frac{1}{4}\right) \sum_{r=n_1/2+1}^{r=n_1} {}_D\alpha_r\right] {}_1 d_d$$

$$+ \frac{1}{4}\left[\sum_{r=n_1/2+1}^{r=n} {}_D\alpha'_r - \sum_{r=0}^{r=n_1/2} {}_D\alpha'_r\right] {}_1 d_d \qquad \text{IV(42)}$$

If $p \geqslant n_1/2$, the standard deviation of ${}_{SM}\Delta_D$ is therefore given by

$$_{SM}(\sigma_\Delta)_D^2 = q\left(1 - \frac{q}{n_3}\right) {}_3 d_d^2 \sigma_\alpha^2 + \frac{n_2}{4}\left\{1 + \frac{2q}{n_3}\left(\frac{q}{n_3} - 1\right)\right\} {}_2 d_d^2 \sigma_\alpha^2$$

$$+ \left[\frac{3}{2}p - \frac{p^2}{n_1} - \frac{3}{8}n_1\right] {}_1 d_d^2 \sigma_\alpha^2 \qquad \text{IV(43a)}$$

If $q = m + n_3/2$, IV(43a) can be written as

$$_{SM}(\sigma_\Delta)_D^2 = \left[n_3\left(\frac{1}{4} - \frac{m^2}{n_3^2}\right) {}_3 d_d^2 + \frac{n_2}{4}\left(\frac{1}{2} + \frac{2m^2}{n_3^2}\right) {}_2 d_d^2\right.$$

$$\left. + \left(\frac{3}{2}p - \frac{p^2}{n_1} - \frac{3}{8}n_1\right) {}_1 d_d^2\right] \sigma_\alpha^2 \qquad \text{IV(43b)}$$

which shows that ${}_{SM}(\sigma_\Delta)_D$ is symmetrical about the centre of the non-diagonal side.

If $p \leqslant n_1/2$, then

$$_{SM}(\sigma_\Delta)_D^2 = q\left(1 - \frac{q}{n_3}\right) {}_3 d_d^2 \sigma_\alpha^2 + \frac{n_2}{4}\left\{1 + \frac{2q}{n_3}\left(\frac{q}{n_3} - 1\right)\right\} {}_2 d_d^2 \sigma_\alpha^2$$

$$+ \left[\frac{p}{2} - \frac{p^2}{n_1} + \frac{n_1}{8}\right] {}_1 d_d^2 \sigma_\alpha^2 \qquad \text{IV(44a)}$$

If $q = m + n_3/2$, this can be written

$$_{SM}(\sigma_\Delta)_D^2 = \left[n_3 \left(\frac{1}{4} - \frac{m^2}{n_3^2} \right) {}_3 d_d^2 + \frac{n_2}{4} \left(\frac{1}{2} + \frac{2m^2}{n_3^2} \right) {}_2 d_d^2 \right.$$

$$\left. + \left(\frac{p}{2} - \frac{p^2}{n_1} + \frac{n_1}{8} \right) {}_1 d_d^2 \right] \sigma_\alpha^2 \qquad \text{IV(44b)}$$

These equations also hold for points on AC and GH.

(*b*)   Uncertainty in height between a point on a diagonal side and a point on EF. This is given by replacing n_3 by n_2, n_2 by n_3, and ${}_3 d_d$ by ${}_2 d_d$ in IV(43) and IV(44).

(6)   *The uncertainty between a point on EF and one on GH*

The height of a point on GH above the datum plane is given by

$$_{SM} z_p = \left[\left(1 - \frac{p}{n_3} \right) \sum_{r=0}^{r=p} {}_{SM} \alpha_r - \frac{p}{n_3} \sum_{r=p+1}^{r=n_3} {}_{SM} \alpha_r \right] {}_3 d_d + c \left(1 - \frac{p}{n_3} \right) + \frac{fp}{n_3}$$

where p is measured from G and $n_3 \geqslant p \geqslant 0$ (see equation IV(20)).

The height of a point on EF above the datum plane is given by

$$_{LM} z_q = \left[\left(1 - \frac{q}{n_2} \right) \sum_{r=0}^{r=q} {}_{LM} \alpha_r - \frac{q}{n_2} \sum_{r=q+1}^{r=n_2} {}_{LM} \alpha_r \right] {}_2 d_d + c \left(1 - \frac{q}{n_2} \right) + \frac{fq}{n_2}$$

where q is measured from E and $n_2 \geqslant q \geqslant 0$ (see equation IV(20)).

The difference in height between a point on GH and one on EF is thus given by

$$_{SM} \Delta_{LM} = \left[\left(1 - \frac{p}{n_3} \right) \sum_{r=0}^{r=p} {}_{SM} \alpha_r - \frac{p}{n_3} \sum_{r=p+1}^{r=n_3} {}_{SM} \alpha_r \right] {}_3 d_d$$

$$- \left[\left(1 - \frac{q}{n_2} \right) \sum_{r=0}^{r=q} {}_{LM} \alpha_r - \frac{q}{n_2} \sum_{r=q+1}^{r=n_2} {}_{LM} \alpha_r \right] {}_2 d_d + (f - c) \left(\frac{p}{n_3} - \frac{q}{n_2} \right)$$

$$\text{IV(45)}$$

where $(f - c)$ is given by IV(24b). Thus

$$_{SM}(\sigma_\Delta)_{LM}^2 = p \left(1 - \frac{p}{n_3} \right) {}_3 d_d^2 \sigma_\alpha^2 + q \left(1 - \frac{q}{n_2} \right) {}_2 d_d^2 \sigma_\alpha^2 + \frac{n_2}{2} \left(\frac{p}{n_3} - \frac{q}{n_2} \right)^2 {}_2 d_d^2 \sigma_\alpha^2$$

$$\text{IV(46)}$$

Practical values

Having established the formulae for the standard deviation of the difference in height between points on the various sight lines, we can now calculate some typical values for these differences. We shall begin by finding the standard deviation of the difference between a point which moves from the centre Q (see Figure IV.1) along a diagonal, thence to an outside line, and thence to a non-diagonal line passing through the centre, and the point Q.

For the first part of this exercise we require equation IV(5), giving the standard deviation of the difference between two points on a diagonal side. The required equation is

$$_D(\sigma_\Delta)_D^2 = \left(1 + \frac{q-p}{n_1}\right)(p-q)\sigma_{\alpha\ 1}^2 d_d^2 \qquad \text{IV(5)}$$

Taking $q = n_1/2$ and $p = n_1/2 + t$, where t goes from 0 to $n_1/2$, we have

$$_D(\sigma_\Delta)_D^2 = \left(1 - \frac{t}{n_1}\right)t\sigma_{\alpha\ 1}^2 d_d^2 \qquad \text{IV(47)}$$

As a typical case we shall assume that a diagonal sight line has 20 points, a long side 16 and a short side 12; further, for the sake of simplicity, we shall assume that $_1d_d = {}_2d_d = {}_3d_d = d, n_1 = 20, n_2 = 16, n_3 = 12$. Thus IV(47) becomes

$$_D(\sigma_\Delta)_D = \sigma_\alpha d\left[t\left(1 - \frac{t}{n_1}\right)\right]^{1/2} \qquad \text{IV(48)}$$

where t goes from 0 to 10, and $n_1 = 20$.

In a similar manner, using IV(38), it can be shown that

$$_L(\sigma_\Delta)_D = \sigma_\alpha d\left[q\left(1 - \frac{q}{n_2}\right) + \frac{n_1 q^2}{2n_2^2} + \frac{n_1}{2}\left(\frac{1}{2} - \frac{q}{n_2}\right)\right]^{1/2} \qquad \text{IV(49)}$$

where q goes from 0 to n_2. $n_2/2$ corresponds to the point G, the centre of AB.

Similarly, using IV(43), it can be shown that

$$_{SM}(\sigma_\Delta)_D = \sigma_\alpha d\left[q\left(1 - \frac{q}{n_3}\right) + \frac{n_2}{4}\left\{1 + \frac{2q}{n_3}\left(\frac{q}{n_3} - 1\right)\right\} + \frac{n_1}{8}\right]^{1/2} \qquad \text{IV(50)}$$

where q goes from 0 to n_3. $n_3/2$ corresponds to the point Q, the centre of GH.

The use of the above three equations enables the standard deviation along the path QBGQ, or for similar paths starting at Q, to be found. For the calculation of the standard deviation along the path QBEQ and similar paths, equation IV(48) and the following two equations IV(51) and IV(52) are required. IV(51) is obtained from IV(49) by writing n_3 in place of n_2, giving

$$_S(\sigma_\Delta)_D = \sigma_\alpha d\left[q\left(1 - \frac{q}{n_3}\right) + \frac{n_1 q^2}{2n_3^2} + \frac{n_1}{2}\left(\frac{1}{2} - \frac{q}{n_3}\right)\right]^{1/2} \qquad \text{IV(51)}$$

where q goes from 0 to n_3, $n_3/2$ corresponds to the point E, the centre of BC. Likewise IV(52) is obtained by replacing n_3 by n_2; and n_2 by n_3 in IV(50), giving

$$_{LM}(\sigma_\Delta)_D = \sigma_\alpha d\left[q\left(1 - \frac{q}{n_2}\right) + \frac{n_3}{4}\left\{1 + \frac{2q}{n_2}\left(\frac{q}{n_2} - 1\right)\right\} + \frac{n_1}{8}\right]^{1/2} \quad \text{IV(52)}$$

where q goes from 0 to n_2. $n_2/2$ corresponds to the point Q, the centre of EF.

The standard deviation of the difference in height between two points on a long outside side (IV(14)) can be written as

$$_{L}(\sigma_\Delta)_L = \left[(p - q)\left(1 - \frac{p - q}{n_2}\right) + \frac{(p - q)^2 n_1}{2n_2^2}\right]^{1/2} d\sigma_\alpha \quad \text{IV(53)}$$

whilst the corresponding standard deviation for a short outside side is given by

$$_{s}(\sigma_\Delta)_S = \left[(p - q)\left(1 - \frac{p - q}{n_3}\right) + \frac{(p - q)^2 n_1}{2n_3^2}\right]^{1/2} d\sigma_\alpha \quad \text{IV(54)}$$

Figures IV.2 and IV.3 show the plotted results of substituting in the above equations. The curves give the relative values of the standard deviation of the difference in height between points on a surface table. For curves originating from Q the ordinates give the relative standard deviation of the difference in height between the point corresponding to the ordinate and the point Q. For curves originating from the points B, with zero ordinates, the relative standard deviation of the difference in height between a point on an outside side and the point B is given by the ordinate corresponding to that point. The ordinates of the curves give values of V, the relative standard

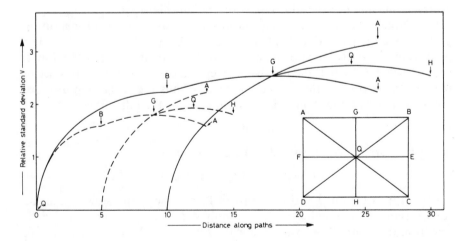

Figure IV.2

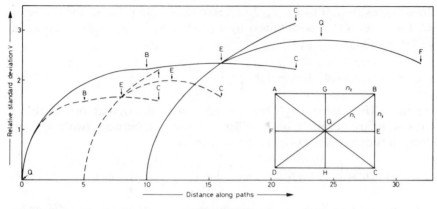

Figure IV.3

deviation. The standard deviation of the difference in height between two points is given by $\sigma = V\sigma_\alpha d$, where σ_α is the standard deviation of the angle measuring device and d is the spacing of the feet of the angle detecting device. The dotted curves have values of n_1, n_2 and n_3, which are half those of the full curves.

Figure IV.2 gives the relative standard deviation of the path QBGH, whilst Figure IV.3 gives the relative standard deviation of the path QBEF. The values given are also true of similar paths QCHG, QCEF, etc.

The following important points should be noted.

(1) The standard deviation of the difference in height between a point on a sight path and the centre Q of the measurements rises rapidly and then flattens out and rises quite slowly. Notice that the points E, H, F and G represent maximum values for points on the outside sides.

(2) Notice that there is a gain in increasing the number of measuring points. Take the point E on the full curves which has a V of about 2.35. The point E on the dotted curves has a value of V of about 1.66. But the dotted curve reaches E by half the number of steps, and thus the d for this set of measurements is twice that of the full curves. Thus if d is the step distance for the full curves, the standard deviation of E for the full curves is

$$\sigma = \sigma_\alpha d\, 2.35$$

whilst the corresponding value for the dotted curves is

$$\sigma = \sigma_\alpha 2d \times 1.66 = \sigma_\alpha d\, 3.32$$

(3) It is to be noted that the value of V increases more slowly than the square root of the number of steps required to reach a point on a sight line from the centre. For instance, for the full curves the value of E is 2.35, whilst the number of steps to E is 18, giving $\sqrt{18} \simeq 4.25$.

(4)   The two sets of curves, full and dotted, with $n_1 = 20$, $n_2 = 16$ and $n_3 = 12$, and $n_1 = 10$, $n_2 = 8$ and $n_3 = 6$ cover most of the cases met with practically, from which it is seen that V has a maximum value of 2.35 for an outside side. The *uncertainty* for such a point is thus $2.35\sigma_\alpha dk_1$, where k_1 is the tolerance factor for the standard deviation $(2.35\sigma_\alpha d)$. The corresponding uncertainty for the dotted curves is $1.66\sigma_\alpha dk_1$.

(5)   It is to be noted that for the full curves the relative standard deviation V along the path QBEQ is about 2.82, whilst along the path QBGQ it is 2.55, but the relative standard deviation V of the difference in height between the two paths on arriving back at Q is, using IV(46), equal to 2.64. This underlines the fact that the uncertainties along the different paths are not independent.

Appendix V

Confidence and Tolerance Limits of y for Given x, that is for $y|x$ for a Mean Line†

The mean line is given by

$$y = \bar{y} + m(x - \bar{x}) \qquad\qquad 10.29(3)$$

where

$$m = \sum \varepsilon_i \tau_i / \sum \varepsilon_i^2 \qquad\qquad 10.29(1)$$

The value of y for given x is given by the right-hand side of 10.29(3). This value of $y|x$ is a mean value, since 10.29(3) is the equation of the mean line. Now $\bar{y}$ is a variable, as is m the gradient, but here $\bar{x}$ is not a variable because all the x values are assumed to be known exactly.

Hence

$$s_{\bar{y}|x}^2 = s_{\bar{y}L}^2 + (x - \bar{x})^2 s_m^2 = s_{\bar{y}L}^2 + (x - \bar{x})^2 s_{yL}^2 / \sum \varepsilon_i^2$$

$$= s_{yL}^2 \left[\frac{1}{n} + (x - \bar{x})^2 / \sum \varepsilon_i^2 \right] \qquad\qquad V(1)$$

where s_{yL} is given by 10.31(2) and s_m by 10.32(6) and $s_{y|x}$ is the estimated standard deviation of the mean value of y for given x, where y is given by equation 10.29(3). The extreme right-hand side of V(1) is obtained by using 10.32(6).

† From p 303.

514

Confidence limits

The confidence limits for $y|x$ are thus given by

$$\bar{y} + m(x - \bar{x}) \pm k_1 s_{\bar{y}|x} \qquad v = n - 2 \geqslant 200$$

or

$$\bar{y} + m(x - \bar{x}) \pm k_2 s_{\bar{y}|x} \qquad v = n - 2 \leqslant 200$$

where k_1 is given by Table II and k_2 by Table IV of Appendix I.

Tolerance limits

The tolerance limits of $y|x$ give the probability range of a single value of y for given x. The estimated standard deviation s_{yL} of y from the mean line is given by 10.31(2); this is independent of x. Since the standard deviation of the mean of $y|x$ for the mean line is given by $s_{\bar{y}|x}$, the total standard deviation for $y|x$ for a single value of y for given x is thus given by

$$s_{y|x}^2 = s_{yL}^2 + s_{\bar{y}|x}^2 = s_{yL}^2 + s_{\bar{y}L}^2 + (x - \bar{x})^2 s_{yL}^2 / \sum \varepsilon_i^2$$

$$= s_{yL}^2 \left[1 + \frac{1}{n} + (x - \bar{x})^2 / \sum \varepsilon_i^2 \right] \qquad \text{V(2)}$$

The tolerance limits for $y|x$ are thus given by

$$\bar{y} + m(x - \bar{x}) \pm k_2 s_{y|x} \qquad v = n - 2 \leqslant 200$$

where k_2 is given by Table IV of Appendix I for the selected probability.

Appendix VI

Computational Formulae

This appendix gives standard statistical formulae in a form suitable for computation, that is in terms of basic parameters, not derived ones. The equations from which the computational equations have been derived may be found in the text by referring to the equation numbers given on the right.

Mean

$$\bar{x} = \sum_1^n x_i / n \qquad\qquad 2.22(1)$$

Weighted mean

$$\bar{x}_w = \sum_1^n w_i x_i \bigg/ \sum_1^n w_i \qquad\qquad 2.58(1)$$

Estimated standard deviation

$$s = \left[\frac{\{ n \sum_1^n x_i^2 - (\sum_1^n x_i)^2 \}}{n(n-1)} \right]^{1/2} \qquad\qquad 2.24(9)$$

Estimated weighted standard deviation

$$s_w = \left[\frac{n \{ \sum_1^n w_i \sum_1^n w_i x_i^2 - (\sum_1^n w_i x_i)^2 \}}{(n-1)(\sum_1^n w_i)^2} \right]^{1/2} \qquad\qquad 2.64(2)$$

Mean gradient of a straight line

$$m = \frac{n \sum_1^n x_i y_i - \sum_1^n x_i \sum_1^n y_i}{n \sum_1^n x_i^2 - (\sum_1^n x_i)^2} \qquad\qquad 10.27(9)$$

Mean weighted gradient of a straight line

$$m_w = \frac{\sum_1^n w_i \sum_1^n w_i x_i y_i - \sum_1^n w_i x_i \sum_1^n w_i y_i}{\sum_1^n w_i \sum_1^n w_i x_i^2 - (\sum_1^n w_i x_i)^2} \qquad\qquad 10.30(1)$$

Mean value of constant c of equation $y = mx + c$

$$c = \frac{\sum_1^n y_i \sum_1^n x_i^2 - \sum_1^n x_i \sum_1^n x_i y_i}{n \sum_1^n x_i^2 - (\sum_1^n x_i)^2} \qquad 10.29(2)$$

Weighted mean value of constant c

$$c_w = \frac{\sum_1^n w_i x_i^2 \sum_1^n w_i y_i - \sum_1^n w_i x_i \sum_1^n w_i x_i y_i}{\sum_1^n w_i \sum_1^n w_i x_i^2 - (\sum_1^n w_i x_i)^2} \qquad 10.30(2)$$

Estimated standard deviation of variable y from mean line

$$s_{yL} = \left\{ \left[n \sum_i^n y_i^2 - \left(\sum_1^n y_i \right)^2 - \frac{(n \sum_1^n x_i y_i - \sum_1^n x_i \sum_1^n y_i)^2}{\{n \sum_1^n x_i^2 - (\sum_1^n x_i)^2\}} \right] \bigg/ n(n-2) \right\}^{1/2}$$

$$10.31(2)$$

Estimated weighted standard deviation of variable y from mean line

$$s_{yLw} = \frac{1}{\sum_1^n w_i} \left[\sum_1^n w_i \sum_1^n w_i y_i^2 - \left(\sum_1^n w_i y_i \right)^2 \right.$$
$$\left. - \frac{(\sum_1^n w_i \sum_1^n w_i x_i y_i - \sum_1^n w_i x_i \sum_1^n w_i y_i)^2}{\{\sum_1^n w_i \sum_1^n w_i x_i^2 - (\sum_1^n w_i x_i)^2\}} \right]^{1/2} \times \left(\frac{n}{n-2} \right)^{1/2}$$

$$10.31(3)$$

Estimated standard deviation of mean of variable y from mean line

$$s_{\bar{y}L} = s_{yL}/\sqrt{n} \qquad VI(1)$$

Estimated weighted standard deviation of mean of variable y from mean line

$$s_{\bar{y}Lw} = s_{yLw}/\sqrt{n} \qquad VI(2)$$

Estimated standard deviation of mean gradient of mean line

$$s_{\bar{m}} = \left\{ \frac{\left[n \sum_1^n y_i^2 - (\sum_1^n y_i)^2 - \frac{(n \sum_1^n x_i y_i - \sum_1^n x_i \sum_1^n y_i)^2}{\{n \sum_1^n x_i^2 - (\sum_1^n x_i)^2\}} \right]^{1/2}}{(n-2)\{n \sum_1^n x_i^2 - (\sum x_i)^2\}} \right\}$$
$$= s_{yL} \left[n \bigg/ \left\{ n \sum_1^n x_i^2 - \left(\sum_1^n x_i \right)^2 \right\} \right]^{1/2} \qquad 10.33(2)$$

Estimated weighted standard deviation of mean gradient of mean line

$$s_{\bar{m}w} =$$
$$\left\{ \frac{\left[\sum_1^n w_i \sum_1^n w_i y_i^2 - (\sum_1^n w_i y_i)^2 - \frac{(\sum_1^n w_i \sum_1^n w_i x_i y_i - \sum_1^n w_i x_i \sum_1^n w_i y_i)^2}{\{\sum_1^n w_i \sum_1^n w_i x_i^2 - (\sum_1^n w_i x_i)^2\}} \right]^{1/2}}{(n-2)\{\sum_1^n w_i \sum_1^n w_i x_i^2 - (\sum_1^n w_i x_i)^2\}} \right\}$$
$$= s_{yLw} \cdot \sum_1^n w_i \bigg/ \left[n \left\{ \sum_1^n w_i \sum_1^n w_i x_i^2 - \left(\sum_1^n w_i x_i \right)^2 \right\} \right]^{1/2} \qquad 10.32(8)$$

Estimated standard deviation of mean value of constant c

$$s_{\bar{c}} = \left\{ \frac{\left[n\sum_1^n y_i^2 - (\sum_1^n y_i)^2 - \frac{(n\sum_1^n x_i y_i - \sum_1^n x_i \sum_1^n y_i)^2}{\{n\sum_1^n x_i^2 - (\sum_1^n x_i)^2\}} \right]}{n(n-2)\{n\sum_1^n x_i^2 - (\sum_1^n x_i)^2\}} \sum_1^n x_i^2 \right\}^{1/2}$$

$$= s_{yL} \left[\sum_1^n x_i^2 \Big/ \left\{ n\sum_1^n x_i^2 - \left(\sum_1^n x_i \right)^2 \right\} \right]^{1/2} \tag{10.34(5)}$$

Estimated weighted standard deviation of mean value of c

$$s_{\bar{c}w} =$$

$$\left\{ \sum_1^n w_i x_i^2 \frac{\left[\sum_1^n w_i \sum_1^n w_i y_i^2 - (\sum_1^n w_i y_i)^2 - \frac{(\sum_1^n w_i \sum_1^n w_i x_i y_i - \sum_1^n w_i x_i \sum_1^n w_i y_i)^2}{\{\sum_1^n w_i \sum_1^n w_i x_i^2 - (\sum_1^n w_i x_i)^2\}} \right]}{(n-2)\sum_1^n w_i \{\sum_1^n w_i \sum_1^n w_i x_i^2 - (\sum_1^n w_i x_i)^2\}} \right\}^{1/2}$$

$$= s_{yLw} \left\{ \sum_1^n w_i x_i^2 \cdot \sum_1^n w_i \Big/ \left[n\left\{ \sum_1^n w_i \sum_1^n w_i x_i^2 - \left(\sum_1^n w_i x_i \right)^2 \right\} \right] \right\}^{1/2} \tag{10.34(6)}$$

Estimated standard deviation of x intercept of mean line, that is of $-c/m = x_0$

$$s_{x_0} = \left[\frac{s_{\bar{y}L}^2}{m^2} + \frac{\bar{y}^2}{m^4} s_{\bar{m}}^2 \right]^{1/2} \dagger \tag{10.36(6)}$$

Estimated weighted standard deviation of x intercept of mean line

$$s_{x_0 w} = \left[\frac{s_{\bar{y}Lw}^2}{m_w^2} + \frac{\bar{y}_w^2 s_{\bar{m}w}^2}{m_w^4} \right]^{1/2} \tag{VI(3)}$$

Coefficient of correlation r

$$r = \frac{n\sum_1^n x_i y_i - \sum_1^n x_i \sum_1^n y_i}{[\{n\sum_1^n x_i^2 - (\sum_1^n x_i)^2\}\{n\sum_1^n y_i^2 - (\sum_1^n y_i)^2\}]^{1/2}} \tag{10.44(3)}$$

Weighted coefficient of correlation

$$r_w = \frac{\sum_1^n w_i \sum_1^n w_i x_i y_i - \sum_1^n w_i x_i \sum_1^n w_i y_i}{[\{\sum_1^n w_i \sum_1^n w_i x_i^2 - (\sum_1^n w_i x_i)^2\}\{\sum_1^n w_i \sum_1^n w_i y_i^2 - (\sum_1^n w_i y_i)^2\}]^{1/2}}$$

$$\tag{10.44(5)}$$

Estimated weighted standard deviation of x for mean line

$$s_{xw} = \frac{\{\sum_1^n w_i \sum_1^n w_i x_i^2 - (\sum_1^n w_i x_i)^2\}^{1/2}}{\sum_1^n w_i} \left(\frac{n}{n-2} \right)^{1/2} \tag{10.44(8)}$$

† Note that

$$s_{x_0} \neq \left[\frac{s_c^2}{m^2} + \frac{c^2}{m^4} s_m^2 \right]^{1/2}$$

since s_c and s_m are correlated, whilst $s_{\bar{y}L}$ and s_m are not. s_{x_0} here has been obtained using 8.04(4). See the section on orthogonal polynomials (Paragraph 10.85 et seq.) which provides uncorrelated coefficients.

Estimated weighted standard deviation of y for mean line

$$s_{yw} = \frac{\{\sum_1^n w_i \sum_1^n w_i y_i^2 - (\sum_1^n w_i y_i)^2\}^{1/2}}{\sum_1^n w_i} \left(\frac{n}{n-2}\right)^{1/2} \qquad 10.44(9)$$

Weighted standard deviation of x and y for mean line $y = mx + c$

$$\sigma_{xw} = \sqrt{\left(\frac{n-2}{n}\right)} s_{xw} \qquad\qquad VI(4)$$

$$\sigma_{yw} = \sqrt{\left(\frac{n-2}{n}\right)} s_{yw} \qquad\qquad VI(5)$$

The σs are defined as

$$\left(\frac{\sum_1^n w_i (x_i - \bar{x})^2}{\sum_1^n w_i}\right)^{1/2}$$

which is only true for *very large n*. Hence for small values of n it is best to use the estimated values of the σs. Generally $\sigma = \sqrt{(v/n)}s$ where v is the number of degrees of freedom for the variable whose standard deviation is computed.

Estimated standard deviation of y|x for mean line

$$s_{y|x} =$$
$$\left\{\frac{\left[n\sum_1^n y_i^2 - (\sum_1^n y_i)^2 - \frac{(n\sum_1^n x_i y_i - \sum_1^n x_i \sum_1^n y_i)^2}{\{n\sum_1^n x_i^2 - (\sum_1^n x_i)^2\}}\right]\left[1 + \frac{1}{n}\left\{1 + \frac{(nx - \sum_1^n x_i)^2}{\{n\sum_1^n x_i^2 - (\sum_1^n x_i)^2\}}\right\}\right]}{n(n-2)}\right\}^{1/2}$$

$$VI(6)$$

Weighted estimated standard deviation of y|x

$$s_{y|xw} = \left[\sum_1^n w_i \sum_1^n w_i y_i^2 - \left(\sum_1^n w_i y_i\right)^2 - \frac{(\sum_1^n w_i \sum_1^n w_i x_i y_i - \sum_1^n w_i x_i \sum_1^n w_i y_i)^2}{\{\sum_1^n w_i \sum_1^n w_i x_i^2 - (\sum_1^n w_i x_i)^2\}}\right]^{1/2}$$

$$\times \left[1 + \left\{\frac{1}{n} + \frac{(x\sum_1^n w_i - \sum_1^n w_i x_i)^2}{\sum_1^n w_i \{\sum_1^n w_i \sum_1^n w_i x_i^2 - (\sum_1^n w_i x_i)^2\}}\right\}\right]^{1/2}$$

$$\times \left(\frac{n}{n-2}\right)^{1/2} \times \frac{1}{\sum_1^n w_i}$$

Estimated standard deviation in the mean of y|x for mean line

$$s_{\bar{y}|x} =$$
$$\frac{\left\{\left[n\sum_1^n y_i^2 - (\sum_1^n y_i)^2 - \frac{(n\sum_1^n x_i y_i - \sum_1^n x_i \sum_1^n y_i)^2}{\{n\sum_1^n x_i^2 - (\sum_1^n x_i)^2\}}\right]\left[1 + \frac{(nx - \sum_1^n x_i)^2}{\{n\sum_1^n x_i^2 - (\sum_1^n x_i)^2\}}\right]\right\}^{1/2}}{n(n-2)^{1/2}}$$

$$VI(7)$$

Estimated weighted standard deviation in the mean of $y|x$ for mean line

$$s_{\bar{y}|xw} = \left[\sum_1^n w_i \sum_1^n w_i y_i^2 - \left(\sum_1^n w_i y_i \right)^2 - \frac{(\sum_1^n w_i \sum_1^n w_i x_i y_i - \sum_1^n w_i x_i \sum_1^n w_i y_i)^2}{\{\sum_1^n w_i \sum_1^n w_i x_i^2 - (\sum_1^n w_i x_i)^2\}} \right]^{1/2}$$

$$\times \left[1 + \frac{n(x \sum_1^n w_i - \sum_1^n w_i x_i)^2}{\sum_1^n w_i \{\sum_1^n w_i \sum_1^n w_i x_i^2 - (\sum_1^n w_i x_i)^2\}} \right]^{1/2} \times \frac{1}{(n-2)^{1/2} \sum_1^n w_i}$$

VI(8)

Appendix VII

A Suite of Formulae for the Computation on a Computer of the Approximate or Exact Values of Various Statistical Functions

Integral of Normal Distribution Between Limits $-k_1$ to k_1

VII(1)   The formula gives the value of the integral

$$\frac{1}{\sqrt{(2\pi)}} \int_{-k_1}^{k_1} e^{-t^2/2} \, dt = \underset{-k \text{ to } k}{P}$$

i.e. the probability of an uncertainty between the limits

$$\bar{x} - k_1\sigma \qquad \text{to} \qquad \bar{x} + k_1\sigma$$

Range $0 \leqslant k \leqslant \infty$.

Approximation

$$\underset{-k_1 \text{ to } k_1}{P} = 1 - \frac{1}{(1 + a_1 k_1 + a_2 k_1^2 + a_3 k_1^3 + a_4 k_1^4 + a_5 k_1^5 + a_6 k_1^6)^{16}}$$

where $a_1 = 0.049\,867\,346\,97$ $a_4 = 0.000\,038\,603\,575$

$\qquad\qquad a_2 = 0.021\,141\,006\,15$ $a_5 = 0.000\,048\,896\,635$

$\qquad\qquad a_3 = 0.003\,277\,626\,32$ $a_6 = 0.000\,005\,382\,975$

Maximum error $\varepsilon = \pm 0.000\,0003$ for $0 \leqslant k_1 \leqslant 5.7$ where $\varepsilon =$ Approximation
– Function

Inverse of the Integral of the Normal Function

VII(2)   The formula gives the value of the limit k where

$$P(k_1) = \int_{-k_1}^{k_1} e^{-c^2/2}\, dc$$

i.e. where k_1 is the semi-range measured as the number of standard deviations corresponding to a probability of $P(k_1)$ covering the range

$$\bar{x} - k_1\sigma \qquad \text{to} \qquad \bar{x} + k_1\sigma$$

Range of $P(k_1)$: $0 \leqslant P(k_1) \leqslant 1$.

Approximation

$$X = \sqrt{[\ln\{2/(1 - P(k_1))\}^2]^{1/2}}$$

$$k_1 = X - \left\{ \frac{a_0 + a_1 X + a_2 X^2}{1 + b_1 X + b_2 X^2 + b_3 X^3} \right\}$$

where $a_0 = 2.515\,517$ $\qquad b_1 = 1.432\,788$

$a_1 = 0.802\,853$ $\qquad b_2 = 0.189\,269$

$a_2 = 0.010\,328$ $\qquad b_3 = 0.001\,308$

Maximum error $\varepsilon = \pm 0.0004$ for $k_1 = 0$ to ∞ where $\varepsilon =$ Approximation
– Function

Formula for Obtaining a Rectangularly Distributed Random Number

VII(3)   If r is a random number between 0 and 1, then if

$$x_{i+1} = [(ax_i + b)/m - \text{Integer}\,((ax_i + b)/m)]/m$$

then

$$r_i = \text{Int}\,(10^5\, x_{i+1}/m)/10^5$$

where Int $\equiv$ Integral part of.

The values of the constants are

$$a = 24{,}298,\; b = 99{,}991,\; m = 199{,}017$$

x_0 is the initial value of x, and is often taken as 0. Any value however may be used as the starting value. It is to be noted that if the same initial seed is used, i.e. x, then the random numbers obtained will always be the same. The above formula gives random numbers r_i to five significant figures.

Formulae for Obtaining a Normally Distributed Random Number (1) using Two Random Numbers Obtained from VII(3) (see paragraph VII(4a)); (2) using One Random Number Obtained from VII(3) (see paragraph VII(4b))

VII(4a)   The required random number R between $-\infty$ to $+\infty$ is given by $R = (-2\ln r_1)^{1/2} \cos(Kr_2)$ where r_1 and r_2 are random numbers obtained from VII(3) and $K = 2\pi$ if the calculator or computer is working in radian mode, or $K = 360$ if they are working in degree mode.

VII(4b)   A random number r is obtained from VII(3) and u derived using $u = r - 0.5$. u is now substituted for $P(k)$ in VII(2) giving the required random Gaussian number as k.

Integral of the Student *t* Distribution

VII(5)   The Student t distribution is given by the integral

$$\int_{-k_2}^{k_2} f(t_v)\, dt_v = \frac{\Gamma((v+1)/2)}{\Gamma(v/2)\sqrt{(v\pi)}} \int_{-k_2}^{k_2} \left(1 + \frac{t_v^2}{v}\right)^{-(v+1)^2} dt_v$$

where $\Gamma(\alpha/2)$ is the gamma function of $\alpha/2$. When α is even

$$\Gamma\left(\frac{\alpha}{2}\right) = \left(\frac{\alpha}{2} - 1\right)\left(\frac{\alpha}{2} - 2\right)\ldots 3, 2, 1$$

valid for $\alpha \geqslant 4$. If $\alpha = 2$ then $\Gamma(1) = 1$.
   When α is odd

$$\Gamma\left(\frac{\alpha}{2}\right) = \left(\frac{\alpha}{2} - 1\right)\left(\frac{\alpha}{2} - 2\right)\cdots \frac{5}{2}, \frac{3}{2}, \frac{1}{2} \sqrt{\pi}$$

valid for $\alpha \geqslant 3$. If $\alpha = 1$ then $\Gamma(\alpha/2) = \sqrt{\pi}$.

Solution of integral

If v is even, write $v/2 - 1 = n$, an integer. Define

$$\theta = \arctan\left(\frac{t_v}{\sqrt{v}}\right).$$

Then

$$\int_{-k_2}^{k_2} f(t_v) \, dt_v = \sin \theta \sum_{i=0}^{i=n} A_i$$

with $A_0 = 1$, and where A_i satisfies the recurrence relation

$$A_i = \frac{(2i-1)}{2i} A_{i-1} \cos^2 \theta$$

Note: Table IV gives values of the inverse of the integral of the Student '*t*' function. The table gives values of the limit k_2 where

$$P(k_2) = \int_{-k_2}^{k_2} f(t_v) \, dt_v$$

for several values of $P(k_2)$ and indexed values of v, the number of degrees of freedom.

If v is odd, write $v/2 - \frac{1}{2} = n$, an integer, and define θ as before. Then

$$\int_{-k_2}^{k_2} f(t_v) \, dt_v = \frac{2}{\pi} \left[\theta + \sin \theta \cos \theta \sum_{i=0}^{i=n-1} B_i \right]$$

where $B_0 = 1$

and

$$B_i = \frac{2i}{2i+1} B_{i-1} \cos^2 \theta$$

and $t_v = k_2$. These formulae are exact, i.e. they are not approximations.

Integral of the *F* Distribution

VII(6)   The integral required is

$$P(F_0) = \int_{F_0}^{\infty} \frac{\Gamma((v_1 + v_2)/2)}{\Gamma(v_1/2)\Gamma(v_2/2)} v_1^{v_1/2} v_2^{v_2/2} \frac{F^{v_1/2-1}}{(v_2 + v_1 F)^{(v_1+v_2)/2}} \, dF$$

where $P(F_0)$ is the probability from F_0 to ∞ for F and

$$F = S_1^2/S_2^2 = \frac{\chi_1^2/v_1}{\chi_2^2/v_2}$$

$P(F_0)$ gives the probability of F_0 being exceeded.

Method 1

Note: The formulae are not valid when v_1 and v_2 are both odd.

Case 1 For v_1 even and v_2 odd or v_2 even with $v_2 > v_1$ write

$$\frac{v_1}{2} = n \text{ (integer)} \qquad \frac{v_2}{2} = m \text{ (integral or fractional)}$$

and

$$t = \frac{v_2}{(v_2 + v_1 F_0)}$$

Then

$$P(F_0) = t^m \sum_{i=0}^{i=n-1} A_i$$

where $A_0 = 1$ and for $i \geqslant 1$

$$A_i = \frac{m + i - 1}{i}(1 - t)A_{i-1}$$

Case 2 For v_2 even and v_1 odd or v_1 even with $v_1 > v_2$, with n, m and t defined as above, but now where m is an integer. Then

$$P(F_0) = 1 - (1 - t)^n \sum_{i=0}^{i=m-1} B_i$$

where $B_0 = 1$ and for

$$i \geqslant 1 \qquad B_i = \frac{n + i - 1}{i} t B_{i-1}$$

Method 2

This method does not limit the values of v_1 and v_2 used. The formula for $P(F_0)$ can be reduced to

$$P(F_0) = \alpha \int_{\theta_1}^{\pi/2} \sin^{v_1 - 1} \theta \cos^{v_2 - 1} \theta \, d\theta$$

where

$$\theta_1 = \text{arc cos}\left(\frac{v_2}{v_2 + v_1 F_0}\right) \qquad \text{and} \qquad \alpha = 2\Gamma\left(\frac{v_1 + v_2}{2}\right)\Big/\Gamma\left(\frac{v_1}{2}\right)\Gamma\left(\frac{v_2}{2}\right)$$

First we evaluate α

$$\Gamma\left(\frac{x}{2}\right) = \left(\frac{x}{2} - 1\right)\left(\frac{x}{2} - 2\right)\left(\frac{x}{2} - 3\right)\ldots 3, 2, 1 \text{ if } x \text{ is even}$$

$$\Gamma\left(\frac{x}{2}\right) = \left(\frac{x}{2} - 1\right)\left(\frac{x}{2} - 2\right)\left(\frac{x}{2} - 3\right)\ldots \frac{5}{2}, \frac{3}{2}, \frac{1}{2} \sqrt{\pi} \text{ if } x \text{ is odd}$$

When x is *even*

(a) $\Gamma\left(\dfrac{x}{2}\right) = G_{i=(x/2)-1}$ where $G_i = G_{i-1}(i)$ and $G_0 = 1$

If x is *odd*

(b) $\Gamma\left(\dfrac{x}{2}\right) = G_{i=(x-1)/2}$ where $G_i = G_{i-1}\left(\dfrac{2i-1}{2}\right)$ and $G_0 = \sqrt{\pi}$

The three values of the Gamma function given in α are calculated with $x = v_1 + v_2$, v_1 and v_2 using either (a) or (b) depending on x being even or odd, whence α. The following three expressions give the solutions to the integral.

Case 1 v_1 odd or even and v_2 even

Let $m = v_2 - 1$ and $n = v_i - 1$. This produces m as odd and n as even or odd. Then

$$P(F_0) = \alpha\left[-\sum_{i=0}^{i=(m-1)/2} A_i + \frac{A_{(m-1)/2}}{(\sin^{n+m}\theta_1)}\right]$$

where

$$A_0 = (\sin^{n+1}\theta_1 \cos^{m-1}\theta_1)/(n+1)$$

and

$$A_i = A_{i-1}\left(\frac{m-2i+1}{n+2i+1}\right)\frac{\sin^2\theta_1}{\cos^2\theta_1}$$

The number of A terms excluding the last term in the square brackets $= \left(\dfrac{m+1}{2}\right)$

The total number of terms $= \left(\dfrac{m+3}{2}\right)$

Case 2 v_1 and v_2 both odd, i.e. m and n both even, $m = v_2 - 1$, $n = v_1 - 1$

$$P(F_0) = \alpha\left[-\sum_{i=0}^{i=(m/2)-1} A_i + \sum_{i=0}^{i=(m+n-2)/2} B_i + B_{(m+n-2)/2}\left(\frac{\pi}{2} - \theta_1\right) \right]$$

where

$$A_0 = (\sin^{n+1}\theta_1 \cos^{m-1}\theta)/(n+1)$$

$$A_i = A_{i-1}\frac{(m-2i+1)\sin^2\theta_1}{(m+2i+1)\cos^2\theta_1}$$

$$B_0 = A_{(m/2)-1}/(n+m)$$

$$B_i = B_{i-1}\frac{(n+m-2i+1)}{(n+m-2i)}$$

The number of A terms $= m/2$
The number of B terms including the last term in the square brackets $=$
$$\frac{m+n+2}{2}$$
The total number of terms $= m + (n+2)/2$

Case 3 v_1 even, v_2 odd, i.e. m even, n odd, $m = v_2 - 1$, $n = v_1 - 1$

$$P(F_0) = \left[-\sum_{i=0}^{i=(m/2)-1} A_i + \sum_{i=0}^{i=(m+n-1)/2} B_i \right]$$

A and B terms as in Case 2
The number of A terms $= m/2$
The number of B terms $= (m+n+1)/2$
The total number of terms $= m + \dfrac{(n+1)}{2}$

Integral of the χ^2 Distribution

VII(7)   The integral required is

$$\int_0^{\chi_0^2} f(x)\,dx \equiv P(\chi_0^2) = \int_0^{\chi_0^2} \frac{e^{-\chi^2/2}\chi^{v/2-1}}{2^{v/2}\Gamma(v/2)}\,d\chi$$

Formulae for $\Gamma(\alpha/2)$ are given in VII(5)

$$P(\chi_0^2) = \frac{c^{-\chi_0^2/2}\chi_0^{v/2+1}}{2^{v/2-1}v\Gamma(v/2)} S$$

where

$$S = 1 + \sum_{i=0}^{i=\infty} A_i$$

where

$$A_i = \frac{A_{i-1}\chi_0^2}{v + 2i}$$

and $A_0 = 1$. The formula given is exact, i.e. not an approximation.

Bibliography

The following list of references includes books and papers that were consulted during the writing of this book, as well as those that might be considered for further reading.

Bowker A H *Ann. Math. Stat.* **17** 1946, pp 238–40. Computation of Factors for Tolerance Limits on a Normal Distribution when Sample is Large

Cadwell J H and Williams D E *Comput. J.* **4** 1961, pp 260–4. Some Orthogonal Methods of Curve and Surface Fitting

Cramér H *Mathematical Methods of Statistics* (Princeton, NJ: Princeton University Press) 1945

Eisenhart C, Hastay W and Wallas W A *Techniques of Statistical Analysis* (New York: McGraw-Hill) 1947

Feller W *An Introduction to Probability and its Applications* vol I (New York: Wiley) 1957

Fisher R A *Statistical Methods for Research Workers* (Edinburgh: Oliver and Boyd) 1970

Fraser D A S *Statistics: An Introduction* (New York: Wiley) 1947

Geigy J R *Documenta Geigy—Scientific Tables* (Geigy Pharmaceutical Co Ltd, Manchester) 1962

Gues P G *Numerical Methods of Curve Fitting* (Cambridge: Cambridge University Press) 1961

Hastings J R *Approximations for Digital Computers* (Princeton, NJ: Princeton University Press) 1955

Hogg R and Craig A *Introduction to Mathematical Statistics* (New York: MacMillan) 1965

Janossy L *Theory and Practice of the Evaluation of Measurements* (Oxford: Oxford University Press) 1965

Kreyszig E *Advanced Engineering Mathematics* (New York: Wiley) 1967

Levy H and Roth L *Elements of Probability* (Oxford: Clarendon) 1951

Lindgren B W *Statistical Theory* (London: Collier-MacMillan) 1968

Lindley D V *Introduction to Probability and Statistics* Parts I and II (Cambridge: Cambridge University Press) 1962

Owen D B *Handbook of Statistical Tables* (Reading, MA: Addison-Wesley) 1962

Parratt L G *Probability and Experimental Errors in Science* (New York: Wiley) 1966

Whittaker E T and Robinson G *Calculus of Observations* (Glasgow: Blackie and Son) 1940

Wilks W S S *Mathematical Statistics* (New York: Wiley) 1962

Index

531